AF709399

Coral Reef Resilience in the Anthropocene

Peter W. Glynn

Coral Reef Resilience in the Anthropocene

A History of Discovery and Research in the Eastern Tropical Pacific Region

Springer

Peter W. Glynn
Department of Marine Biology and Ecology
University of Miami
Miami, FL, USA

ISBN 978-3-031-81420-4 ISBN 978-3-031-81421-1 (eBook)
https://doi.org/10.1007/978-3-031-81421-1

© The Editor(s) (if applicable) and The Author(s), under exclusive license to Springer Nature Switzerland AG 2025

This work is subject to copyright. All rights are solely and exclusively licensed by the Publisher, whether the whole or part of the material is concerned, specifically the rights of translation, reprinting, reuse of illustrations, recitation, broadcasting, reproduction on microfilms or in any other physical way, and transmission or information storage and retrieval, electronic adaptation, computer software, or by similar or dissimilar methodology now known or hereafter developed.
The use of general descriptive names, registered names, trademarks, service marks, etc. in this publication does not imply, even in the absence of a specific statement, that such names are exempt from the relevant protective laws and regulations and therefore free for general use.
The publisher, the authors and the editors are safe to assume that the advice and information in this book are believed to be true and accurate at the date of publication. Neither the publisher nor the authors or the editors give a warranty, expressed or implied, with respect to the material contained herein or for any errors or omissions that may have been made. The publisher remains neutral with regard to jurisdictional claims in published maps and institutional affiliations.

Cover Image: Juan L. Maté surveying *Pocillopora* condition, Isla del Canal de Afuera coral reef, Gulf of Chiriquí, Panama, 10 April 2024. Credit: Bernhard Riegl.

This Springer imprint is published by the registered company Springer Nature Switzerland AG
The registered company address is: Gewerbestrasse 11, 6330 Cham, Switzerland

If disposing of this product, please recycle the paper.

With affection and gratitude to my wonderfully nurturing grandparents, Emma J. and Peter J. Glynn, who rescued me and then became my adoptive parents, making my life's profession and this book possible. Also, to my wife, Carmen Soé, and children, Mildred J. and Peter J. Glynn, who patiently accepted my passion for coral reef research and endured the many absences that this required.

To my South, Central, and North American students and colleagues who provided valued succor, camaraderie, inspiration, and insightful contributions to the field of coral reef science.

And to Chuck (Charles G.) Messing, who initially was meant to help illustrate this book. We blissfully worked together in the beginning months but soon health problems befell Chuck, sadly cutting his life and our collaboration short. I express gratitude to Chuck, a marine biologist extraordinaire, artist, musician, and thespian, for his good cheer, insights, and encouragement on this project.

Acknowledgments

The exploration, research, miscellaneous activities, and information herein presented have benefited immensely from the help of numerous people. Although several people assisted in multiple areas, for the sake of brevity they are generally mentioned where they contributed most to offering help and advancing coral science. My sincere apologies for any omissions.

I am honor-bound to acknowledge my science teachers, from high school through graduate school, who inspired, imparted knowledge, and helped me understand the structure and philosophy of the scientific method: Curtis R. Yeager, Marvin A. Nottingham, Robert D. Harwood, Martin R. Brittan, Benton Buttrey, Charles M. Vaughan, Arthur M. Pardee, Richard H. Rosenblatt, Donald P. Abbott, Cornelius B. van Niel, Lawrence R. Blinks,[1] Rolf L. Bolin, and Arthur Giese.

Thanks are due to the following for assistance in **coral reef drilling and the study of the physical and chemical features of coral skeletons (sclerochronology)**: James G. Babb, James L. Campbell, Todd M. Campbell, Julie Cole, Mitchell Colgan, H. S. Conroy, Jorge Cortés, Marsha S. Cox, Ellen M. Druffel, Robert B. Dunbar, René Espinosa, B. Hitchko, David Hollander, Ronald J. Lauzon, Braddock K. Linsley, Ian G. Macintyre, David Malmquist, James E. McRae, David A. Mucciarone, William F. Precht, Michael J. Risk, Glen T. Shen, Tom Smalley, B. Smith, Robert H. Stewart, Robert Stuckenrath, Aníbal Velarde, J. Waters, Gerard M. Wellington, D. Wootan, and Aaron Yedid; **for help with coral reef monitoring, collections, species identifications, technical, and scientific advice**: Peter A. Abrams, Juan J. Alvarado, Jerald S. Ault, Stuart Banks, Iliana B. Baums, Charles

[1] Prof. Blinks, director of the Hopkins Marine Station (HMS), performed an unexpected and memorable act of kindness. The HMS administrative office had just informed me that I was $65.00 short in completing payment for tuition. Sitting on the stairs to the Loeb building's second floor, pensive and a bit worried on how I could raise this amount of money, Prof. Blinks hurriedly brushed by me on his way upstairs. He then circled back, tapped me on the arm and asked, "Is everything alright Mr. Glynn?" Upon learning of my payment problem, he immediately opened his wallet, counted out $65.00, and told me to pay him back when possible. How can such a kind gesture ever be forgotten?

H. Baxter, Billy Benstedsmith, Paul Best, Charles Birkeland, Lucy Bunkley-Williams, Mitchell W. Colgan, Susan B. Colley, Jorge Cortés, Ross Cunning, Michael De Alessi, Mark W. Denny, C. Mark Eakin, Ian C. Enochs, Tom Eelkema, René Espinosa, Joshua S. Feingold, Frank J. Fernández, Peggy Fong, Peter J. Glynn, Jeffrey B. Graham, Richard W. Grigg, Héctor Guzmán, Kenneth L Heck, Raymond C. Highsmith, Bert Hoeksema, Carlos Jiménez, Kirstie L. Kaiser, Jeremy M. Kerr, Samuel J. Leonard, Joshua Levy, Priscilla Martínez, Juan L. Maté, Alissa B. Mones, Leonard Muscatine, Ranjan Muthukrishnan, Sergio Navarrete, David Obura, Donald B. Olson, Gustav Paulay, John Pearse, Esther C. Peters, Michel Pichon, Ellen J. Prager, Samuel Purkis, Diomedes Quintero, Marjorie L. Reaka, Héctor Reyes-Bonilla, Fernando Rivera, Juan B. del Rosario, Bernhard M. Riegl, D. Ross Robertson, Gary Robinson, David B. Smith, Tyler B. Smith, Bernardo Vargas-Ángel, Aníbal Velarde, J. E. N. Veron, Robert R. Warner, Ernesto Weil, Gerard M. Wellington, John W. Wells, Evie A. Wieters, Bert Williams, Henk Wolda, Janie L. Wulff, and Fernando A. Zapata. Special thanks are due to Aníbal Velarde and Susan Colley for their many years of faithful service in the field, laboratory, and at localities far from Miami and Panama.

Research vessel support was expertly provided by the crews and the following ship masters: Steve Breen, Jack N. A. Bryan, Lenín Cruz, Tim Ekstrom, Michel Garcia, Kevan Mantell, Lt. Cmdr. D. L. Markowski, Bernhard Schreyer, David West, and Paul Wolter. Air transportation from mainland Ecuador to the Galápagos Islands was subsidized by TAME (Transporte Aéreo Militar Ecuatoriano), an Ecuadorian national airline.

I am grateful for the excellent literature research and acquisition support provided by the RSMAES library staff: Helen D. Albertson, Annie Campbell, Angela Clark, Lisa Fish, and Kay K. Hale. Jack Marquardt, librarian at the Museum of Natural History, Smithsonian Institution, faithfully supplied literature in the early years of our research in Panama.

I thank those who generously offered photographs and help with creating figures; these people are acknowledged in the figure captions. Douglas Tyrrell and Blendia Yanina Tyrrell kindly archived and digitally prepared the figures included in the book.

For energizing research and assistance in coral studies, I thank the cadre of Panamanian students who frequented my laboratory in the early stages of eastern Pacific investigations: Juan Achurra, Damaris P. Ávila H., Tayra E. Barsallo M., Magnolia O. Calderón Mendoza, Francia B. Cedeño M., Fernando Crastz, José Manuel Gil Lásso, Eric Edgardo González C., Victor Martínez, Juan Laurentino Maté Touriño, Eira Mercedes Polo G., Antonio Montaner, Félix A. Oramas N., Miguel A. Perez P., Arcadio Rodaniche, Guadalupe Scott, Rosemary E. Segistán, and Eduardo Valdés Araúz. Luis D'Croz and Juan B. del Rosario, Departamento de Biología Acuática, Universidad de Panamá, are thanked for their help in student mentoring and sharing laboratory space.

My involvement with teaching and mentoring student research was elevated to a new level when I joined the RSMAES faculty in 1983. **The 1983–2016 period:** I thank the following individuals (then students in several cases) for their scientific contributions and collaboration during this intellectually stimulating 30 + year period: Jamie C. Afflerbach, Rafael J. Araújo, Kelly Attridge, Jerald S. Ault, Andrew C. Baker, Heather Balchowsky, Rebecca L. Bal Bailey, O. Barrio, Margaret Basil, Iliana B. Baums, Tracy Baynes, Bernadette Bezy, Kathy Black, Celina Bourbonierre, Viktor W. Brandtneris, Kathryn A. Brown, William E. Browne, Carlos Carbo, Robert W. Carter III, Henry Chaney, Nicola Chapman, Jorge Cortés, David Day, Juan B. Del Rosario, Kyle Dettloff, Helga Dienes-Diaz, C. Mark Eakin, June M. Eakin, Ian C. Enochs, Eduardo Espinosa, René Espinosa, Joshua S. Feingold, Julie C. Firman, Laurel Flebotte (Feingold), Linda Flores, Nohora Galvis, Nancy J. Gassman, Phillip Gillette, Jorge Gomezjurado, Benjamin Grassian, Héctor M. Guzmán, Katherine Hendrickson, James J. Herlan, Daniel M. Holstein, Joy Holtvluwer Ting, Christiane Hueerkamp, Michael Hyatt, J. Jara, Carlos Jiménez, Cindy Johnson, Eric Johnson, Erin Kapostasy, Susan Laessig, Todd C. LaJeunesse, Amber Little, Edward Major, Derek P. Manzello, Carolyn Margolin, Priscilla Martinez, Juan L. Maté Touriño, Anita S. May, Sheila McKenna, Donald F. McNeill, Rafael Menoscal, Godfrey Merlen, Monica Medina, Michael D. Moore, Alison L. Moulding, Alejandro Muentes, David O. Obura, Mauricio Ortiz, Ana M. Palacio-Castro, John E. Parkinson, Nicole Paul, John and Vicki Pearse (Buchsbaum), Jimmy Pefiaherrera, Juan Peña, Douglas Perrine, Ellen J. Prager, Héctor Reyes-Bonilla, Jill L. Richardson (formerly Borger), Robert H. Richmond, Fernando Rivera, Adrienne S. Romanski (now Correa), J. Sabater, Graham Saunders, J. Sabater, Sergio Salazar, Joseph E. Serafy, David B. Smith, Tyler B. Smith, J. H. Stout, Alina M. Szmant, Susan Thornton, Lauren Toth, Bonnie C. Tucker, Johnny Valenzuela, Bernardo Vargas-Ángel, Aníbal Velarde, Geerat J. Vermeij, Grazia Walker, Tiffani Woodworth, Joyce A. Yager, Kiyoshi Yamazato, and Aaron Yedid. **Post 2016. I thank the new cadre of students and established researchers who joined my retirement studies conducted primarily in the waters of southeastern Florida, Students:** Carl Anderson, Nathan J. Baldwin, Rachel N. Barrales, Brian Coffman, Philip L. Davidson, Joshua H. Dominguez, Diana W. Dowd, Brooke Enright, Tayla N. Fortson, Michael P. C. Fuller, Benjamin Grassian, Julie M. Gross, María José Guarderas, Rachel L. Ionata, Nicholas P. Jones, Joshua Levy, Nicolas Martinez, Samuel May, Shannon G. Moorhead, David E. Newhard, Marc G. Palmisano, Jason S. Presnell, Karim D. Primov, Bernhard M. Riegl, Bernhard M. Riegl Jr., Robert P. Roemer, Olivia Schuitema, Nash Soderberg, Jeongran Vanderwoude, Alexandra Wilde, Megan K. Williams, and Olivia Williamson. **Faculty, colleagues, technicians, volunteers, and others:** Renata Alitto, Claudia Alvarez, Richard Aronson, Fred Aschauer, Stuart Banks, Iliana B. Baums, B. Beck, Daniel Benetti, Patricia Blackwelder, Larry E. Brand, William E. Browne, Richard C. Brusca, William L. Bullock, D. Cadien, Dale R. Calder, Tom Carpenter, Carlo Cerrano, Ana B. Christensen, Amy Clement, Julia Cole, María M. Criales, Evan D'Alessandro, N. DeLoach, Kyle Dettloff, Cristina Diaz, Mara R. Diaz, Tom D. Dillehay, D. Drumm, Eliseo Duarte, Kevin Eckelbarger, Joshua S. Feingold, Kirk Fitzhugh, Brian Francisco, Phillip R. Gillette, Walter Goldberg, Terrence Gosliner, Edgar Graham, Martin Grosell, J. M. Guerra-García, Michael Hart, James C. Hendee, Gordon Hendler, Lea-Anne Henry, Amy Hiron, Ronald Hoenig, Tom Hughes, Pat

A. Hutchings, Inti Keith, James Klaus, Chris Langdon, Howard Lasker, Sam Maddingly, Kenneth Marks, George Matsumoto, Robert McQueeny, Serag S. Mehterian, Brendyn Meisinger, Liza Merly, Charles Messing, Christopher Meyer, Claudia E. Mills, Amy Moran, Peter J. Mumby, Patricia J. Netherly, Markus Niemeyer, Louise Page, Mark R. Patterson, Jorge A. Quintela Fernandez, Marjorie Reaka, D. Abigail Renegar, Bernhard M. Riegl, Bernhard M. Riegl, Jr., Amanda Robinson, Robert P. Roemer, Benjamin I. Ruttenberg, Peter F. Sale, Alonso Eduardo Sanchez, Michael C. Schmale, Joseph Serafy, Geoffrey Shideler, Erik Stabenau, Sascha C. C. Steiner, Marjorie Suarez, Daniel O. Suman, Mariana Tovar, Carlos Tudela, Victor Vásquez, Mariana Vera Zambrano, Chantal Victor, Nancy A. Voss, Regina Wetzer, and Morgan Will.

I thank the following persons who contributed to and encouraged Caribbean and eastern Pacific coral research at various times over the years: Andrew Bruckner, Eugenio Carpizo Ituarte, Ángel Chiriboga, Mitchell W. Colgan, Joseph H. Connell, Adrienne M. S. Correa, Robert Endean, Peggy Fong, Julio Garcia, Nora I. Goreau, Thomas J. Goreau, Malcolm J. Graham, Leslie Harris, William Newman, Juan Carlos Moncayo, John Ogden, Roberto Pépolas, Morgan Pratchett, D. P, G. Renaud, Alma Paola Rodríguez Troncoso, Sergio Salazar, Nancy A. Voss, and Herman Wirshing. I thank my family for moral support and cheering me on: Melissa L. and Joey Dyer, Bari R. Glynn, Olivia A. Glynn, Peter J. Glynn, Peter D. Marquez, Natasha G. Tupinamba', and Patrick G. Wilson.

Small and large parts of this narrative were reviewed and commented on by numerous individuals who offered suggestions, information, and corrected errors: Falk C, Amelung, Richard R. Aronson, Andrew C. Baker, Iliana B. Baums, Charles Birkeland, Marilyn Brandt, Stephen D. Cairns, Peter Castro, Henry W. Chaney, Mitchell Colgan, Susan B. Colley, Jorge Cortés, Amílcar Cupul-Magaña, Caroline Dennison, Mark W. Denny, Tui De Roy, C. Mark Eakin, Ian C. Enochs, Joshua S. Feingold, Paul C. Fiedler, Vincent Flynn, Colin Foord, Nohora Galvis, Peter J. Glynn, Victoria M. Glynn, Walter M. Goldberg, Bert W. Hoeksema, Kristopher Kernauskas, Todd C. LaJeunesse, Haris A. Lessios, Andrés López-Pérez, Derek P. Manzello, Juan L. Maté, Donald F. McNeill, Charles G. Messing, Carlos G. Muñoz, David O. Obura, John C. Ogden, Marjorie F. Oleksiak, Ana M. Palacio-Castro, Esther C. Peters, Guillermo P. Podestá, James W. Porter, William F. Precht, Marjorie L. Reaka, Héctor Reyes-Bonilla, Bernhard M. Riegl, Bernhard M. Riegl Jr., Fernando Rivera, Ana Paola Rodríguez-Troncoso, Tyler B. Smith, George D. Stanley, Jr., Alina M. Szmant, Lauren T. Toth, Erwan Deirieu Trottin, Michael Vecchione, Janet Voight, Robert Voss, Joerg Wiedenmann, and Fernando A. Zapata. I apologize and take responsibility for any omissions.

Laboratory directors, Juan Rivero and Jack Randall in Puerto Rico, and Martin Moynihan and Ira Rubinoff in Panama, are acknowledged for enthusiastically encouraging coral reef research at their research institutes. Our research in Ecuador benefitted from help offered by the Charles Darwin Research Station (CDRS), the Charles Darwin Foundation, the Galápagos National Park, and the Lenín Cruz family of Floreana Island. The many CDRS directors, including Craig MacFarland and Günther Reck in the early years, provided essential assistance in launching our research program in the Galápagos Islands. Financial support from the Smithsonian Institution's Scholarly Studies Program, the U.S. National Science Foundation, the National Geographic Society, and the MacArthur Foundation are duly recognized for providing research funding and student support.

In the final submission process, I thank Éva Lörinczi, Publishing Editor, Springer Nature Group, for her guidance, efficiency, and encouragement. For the book formatting and printing, I thank Ramya Ravichandran, Production Editor, Kali H. Gayathri, Production Supervisor, and their teams. Peter J. Glynn, Bernhard M. Riegl, B. M. Riegl Jr., and Ana Palacio-Castro greatly assisted in the final manuscript submission.

Introduction

Coral reefs are undeniably in great peril, on the verge of collapse and loss globally. This book spans six decades of discovery and ecological investigations of eastern Pacific coral reefs, a long overlooked, vibrant, and diverse ecosystem in a marginal tropical Pacific region. It begins with the discovery of pristine coral reefs in Panama in the 1960s, documents stressful El Niño warming events causing widespread coral bleaching and mortality during the 1980s–1990s, coral recovery in the 1990s–early 2000s, and presently (2023–2024) renewed extreme El Niño warming and region-wide heat waves, continuing coral mortality, and setbacks in reef recovery.

Basic coral reef terminology, reef structures, geomorphological reef types, and ecological interactions and functions relating to eastern Pacific coral reef ecosystems are introduced in Chapter 1 ("A Primer for Eastern Pacific Coral Reefs") as background for discussions in Chaps. 2-13. In Chap. 2 ("Coral Reef Discoveries"), discoveries and adventures relating to eastern Pacific coral reefs in the equatorial eastern Pacific (Panama, Costa Rica, Colombia, Ecuador, and Nicaragua) and biogeographically related areas (Mexico, Clipperton Atoll, and Easter Island) are described as explorations and surveys ranged across the eastern tropical Pacific. Chuck Messing and I illustrated several key species with sketches discussed here and in other chapters. Personal anecdotes relating to our research experiences, some amusing and instructive, are presented. These stories, ranging from frightening to pleasant, and occasionally downright enjoyable, are normally off limits for publication in scientific journals. Some of these anecdotes may be useful in offering insights into potential problems and pitfalls connected with research and travels across remote, often little known, culturally diverse, and occasionally hazardous terrain.

In Chap. 3 ("Origin of the Eastern Pacific Coral Reef Province"), it is hypothesized that nearly all reef-building coral species, including numerous reef-associated metazoans, originated from central/western Pacific genetic lineages via transpacific long distance dispersal. Historically, early coral reef scientists argued that coral reefs do not occur in the eastern Pacific because of cold upwelling conditions. A new understanding of the environmental challenges affecting corals in this region includes aperiodic stressful warming (El Niño events) and the threat of low aragonite saturation (ocean acidification) accompanying upwelling and increasing CO_2

emissions. These conditions are examined in Chap. 4 ("Physical Environmental Challenges"). The wide-ranging and multifarious direct and indirect impacts of El Niño and La Niña (ENSO) are discussed in Chap. 5 ("El Niño Southern Oscillation").

My team's main research activities were centered on five ecological topics: "Coral Reproduction and Recruitment" (Chap. 6), "Corallivores and Reef-Dwelling Predators" (Chap. 7), "Bioerosion" (Chap. 9), "Biodiversity, the Seen and Unseen Biota" (Chap. 10), and "Symbiosis, Microorganisms to the Megafauna" (Chap. 11). Large animal encounters, including a variety of aquatic vertebrates, e.g., bony fishes, sharks, dolphins, whales, reptiles, and birds (some peripheral to coral reefs) are presented in "Megafauna" in Chap. 8 because of their iconic status and intriguing behaviors.

"Coral Reproduction and Recruitment" (Chap. 6). Contradicting the early notion that eastern Pacific corals were relatively sterile or non-reproductive, all species investigated were highly fecund, with most demonstrating both sexual and asexual reproductive activity. Thus, corals experiencing severe mortality events have the potential to self-seed and recover from surviving populations instead of solely by distant propagule dispersal across the vast West/Central Pacific Ocean.

"Corallivores and Reef-Dwelling Predators" (Chap. 7). Despite remarkably high coral consumption by diverse predators of corals (invertebrate and vertebrate corallivores), eastern Pacific reef accretion is high and on par with carbonate production rates in other tropical regions. Coral reefs as welcome shelter for potential reef prey are questioned considering the abundance and diversity of cryptic predators.

"Bioerosion" (Chap. 9). Biological destructive processes are highly evident on eastern Pacific coral reefs, particularly following severe El Niño-induced mortality events. A particularly destructive sea urchin bioeroder and the methods employed to measure its impact are presented. It is concluded that bioerosion is not totally negative because the production and perpetuation of reef carbonate sediment habitats and coral asexual reproduction depend significantly on this process.

"Biodiversity, the Seen and Unseen Biota" (Chap. 10). The biodiversity of coral reefs, the tropical rainforests of the sea, is examined regarding the highly visible macrobiota (macroalgae, corals, mollusks, echinoderms, fishes *inter alia*) as well as the invisible cryptozoic biota and microscopic zooxanthella endosymbionts of corals. Loss of coral reef structural complexity, due to coral mortality and bioerosion, can result in catastrophic biodiversity declines.

"Symbiosis, Microorganisms to the Megafauna" (Chap. 11). The three general types of symbioses—commensalism, mutualism, and parasitism—are abundantly represented on coral reefs. The very existence of coral reefs depends on the mutualism between coevolved coral hosts and endosymbiotic zooxanthellae (and innumerable other microorganisms). Several notable symbiotic relationships are discussed involving coral crustacean guards, boring bivalves, fish cleaners, and microorganisms (including pathogens). A hypothetical case of the origin and evolution of a coral-crab guard mutualism is proposed.

"Retirement, A Subtropical Research Addendum" (Chap. 12). Notwithstanding my official retirement in May 2016, I could continue with teaching, research, and student mentoring for several more years. Teaching marine biology and mentoring were happily extended to college-bound high school students (Summer Scholars Program, University of Miami) and a graduate course, "Coral Reef Ecology" (Nova Southeastern University). Continuation of coral-related research initiatives allowed me to guide and mentor students interested in developing research skills. This work involved benthic ctenophores and an exotic brittle star species, both epizoites (host surface dwellers) of octocorals. Other research projects with colleagues involved (a) coral "rolling stones" (coralliths), (b) the occurrence of ancient *Pocillopora* coral offerings in northern Peru (13,300-year-old specimens in a ceremonial mound at Huaca Prieta). (c) publication of a review article on the state of coral reefs in the Galápagos Islands, and (d) work on a research manuscript of long-term biodiversity changes in an intertidal marine community in Pacific Grove, California. This last project, still in progress with Jamie Leonard and colleagues, takes me back to my doctoral research days in 1959–62. Chap. 12 summarizes these activities.

"Uncharted Waters, The Fate of Coral Reef Ecosystems" (Chap. 13). The present state of eastern Pacific and global coral reefs and climate-related extinctions of coral reef biota are summarized in Chap. 13. Climate change projections are noted with particular emphasis on global warming, heat domes, heat waves, and coral bleaching/mortality related to recurring El Niño events. The development and effects of the 2023–2024 El Niño event relating to eastern Pacific corals are presented (up to May 2024).[1] Coral reef protection and restoration initiatives are discussed. I advocate for supporting scientifically based efforts to move away from (reduce) the root cause of climate change—excessive use of fossil fuels—and not to abandon hope of losing Planet Earth's ecosystems, including the quintessentially biodiverse coral reefs. This book ends on a tenuous hopeful note that our youthful generation Z (and their descendants) will find ways to retain some semblance of coral reefs into the future.

How did I develop a passion for researching eastern Pacific corals and coral reefs? Coral reefs, diverse, vibrant, and complex ecosystems, captivated my interest at an early age. Marine life in general fascinated me, largely encouraged by my great aunt, Dora Conrad, a lifelong schoolteacher. Her love of the natural world was unbounded. "High ho, a rendezvous with Nature we shall go" was a common refrain. She walked the beaches and explored tide pools with me in Coronado, California, an island connected to the mainland by a long sandy strand. She further cultivated my interests in the natural sciences by taking me to the San Diego Museum of Natural History in Balboa Park and the marine aquarium at the Scripps Institution of Oceanography in La Jolla. These were memorable and exciting times of my childhood. I was truly awe-struck by the museum's marine life exhibits and the sea life at the Scripps aquarium. Aunt Dora encouraged me to make a collection of the seashells and biological oddities we found on our beach walks. Soon, several specimen cabinets were filled and began overflowing from my bedroom. My grandmother, Emma Glynn, encouraged me to move the growing collections, by

[1] Stressful El Niño warming was still persistent in March 2024, with projections indicating a quick transition to neutral conditions by April–June 2024 (NOAA Climate Prediction Center, NOAA ENSO blog, accessed in March 2024).

now also including alcohol-preserved specimens, to a vacant guest cottage in our backyard. This was to become my "laboratory," where I identified, examined, and dissected various curated specimens. My Great Uncle, Master Sargeant Slick Taylor, on medical leave from American Samoa during World War II, brought home assorted reef invertebrates, which I excitedly added to my collections. The assortment of specimens, however, was without corals.

Coronado's geographic location is temperate, north of subtropical climes that support reef-building corals. This location, however, is very near the international border with Baja California, Mexico, and the Sea of Cortez, of Steinbeck and Ricketts' (1941) fame. In the late 1940s and early 1950s, several of my friends and I—John Elwell, Stephen and Patrick Flynn, Thomas Carlin, Jr., and Edward Barnard—began making excursions into Baja California to explore and dive along the Pacific coast. Since this coastal stretch is under the influence of the California Current, the waters were cool to cold no matter how far south we traveled. We needed to explore the gulf coast of Baja California, the Sea of Cortez, where waters were seasonally warmer, supporting a subtropical biota, including corals. We ventured across the Baja California peninsula to San Felipe and a few kilometers south where we found small coral nubbins. Roads and trails to the south were non-existent or very difficult to navigate. Even the roads to San Felipe sometimes presented challenges. While driving late one night to San Felipe, while my companions slept, I began encountering large boulders on the road. Unaware that this was a warning, I weaved in and around them and finally came to an abrupt stop when I lost sight of the road immediately ahead. Exiting our car and walking toward a deafening, roaring sound, the road abruptly disappeared in front of me. About 5 meters below was a rushing turbulent river! A distant rainstorm had washed away this section of the road. I returned to our car and awakened my companions. Walking to the edge of the precipice, we stood looking down at the raging torrent, stunned, and fortunate to abandon this excursion to San Felipe.

We later decided to explore the eastern shores of the upper Gulf of California, the coastal areas and offshore islands of Mexico's Sonora state. These areas included San Carlos Bay (near Guaymas) and surroundings, and Tiburón Island. We towed a skiff from Coronado to Kino, carefully crossed the channel (Canal del Infiernillo = "Tiny Hell's Channel") to Tiburón Island and camped on the south end of the island. This area marked our first enjoyable encounter with Native Americans, the stately Seri people. Corals were present at these localities, but in low abundance and variety. Traveling by bus and exploring farther south into Mexico, to Mazatlán and Manzanillo, revealed increasing coral abundance and diversity, but no structural coral reef development.

After graduating from the University of South Dakota, I was offered a position by Donald P, Abbott (DPA)[2] to engage in graduate studies in his laboratory at the

[2] DPA was a stellar advisor who was greatly admired by his students. His guidance in conducting research was unparalleled. He was also an expert lecturer in the classroom and an inspiring instructor in the laboratory and field, always collaborating with students on the practical assignments of the day. Don's 2–3 hour long lectures, following early morning collections in the intertidal zone,

Hopkins Marine Station (HMS), Pacific Grove, California, the marine laboratory of Stanford University. Although my doctoral research centered on the ecology of an intertidal algal-barnacle community (Glynn 1965), conversations with Dr. Abbott about his research experiences on Ifaluk Atoll (Caroline Islands) kept my hopes alive that one day I would be able to engage in coral reef research. I relished reading his book, *Coral Island*, co-authored with Marston Bates, relating studies of Ifalukian culture and survival on a remote and tiny coral island (Bates and Abbott 1958).

Thanks to a fellow graduate student and friend, Bernie (Bernard D.) Fink, during my final year at HMS I met Carmen S. Quiñones Acosta, who was teaching Spanish at Pacific Grove High School. Bernie kindly invited me to a spaghetti dinner party where I first met Carmen. I was instantly smitten by her charm, energy, and intellect. We were married in Pacific Grove a short time later. I knew we would make a perfect match, and we have—64 years later and still counting. I began inquiring, "There are coral reefs in Puerto Rico, right?" "Yes, indeed." and Carmen learned, through inquiries of her father, Rubén, that there was an opening for a marine biologist at the Institute of Marine Biology (IMB), University of Puerto Rico, Mayagüez—in her hometown, no less. Perfect! Now to convince my advisor, DPA, that I could complete my research (analyses and writing) in Puerto Rico, return to Stanford and defend all dissertation requirements and meet the graduation deadlines. Fortunately, the stars were aligned, I was offered the position at IMB, and my wishes and plans were realized.

My seven years of teaching and research in Puerto Rico were most enjoyable and rewarding. I finally became intimately familiar with Caribbean coral reefs, principally in southwestern Puerto Rico off the fishing village of La Parguera. Johnny (Juan A.) Rivero offered me a position as a research scientist in the IMB, allowing me time to pursue reef studies. I have fond memories of several colleagues who assisted in my research in several ways, including Luis Almodóvar, Charles E. Cutress, Juan Gerardo González, Jack (John E.) Randall, and Fermín Sagardía. I am grateful to Frank Fernández for his assistance, especially his endurance and good nature in monitoring 24- and 48-hour plankton surveys on offshore coral cays. Numerous shallow-shelf and deep shelf-edge reefs were accessible for study near the marine laboratory on Magueyes Island, La Parguera (Ballantine et al. 2008; Hubbard et al. 2008). During this time, coral reefs were thriving and actively accreting with large areas, 10 s of hectares, of live *Acropora palmata* (Elkhorn Coral) on exposed reef fronts with no signs of coral bleaching or diseases. The variety of reef life was remarkable. A symbiotic association of a chiton species with isopods fascinated me and set me on a sidetrack of isopod systematics and ecology (Glynn 1965; Menzies and Glynn 1968) that eventually spilled over into the eastern Pacific (e.g., Glynn and Glynn 1974).

On one of my coral reef field trip classes off La Parguera, I first met Ira Rubinoff. Jack Randall, director of the IMB, asked if Ira could join us on our field trip. Ira was

---

were legendary. His laboratory drawings and notes were exquisite and published posthumously with a five-star review rating (Abbott, 1987). He was especially careful in not taking any credit for his students' work, never publishing with them on their research degree requirements.

a recent doctoral graduate from Harvard University, en route to Panama and the Smithsonian Tropical Research Institute (STRI), where he was recently appointed to expand the nascent marine program. A few months later, I received a letter from Ira asking if I would be interested in applying for a position at STRI. After a site visit in Panama, including both Caribbean and Pacific environments and laboratories, separated by only 60 km across the Panamanian isthmus, I enthusiastically accepted a research biologist position at STRI. This was an opportunity to continue investigating Caribbean coral reefs and to begin exploring the Pacific coast of Panama and other nearby equatorial Pacific areas for the presence of coral reefs. Carmen and I left Puerto Rico, *La Perla del Caribe*, with heavy hearts and tears but with hopes for a new and exciting future.

As in all scientific studies, one of the first essential tasks is to review the relevant literature of prior research, in my case coral studies conducted in the Pacific waters of Panama. Aside from the earliest taxonomic work of Verrill (1866), who named coral species collected in Panamanian waters (by field assistants), a study of special interest was that of the S. Y. *St. George* Expedition. This was an expedition led by Cyril Crossland, a renowned coral reef scientist, who spent considerable time investigating the "marine ecology and coral formations" off the Pacific coasts of Panama and Colombia. The results of Crossland's (1927) surveys were most discouraging, concluding his eastern Pacific observations as follows, "... without coral deposits of any considerable size, and certainly no true reefs." We later discovered numerous coral reefs surrounding Coiba Island (Gulf of Chiriquí, Panama) and at Gorgona Island (Colombia), localities surveyed by Crossland.

Soon after arriving in Panama, I was introduced to Bob (Robert) Stewart in the engineering division of the Panama Canal Company. Bob was a geologist and scuba diver who frequented the Pearl Islands, 40 km south of my new laboratory at Naos Island. I immediately engaged him in discussing coral reefs and my disappointment with Crossland's conclusion on the absence of reefs in Panama. "Not true!" Bob exploded with disagreement over Crossland's assessment. Not long afterwards, I was diving with Bob and gleefully surveying coral reefs at Contadora and Saboga Islands in the northern Pearl Islands Archipelago. This experience with Bob was exhilarating. Yes, there were abundant corals and from all appearances, there were likely true structural coral reefs in the eastern Pacific (Glynn et al. 1972; Glynn and Stewart 1973).[3] I immediately pinned nautical charts of Panama and adjacent countries onto my laboratory's bulletin board to highlight sites suspected of supporting reef growth and therefore needing study. Beyond Panama, and logistically and politically within reach, were the coastal areas of Costa Rica, Colombia, Ecuador, and some offshore islands such as the Galápagos Islands.

This book chronicles the discovery and dynamic changes of eastern Pacific coral reefs and identifies the researchers who observed and studied them since their entry into the Anthropocene Age. It offers an overview of undisturbed and vibrant reef

[3] Subsurface drilling and radiometric dating of corals were later performed to document the structure, accretion rates, and ages of the newly discovered coral reefs (Glynn and Macintyre, 1977).

growth to a region-wide transition of coral bleaching and mortality initiated by stressful El Niño warming events and heat waves. With the accelerating increase in reef researchers across the eastern Pacific reef tract in recent decades, and the development of novel study methods, this has resulted in a comprehensive understanding of the responses of corals to diverse disturbances. With hopeful anticipation, this history of reef studies and newly acquired knowledge will benefit coral protection, recovery, and reef restoration.

References

Abbott DP (1987) Observing marine invertebrates: drawings from the laboratory. Stanford University Press, Stanford, 380 p

Ballantine DL, Appeldoorn RS, Yoshioka P, Weil E, Armstrong R, Garcia JR, Otero E, Pagan F, Sherman C, Hernandez-Delgado EA, Bruckner A, Lilyestrom C (2008) Biology and ecology of Puerto Rican coral reefs. Pp. 375–406, In: Riegl BM, Dodge RE. (eds.) Coral reefs of the USA. Springer, Cham

Bates M, Abbott DP (1958) Coral Island, Portrait of an Atoll. Charles Scribner's Sons, New York, 254 p

Crossland C (1927) The expedition to the South Pacific of the SY 'St. George'. Marine ecology and coral formations in the Panama region, the Galapagos and Marquesas Islands, and the Atoll of Napuka. Trans R Soc Edinb 55:531–554

Glynn PW (1965) Community composition, structure, and interrelationships in the marine intertidal *Endocladia muricata—Balanus glandula* association in Monterey Bay, California. Beaufortia 12 (148):1–198

Glynn PW, Stewart RH (1973) Distribution of coral reefs in the Pearl Islands (Gulf of Panama) in relation to thermal conditions. Limnol Oceanogr 18:367–379

Glynn PW, Glynn CS (1974) On the systematics of *Ancinus* (Isopoda, Sphaeromatidae), with the description of a new species from the tropical eastern Pacific. Pac Sci 28(4):401–422

Glynn PW, Macintyre IG (1977) Growth rate and age of coral reefs on the Pacific coast of Panama. In: Proceedings of the third international coral reef symposium, Miami, vol 2, pp. 251–259

Glynn PW, Stewart RH, McCosker JE (1972) Pacific coral reefs of Panama: structure, distribution and predators. Geologische Rundschau 61(2):483–519

Hubbard DK, Burke RB, Gill IP, Ramirez WR, Sherman C (2008) Coral-reef geology: Puerto Rico and the US Virgin Islands. In: Riegl BM, Dodge RE (eds.) Coral reefs of the USA. Springer, Cham, pp 263–302

Menzies RJ, Glynn PW (1968) The common marine isopod Crustacea of Puerto Rico: a handbook for marine biologists. Stud. Fauna Curaçao Other Carib. Islands 27.1–133

Steinbeck J, Ricketts EF (1941) Sea of Cortez, a leisurely journal of travel and research. The Viking Press, New York. 598 p

Verrill AE (1866) On the polyps and corals of Panama with descriptions of new species. Proc Boston Soc Nat Hist 10:325–357

Contents

Chapter 1
A Primer for Eastern Pacific Coral Reefs

Abstract Because coral reef terminology is so highly specialized, commonly encountered terms are introduced here, mainly with reference to eastern Pacific corals and reefs, to facilitate understanding of topics in later chapters. The distinction between coral reefs and coral communities is examined with particular attention to the coral holobiont, that is the coevolved mutualism of photosynthetic algal endosymbionts (zooxanthellae) and their scleractinian coral hosts. This ancient, coevolved symbiosis is the fundamental driver of reef development and growth. Basic coral biology and coral reef ecological processes are introduced, including growth, reproduction, predation, bioerosion, and potential environmental disturbances (e.g., ENSO anomalous warming and cooling events, upwelling, low carbonate saturation state, low tidal exposure of shallow-occurring corals, and tectonism). Reef types, their morphology, biotic zonation, and occurrence within the eastern Pacific Province are noted as well as prominent zooxanthellate coral taxa. Reef development and the characteristics of eastern Pacific reefs are briefly compared with other global coral reef regions.

Introduction

Considering the vastness and massiveness of coral reefs, which are visible from outer space, and the intense interest in the geology and biology of reefs in the 19th and 20th centuries, it is surprising that so much confusion has surrounded their true structural nature and development. Over the years, amateur naturalists and specialists devoted to coral reef studies have confused **coral communities** with structural **coral reefs**. Steven Wainwright's (1965)[1] study of coral reefs in the Red Sea

[1] I reference both early and recent studies because it is crucial to know the primary sources of information. Whenever someone observes something for the "first time" or arrives at a "novel" hypothesis, it is possible these subjects were noted by earlier investigators, and later discarded, modified, or accepted. Knowledge of early studies can often lead to a healthy change of attitude, from hubris to humility.

© The Author(s), under exclusive license to Springer Nature Switzerland AG 2025

P. W. Glynn, *Coral Reef Resilience in the Anthropocene*,
https://doi.org/10.1007/978-3-031-81421-1_1

advanced a careful distinction between coral communities, i.e., populations of coral colonies growing on predominantly firm and often non-coral substrates, and coral reefs, biogenic structures composed primarily of coral colonies growing on and contributing to the building of skeletal frameworks. Coral reefs are enduring constructions that can grow continuously for hundreds and even thousands of years whereas coral communities are more ephemeral; their colonies grow, die, and are replaced by new colonies. I contrast coral communities as re-forming, multigenerational colony aggregations from coral reefs which are growing (accreting) formations, producing carbonate deposits capable of modifying surrounding environmental conditions. The structure of coral communities does not change appreciably over time, but coral reefs undergo notable changes in three dimensions. In the eastern Pacific, coral communities do not reveal appreciable changes in their overall appearance over multiple decades. Coral reefs do change, however. They often grow vertically, building massive frameworks that form and re-shape different reef zones (see below). David Stoddart (1969) elaborated on this important distinction in the context of geologically Recent global coral reef morphologies and distributions. **Recent**, in this context, is synonymous with the Holocene epoch, which began about 11,700 years ago at the end of the last glacial advance and continues today. Sea-level rise slowed at this time and continued to slow to the current global annual average of 3.6 mm. Global sea level has risen between 21–24 cm since 1880 and is presently accelerating because of human-induced climate change. We have entered the **Anthropocene Epoch**, a proposed interval of geologic time in which the activities of humankind began to substantially alter Earth's geology, atmosphere, oceans, and climate with a transformative impact on the biosphere.[2]

Shortly after Stoddart's informative review, Jones and Endean (1973, 1977) published an edited, four-volume comprehensive treatise on the geology and biology of the world's coral reefs. Despite their thoroughness, I was surprised at how little information was presented on eastern Pacific coral reefs. Predator-prey relationships of corallivores (coral-eating animals) and the geology of Clipperton Atoll were mentioned briefly. Indeed, the tropical eastern Pacific region needed attention[3].

---

[2]When did the Anthropocene Epoch begin? Many authorities suggest that it started in the 1950s with the addition of radioactive isotopes and plastics in the geologic rock record. Indeed, microplastics (Hieri et al. 2021) and radioactive isotopes (Guan et al. 2021) are readily incorporated in coral skeletons.

[3]Bob Endean invited me to write a review of the ecology of Caribbean coral reefs following the very unfortunate and premature death of Thomas (Tom) F. Goreau in 1970. My first investigations of the ecology of coral reefs began in Puerto Rico in 1960. I moved to the Smithsonian Tropical Research Institute (STRI) in Panama in late 1967 and began studies of eastern Pacific coral reefs in 1970 (Glynn 1972; Glynn et al. 1972). I continued with Caribbean reef studies on the north coast of Panama in the early years of my move to Panama, primarily geomorphology (Glynn 1973a) and Holocene reef history (Macintyre and Glynn 1976). Two related studies of the trophic structure of a coral reef in Puerto Rico, Laurel Reef, required further work (Glynn (1973b, c). Analyses completed in my laboratory in Panama revealed evidence of significant plankton depletion in water streaming over the Laurel coral reef flat community. The filtering efficiency of the reef community amounted to 91% and 60% removal of phytoplankton and zooplankton respectively, indicating the importance of an allochthonous food source. This result was gratifying as it was in line with the classic studies of C. M. Yonge and co-workers on Australia's Great Barrier Reef. Further, it pro-

Reef-Building Organisms

The great majority of organisms that secrete calcium carbonate ($CaCO_3$) or limestone skeletons as they grow, and thus potentially contribute to reef building, are stony, hard, or scleractinian corals (Phylum Cnidaria, Order Scleractinia) and calcareous, or coralline, red algae (Phylum Rhodophyta, Family Corallinaceae). Although all red algae are photosynthetic, only about half of the over 1600 living species of stony corals are autotrophic. Many azooxanthellate corals live in the deep sea in total darkness. Photosynthetic corals are often referred to as **phototrophic**, **symbiotic**, or **zooxanthellate**, after the presence of microscopic **zooxanthellae** that reside in the cells that line their digestive cavity and perform photosynthesis. Zooxanthellae are single-celled organisms called dinoflagellates, which include important members of the plankton and are responsible for red tides, bioluminescence, and paralytic shellfish poisoning. Another term applied to symbiotic corals is **holobiont**, which refers to the coral host and all its microbial associates, including zooxanthellae, bacteria, and viruses (Rohwer and Youle 2010).

The numerous stony corals without photosynthetic symbionts are **azooxanthellate** and generally exhibit slow growth compared to zooxanthellate species. Many azooxanthellate species occur in shallow reef communities, but usually under low light conditions such as in cavities or caves. The majority of azooxanthellate species inhabit deep, aphotic ocean depths[4]

The algal taxon, Corallinaceae, consists of two groups, the articulated corallines, and **crustose coralline algae** (**CCA**). Articulated species are branching, tree-like forms that fall apart into separate calcareous pieces after death whereas crustose species form thin calcareous crusts and cement-like accretions that can aid in binding together coral and other skeletal structures. If stony corals can be considered the bricks of a reef, crustose red algae are the mortar. These photosynthetic corals and algae typically calcify relatively rapidly in sunlit tropical environments, thus supplying essential building materials for the formation of limestone structures — coral reefs. The coral reef **framework** or scaffolding, that forms over time, builds vertically upward. Extreme low tide limits vertical growth. Above this tidal level, prolonged exposure to the warm/hot and dry atmospheric conditions results in coral bleaching, desiccation, and death.

---

vided an additional example of the vital role of subsidized trophic resources in diverse food webs (Polis and Hurd 1995; Polis et al. 1997).

[4] One of the world's largest known deep sea coral provinces is located on the Blake Plateau off the southeast coast of the USA. The coral buildups or mounds, composed dominantly of the delicately branching colonies of *Lophelia pertusa*, were first detected by echo sounding in 1956 and authenticated to be living coral banks and mounds by Stetson et al. (1962). These coral communities, referred to as the "million mounds" region, occur at depths of 800 to 1000 m and are present over 26,000 km² of sea floor according to recent NOAA surveys (Windows to the deep 2019 Expedition). Some coral mounds, reaching 100 m in height, may well be hundreds of thousands of years old. For a recent account of this deep sea biodiverse hotspot see the NRDC Expert Blog, Natural Resources Defense Council blog, <nrdc.org/bio/alison-chase/scientists-discover-worlds-largest-deep-sea-coral-reef-habitat> January 23, 2024.

Reef-building corals and calcareous algae are the **primary frame builders**. Many other carbonate-secreting organisms, known as **secondary frame builders,** also contribute to reef growth; these include non-scleractinian fire corals and octocorals, sponges, polychaete tubeworms, bryozoans, mollusks, echinoderms, and some green and brown algae. These organisms precipitate calcium carbonate, usually in lesser amounts, and infill reef cavities with their skeletal remains packing and consolidating the reef frameworks. Several other kinds of organisms that do not deposit skeletons, such as microalgae, bacterial mats, fleshy algae, sea anemones, and sponges, stabilize the framework and contribute to reef growth. Wulff and Buss (1979) demonstrated that sponges in the Caribbean could bind and stabilize corals, an effect also observed on eastern Pacific coral reefs. Under optimal levels of sea water alkalinity, cementation of limestone skeletons contributes significantly to the consolidation and stabilization of reef structures (Manzello 2010; Manzello et al. 2008). In addition to the primary and secondary calcifying frame builders, Dennis Hubbard (2011) has emphasized the critical roles of bioerosion, cementation and encrustation in reef accretion. His extensive investigation of the internal structure of Caribbean coral reefs led him to facetiously (but seriously) refer to coral reefs as "garbage piles"—50% reef builders/50% detritus.[5]

In the eastern Pacific, branching corals in the genus *Pocillopora* are the dominant reef builders (Fig. 1.1). They form vertically-oriented frameworks whose branches are often joined by crustose coralline algae and sponges below the zone of live coral tissue. *Pocillopora* patches on some eastern Pacific reefs are chiefly clones, likely

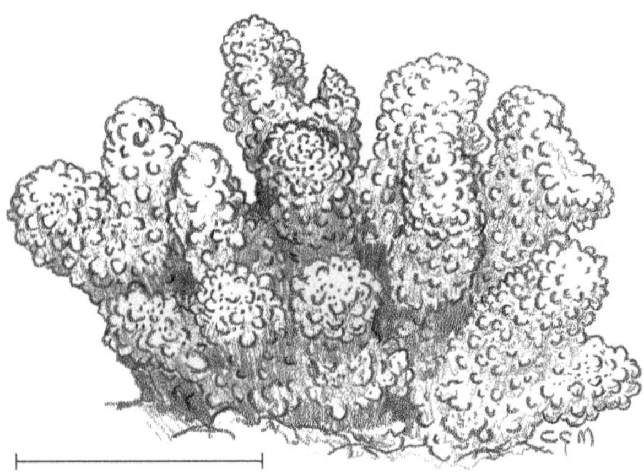

Fig. 1.1 *Pocillopora* sp., a principal eastern Pacific reef-building branching coral species, scale bar = 10 cm. This colony is morphologically similar to *Pocillopora* type 1, *Pocillopora grandis* (see Chap. 10). Drawing courtesy of Charles Messing

[5]A US Army colonel in Panama who supported our reef studies once asked if I would give him an attractive coral for his desk. In one of our surveys, I spotted a nice, clean, recently dead colony. I collected the coral, sent it to his office and received a thank you note. About a week later, two junior officers of hostile bearing arrived at my laboratory with the coral, pointing out that it was growing on a live .50 caliber bullet. They were investigating whether this potentially hazardous gift was intentional or accidental. Accidental, of course, but an example of reef "trash" and the ability of corals to settle, grow and incorporate all manner of available objects into the fabric of reef structures.

formed asexually when colonies fragment and then grow separately (Pinzón et al. 2012; Baums et al. 2014). Such clones are thought to be more prevalent in marginal coral communities where corals are less abundant due to unfavorable environmental conditions and thus less likely to reproduce sexually (**Allee effect**, low chance of gamete encounters). Massive or mound-forming colonies, such as *Porites lobata, Pavona clavus, Pavona gigantea,* and *Gardineroseris planulata,* also construct structural reefs in the eastern Pacific (Fig. 1.2). For example, *P. lobata* dominates reefs along southern mainland Costa Rica and at Cocos Island, and *P. clavus* on northwestern Costa Rican coastal reefs (Cortés and Jiménez 2003).

Unlike coral reefs in the Indo-Pacific region, where crustose coralline algae (CCA) are major framework builders forming **algal ridges**—elevated features on reef fronts subject to powerful wave action. Some eastern Pacific reefs in areas of high wave assault, for example in Panama, Colombia, and at Clipperton Atoll, have incipient algal ridges or slightly elevated rims in areas of high turbulence. Bill (William) Newman, of the Scripps Institution of Oceanography, and his doctoral student, Tom Dana, noted the paucity of calcareous algal buildups during reef surveys in Panama in the 1970s. They also remarked on how eastern Pacific reef-building occurs primarily in sheltered areas, such as coves and shore areas protected from strong wave action. In addition, they observed that white sandy beaches composed of broken carbonate skeletons were reliable indicators of patch reefs growing just offshore.

Jorge Cortés of the University of Costa Rica summarized other characteristics of eastern Pacific coral reefs in two publications (Cortés 1997, 2011): (a) relative youth and small size, (b) discontinuous distribution, (c) low species diversity, particularly of reef-building corals, (d) occurrence at shallow depths, (e) subject to frequent disturbances, (f) high bioerosion, (g) uncemented structures, and (h) slow recovery. Recent studies have demonstrated rapid recovery of bleached corals that have survived extreme thermal conditions (Chap. 13).

Unquestionably, most eastern Pacific coral reefs are youthful and small compared with other reef regions. This observation may have contributed to the

Fig. 1.2 *Porites lobata*, a principal eastern Pacific reef-building massive coral species, scale bar = 25 cm. Drawing courtesy of Charles Messing

confusion over whether true coral reefs exist in the eastern Pacific and in other marginal environments such as the Arabian Gulf and the Gulf of Oman.

Besides the attractive biogenic structures formed, coral reefs provide numerous benefits to the natural world and humankind, e.g., (a) capability to alter currents, dampen turbulence and protect shorelines, (b) providing habitat niches (shelter and substrates) that promote biodiversity and increase biomass, (c) nourishment for coastal populations, and (d) the revelation of historical and ancient environmental conditions. The latter can often be ascertained from carbonate skeletal remains through physico-chemical analysis.

The eastern tropical Pacific is home to relatively few zooxanthellate reef-building coral species, from nearly 30 along the coasts of Mexico, Costa Rica, Panama, and Colombia to about 15 in El Salvador and Nicaragua (Figs. 1.3 and 1.4). Coastal Ecuador and the Galápagos Islands have 20–23 species with only two reported from Peru (Figs. 1.4 and 1.6). By contrast, their numbers increase dramatically across the central and western Pacific, from about 60 species in Hawaii to 160 at Palmyra Atoll, and 350 on Australia's Great Barrier Reef. These numbers are only approximations because of shifting distributions, discoveries, local extinctions and taxonomic revisions resulting from better understanding of coral species boundaries

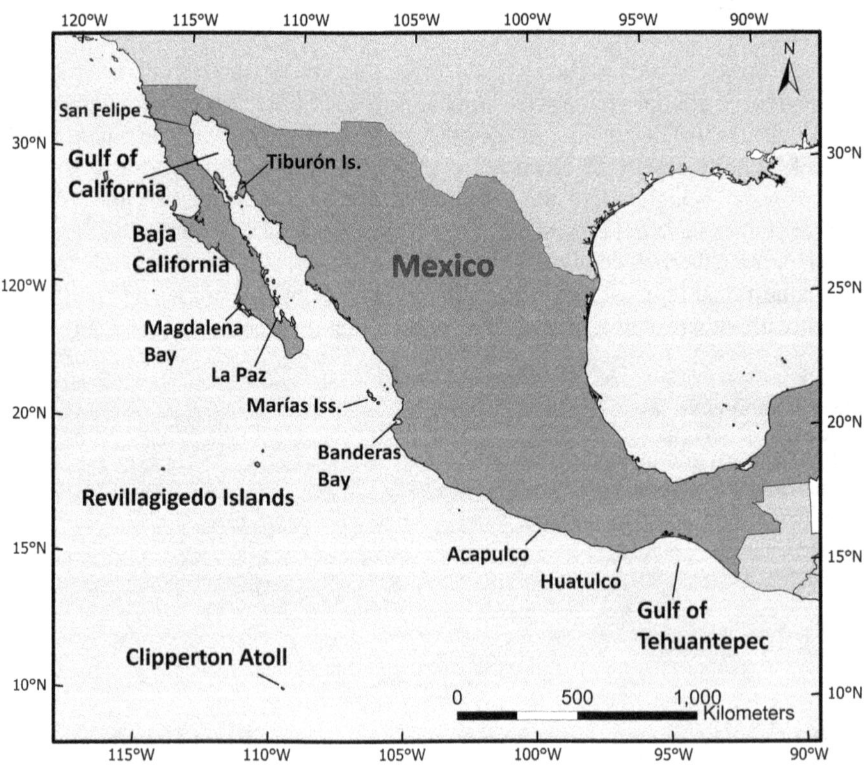

Fig. 1.3 Mexican Pacific coastal coral sites, highlighting major coastal regions and islands. Map adapted and edited by Diana Moanga

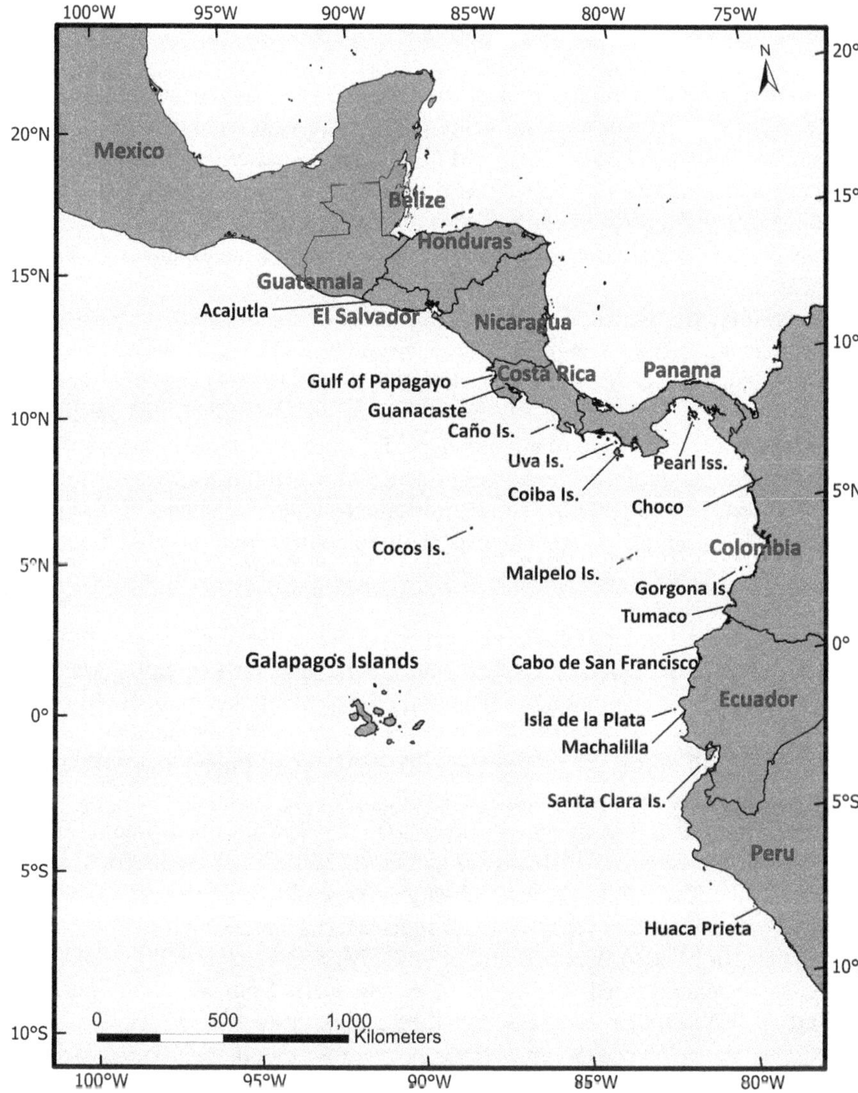

Fig. 1.4 Equatorial eastern Pacific coral sites including the Huaca Prieta prehistoric settlement in Peru where the archaeological specimen of *Pocillopora* was discovered (Glynn et al. 2022). Map adapted and edited by Diana Moanga

(thanks to molecular genetic studies). Four species of fire corals, Family Milleporidae, which are distant zooxanthellate relatives of stony corals (related to freshwater *Hydra*), also occur in the eastern Pacific at Panama and Clipperton Atoll. They have undergone changes in occurrence in recent years, however, due to local and regional extinctions (Glynn et al. 2017). Unlike species rich and complex coral ecosystems of the western Pacific, the few coral species present in the eastern Pacific reflect a relative simplicity that offers advantages in the investigation of ecological processes.

Destructive Reef Processes

Corallivores, reef animals that prey on corals, are diverse and abundant on eastern Pacific reefs; these include species of mollusks, crustaceans, echinoderms, and several fish families (Enochs and Glynn 2017). Although corallivory is common-place, outbreak predation events leading to reef devastation have not been reported. Intense sea urchin grazing following severe El Niño bleaching and mortality, however, has resulted in the degradation and erosion of reefs in some equatorial eastern Pacific areas.

Bioerosion, the breakdown of hard substances by organisms, is particularly prevalent in the eastern Pacific marine environment. On coral reefs, bioeroding organisms cause the dissolution, excavation, and abrasion of carbonate skeletons and limestone deposits via boring, burrowing, drilling, scraping, and grazing by **microborers** (such as algae, fungi, and bacteria), and **macroborers** (e.g., sponges, polychaete worms, mollusks, crustaceans, sea urchins and fishes). Ray (Raymond C.) Highsmith (1980) highlighted how the nutrient-rich environments of the eastern tropical Pacific promote bioerosion, especially in contrast to the low-nutrient, gin-clear waters of many other coral environments that typically occur on the western sides of ocean basins.

The serious decline of global coral reefs over the last several decades has been attributed to a combination of effects, often interacting, such as global warming, coastal development, and pollution (Pandolfi et al. 2005; Baker et al. 2008; De'ath et al. 2012; Hughes et al. 2018). Coral cover has declined precipitously in some places, such as Florida and the Great Barrier Reef. Interestingly, however, several coral reefs that have been monitored for decades at Cocos Island and the Guanacaste district of Costa Rica (Guzmán and Cortés 2001, 2007; Cortés and Jiménez 2003), Panama, and the Galápagos Islands (Glynn et al. 2015; Fong et al. 2017), and at Gorgona Island, Colombia (Zapata and Vargas-Ángel 2003; Zapata 2017), have all regained high live coral cover in 10 to 20 years after severe bleaching disturbances. Two studies, in Mexico (Cruz-García et al. 2020) and Panama (Glynn and Fong 2006), have demonstrated live coral tissue recovery from **El Niño Southern Oscillation (ENSO)** thermal stress in as little as months to a few years.

ENSO is a coupled ocean-atmospheric system resulting in the irregular periodic occurrence of warm El Niño or cool La Niña waters in the eastern tropical Pacific region. A common metric used to gauge the **recovery** of coral reefs is live coral cover, i.e., a return of living corals to pre-disturbance abundance[6]. A coral community that can absorb environmental stress and still return to or retain its biological composition and function is considered ecologically **resilient**. Branching and massive coral colonies that suffered partial mortality associated with ENSO events in

[6]A more realistic measure of ecosystem recovery should consider the return of the structure and function of coral reef communities to pre-disturbance conditions after having been subject to a physical or biotic stressor or multiple stressors.

Panama revealed rapid tissue regrowth in one year, resulting in 37.9% and 21.5% live tissue recovery, respectively (Glynn and Fong 2006). In addition, Romero-Torres et al. (2020) found that coral reefs recovered regionally across the eastern Pacific over periods of 10–15 years. Considering that coral reef recovery requires some combination of successful sexual reproduction, asexual regeneration, recruitment of larvae (which is generally low in the eastern Pacific), and escape from a gauntlet of competitors, diseases, corallivores, and resistance to recurrent thermal disturbances, 15, 20 or even 25 years would seem reasonable periods for community recovery in areas subject to extreme and recurring environmental stressors.

Plant Communities

Red, green, and brown fleshy **macroalgae** occur on coral reefs in the eastern Pacific but are less diverse and abundant than in the Caribbean and Indo-Pacific regions. The species-poor flora reported for some areas, however, is probably more apparent than real (Earle 1972). In a recent assessment, the species richness of macroalgae ranges from 16 to 24 in El Salvador and Guatemala, respectively, to 232 to 316 in Mexico and the Galápagos Islands, respectively (Cortés et al. 2017). Prominent macroalgal communities exist in the Galápagos Islands, even species of kelps (e.g., *Eisenia galapagensis*), but in nutrient rich upwelling environments, such as Caleta Iguana, far from coral reefs (Glynn and Wellington 1983). An intriguing mixture of macroalgae and reef-building corals occurs along the seasonal upwelling coast of Dhofar in southern Oman (Glynn 1993; Sheppard 2000). Kelps are prominent during the monsoonal upwelling season, followed by corals when upwelling ceases, waters warm and algal populations decline. Corals are abundant and reef development common in the Gulf of Oman in the lee of monsoonal upwelling (Fig. 1.5).

As noted earlier, coralline algal taxa occur on eastern Pacific reefs (e.g., *Lithophyllum*, *Lithothamnion*, and *Porolithon*) but are not prominently involved in reef building as in other coral reef regions. Coralline algae that can contribute importantly to reef building globally can add as much to reef carbonate production as corals (Cornwall et al. 2023). The calcified, segmented, green algal genus *Halimeda* is omnipresent in most coral reef regions but is rarely encountered in the eastern Pacific.

Algal turfs and **algal lawns,** low-lying (mm-cm high) patches of filamentous and microalgae are commonly present on shallow coral reefs worldwide, including eastern Pacific reefs. Intense grazing by reef herbivores is largely responsible for these patches as well as damselfish that "cultivate" their territories and consume the algae (Eakin 1987). These damselfish patches are sometimes called "algal gardens" and damselfish intensely defend them against fish and invertebrate herbivores and other intruders. Territorial damselfish are highly influential in the eastern Pacific where they have been shown to control coral zonation (Wellington 1982) and limit bioerosion (Glynn 1988; Eakin 1992) and corallivory (Glynn and Colgan 1988).

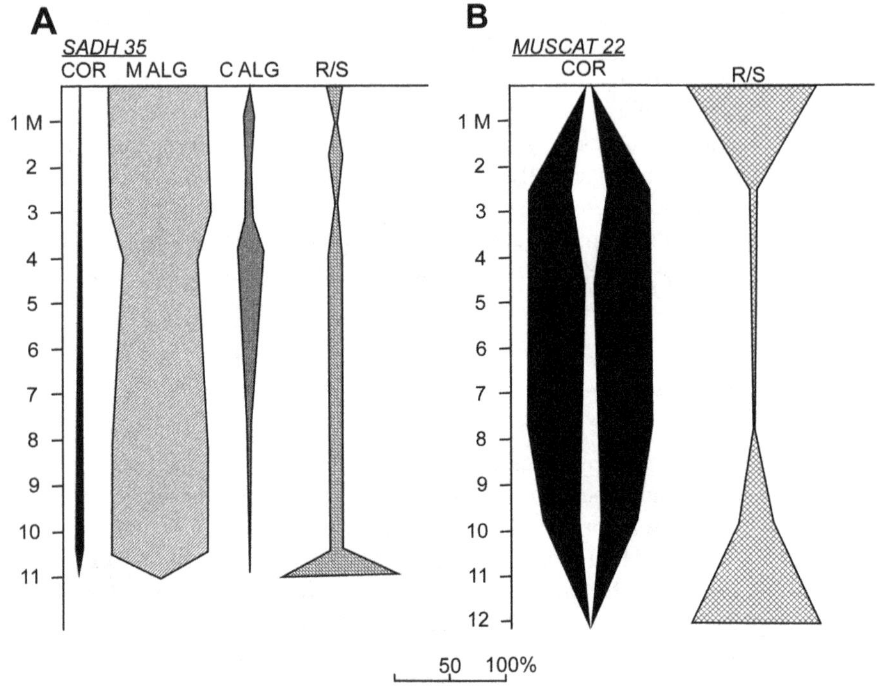

Fig. 1.5 Percent cover with depth of major epibenthos at an Arabian Sea upwelling site (A) and reef site in the Gulf of Oman (B). Sampled at the end of upwelling in October 1982. COR, reef-building corals (solid = live, clear = dead);M-ALG, macroalgae; C-ALG, coralline algae; R/S, rock and/or sand. Modified after Glynn (1993)

Eastern Pacific coral reefs differ from reefs in other biogeographic regions by the virtual absence of associated seagrass ecosystems. Sparse shoal grass (*Halodule* sp.) is present on some shallow sand banks adjacent to coral reefs in the Gulf of Chiriquí, Panama, but never the luxurious sea grass communities characteristic of the Caribbean north coast of Panama and elsewhere in tropical reef waters. Coral reefs and seagrass ecosystems together share important ecological interactions, such as buffering negative terrestrial influences and serving as nurseries (Ogden 1997). Instead of sea grasses, eastern Pacific coral reefs often grade into **rhodolith** communities, aggregations of fist-size, unattached, free-living coralline algae and diverse invertebrates, usually at slightly greater (10 s of meters) depths (Steller et al. 2009; Cortés et al. 2017). Like seagrass communities, rhodolith communities provide structure and shelter for numerous species, which contribute to local biodiversity.

Environmental Conditions

Of the environmental conditions promoting coral reef growth, warm sea water (not abnormally warm) is critical and has been emphasized since the early days of coral reef science. The global distribution of coral reefs is roughly bound by the 18 °C minimum monthly isotherm, a relationship supported by T. Wayland Vaughan (1919). The influence of sea water temperature is especially evident in the eastern tropical Pacific region where the eastern boundary currents (California and Peru-Chile currents) and upwelling centers exercise a strong influence on thermal conditions. Also, ENSO events can generate stressful high and low anomalous thermal conditions over periods of months. The 1982–83 ENSO event enormously impacted eastern Pacific marine and terrestrial ecosystems and was a harbinger of global coral reef decline (Glynn 1984, 1990; Brown 1987). This severe disturbance immediately changed the direction of my/our[7] research focus.

Unlike temperate and polar latitudes where seasonal changes are marked by day length and temperature differences, the tropics experience slight changes in day length but notable contrasts in the wet and dry seasons. These seasonal differences in rainfall are associated with the north-south migration of the **Intertropical Convergence Zone (ITCZ)**, a virtually windless belt that circles the earth near the equator and is known by mariners as the Doldrums. Here, the northeast and southeast trade winds converge, tropical temperatures cause the moist air to rise (convection) and eventually cool, causing heavy rainfall. In the eastern Pacific, this heavy rainfall zone reaches its northern-most location at El Salvador, near 12°N, in the northern summer (July–September). Its southern-most position, close to Ecuador's equator (1–2°N), occurs in February (Fiedler and Lavin 2017). The ITCZ thus affects different portions of Central and South American coastal areas at different times of year potentially causing flooding, excessive river discharge, siltation, and high cloud cover.

In contrast with the western Pacific, where the Kuroshio and East Australian Ocean currents transport warm waters from the tropics to northern and southern higher latitudes, respectively, the major currents that flow along the eastern Pacific margin bring cooler waters from higher latitudes toward the equator. Here, the southbound California Current and northbound Peru, or Humboldt Current, result in a much narrower latitudinal band of warm water conditions. In the western Pacific, corals range from Tokyo Bay (35.5°N) to slightly below the Tropic of Capricorn (23.5°S) in Australia, a span of 59 degrees of latitude covering about 6550 km. In contrast, coral communities and coral reefs in the eastern Pacific range from near the Tropic of Cancer in Baja California (23.5°N) to the Gulf of Guayaquil in Ecuador (4°S), a span of 27.5 degrees of latitude covering about 3050 km. As these major

[7] When I refer to 'we', 'our', or 'team' in this narrative I am acknowledging the research efforts of my students, colleagues, and myself. It is necessary to emphasize the significance of 'we', 'our', or 'team' instead of 'I', 'me', and 'my'. Advances in most areas of coral reef research are clearly best achieved by a team effort, utilizing the insights and skills of different talented individuals.

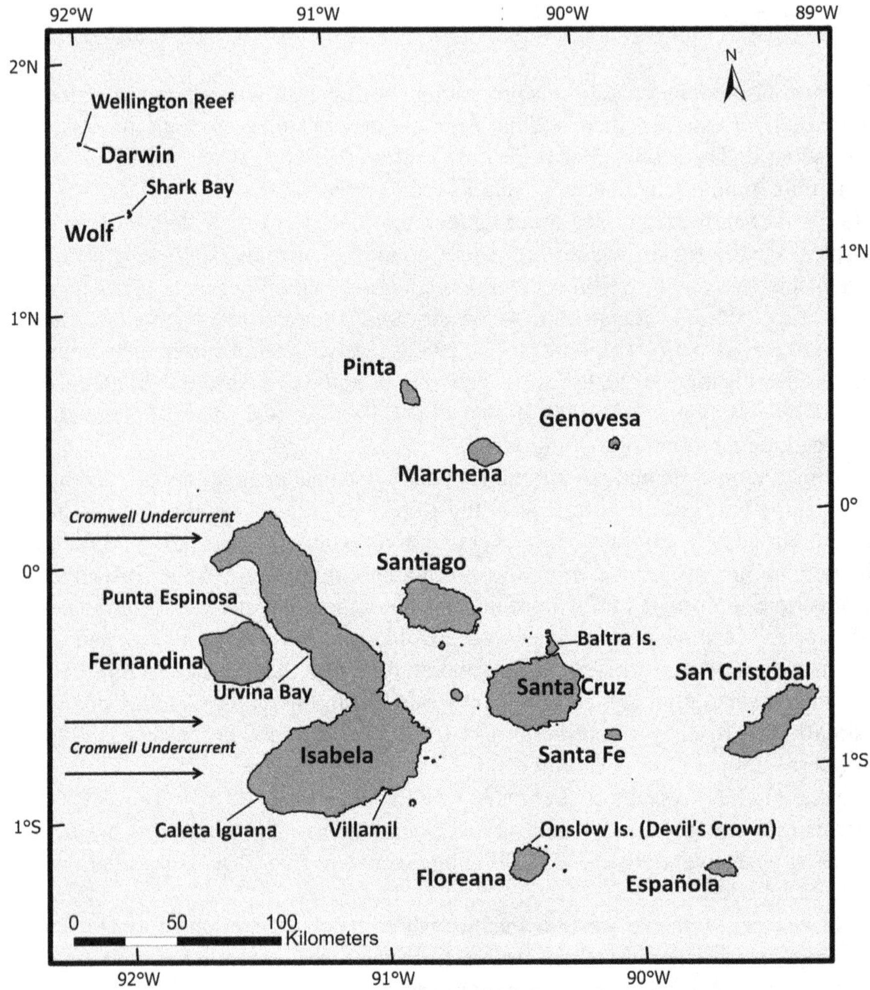

Fig. 1.6 Galápagos Archipelago showing location of 14 larger island sites with pre-1982-83 El Niño coral communities. Map adapted and edited by Diana Moanga

currents flow south and north toward the equator in the eastern Pacific, they veer to the west due to the earth's rotation (**Coriolis effect**) and are replaced by cooler and nutrient-rich water that rises from deeper levels of the water column (upwelling), even at the equator. In addition, three areas along the Mesoamerican isthmus are subject to seasonal upwelling of cooler water. The Gulf of Tehuantepec in southern Mexico, the Gulf of Papagayo spanning the Nicaragua/Costa Rica border, and the Gulf of Panama/Panama Bight (Fiedler and Lavin 2017) are all low terrain areas where wind jet streams arising from the Caribbean trade winds cross the isthmus and move surface waters offshore by means of advection. In addition, the Galápagos

Archipelago, especially the western and southern/central islands, are subject to strong equatorial upwelling, a result of the eastward moving and shoaling **Cromwell Undercurrent** (Fig. 1.6). In August 1975, we installed a temperature recorder at 2 m depth at one of the upwelling sites (Caleta Iguana, Isabela Island); when it was later retrieved in September it showed temperatures ranging between 14 to 17 °C (Glynn and Wellington 1983). When Jerry (Gerard) Wellington, Chuck (Charles) Birkeland, John Wells and I surveyed this site in 1975, it was a foggy morning, the water was frigid — bone chilling — and of low visibility, but the marine community was fascinating. A luxuriant kelp community, azooxanthellate corals, tiny abalone (*Haliotis dalli*) and a diverse temperate fish fauna were present just 110 km south of the equator, all characteristic of much higher latitudes.

In addition to sea temperature, several other environmental factors can have important effects on eastern Pacific zooxanthellate corals. Joanie (Joan Ann) Kleypas, National Center for Atmospheric Research, Boulder, Colorado, and colleagues analyzed the limiting environmental conditions in the context of global coral reef distributions and paid special attention to "marginal" reef environments (Kleypas et al. 1999). Photosynthetic corals are very sensitive organisms; they respond to changing light conditions, salinity, and the concentrations of carbonate ions and nutrients dissolved in sea water. Since coral holobionts are autotrophic their symbiotic algae require adequate light, photosynthetic active radiation (PAR), to perform the essential metabolic function of photosynthesis. Both light levels (irradiance) and salinity fluctuate substantially along the Pacific coast of Colombia and other eastern Pacific areas of high rainfall. The Chocó coast of Colombia boasts one of the highest annual rainfall totals in the world, with values ranging from 7000 to 9000 mm per year and reaching as high as 13,000 mm in some areas. The resulting river discharges cause marked reductions in salinity and high siltation. Rivers also introduce nutrients to coastal areas, and the frequent high nutrient loads can stimulate phytoplankton blooms, which together with rainfall-associated cloud cover greatly reduce light penetration into sea water. Nutrient pulses from natural runoff and anthropogenic pollution sources, such as fertilizers and sewage, can also lead to the proliferation of competing benthic algae, and bacteria and viruses. Virulent microorganisms cause diseases and death of corals, in addition to negative impacts on the physiology of reef biota and slowing accretion of limestone frameworks (Koop et al. 2001; Rohwer and Youle 2010; Silbiger et al. 2018).

Once while anchored at Punta Entrada,[8] Gulf of Chiriquí, a major river drainage area in southwestern Panama, our team was surprised to see several small pocilloporid patch reefs just inside the estuary. These reefs were exposed on an outgoing spring (extreme) tide. A quick scuba inspection, our boat captain was anxious to exit the estuary, revealed that the reefs were constructed nearly entirely of pocilloporid

[8] Punta Entrada is a sheltered anchorage site and frequent entry point for boats of shallow draft. We often used this estuarine waterway to transport supplies and personnel to our offshore island study sites.

corals with framework heights between 1 to 2 m. Unsurprisingly, the water column visibility was low, ranging between 1–2 meters. The coral community consisted of several species typically associated with pocilloporid reefs on the offshore islands: the Guineafowl Pufferfish *Arothron meleagris*, and coral crustacean symbionts (crabs and shrimp) and hermit crab corallivores (coral predators). We speculated that a **saltwater wedge**, a high-density saline layer that forms along the bottom of estuaries, might provide suitable conditions for these corals in an otherwise hostile brackish water environment. It would seem worthwhile to explore the extent of this coral development because such isolated reef patches, shielded perhaps from offshore/oceanic stress effects, might serve as refugia during ENSO bleaching events.

Aragonite saturation state, expressed as $\mathbf{\Omega_{arag}}$, is a measure of a solution's tendency to precipitate or dissolve aragonite (a specific crystal form of calcium carbonate). Ω is the Greek letter omega, and the "arag" subscript refers to aragonite that corals use to build their skeletons. Modern-day coral reefs have grown and developed under values of $\Omega_{arag} > 3.3$, with most reefs experiencing Ω_{arag} values of 4 or greater during preindustrial times (Hoegh-Guldberg et al. 2007). The oceans have absorbed about 1/3 of the anthropogenic carbon dioxide (CO_2) produced from the burning of fossil fuels (Sabine et al. 2004). As CO_2 dissolves in sea water, it reacts with water to produce hydrogen ions, thus lowering the pH, a process termed **ocean acidification** (Caldeira and Wickett 2003). The addition of dissolved CO_2 alters the seawater carbonate equilibrium and depresses both pH and Ω_{arag}. As Ω_{arag} declines, coral growth rates decrease while abiotic and biologically-mediated dissolution by endolithic algae and clionaid sponges increase (Glynn and Manzello 2015). In marginal reef areas such as the Galápagos Islands, Ω_{arag} is slightly higher than 3 (Manzello et al. 2008, 2014). Kleypas et al. (1999) drew attention to this factor—ocean acidification — in general terms and to its potential increasing importance due to the rising levels of anthropogenic greenhouse gasses. In summary, six physical factors affect healthy coral growth: temperature, light, salinity, nutrients, siltation, and carbonate saturation state.

Tectonic events that cause earthquakes, landslides and tsunamis occur relatively frequently in the equatorial eastern Pacific. Few such disturbances, however, have been documented to affect coral reefs. The Tumaco earthquake of 1979, magnitude 7.9–8.1, caused little damage to an area supporting abundant corals and a small patch reef at the south end of Gorgona Island, Colombia (Zapata and Vargas-Ángel 2003). A tsunami inundated the site and the local sea floor subsided 0.8 m, but the corals suffered no negative impacts (Herd et al. 1981). Uplift of reefs can be dramatic, however, as occurred on the Caribbean coast of Costa Rica (Cortés et al. 1992, 1994a). Coral reefs and limestone platforms were suddenly raised from 50 to 190 cm above sea level along 85 km of the Costa Rican coastline. These exposures resulted in mass mortalities of marine biota. Temblor-induced landslides along steep coastal areas can result in high siltation and burial of nearshore ecosystems (Garwood et al. 1979). In 1954, the sudden upward movement of magma along the Urvina Bay shelf, located on the Alcedo Volcano coast (Isabela Island, Galápagos Islands) resulted in several km^2 of coral-rich communities being raised 7 m above sea level (Couffer 1956; Colgan and Malmquist 1987) (Figs. 1.7 and 1.8). The

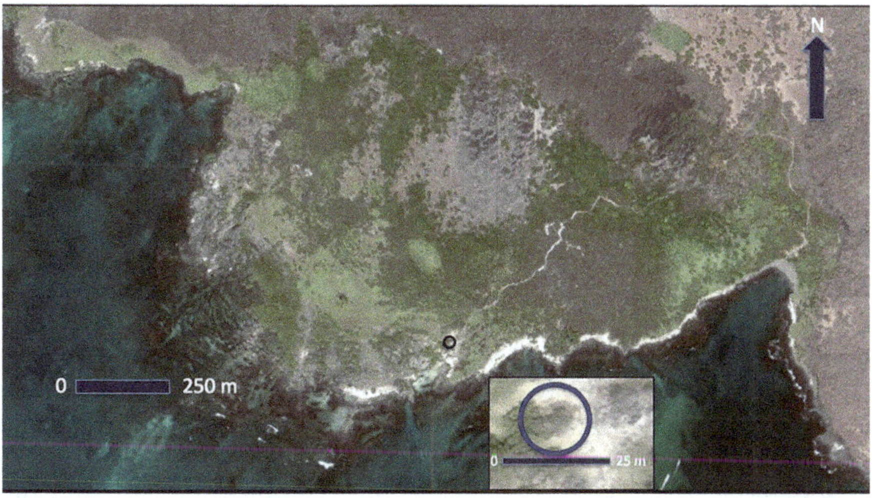

Fig. 1.7 Google Earth satellite image, from 24 December 2014, of the Urvina Bay uplift, Isabela Island, Galápagos Islands. The light green areas are vegetated and were uplifted in 1954. The inserted circle surrounds the large *Pavona clavus* colony. Located short distances north and east, respectively, are Volcano Darwin and Volcano Alcedo

Fig. 1.8 Mitchell Colgan inspecting a single microatoll colony of *Pavona clavus* at the Urvina Bay uplift site, Isabela Island (February 1986). From a detailed study of the skeletal growth record, this colony was determined to have an estimated age of 367 years in 1954 (Dunbar et al. 1994). The Darwin Volcano is visible in the distant background

uplifted shelf was investigated extensively by Mitchell Colgan and assistants in 1986–87 who produced detailed maps of the location and nature of pre-ENSO 1982–83 coral communities in this remote area of the Galápagos Islands (Colgan 1990).

Coral Reproduction

As noted above, corals can recover population losses from a variety of disturbances via sexual and asexual reproduction. Corals reproduce sexually in two ways: **spawning** and **brooding**. Coral spawners release gametes (eggs and sperm) into surrounding waters. Fertilization occurs externally and the fertilized egg, or zygote, then develops into a **planula** larva. Planulae remain in the plankton for hours, days or weeks depending on the species and environmental conditions. In brooding corals, released sperm find mature eggs either inside or on the surface of polyps. The fertilized eggs are retained for varying periods (weeks to months) in or on the maternal polyp.[9] The zygotes develop into planula larvae and are eventually released into the water column. Most zooxanthellate corals are broadcast spawners, including 13 of the 14 eastern Pacific species studied (Glynn and Colley 2008). *Porites panamensis* is the only zooxanthellate coral in the eastern Pacific that broods, but it is not known how long it retains developing larvae before they are released.

Coral Reef Classification, Morphology and Zonation

Although several major coral reef types have been described (e.g., Ladd 1973; Hopley 2011), three are widely recognized, namely **fringing**, **barrier**, and **atoll** coral reefs. Darwin (1842) described all three in his elegant and insightful hypothesis of reef development during the progressive subsidence of Pacific oceanic volcanoes. Fringing reefs, as the name implies, develop along shorelines around islands or continental land masses, often consisting of narrow, shallow lagoons and **reef flats** with scattered pools. The leeward reef flats, seldom exposed to strong turbulence, can support diverse communities of algae and metazoans that require some degree of shelter. Fringing reefs are common in the eastern Pacific and are constructed of branching pocilloporid corals and massive poritid or agariciid corals. The internal structure of one of several well-developed fringing reefs at Gorgona Island, Colombia, consists of a mix of branching and massive coral colonies, voids (empty spaces), and compacted sand sediments (Fig. 1.9).

[9] The retention of zygotes is probably highly variable among species and has not been adequately studied. The time that developing larvae are retained before release in *Porites panamensis*, the single zooxanthellate eastern Pacific brooder, is unknown. An azooxanthellate brooder in Pacific Panama, *Tubastraea coccinea*, retains planulae for about 6 weeks (Glynn et al. 2008).

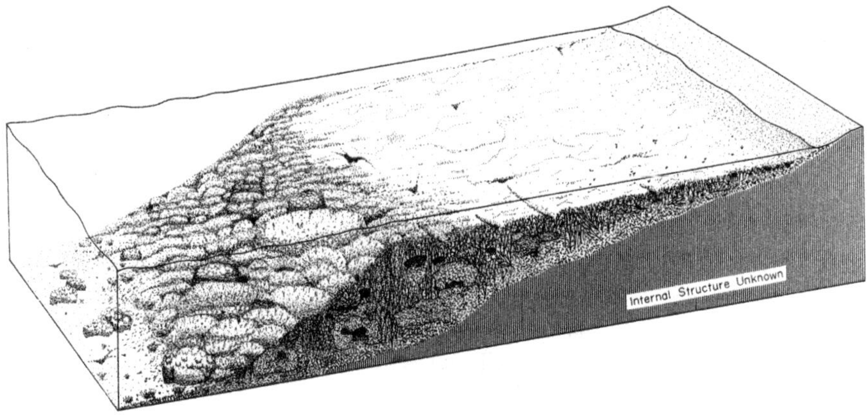

Fig. 1.9 Three dimensional view of a *Pocillopora* fringing reef (La Azufrada Reef) at Gorgona Island, Colombia (May 1979). The shore to reef-front section showing presumed internal structure is based on probing and retrieval of cored sediments. Approximate reef width (shoreline to reef base) ~ 120 m; maximum vertical framework development = 7–8 m. More detailed study of La Azufrada reef morphology has revealed five depth-related zones (from shoreline to seaward): back reef, reef flat, reef crest, reef front, and upper reef slope (Zapata and Vargas-Ángel 2003). Drawing by P. W. Glynn, modified after Glynn et al. 1982

Barrier reefs generally occur farther offshore and have a relatively deep lagoon between the back or leeward side of the reef and the shoreline. No barrier reefs have been reported in the eastern Pacific region. The near absence of broad continental shelves and the presence of steep drop offs of nearshore terrain are probably responsible for this absence (Glynn and Ault 2000). An atoll is a roughly circular ring of islets and reefs surrounding a central lagoon, although it may also be irregularly oval, elongated, or even polygonal. Breaks or passes across the reef allow an interchange of water between the lagoon and open sea. According to Darwin's model, the vertical growth of a reef around an island leads to the formation of a relatively deep lagoon as the island subsides. With continuing subsidence, vertical coral growth near the sea surface forms the atoll ring-like structure. Clipperton Island, about 1300 km SW of the Mexican mainland (Acapulco), is the only known atoll in the eastern Pacific (Charpy 2009; Glynn et al. 2017), although, strictly speaking, it is an "almost atoll", as a volcanic rock remains near an old passage on the southeast side of the reef (Trichet 2009).

In addition to these three reef types, **patch reefs** are relatively small and isolated coral buildups surrounded by sediments, often present within barrier reefs and atoll lagoons. In the eastern Pacific, pocilloporid patch reefs often build vertically to the extreme low tidal datum. Both local reef types, fringing and patch reefs, consist of coral frameworks ranging from 1 to 5 m in height, more rarely 10 to 13 m. Ian Macintyre and I probed and core-drilled a patch reef at Uva Island, Panama and found that it had attained at least 11 m in height.[10] Radiometric dating of a coral fragment recovered from the reef base, where reef growth was initiated, revealed a date of 5585 ± 75 years BP (before present). This pocilloporid patch reef began accreting near the end of the Stone Age (between 4000 BC and 2000 BC) and the beginning of the Egyptian Dynasties. Comparable reef growth initiation of a fringing, massive, poritid reef in Costa Rica began about 5500 years BP (Cortés et al. 1994b). Compared to coral reefs in the Indo-Pacific and wider Caribbean regions, which are constructed of numerous coral species (10 s to 100 s), eastern Pacific reefs consist of only five to <10 species in the genera *Pocillopora*, *Porites*, and *Pavona* (Cortés 1997; Glynn 2001).

Sealevel is continually changing, and the rate of change is critically important to reef growth. At the end of the Last Glacial Period (Pleistocene Epoch) sea level rise was so rapid that reef growth could not keep pace with it. Beginning approximately 11,500 years ago (Holocene Epoch) the advancing sea began to decelerate and by 6–7 thousand years ago reef growth could keep up with rising sea level. At the Fifth International Coral Reef Congress in Tahiti in 1985, Conrad Neumann and Ian Macintyre proposed a simplified but novel scheme to explain reef growth in relation to rates of sea level rise. Their presentation created a palpable buzz of excitement at this meeting. Keep-up, catch-up, give-up was the circulating refrain among many of the congress's attendees. The age estimates of the reefs in Panama and Costa Rica suggested keep-up reef development. If current projections for global warming and accelerated sea level rise are realized, eastern Pacific reefs (and global reefs) will soon enter the give-up growth mode (Neumann and Macintyre 1985).

I suggest the name **bank reef** for an eastern Pacific reef structure that does not correspond well with any of the aforenamed reef types. This reef type consists chiefly of massive coral species such as *Porites lobata* and *Pavona* spp., lies at various distances from the shoreline, and is rarely subaerially exposed. Bank reefs also occur in deeper and more exposed open water environments and are often linear in shape. The longest known eastern Pacific bank reef is located at Cocos Island (Costa Rica) and is slightly more than 2 km in length (Guzmán and Cortés 1992, 2007; Glynn et al. 2017). The Wellington Reef at Darwin Island, Galápagos Islands, just under 1 km in length, is another example of an eastern Pacific bank reef (Glynn et al. 2015). The incipient bank reef structure overtopping a lava tube off Motu Tautara at Easter Island suggests how a reef might orient linearly as coral growth advances along an elongated basalt rock basement (Hubbard and Garcia 2003).

[10] Due to the relative ease of penetrating pocilloporid reef frameworks, we developed a black-iron pipe probe that could be hammered and rotated into reefs to determine their thicknesses (Glynn and Macintyre 1977). The individual pipe sections were one meter in length and could be screwed together to create pipe strings up to 10 meters in height or higher. On one occasion, when Aníbal Velarde and I were probing a pocilloporid fringing reef at low tide on Pedro González Island (Pearl Islands, Panama) the pipe sections thankfully thwarted an assault. Four thugs began to approach us, and Aníbal alerted me that we were about to be attacked. The nearby village had a tarnished reputation for supporting robberies targeting outsiders. We each brandished a section of iron pipe and admonished the gang to cease and desist. Our threat had succeeded; we were able to finish the reef probing before the rising tide (Mother Nature) forced us off the reef flat!

To characterize and compare reef structures and habitats, reefs can be subdivided into different zones, distinguished by physical features such as depth, wave energy, and the underlying topography. Three major zones have long been recognized—**reef flat**, **forereef slope**, and **reef base**. Reef flats of fringing reefs lie near the shoreline. They can vary from a few meters to 50–100 meters in breadth (from inshore to offshore), and as reef flats are often exposed at extreme low tides, they represent the upper limit of coral growth. Pocilloporid corals are usually predominant on eastern Pacific reef flats, and their live cover is closely related to extreme low tidal exposures. A series of extreme mid-day exposures can cause high coral mortality and low live coral cover, a common occurrence on reef flats across the equatorial eastern tropical Pacific (Glynn 1976; Eakin and Glynn 1996; Cortés and Jiménez 2003; Maté 2003; Zapata and Vargas-Ángel 2003). Massive corals, especially *Porites lobata*, are also present on some reef flats. Coral tissues at the uppermost surfaces of a colony are killed due to tide-related exposure; the lower submerged edges survive and continue to grow, producing a raised living coral rim surrounding a depressed, often sediment filled center that resembles a miniature atoll (Fig. 1.10). This type of altered colony is termed a **microatoll**. Microatolls on the Caño Island reef flat include several colonies nearly 2 m in diameter and 200 to 300 years in age (Guzmán and Cortés 1989; Macintyre et al. 1992). The exposed windward or seaward edge of a reef flat is often slightly elevated and termed the **reef crest**. Crustose coralline algae that flourish in this wave-exposed environment are responsible for the slight increase in elevation. Algal ridges on fringing reefs occur at several eastern Pacific sites, for example on the western, exposed side of Gorgona Island (Glynn et al. 1982; Zapata and Vargas-Ángel 2003). One must exercise caution when navigating this feature at low tide, due to the fragility of the coral/algal ridge, and the presence of long spined sea urchins (*Diadema mexicanum*) and calcareous vermetid tubes (a gastropod mollusk) with razor-sharp rims.

The seaward slopes of fringing reefs commonly descend into deeper water at angles of 10° to 30° and generally support high pocilloporid cover. Massive corals are also often present in this zone, such as at Punta Islotes, Golfo Dulce, Costa Rica (Cortés 1990). Coral cover is generally high, the angle of decline lessens with depth and the slope zone merges into the reef base at 5 to 15 m depth. Coral framework blocks and massive overturned colonies, calved from steep reef slopes, are sometimes scattered about the lower reef slope and reef base, the result of dislodgement and down-slope slumping due to storms, seismic activity, and bioerosion (Lirman et al. 2001; Alvarado et al. 2017). Coral abundance tends to diminish at the reef base, commonly between 5–15 m depth, depending on water transparency, light penetration, and turbulence. Carbonate rubble and sand also increase in the reef base zone. Coral **richness**, the number of species in a habitat, is usually greatest in the lower seaward slope and reef base zones and may include mound, platy, nodular, and encrusting colony forms (Reyes-Bonilla 2003; Zapata and Vargas-Ángel 2003). This may be due to lessened interspecific competition among corals for substrate space (Glynn et al. 1972).

Clipperton Atoll has the steepest reported reef slope in the eastern Pacific of 80° + and extends from about 20 to 60 m depth (Glynn et al. 1996; Glynn 2001;

Fig. 1.10 *Porites lobata* microatolls on the northwestern reef flat of Caño Island, Costa Rica (January 1986). Extreme low tidal exposure with numerous 1–2 m diameter microatolls in view. A, Héctor M. Guzmán (left) and Peter W. Glynn holding a 1 m^2 quadrat. B, A 2+ diameter microatoll with an eroded central area surrounded by live submerged tissues which accreted laterally. Colonies 1.4–1.8 m in diameter revealed C-14 ages ranging from 140 to 240 ± 60 yrs. B. P. (before present) (Macintyre et al. 1992). Photographs courtesy of I. G. Macintyre

Charpy 2009). Terraces at 15–18 m and 50–58 m depth are present on the seaward slope likely reflecting erosion that occurred when sea level was lower (e.g., Chevalier 1973; Ladd 1973). Here four coral colony forms are common: massive, massive with skirt-like plates, plates, and foliose. Patch reef and bank reef slopes can also support a combination of branching and massive/platy coral species, with the latter two forms prevalent on bank reefs. Although reef base coral zones often exhibit the greatest coral species richness, they do not necessarily support the greatest density or abundance of live coral, i.e., coral cover. Generally, most reef base zones are shallowest (2–5 m) at near-shore sites and increase in depth at offshore islands (10–15 m) where water transparency is higher. Clipperton Atoll supports deep, reef base zooxanthellate corals, for example abundant assemblages of *Pavona varians* at 70 m depth (Glynn et al. 1996).

Recent ROV (remotely operated vehicle) surveys of deep, off-reef habitats have revealed zooxanthellate coral patches at depths exceeding 100 m at Rapa Nui (Easter Island) (see Chaps. 2 and 13). These deep coral assemblages are termed **mesophotic coral ecosystems**; note the term "coral ecosystems" instead of "coral reefs." This is an accurate distinction because deep-occurring corals are not involved in reef building. They can, however, form low profile patches ~1 m in height and consist of dense coral populations. Rapa Nui is an oceanic island with clear waters allowing for deep light penetration. In addition, it is surrounded by a deep surface mixed layer ranging between 80–115 m in depth (Moraga et al. 1999). Hoeksema et al. (2019) reported mesophotic zooxanthellate mushroom coral populations of approximately 500 individuals m^{-2} at Rapa Nui. The coral species at these greater depths belong to scleractinian taxa, both symbiotic (e.g., *Leptoseris*, *Agaricia*, and *Madrepora*) and asymbiotic species known to inhabit dimly lit environments. Genomic studies indicate that there may exist many more mesophotic zooxanthellate coral taxa globally than currently recognized (Gijnbers et al. 2023). Jorge Cortés and colleagues performed extensive submersible surveys at Cocos Island, but did not encounter deep coral communities (Cortés 2016). Based on the absence of deep-occurring zooxanthellate corals at Cocos Island and knowledge of water column characteristics in the equatorial eastern Pacific, Jorge concluded that mesophotic coral ecosystems are probably absent from this region due to the following limiting conditions (Cortés 2019): (a) a shallow thermocline (~ 50 m depth), with reduced light levels at the thermocline transition depth, (b) relatively low temperatures, (c) low oxygen concentration, and (d) a low carbonate saturation state (Ω_{arag}).

References

Alvarado JJ, Grassian B, Cantera-Kintz JR, Carballo JL, Londoño-Cruz E (2017) Coral reef bioerosion in the eastern tropical Pacific. In: Glynn PW, Manzello DP, Enochs IC (eds) Coral reefs of the eastern tropical Pacific: persistence and loss in a dynamic environment. Coral reefs of the world 8. Springer, Dordrecht, pp 369–403

Baker AC, Glynn PW, Riegl B (2008) Climate change and coral reef bleaching: an ecological assessment of long-term impacts, recovery trends and future outlook. Est Coast Shelf Sci 80:435–471. https://doi.org/10.1016/j.ecss2008.09.003

Baums IB, Devlin-Durante M, Laing BAA, Feingold J, Smith T, Bruckner A, Monteiro J (2014) Marginal coral populations: the densest known aggregation of *Pocillopora* in the Galápagos archipelago is of asexual origin. Front Mar Sci 1:59. https://doi.org/10.3389/fmars.2014.00059

Brown BE (1987) Worldwide death of corals: natural cyclical events or man-made pollution? Mar Pollut Bull 18:9–13

Caldeira K, Wickett ME (2003) Anthropogenic carbon and ocean pH. Nature 425:365

Charpy L (2009) Clipperton, environnement et biodiversité d'un microcosme océanique. Muséum national d'Histoire naturelle, Paris. IRD, Marseille, p 420 (Patrimoines naturels 68)

Chevalier JP (1973) Coral reefs of New Caledonia. In: Jones OA, Endean R (eds) Biology and geology of coral reefs, geology 1. Academic, New York, pp 143–167

Colgan MW (1990) El Niño and the history of eastern Pacific reef building. In: Glynn PW (ed) Global ecological consequences of the 1982–83 El Niño-Southern Oscillation. Elsevier oceanography series 52. Elsevier, Amsterdam, pp 183–232

Colgan MW, Malmquist D (1987) The Urvina Bay uplift: a dry trek through a coral community. Oceanus 30:61–66

Cornwall CE, Carlot J, Branson O, Courtney TA, Harvey BP, Perry CT, Andersson AJ, Diaz-Pulido G, Johnson MD, Kennedy C (2023) Crustose coralline algae can contribute more than corals to coral reef carbonate production. Comm Earth Environ 4:105. https://doi.org/10.1038/s43247-023-00766-w

Cortés J (1990) The coral reefs of Golfo Dulce, Costa Rica: distribution and community structure. Atoll Res Bull 344-1-37:1–37

Cortés J (1997) Biology and geology of eastern Pacific coral reefs. In: Proceedings of the eighth international coral reef symposium, Panama, vol 1, pp 57–64

Cortés J (2011) Eastern tropical Pacific coral reefs. In: Hopley D (ed) The encyclopedia of modern coral reefs: structure, form and process. Springer, Berlin, pp 351–358

Cortés J (2016) Isla del Coco: coastal and marine ecosystems. In: Kappelle M (ed) Costa Rican ecosystems. University of Chicago Press, Chicago, pp 162–191

Cortés J (2019) Isla del Coco, Costa Rica, eastern tropical Pacific. In: Loya Y, Puglise KA, Bridge T (eds) Mesophotic coral ecosystems. Coral reefs of the world 12. Springer, Berlin, pp 465–475. https://doi.org/10.1007/978-3-319-92735-0-26

Cortés J, Jiménez C (2003) Corals and coral reefs of the Pacific coast of Costa Rica: history, research and status. In: Cortés J (ed) Latin American coral reefs. Elsevier Science, Amsterdam, pp 361–385

Cortés J, Soto R, Jiménez C (1994a) Efectos ecológicos del Terremoto de Limón. Revista Geológica de América Central, Volumen Especial sobre el Terremoto de Limón, pp 187–192

Cortés J, Macintyre IG, Glynn PW (1994b) Holocene growth history of an eastern Pacific fringing reef, Punta Islotes, Costa Rica. Coral Reefs 13:65–73

Cortés J, Soto R, Jiménez C, Astorga A (1992) Earthquake associated mortality of intertidal and coral reef organisms. In: Proceedings of the seventh international coral reef symposium, Guam, vol 1, pp 235–240

Cortés J, Enochs IC, Sibaja-Cordero J, Hernández L, Alvarado JJ, Breedy O, Cruz-Barraza JA, Esquivel-Garrote O, Fernández-García C, Hermosillo A et al (2017) Marine biodiversity of eastern tropical Pacific coral reefs. In: Glynn PW, Manzello DP, Enochs IC (eds) Coral reefs of the eastern tropical Pacific, persistence and loss in a dynamic environment, coral reefs of the world 8. Springer, Cham, pp 203–250

Couffer JC (1956) The disappearance of Urvina Bay. Nat Hist 65:78–83

Cruz-García R, Rodríguez-Troncoso AP, Rodríguez-Zaragoza FA, Mayfield A, Cupul-Magaña AL (2020) Ephemeral effects of El Niño Southern Oscillation events on an eastern tropical Pacific coral community. Mar Freshw Res 71:1259. https://doi.org/10.1071/MF18481

Darwin CR (1842) The structure and distribution of coral reefs. Being the first part of the geology of the voyage of the *Beagle*, under the command of Capt. Fitzroy, R. N. during the years 1832 to 1836, vol 12. Smith Elder, London, p 115

De'ath G, Fabricius KE, Sweatman H, Puotinen M (2012) The 27-year decline of coral cover on the Great Barrier Reef and its causes. Proc Nat Acad Sci USA 109(44):17995–17999. https://doi.org/10.1073/pnas.1208909109

Dunbar RB, Wellington GM, Colgan MW, Glynn PW (1994) Eastern Pacific sea surface temperature since 1600 A. D.: the $\delta^{18}O$ record of climate variability in Galápagos. Paleoceanography 9:291–316

Eakin CM (1987) Damselfishes and their algal lawns: a case of plural mutualism. Symbiosis 4(1–3):275–288

Eakin CM (1992) Post-El Niño Panamanian reefs: less accretion, more erosion and damselfish protection. In: Proceedings of seventh international coral reef symposium, vol 1, pp 387–396

Eakin CM, Glynn PW (1996) Low tidal exposures and reef mortalities in the eastern Pacific. Coral Reefs 15:120

Earle SA (1972) A review of the marine plants of Panamá. In: Jones ML (ed) The Panamic biota: some observations prior to a sea-level canal. Bull Biol Soc Wash 2:69–87

Enochs IC, Glynn PW (2017) Corallivory in the eastern Pacific. In: Glynn PW, Manzello DP, Enochs IC (eds) Coral reefs of the eastern tropical Pacific: persistence and loss in a dynamic environment. Coral reefs of the world 8. Springer, Dordrecht, pp 315–337. https://doi.org/10.1007/978-94-017-7499-4_11

Fiedler PC, Lavin MF (2017) Oceanographic conditions of the eastern tropical Pacific. In: Glynn PW, Manzello DP, Enochs IC (eds) Coral reefs of the eastern tropical Pacific: persistence and loss in a dynamic environment. Coral reefs of the world 8. Springer, Dordrecht, pp 59–83

Fong P, Smith TB, Muthukrishnan R (2017) Algal dynamics: alternate stable states of reefs in the eastern tropical Pacific. In: Glynn PW, Manzello DP, Enochs IC (eds) Coral reefs of the eastern tropical Pacific: persistence and loss in a dynamic environment. Coral reefs of the world 8. Springer, Dordrecht, pp 339–367. https://doi.org/10.1007/978-94-017-7499-4_11

Garwood NC, Janos DP, Brokaw N (1979) Earthquake caused landslides: a major disturbance to tropical forests. Science 205:997–999

Gijnbers JC, Englebert N, Prata KE, Pichon M, Dinesen Z, Brunner R, Eyal G, Gonzalez-Zapata FL, Kahng SE, Latijnhouwers KRW, Muir P, Radice VZ, Sanchez JA, Vermeij MJA, Hoegh-Guldberg O, Jacobs SJ, Bongaerts P (2023) Global phylogenomic assessment of *Leptoseris* and *Agaricia* reveals substantial undescribed diversity at mesophotic depths. BMC Biol 21:147. https://doi.org/10.1186/s12915-023-01630-1

Glynn PW (1972) Observations on the ecology of the Caribbean and Pacific coasts of Panamá. In: Jones ML (ed) The Panamic biota: some observations prior to a sea-level canal. Bull Biol Soc Wash 2:13–30

Glynn PW (1973a) Aspects of the ecology of coral reefs in the western Atlantic region. In: Jones OA, Endean R (eds) Biology and geology of coral reefs, vol II, Biology 1. Academic Press, New York, pp 271–324

Glynn PW (1973b) Ecology of a Caribbean coral reef. The *Porites* reef-flat biotope: part I. Meteorology and hydrography. Mar Biol 20(4):297–318

Glynn PW (1973c) Ecology of a Caribbean coral reef. The *Porites* reef flat biotope. Part II. Plankton community with evidence for depletion. Mar Biol 22:1–21

Glynn PW (1976) Some physical and biological determinants of coral community structure in the eastern Pacific. Ecol Monogr 46(4):431–456

Glynn PW (1984) Widespread coral mortality and the 1982-83 El Niño warming event. Environ Conserv 11(2):133–146

Glynn PW (1988) El Niño warming, coral mortality and reef framework destruction by echinoid bioerosion in the eastern Pacific. Galaxea 7(2):129–160

Glynn PW (1990) Global ecological consequences of the 1982–83 El Niño-Southern Oscillation. Elsevier oceanography series 52, Elsevier, Amsterdam, p 563

Glynn PW (1993) Monsoonal upwelling and episodic *Acanthaster* predation as probable controls of coral reef distribution and community structure in Oman, Indian Ocean. Atoll Res Bull 379:1–66

Glynn PW (2001) Eastern Pacific coral reef ecosystems. In: Seeliger U, Kjerfve B (eds) Coastal marine ecosystems of Latin America. Ecological studies, vol 144. Springer, Berlin, pp 281–305

Glynn PW, Colgan MW (1988) Defense of corals and enhancement of coral diversity by territorial damselfishes. In: Proceedings of the 6th international coral reef symposium, Australia, vol 2, pp 157–163

Glynn PW, Ault JS (2000) A biogeographic analysis and review of the far eastern Pacific coral reef region. Coral Reefs 19:1–23

Glynn PW, Colley SB (2008) Survival of brooding and broadcasting reef corals following large scale disturbances: is there any hope for broadcasting species during global warming? In: Proceedings of 11th international coral reef symposium, Ft. Lauderdale, vol 1, pp 368–372

Glynn PW, Fong P (2006) Patterns of reef coral recovery by the regrowth of surviving tissues following the 1997-98 El Niño warming and 2000, 2001 upwelling cool events in Panamá, eastern Pacific. In: Proceedings of 10th international coral reef symposium, Okinawa, pp 624–630

Glynn PW, Wellington GM (1983) Corals and coral reefs of the Galápagos Islands. With an annotated list of the scleractinian corals of the Galápagos by John W. Wells. University of California Press, Berkeley, p 330

Glynn PW, Macintyre IG (1977) Growth rate and age of coral reefs on the Pacific coast of Panama. In: Proceedings of 3rd international coral reef symposium, Miami, vol 2, pp 251–259

Glynn PW, Manzello DP (2015) Bioerosion and coral reef growth; a dynamic balance. In: Birkeland C (ed) Coral reefs in the Anthropocene. Springer, Dordrecht, pp 67–97. https://doi.org/10.1007/978-94-017-7249-5_4

Glynn PW, Stewart RH, McCosker JE (1972) Pacific coral reefs of Panama: structure, distribution, and predators. Geol Rundsch 61(2):483–519

Glynn PW, von Prahl H, Guhl F (1982) Coral reefs of Gorgona Island, Colombia, with special reference to corallivores and their influence on community structure and reef development. Anales del Instituto de Investigaciones Marinas de Punta de Betín 12:185–214

Glynn PW, Veron JEN, Wellington GM (1996) Clipperton Atoll (eastern Pacific): oceanography, geomorphology, reef-building coral ecology and biogeography. Coral Reefs 15:71–99

Glynn PW, Riegl B, Purkis S, Kerr JM, Smith TB (2015) Coral reef recovery in the Galápagos Islands: the northernmost islands (Darwin and Wenman). Coral Reefs 34:421–436

Glynn PW, Colley SB, Maté JL, Cortés J, Guzmán HM, Bailey RL, Feingold JS, Enochs IC (2008) Reproductive ecology of the azooxanthellate coral *Tubastraea coccinea* in the equatorial eastern Pacific: part V. Dendrophylliidae. Mar Biol 153:529–544

Glynn PW, Alvarado JJ, Banks S, Cortés J, Feingold JS, Jiménez C et al (2017) Eastern Pacific coral reef provinces, coral community structure and composition: an overview. In: Glynn PW, Manzello DP, Enochs IC (eds) Coral reefs of the eastern tropical Pacific: persistence and loss in a dynamic environment. Coral reefs of the world 8. Springer, Dordrecht, pp 107–176

Glynn PW, Dillehay TD, Netherly PJ, Riegl BM (2022) Prehistoric reef-building coral occurrence in North Peru J Island Coastal Archaeol 19:420–432. https://doi.org/10.1080/15564894.2022.2072026

Guan Y, Mai J, Wang H, Zhang P, Huang C, Liu Z, Zhan X, De Cesare M, He X, Wang X, Ye M (2021) Plutonium isotopes and radionuclides in corals around Weizhou land in Beibu Gulf, China. Appl Radiat Isot 176:109873. https://doi.org/10.1016/j.apradiso.2021.109873

Guzmán HM, Cortés J (1989) Coral reef community structure at Caño Island, Pacific Costa Rica. Pubbl Staz Zool Napoli (I: Marine Ecology) 10:23–41

Guzmán HM, Cortés J (1992) Cocos Island (Pacific of Costa Rica) coral reefs after the 1982-83 El Niño disturbance. Rev Biol Trop 40:309–324

Guzmán HM, Cortés J (2001) Changes in reef community structure after fifteen years of natural disturbances in the eastern Pacific (Costa Rica). Bull Mar Sci 69:133–149

Guzmán HM, Cortés J (2007) Reef recovery 20 years after the 1982-1983 El Niño massive mortality. Mar Biol 151:401–411

Herd DG, Youd L, Meyer H, Arango JL, Pearson WJ, Mendoza C (1981) The great Tumaco, Colombia earthquake of 12 December 1979. Science 211:441–445

Hieri F, Wu HC, Westphal H (2021) Scleractinian corals incorporate microplastic particles: identification from a laboratory study. Environ Sci Pollut 28(28):37882–37893. https://doi.org/10.1007/s11356-021-13240-x

Highsmith RC (1980) Geographic pattern of coral bioerosion: a productivity hypothesis. J Exp Mar Biol Ecol 46:177–196

Hoegh-Guldberg O, Mumby PJ, Hooten AJ, Steneck RS, Greenfield P, Gomez E, Harvell CD, Sale PF, Edwards AJ, Caldeira K, Knowlton N, Eakin CM, Iglesias-Prieto R, Muthiga N, Bradbury RH, Dubi A, Hatziolos ME (2007) Coral reefs under rapid climate change and ocean acidification. Science 318:1737–1742. https://doi.org/10.1126/science.1152509

Hoeksema BW, Sellanes J, Easton EE (2019) A high-latitude, mesophotic *Cycloseris* field at 85 m depth off Rapa Nui (Easter Island). Bull Mar Sci 95(1):101–102. https://doi.org/10.5343/bms.2018.0053

Hopley D (2011) Reef classification by Hopley (1982). In: Hopley D (ed) Encyclopedia of modern coral reefs: structure, form and process. Springer, Dordrecht, pp 849–854

Hubbard DK (2011) Reef drilling. In: Hopley D (ed) The encyclopedia of modern coral reefs: structure, form and process. Springer, Berlin, pp 856–868

Hubbard D, Garcia M (2003) The corals and coral reefs of Easter Island—a preliminary look. In: Loret J, Tanacredi JT (eds) Easter Island: scientific exploration into the world's environmental problems in microcosm. Kluwer Academic, New York, pp 53–77

Hughes TP, Kerry JT, Baird AH, Connelly SR, Dietzel A, Eakin CM, Heron SF, Hoey AS, Hoogenboom MO, Liu G, McWilliam MJ, Pears RJ, Pratchett MS, Skirving WJ, Stella JS, Torda G (2018) Global warming transforms coral reef assemblages. Nature 556(7702):492–496. https://doi.org/10.1038/s41586-018-0041-2

Jones OA, Endean R (eds) (1973) Biology and geology of coral reefs, vol 2, Biol. 1. Academic Press, New York

Jones OA, Endean R (eds) (1977) Biology and geology of coral reefs, vol 4, Geol. 2. Academic Press, New York

Koop K, Booth D, Broadbent A, Brodie J, Bucher D, Capone D, Coll J (2001) Encore: the effect of nutrient enrichment on coral reefs, synthesis of results and conclusions. Mar Pollut Bull 42:91–120

Kleypas JA, McManus JW, Meñez LAB (1999) Environmental limits to coral reef development: where do we draw the line? Am Zool 39:146–159

Ladd HS (1973) Bikini and Eniwetok atolls, Marshall Islands. In: Jones OA, Endean R (eds) Biology and geology of coral reefs, geology 1. Academic Press, New York, pp 93–112

Lirman D, Glynn PW, Baker AC, Leyte Morales GE (2001) Combined effects of three sequential storms on the Huatulco coral reef tract, Mexico. Bull Mar Sci 69(1):267–278

Macintyre IG, Glynn PW (1976) Evolution of modern Caribbean fringing reef, Galeta Point, Panamá. Am Assoc Pet Geol Bull 60(7):1054–1072

Macintyre IG, Glynn PW, Cortés J (1992) Holocene reef history in the eastern Pacific: mainland Costa Rica, Caño Island, Cocos Island, and Galápagos Islands. In: Proceedings of the 7th international coral reef symposium, Guam, vol 2, pp 1174–1184

Manzello DP (2010) Ocean acidification hotspots: spatiotemporal dynamics of the seawater CO_2 system of eastern Pacific coral reefs. Limnol Oceanogr 55:239–248

Manzello DP, Kleypas JA, Budd DA, Eakin CM, Glynn PW, Langdon C (2008) Poorly cemented coral reefs of the eastern tropical Pacific: possible insights into reef development in a high-CO_2 world. Proc Nat Acad Sci USA 105(30):10450–10455. https://doi.org/10.1073/pnas.0712167105

Manzello DP, Enochs IC, Bruckner A, Renaud PG, Kolodziej G, Budd DA, Carlton R, Glynn PW (2014) Galápagos coral reef persistence after ENSO warming across an acidification gradient. Geophys Res Lett 41:9001–9008. https://doi.org/10.1002/2014GL062501

Maté JL (2003) Corals and coral reefs of the Pacific coast of Panamá. In: Cortés J (ed) Latin American coral reefs. Elsevier Science, Amsterdam, pp 387–417

Moraga J, Valle-Levinson A, Oliveres J (1999) Hydrography and geostrophy around Easter Island. Deep-Sea Res I Oceanogr Res Pap 46(4):715–731

Neumann AC, Macintyre IG (1985) Reef response to sea-level rise: keep-up, catch-up, or give-up. In: Proceedings of fifth international coral reef congress, Tahiti, vol 3, pp 205–210

Ogden JC (1997) Ecosystem interactions in the tropical coastal seascape. In: Birkeland C (ed) Life and death of coral reefs. Chapman and Hall, New York, pp 288–297

Pandolfi JM, Jackson JBC, Baron N, Bradbury RH, Guzman HM et al (2005) Are US coral reefs on the slippery slope to slime? Science 307(5716):1725–1726

Pinzón JH, Reyes-Bonilla H, Baums IB, LaJeunesse TC (2012) Contrasting clonal structure among *Pocillopora* (Scleractinia) communities at two environmentally distinct sites in the Gulf of California. Coral Reefs 31:765–777. https://doi.org/10.1007/s00338-012-0887-y

Rohwer F, Youle M (2010) Coral reefs in the microbial seas. Plaid Press, Granada Hills, p 201

Polis GA, Hurd SD (1995) Extraordinarily high spider densities on islands: flow of energy from the marine to terrestrial food webs and the absence of predation. Proc Nat Acad Sci USA 92(10):4382–4386

Polis GA, Anderson WB, Holt RD (1997) Toward an integration of landscape and food web ecology: the dynamics of spatially subsidized food webs. Annu Rev Ecol Syst 28(1):289–316. https://doi.org/10.1146/annurev.ecolsys.28.1.289

Reyes-Bonilla H (2003) Coral reefs of the Pacific coast of México. In: Cortés J (ed) Latin American coral reefs. Elsevier, Amsterdam, pp 331–349

Romero-Torres M, Acosta A, Palacio-Castro AM, Treml EA, Zapata FA, Paz-García DA, Porter JW (2020) Coral reef resilience to thermal stress in the eastern tropical Pacific. Glob Chang Biol 26:3880–3890. https://doi.org/10.1111/gcb.15126

Sabine CL, Feely RA, Gruber N, Key RM, Lee K, Bullister JL, Wanninkhof R, Wong CS, Wallace DWR, Tillbrook B, Millero FJ, Peng T-H, Kozyr A, Ono T, Rios AF (2004) The oceanic sink for anthropogenic CO_2. Science 305(5682):367–371

Sheppard CR (2000) Coral reefs of the Western Indian Ocean: an overview. In: McClanahan TR, Sheppard CRC, Obura DO (eds) Coral reefs of the Indian Ocean: their ecology and conservation. Oxford University Press, Oxford, pp 3–38

Silbiger NJ, Nelson CE, Remple K, Sevilla JK, Quinlan ZA, Putnam HM, Fox MD, Donahue MJ (2018) Nutrient pollution disrupts key ecosystem functions on coral reefs. Proc R Soc B 285(1880):20172718

Steller DL, Foster MS, Riosmena-Rodríguez R (2009) Living rhodolith bed ecosystems in the Gulf of California. In: Johnson JM, Ledesma-Varguez J (eds) Atlas of coastal ecosystems in the Gulf of California: past and present. University of Arizona Press, Tucson, pp 72–82

Stetson TR, Fleming D, Squires D, Pratt R (1962) Coral banks occurring in deep water on the Blake plateau. Am Mus Novit 2114:39

Stoddart DR (1969) Ecology and morphology of recent coral reefs. Biol Rev 44:433–498

Trichet J (2009) Origine, situation et traits morphologiques généraux de I'ile. In: Charpy L (ed) Clipperton, environnement et biodiversité d'un microcosme océanique, Muséum national d'Histoire naturelle, Paris, IRD, Marseille, Patrimoines naturels, vol 68, pp 23–28

Vaughan TW (1919) Corals and the formation of coral reefs. Annu Rep Smithson Inst 17:189–238

Wainwright SA (1965) Reef communities visited by the South Red Sea expedition 1962. Bull Sea Fish Res Sta Israel

Wellington GM (1982) Depth zonation of corals in the Gulf of Panama: control and facilitation by resident reef fishes. Ecol Monogr 52:223–241

Wulff JL, Buss LW (1979) Do sponges help hold coral reefs together? Nature (London) 281:474–475

Zapata FA (2017) Temporal dynamics of coral and algal cover and their drivers on a coral reef of Gorgona Island, Colombia (eastern tropical Pacific). Rev Acad Colomb Cienc Ex Fis Nat 41:298–310. https://doi.org/10.18257/raccefyn.486

Zapata FA, Vargas-Ángel B (2003) Corals and coral reefs of the Pacific coast of Colombia. In: Cortés J (ed) Latin American coral reefs. Elsevier, Amsterdam, pp 419–447

Chapter 2
Coral Reef Discoveries

Abstract Exploration and adventures of the little known reef-building coral fauna in the equatorial eastern Pacific (Panama, Costa Rica, Colombia, Ecuador, including the Galápagos Islands) in the 1960s and 1970s revealed abundant coral communities and coral reefs, and noteworthy experiences. Aerial photography performed during midday extreme low tides followed by ground truth scuba diving were effective means in locating coral formations. Among the few reef-building coral species encountered, taxa in the genera *Pocillopora*, *Porites*, and *Pavona* were predominant. Reef framework accumulation (vertical thickness) ranged from 1–2 m to 10–12 m. Reefs constructed by species of the three genera mentioned above were present in upwelling and non-upwelling environments, mostly in sheltered bays and on leeward shores. Small patch reefs were observed in estuarine environments in Panama. Robust coral community and reef development occurred in the gulfs of Chiriquí and Panama (Panama), the Guanacaste Province, Caño Island and Golfo Dulce (Costa Rica), Gorgona Island (Colombia), Machalilla and Isla de La Plata (Ecuador), and Wolf (Wenman) and Darwin islands (northern Galápagos Islands, Ecuador). Beginning in the 1980s, robust coral community and reef development have been documented at eastern Pacific oceanic islands (e.g., Clipperton Atoll, Revillagigedo Islands, and Cocos Island), coastal Mexico, Nicaragua, and distant Rapa Nui (or Easter Island, Chile).

Introduction

From my newly assigned laboratory, firmly affixed to concrete pilings alongside the Pacific terminus of the Panama Canal, the Taboga Islands were in plain view to the southwest over the Pacific Ocean (Fig. 2.1). During World War II this building served as a degaussian station for ships transiting the canal. Lines carrying powerful electromagnetic currents were attached to ships to counteract their magnetic fields in order to prevent the detonation of mines protecting the canal.

In my first conversations with Bob (Robert H.) Stewart and Tony (Anthony) Mann, Engineering Division, Panama Canal Company, these diving enthusiasts

© The Author(s), under exclusive license to Springer Nature Switzerland AG 2025
P. W. Glynn, *Coral Reef Resilience in the Anthropocene*,
https://doi.org/10.1007/978-3-031-81421-1_2

Fig. 2.1 P. W. Glynn's laboratory in Panama while employed by the Smithsonian Tropical Research Institute (1967–1983). This building, located alongside the Pacific entrance to the Panama Canal during World War II, served as a degaussian station. The pier structure was refurbished and now serves as a public fishing platform

excitedly described the presence of corals at Taboga Island and coral reef-like formations in the Pearl Islands, farther south in the Gulf of Panama. According to opportunities to travel and engage in marine surveys, I began investigations on both the north Caribbean and south Pacific coasts of Panama. In approximately chronological order, coral studies were extended to Costa Rica, Colombia, Ecuador (Galápagos Islands), and Chile (Rapa Nui or Easter Island; Fig. 2.2).

Panama

The mystery of the existence of eastern Pacific coral reefs began when I visited Panama in 1968, during an interview for a position with the Smithsonian Tropical Research Institute (STRI). Dr. Ira Rubinoff was my host, tasked with introducing me to the research laboratories and marine environments of the Isthmus of Panama. Coming from the vibrant coral reefs of southwestern Puerto Rico Ira did not want to disappoint, so he introduced me to the Caribbean coral reefs of Panama, which were remarkable. Since there was great interest at STRI in transisthmian ecosystem comparative studies, I wondered about the occurrence of coral reefs on the Pacific side of Panama. The only marine environments I observed during my visit were mangrove water ways and mud flats, sandy beaches, and rocky shores near the Panama

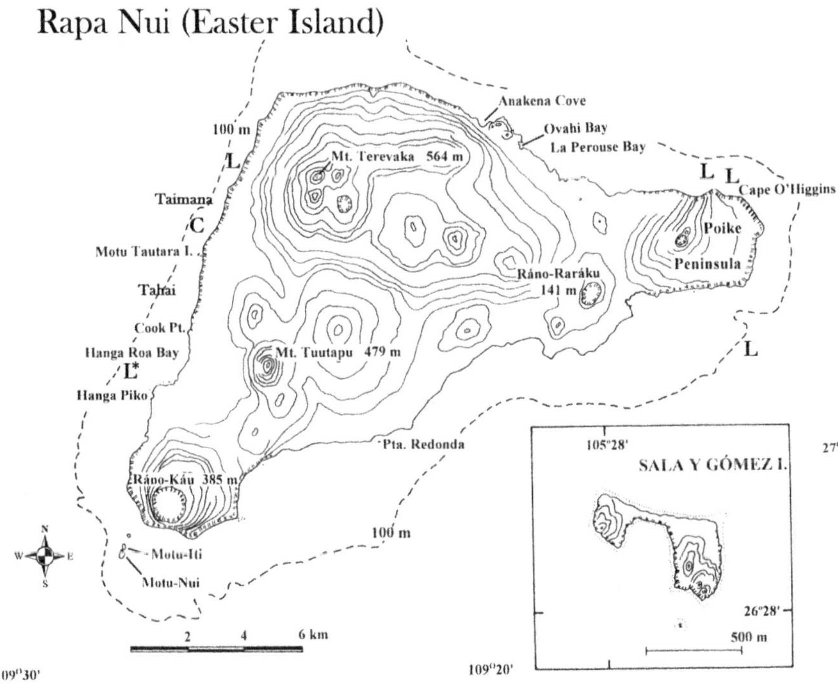

Fig. 2.2 Rapa Nui (Easter Island) and Sala y Gómez Island (inset map), southern-most outposts of the eastern Pacific coral fauna. Mesophotic coral community locations are denoted by **C**, *Cycloseris* (from Hoeksema et al. 2019) and **L**, *Leptoseris* (from Sellanes et al. 2021). **L*** denotes location of the mostly dead coral community. Map modified after Hubbard and Garcia (2003), courtesy of Diana A. Moanga

Canal (Punta Paitilla). Only vague answers were offered when I asked about Pacific coral reefs. In a literature search I found that some early workers remarked that zooxanthellate corals were abundant in the Gulf of Panama, including several named species (Verrill, 1864, 1866; Vaughan 1906). Extensive surveys by the St. George Expedition, however, reported by Crossland (1927), failed to reveal any coral reef development in Panama, Colombia, or the Galápagos Islands. But then, in a more recent publication, J. Wyatt Durham offered up a ray of hope, "With respect to the absence of coral reefs on the Pacific shores of the Americas and the adjacent islands, it may possibly be more apparent than real." (Durham and Barnard 1952). The mystery continued.

Our first coral surveys in Panama, in the late 1960s, were conducted among the numerous nearby islands in the Gulf of Panama—the Taboga Islands group and the Pearl Islands, the latter consisting of 39 main islands and numerous islets and shoals. The waters near Naos Island and my laboratory were usually turbid (laden with suspended sediments), and not conducive to zooxanthellate or phototrophic coral growth. Fist-sized and smaller azooxanthellate corals were abundant, but only once did we find a living colony of *Pocillopora*. My faithful companion and long-term assistant, Aníbal Velarde, began employment as a custodian at the Pacific marine laboratory of the Smithsonian Tropical Research Institute (STRI). He soon began to show a strong interest and talent in marine investigations. This, combined with his water skills and scuba diving proficiency soon resulted in his promotion and transfer to my laboratory as a biological assistant. He became my trusted right-hand man and dear friend with a passion for coral reef research.

The nearest corals for study occurred at Taboga Island, about 20 km southwest of my laboratory. We established a study site on the north side of Taboga where species of *Pocillopora*, *Pavona* and *Porites* were abundant and demonstrating incipient reef growth.

Our studies in the Pearl Islands were concentrated in the northern sector of the archipelago, at Contadora, Saboga, and Chapera islands. Several coral species were abundant at these islands, as well as structural patch and fringing coral reefs. Today these three islands are popular tourist destinations, but they were not always friendly places to visit in the 1960s and earlier. We were told that brigands operated out of the Saboga village in the late 1800s and early 1900s. We also heard that the villagers, mostly former pearl divers, were occasionally visited by pirates. There was supposedly a special hiding place for valuables and maidens in the crypt below the foundation of the village church. In the late 1970s I walked through the village with the Wellington family, and we immediately perceived an unwelcome demeanor from the Saboga folk. Jerry (Gerard) Wellington, his wife Patricia, and daughter Simone, were searching for a house to rent.

One of the difficulties in observing coral reefs in the eastern Pacific is the common occurrence of turbid water (with suspended sediments and plankton) which obscures subsurface features. We found, however, that at extreme low tides and during daylight hours, especially in the relatively cloudless dry season (mid-December through April), coral reef flats are subaerially exposed and clearly visible. We began to take advantage of these exposure events by means of aerial reconnaissance, i.e., overflights of suspected reef sites in a fixed wing Cessna 176 aircraft. Thanks to Captain Dennis Cismowski, a helicopter pilot in the U.S. Army (still then on active duty and on leave from the Viet Nam war zone), we discovered numerous coral reefs in this manner. In order to capture planar views of reefs, generally flying between 1000–3000 m elevation, Dennis would dip the plane to starboard so that I could photoshoot nearly vertically. I attached a polarizing filter over the camera lens to reduce sunlight reflection and glare. Reef areas, shapes, and orientation relative to coastal features could then be determined from the aerial photographs (see Glynn and Stewart 1973, Figs. 2.2 and 2.3; Glynn and Macintyre 1977, Figs. 2.5 and 2.6).[1]

[1]Aerial photography also was performed in the 1970s with other pilots, for whom I am grateful: Dennis Calvo (Costa Rica), William S. Cowles and Kenneth Day (Panama), U.S. Army Southern Command, 352nd Aviation Detachment, Maj. D. S. Tracey, and others (Panama, and Malpelo Island, Colombia). The U.S. Army flights were helpful, but not ideal as I was confined to a back seat and could not open a window for better views and to reduce glare. On one flight, I heard the pilot communicating with the co-pilot over the 'intercom' to test the endurance of their science passenger to g-forces, by maneuvering the plane into sudden dives and turns. I kept silent because

Fig. 2.3 Jorge Cortés transcribing field notes and identifying corals at the Casa Verde field house in Cuajiniquil, Guanacaste, Costa Rica (May 2019). This center serves as a base of operations for the Marine Biodiversity, Área de Conservación Guanacaste Project (BioMar-ACG). Photograph courtesy of Rocío Córdoba

The next step involved in these reconnaissance missions was "ground truthing" or *in situ* verification of coral species, associated reef biota, and coral zonation. Occasionally we could land our plane close to recently discovered reefs and survey them right away. We always carried dive gear and jerry cans of fuel in the plane's back seat. More often, however, it was necessary to reach the reefs at some future time by automobile or boat.

This method of reef discovery was not without risks. During one aerial reconnaissance mission with Dennis in the Gulf of Chiriquí, we spotted several new reefs that I was anxious to photograph and document. This mission resulted in considerable anxiety. After arriving over the Gulf of Chiriquí from our takeoff field near Panama City, I urged Dennis to continue circling some of the newly found reefs in order to capture high quality images. After an hour or so Dennis remarked, "We are getting very low on fuel." I glanced at our fuel gauges and saw that one tank was empty and the needle on the second tank was nearly on the empty mark! "Dennis, you should have mentioned this to me earlier." Since we were not far (about 70 km) from an airport in the city of David, in western Panama, Dennis decided to try landing there. We prepared for a "pancake landing" in the sea if our fuel ran out. The idea behind this type of landing is to cause the aircraft to stall, i.e., to decrease the forward air speed and lift as much as possible and then land "softly" on the water, belly flat on the surface. We opened our doors and held seat cushions against the cockpit console to protect our heads from impact. We cleared the open sea, surf zone and beach — still no sputtering—glided over a fringe mangrove forest, and *voilá*, there was the landing field dead ahead. Dennis landed the plane safely with both fuel tanks registering empty!

I was not prone to vertigo. The violent maneuvering soon began but did not last long because the co-pilot became nauseous. I offered a wry smile to the pilot when he glanced back to gauge my condition.

Our early coral reef "ground truthing" surveys benefitted from the help of many individuals. United States military diver volunteers, Special Forces Sergeants Jim (James G.) Babb, "Hogjaw" (James) Campbell, Bud (G. W.) McLeod, and Chief Warrant Officer and Marine Gunner Ron Lauzon, were especially helpful in reef drilling operations. After completing each research task, Bud would take great pleasure in throwing the chief scientist (usually me) overboard. This was a small price to pay for his considerable help in our studies. Ron's encomium of his participation in our investigations was especially appreciated, paraphrasing "My worst day assisting in Glynn's coral projects was more pleasurable than the best day involved in military diving operations." Our boat crew support team, Jack (J. N. A.) Bryan and David West, both Boatmasters, and Lucho (Luis) Cruz, crew and cook, always transported us to and from our reef sites safely. Jack was an impressively strong man. He would easily lift single handedly a 25 horsepower outboard motor.

From the late 1960s to the early 1970s, coral communities and structural reefs had been reported from numerous sites in the Gulfs of Panama and Chiriquí (Glynn 1972; Glynn and Stewart 1973; Glynn et al. 1972; Porter 1972). These findings triggered unprecedented interest in the newly discovered reefs in Panama, which resulted in a surge of coral reef investigations at STRI. Once a lonely place for coral reef researchers in the 1960s, STRI became a focus of activity in the 1970s and later. A welcome uptick in all manner of coral studies occurred at STRI which involved graduate students, postdoctoral fellows, and established coral reef researchers. These were heady times; several of these early studies are discussed in the following chapters of this book.

Costa Rica

As the magnitude of Panama's coral reef development came into focus, we were naturally curious about the occurrence of reef building elsewhere in the eastern Pacific. Our next destination of interest was along the shores of Costa Rica. Fernando Crastz, a University of Panama student, two student volunteers, and I crossed the Costa Rican border in the early 1970s, in a four-wheel drive jeep loaded down with camping and diving gear. On our way to the Pacific coast, we stopped at the University of Costa Rica (UCR) in San Jose to pay our respects to Prof. Manuel Murillo and to inform him of our plans. His interest in corals and marine life in general was well known. At that time, Manuel was a co-founder and director of the marine science research center at UCR, which today is known as the Centro de Investigación en Ciencias del Mar y Limnología (CIMAR). Our meeting with Manuel was cordial and informative and he encouraged us to proceed with our survey plans and gave us his blessings.

After arriving in the Guanacaste region of Costa Rica, we established our camp site at Ocotal Beach in Ocotal Bay. Guanacaste is a province in northwestern Costa Rica with numerous protected bays and islands. The name 'ocotal' derives from the Spanish word 'ocote', which is a species of pine tree. We surveyed several sites around Culebra Bay where roads and trails allowed access to the shoreline. While we found some coral communities, this exploratory excursion was somewhat disappointing because no structural coral reefs were observed. The tropical dry forest biome (a widespread type of biotic community) behind our beach camp site, however, did reveal some surprises regarding its biota. Unlike the verdant shoreline near Panama City, the landscape appeared dry with stubby bushes and abundant prickly pear cactus plants. Cottontail rabbits (*Sylvilagus brasiliensis*) were abundant as was a tropical rattlesnake (*Crotalus simus*). One morning departing our camp site we needed to abruptly stop our jeep to allow a surprisingly large (1.5 m long) rattlesnake to cross the road. This reminded me of some of my early boyhood days when living in Dulzura, California. While hiking to and from school, I always exercised caution and carried a willow switch to coax snakes off the trail.

Since our first Costa Rican reef reconnaissance venture was not very successful, I decided to take up Dennis Cismowski's good offer to perform overflights of various sectors of the shoreline. We conducted aerial surveys in Costa Rica during the early to mid-1970s. In the extreme northwestern part of Costa Rica, near the Nicaraguan border, we obtained photographs showing reef-like structures at the north and south sides of the Santa Elena Peninsula, the Islas Murciélago (Bat Islands), and the Gulf of Papagayo (named after a multicolored macaw parrot) south to Punta Gorda along the north shoreline of the Nicoya Peninsula. Several sites of potential interest were located. Many of these areas were later visited and found to support living (and some dead) patch and fringing pocilloporid coral reefs. Precise locations and general information on these coral formations are published in Glynn et al. (1983), Cortés and Jiménez (2003), Jiménez et al. (2010), and Glynn et al. (2017). Two dead coral reefs, one near the entrance of Santa Elena Bay, and the Matapalo Reef, on the leeward side of Punta Gorda, have revealed intriguing evidence on the causes of their death, which is discussed in later chapters.[2]

Nearly all of our surveys in Costa Rica were pleasant and uneventful except for one incident at Santa Elena Bay in 1978. This occurred during the Sandinista revolution in Nicaragua when armed clashes spilled across the Nicaraguan/Costa Rican border. On our way to the bay, with a trailer and boat in tow, we were stopped at a check point by Costa Rican guards who warned us to exercise caution in the vicinity of the Somoza ranch on the Santa Elena Peninsula. General Somoza, then president of Nicaragua, was a prime target of the Sandinista National Liberation Front. We quickly skirted the ranch, located on an isolated stretch of the peninsula, and parked under a canopy of trees on the bay's shore where we could launch our dive boat. After completing our work on a dead reef outside of the bay, in the Gulf of Santa Elena, we surfaced and were boarding our diving support boat. One of us caught a glimpse of a bright reflection from a far off object on the horizon to the north. It appeared to be approaching us. As it neared, we could see that it was a dark-colored gun boat. We immediately pulled anchor and headed for the bay at full speed. The

[2] Various lines of evidence suggest that the dead pocilloporid reef just outside of Santa Elena Bay succumbed to sea water chilling during the Little Ice Age, early fourteenth century through the mid-nineteenth century (Glynn et al. 1983).

gunboat turned and began to follow us but did not enter the bay. We quickly loaded our boat onto the trailer and departed the peninsula post haste. We suspected this was a Sandinista incursion into Costa Rican territorial waters but were never certain. Nonetheless, this was quite an unsettling event.

Our later coral reef surveys in Costa Rica were concentrated along mid- and south-coastal areas and at Caño Island. Numerous coral communities were observed at Quepos and Punta Uvita. Caño Island exhibited an impressive reef flat on its north shore (Glynn et al. 1983). Héctor Guzmán described in detail the Caño Island reef formations and coral communities for his Master's degree research requirement under Jorge Cortés' guidance (Guzmán 1986). Héctor's research was later expanded and documented the occurrence of several coral reefs surrounding the island (Guzmán and Cortés 1989, 2001). In many ways, coral research at Caño Island was very pleasant in spite of incessant battles with cockroaches and warding off pesky land crab encounters at night. The incredibly strong work ethic, dedication to coral research, and good cheer of the Costa Rican coral team led by Jorge, effectively suppressed the minor pest annoyances. Further, Héctor made a delightful discovery that was enjoyed by all. He discovered a heavenly freshwater pool, hidden in the island's forest, that he named "La Poza de Venus." It was a joyous occasion to bathe in the cool and limpid waters of this shaded pond after days of diving in the sea.

During one of our early road trips into Costa Rica we detoured from the InterAmerican Highway to visit the settlement of Golfito at Golfo Dulce. Golfo Dulce is a large gulf, about 40 km in length and 10 to 15 km wide. Hydrographic charts show extensive shallow shelf areas that could support coral growth and was therefore worth exploring. Golfito was a very busy port, a United Fruit Company enclave surrounded by extensive banana plantations. In many ways this small town was similar to Balboa of the Panama Canal Company with a strong Yankee or United States presence. There was a Zona Americana with company controlled housing, stores, schools, an infirmary, theater, and more, all the amenities of a town in the USA. There was also a guard house through which we needed to pass. Our STRI (Smithsonian) travel office had called ahead to inform the fruit company that we would be paying them a visit. We were politely received and put in touch with Donald McGuiness who would offer his boat service for our coral surveys. Observations along the north-east shore of Golfito revealed pocilloporid communities and small pocilloporid fringing reefs. Flourishing clusters of a small knobby coral, *Psammocora stellata*, dominated the reef flats.

On one occasion, when Dennis and I were returning to Panama after flying over Caño Island, he remarked that we were running low on fuel but had a sufficient amount to fly over the northern or inner-most reaches of Golfo Dulce. Considering our previous survey results at Golfito, amounting to the occurrence of only modest coral patches, I was not in favor of Dennis' suggested flight path but consented due to our low fuel supply and his best judgment. We observed Golfo Dulce ahead as we cleared the forested terrain below. Two large, lightly colored, shallow bottom areas immediately caught our attention. These were unmistakably two well developed coral reefs, separated by only a few km, one at Punta Islotes and the other at Punta Bejuco. The occurrence of these reefs was entirely unexpected because of the relatively meager coral presence observed during our earlier Golfito survey. Jorge

Cortés, in the late 1980s, investigated the structure and growth history of the Punta Islotes reef complex which resulted in key publications that satisfied his doctoral research requirements (Cortés 1990a, b, 1991; Cortés et al. 1994) (Fig. 2.3).[3]

Colombia

The numerous coral and reef discoveries in Panama and Costa Rica were truly remarkable, exceeding all expectations. The discovery and mapping of eastern Pacific coral formations, virtually unknown to the scientific world, were perhaps the most exciting days of my research career. Martin Moynihan, the director of STRI, and Ira Rubinoff, associate director of the marine program, were equally excited with these discoveries and offered our coral team abundant encouragement and support. Could the occurrence of coral formations continue south of Panama, along the South American coasts of Colombia and Ecuador?

When I began searching for coral reefs in the equatorial eastern Pacific, so little was known about their presence in this region I was unsure of where to begin. The best guide available was Joubin's (1912) atlas of the known distribution of coral reefs of the world, with additions by John W. Wells (1957). I was fortunate in acquiring Joubin's original publication, illustrating the locations of global coral reefs, including those of the eastern Pacific region. Coral reef presence was denoted by red markings on large scale maps. I pinned the map page of tropical western America to my laboratory's bulletin board as a guide. As our surveys progressed in Panama and Costa Rica, we found that some reef sites were properly marked on Joubin's map, but many were not; several rocky promontories were misidentified as coral reefs. For example, we are now aware that Joubin's relatively large red patch at Cabo San Francisco, Ecuador, represents a long stretch of rocky outcrops with few if any reef-building corals. Possibly the nature of this coastal stretch was taken from a mariner's log marked *rif* or *riff*, a medieval European word meaning "a rocky relief feature close to the sea surface to be of danger to sailing vessels." Two coral rich areas occur farther south in Ecuador, at El Faro and the Surrones Bank (sites discovered by Fernando Rivera and Priscilla Martinez and noted in Glynn et al. 2017). Early in our aerial surveys we were especially anxious to explore the Cabo San Francisco region, because of Joubin's prominent red markings in that area.

Again, I approached Dennis for his advice and willingness to fly south along the Colombian coast to northern Ecuador. Without hesitation he approved the idea and

[3] Jorge was one of my first doctoral students, arriving at RSMAES in 1985 on a Fulbright Scholarship. Upon completion of his graduate studies, he returned to the University of Costa Rica, San Pedro, and rose to the rank of Professor, a resolute scholar, teacher, and mentor. Jorge's pioneering research, and that of his students, has resulted in the discovery of numerous reefs and a plethora of ecological studies of Costa Rica's coral reef ecosystems. With an appreciation of the fragility and value of coral reefs, Jorge became a steadfast advocate for the protection of this natural resource. He received the National Science Prize, Clodomiro Picado Twight, of Costa Rica in 1982, the prestigious Eminence of Research award of the International Coral Reef Society in 2021, and became a Professor Emeritus in 2022, still actively engaged in coral research.

considered it feasible and efficient in terms of cost and time saving. Two reconnaissance missions could be completed during a single flight, i.e., aerial photography along the Colombian coast and the northern coast of Ecuador. So, early one morning in the dry season (March 1975), Dennis and I were flying south along the Panamanian coast in our trusty Cessna, with diving gear and Jerry cans of aviation fuel stowed in the back seat. Several weeks before our departure I requested permission, through our Smithsonian international travel office, to enter Colombia, refuel, and continue our flight to Ecuador; permission was also requested from the Colombian and Ecuadorian embassies in Panama (and granted by both countries) to fly into Esmeraldas, Ecuador, the nearest landing site from Cabo San Francisco (Fig. 1.4).

From the Panamanian border south to Buenaventura, Colombia, we spotted and photographed coral reefs (at Punta Tebada, Gulf of Cupica, and Ensenada de Utría), which were later investigated by Bernardo Vargas-Ángel (1988, 1989). Bernardo first appeared in my office at RSMAES in 1990 when he modestly proposed researching coral reefs along the Colombian Chocó coast (Fig. 1.4). I do not believe he realized my delight, but I was greatly pleased that this aspiring graduate student proposed conducting reef studies in this challenging environmental setting of the Chocó. As it turned out, Bernardo demonstrated that pocilloporid reef development did occur in this area. He provided detailed descriptions of coral community structure, coral growth rates, the geomorphology of reefs, and documented local disturbances and reef recovery (Vargas-Ángel, 1996, 2001; Zapata and Vargas-Ángel 2003). His principal reef study site, the 10.5 ha (hectare) La Chola reef, is one of the largest pocilloporid fringing reefs on eastern Pacific continental shores. The occurrence of this reef was unexpected considering the extreme climate conditions of the Chocó sector of the Colombian coast. The annual rainfall ranges from 5000 to 7000 mm and more, one of the world's wettest regions. This high rainfall contributes to several environmental factors that are inimical to coral growth, namely high cloud cover causing reduced sunlight, low salinity sea water, and elevated sedimentation from river drainage; the latter condition is exacerbated sporadically by landslides resulting from frequent and severe seismic events. There were (and still are) other concerns of personal safety, namely tropical diseases such as yellow fever and malaria, and regular drug trafficking in this area.

As Dennis and I were approaching foul weather in our flight south, an ominous dark storm just north of Buenaventura, he handed our plane's controls to me to evaluate and sharpen my flying skills under extreme turbulent conditions. We flew directly into a severe thunderstorm with frequent lightning strikes, buffeting winds, heavy rain, and zero visibility. After a short while the plane began a yawing motion, which I could not get under control. I was failing at flying on instruments. Had I not been able to hand the controls back to Dennis, who quickly made the necessary adjustments to regain control, I am certain the outcome would have been disastrous.

After a brief stopover and refueling at the Buenaventura airport we headed to the Cali international airport to clear customs and file our flight plan to continue to Esmeraldas. This is where the trouble began with the Cali airport administrator. He told us that our papers were bogus, not in order, and that we needed to return to Panama immediately! We suspected he was up to bribery, but we complied, boarded our Cessna airplane, and headed on a northwest course back to Panama. In a short while we were approaching a low, coastal mountain range. Dennis gave me a knowing glance and said, "Peter, you really wanted to survey that suspected coral reef coastal stretch in Ecuador, didn't you?" "Yes, I did," I replied, "a great disappointment." Dennis continued, "Well, once we reach the coast, I can fly low, in the shadow of the mountain range, thus avoiding radar detection." "What do you say, Peter?" A little while later Dennis banked the plane and put us on a southerly flight path to Ecuador.

On our way south, we circled Gorgona Island, Colombia (Fig. 1.4) and obtained a nice series of photographs of several fringing reefs around the island, thus supporting Joubin's red markings for this area (von Prahl et al. 1979; Glynn et al. 1982). Henry von Prahl (deceased[4]), Fernando A. Zapata and students at the Universidad del Valle (UV, Cali) have investigated these coral reefs extensively and continue their studies at this former prison island, which is now a protected national reserve (Fig. 2.4). The present Gorgona National Natural Park, established in 1984, serves

Fig. 2.4 Fernando Zapata and his student monitoring team at Gorgona Island Natural National Park, Colombia (October 2005). From left to right, port side of dive boat, Fernando Zapata and Raúl Navas, starboard side, Luis Chasqui, Tania Llanes and Adriana Alzate. This field sampling exercise involved the annual monitoring of coral reefs under the Colombian government sponsored SIMAC program. SIMAC is an acronym for Sistema Nacional de Monitoreo de Arrecifes Coralinos en Colombia (National Monitoring. System of Coral Reefs in Colombia)

[4]While lecturing to a class at RSMAES, I was stunned in hearing from one of my students of Henry's death, my esteemed Colombian colleague and friend. Tragically, Henry was among the 111 passengers aboard an Avianca airliner who were killed on 27 November 1989 by a terrorist's explosive device (Glynn 2001a). Due to Henry's pioneering coral research in Colombia, we dedicated a special issue of the Bulletin of Marine Science on Eastern Pacific ENSO coral studies to his memory (Glynn and Colley 2001)

as a prime coral reef study site for the Coral Reef Ecology Research Group (UV) and other investigators in Colombia.

A relatively short time following our overflight of Gorgona Island Dennis remarked that we were running low on fuel, again! This fuel problem was usually my fault for requesting repeated passes over reef areas of interest. We needed to land, but all we could see below at ground level were vast tracks of thick vegetation and marshy terrain. Dennis found a possible landing site on his aeronautical chart, a small airfield at the port city island of Tumaco, about 40 km from the Colombia/Ecuador border (Fig. 1.4). He was aware of an internationally recognized agreement that if it were necessary to make an emergency landing, one would be given a green light signal to land. Dennis circled the landing field, faked engine problems, and then we received a green light from the tower. Immediately upon landing, our plane was surrounded by about 10 armed soldiers, with M1 rifles, who demanded that we de-plane. They then escorted us to their headquarters. *En route*, Dennis startled me by saying that he had a loaded 45 pistol in his flight boot and thought we could shoot our way out of this situation. I considered this the height of recklessness and told him I wanted no part of this plan and would fall flat on the tarmac, arms and legs spread-eagle, if he reached for his gun. We are not in a Vietnamese war zone! We continued walking under guard and to the post's headquarters where we faced a lieutenant with a stern demeanor who demanded to see our identification documents. I handed my U.S. government official passport to the officer; he examined it and kept it. Dennis offered up his identification papers (an active duty army captain) and the lieutenant immediately came to attention and snapped him a clean salute. A matter of a captain outranking a lieutenant and, as we learned later, a Colombian officer who spent time training in the Panama Canal Zone and who greatly admired the U.S. Army. The lieutenant quickly adopted an amiable attitude. He apologized for the arrest, but said he needed to detain us for two days in order to satisfy the Cali airport administrator's complaint. The administrator, who suspected we would attempt to fly south to Ecuador, radioed ahead to Tumaco to be on the lookout for two Norte Americanos in a Cessna 176. Instead of the grimy jail cell, the kindly lieutenant offered us his quarters, and instructed us to shower and rest after our long flight.

Waking up the next morning, after a pleasant evening and repast with the lieutenant, we joined him for breakfast, and then he asked if we would like to accompany him on a sightseeing tour of Tumaco. "Claro que sí", of course we would; we climbed into his jeep and off we went. We drove through the village, past the marketplace and port, greeting the town's folks as we proceeded. We also visited the town's most celebrated site, the church. The church was Tumaco's *piece-de-resistance*, whose early pastor, Friar Gerardo Larrondo, led a procession to the sea in a time of crisis that became celebrated worldwide as the 1906 Miracle of Tumaco. With the succor of the Eucharistic Christ, it was purported that a tsunami was prevented from inundating and destroying the village. The magnitude of the earthquake that generated this disturbance was estimated to be 8.4–8.8 on the Richter scale. This is yet another example of a significant disturbance event that could potentially affect coral reefs (e.g., through structural damage or uplift and subaerial exposure), in this case along the Chocó coast.

After two days of respectful treatment and leisure at the military base, involving many games of volleyball, table tennis, and billiards, our lieutenant host told us that we would be allowed to "escape" early in the morning of the third day. Our Cessna would be fully fueled, and our escape would not be announced until noontime, giving us time to land and refuel at Buenaventura and then cross the Panamanian/Colombian border before dark. After flying all day we finally landed on the unlit air strip in Panama in full darkness, having again photographed many of the coral reefs of the Colombian Chocó region. Verification of Joubin's red markings at Cabo San Francisco still beckoned.

My first on-site surveys in Colombia began in 1979 when I was invited by Henry von Prahl to join the Sula III Expedition to investigate Gorgona Island's flora and fauna. An extraordinarily gifted naturalist and artist, with contributions in the fields of marine biology, aquaculture, mangrove ecology, ornithology, *inter alia,* Henry was an innocent casualty of drug war violence in 1989 [Henry von Prahl (1948–1989)]. The Sula expedition consisted of a large contingent of Colombian and foreign naturalists with specialties in botany and zoology. Funding was provided by COLCIENCIAS, an agency of the Colombian government supporting scientific studies. All participants met at the Universidad del Valle in Cali, and then traveled by bus to Buenaventura where we boarded a motor sailing boat for travel to Gorgona Island. We departed Buenaventura at 7:00 PM, slept on the open deck during passage, and arrived at Gorgona Island at 7:00 AM the next morning. Our stay on the island was planned for 6 full working days (18–23 May). I was fearful that our equipment and supplies (e.g., microscopes, field journals, literature, and clothing) might be drenched by a passing rainstorm since we had entered the ITCZ (Inter Tropical Convergence Zone) wet season. We were lucky; we arrived at Gorgona Island a bit drowsy but dry.

At the time of our visit, Gorgona Island was still a penal colony, holding some of the most treacherous criminals in Colombia. We all felt well protected, however, because the prison's guards were in full force and effectively managing the facility. Each research team was assigned an inmate, presumably from an incarcerated record of good behavior, to help and facilitate our needs. The coral team was assisted by several inmates, one with the moniker El Diablo (the devil). His visage had an uncanny resemblance to a common caricature of Mephistopheles—a pointed, long, narrow nose and chin, and sinister penetrating eyes. He appeared to be in his fifties, and was amicable, respectful and helpful. No special favors were offered to us regarding our accommodations and meals. Our living conditions were spartan but clean. The usual evening meals consisted of fish (bony green garfish) served with rice. It was a treat to be served a bread roll. These always seemed to be in short supply, however, El Diablo sensed that we enjoyed the freshly baked rolls so found a way to make them more available.

I befriended El Diablo and chatted with him in the evenings. Regrettably, I never learned his surname. He seemed such a gentle and kind soul to be living out his years in the infamous Gorgona Island prison. One evening I ventured to ask him about the legal charges and conviction that brought him to this maximum-security facility. He was anxious to tell his story, which began on a fishing venture with seasoned fishers when he was still a teenager. They were dynamite fishing (which is prohibited in Colombia) in a coastal lagoon, lighting and throwing sticks of dynamite into the water to stun and kill fish. They were reaping a bountiful catch until they spotted a police patrol boat rounding a point and heading in their direction at full speed. All on board the fishing boat jumped into the water, except for El Diablo, and swam to the mangrove thickets to seek cover. El Diablo did not follow them, possibly because he could not swim, but remained alone in the boat. As the patrol boat pulled alongside, El Diablo panicked, lit a stick of dynamite, and threw it at the police boat. Tragically, this resulted in the death of one of the patrol officers and a life sentence in prison for this teenager's audacious attack. Executions are not performed in Colombia. I sympathetically remarked, "Ay señor, if only you just raised your arms and politely surrendered your sentence would have amounted to no more than a couple of years of imprisonment." El Diablo, sighed, "I know, I was stupid like a burro (donkey)."

Our prisoner assistants with boating skills were very helpful during diving activities. They provided *cayuco* (dugout canoe) support, and on one occasion first aide assistance. In surfacing after a scuba dive, I experienced intense pain in one arm. A tentacle of *Physalia physalis*, the Portuguese man-of-war jellyfish (technically a cnidarian siphonophore), was draped over my arm. With a mind not to panic, I reasoned that by rotating my arm above water I could uncoil the tentacle. To my dismay, a wind gust caught the tentacle and double wrapped it around my arm. My *cayuco* support crew recognized this problem, grabbed hold of me, and held me firmly to the gunwales as they paddled quickly to shore. With machete in hand, one of the crew found a papaya tree, climbed the tree, cut loose a fruit, sliced it and then applied the oozing white sap to the swelling lines of welts on my arm. In a matter of minutes, the excruciating pain waned. The papain enzyme in the sap neutralized the venom of the stinging nematocysts. I was amazed at how quickly our prisoner assistants recognized and resolved this medical emergency. I thanked them profusely for their care.

From surveys of numerous eastern Pacific localities, it was soon realized that one coral genus in particular was overwhelmingly prominent in reef building, namely *Pocillopora*. The recognized morphospecies of *Pocillopora* contributing most commonly to framework structures were *P. damicornis*, *P. elegans*, and *P. verrucosa.*[5] These corals form relatively fragile frameworks of vertically oriented branches

[5]According to a study conducted in Australia, based on molecular and morphometric evidence, *P. damicornis* was determined to be a variant of *P. verrucosa* (Schmidt-Roach et al. 2014). Molecular evidence is derived from a coral's genome, and morphometric evidence from anatomical traits (mostly skeletal structures and colony form). More recent studies, based on the genetic structure of the coral host and the fidelity of zooxanthella endosymbionts with particular coral species, have revealed three additional morphospecies that are genetically equivalent to *Pocillopora damicornis*. The reader is referred to Fig. 10.2 in Chap. 10 for more detailed information on this topic.

capable of rapid growth. The reader is referred to Veron (2000) for morphological descriptions of these species and their global distributions. Similar pocilloporid reef development, referred to as 'reefs' or 'large monospecific stands', has been reported at Rottnest Island, SW Australia (Veron and Marsh 1988; Ross et al. 2015), and in the Gulf of Oman (Glynn 1983, 1993; Sheppard and Salm 1988; Salm 1993). This raises the intriguing question, "What is the minimum number of coral species necessary for reef building?" Interestingly, the single genus *Pocillopora* has the capacity to construct reefs, creating well defined morphological features such as reef flats, reef slopes and reef base zones. Jorge Cortés and co-workers in Costa Rica have observed that *Porites lobata* and *Pavona clavus,* each alone, can also construct coral reefs. Contemporary paucispecific reef building is not unprecedented since coral reefs were constructed by *Porites* species alone in marginal Late Miocene environments in the Mediterranean Sea (Perrin and Bosellini 2013).

Ecuador

In late 1974, I was informed that Jerry (Gerard M.) Wellington, then a U.S. Peace Corps volunteer, was charged with mapping shallow marine communities in the Galápagos Islands for a development plan to aid in the creation of an Ecuadorian national park (Wellington 1975). I also received information that Jerry was reporting the presence of reef-building corals in his surveys. With Jerry's advanced knowledge of the marine biota, this would be an advantageous time to perform a comprehensive, island-wide survey of coral development. The absence of coral reefs in the Galápagos Islands, the condition advanced by Darwin (1842) and Dana (1890), or their presence according to Joubin's red markings (1912), could be determined by surveys across the archipelago (Fig. 1.6). I proposed mounting an exploratory expedition to Ira Rubinoff, who enthusiastically supported the idea and obtained further support from S. Dillon Ripley (Secretary of the Smithsonian Institution) and the Charles Darwin Research Station (CDRS), Puerto Ayora, Santa Cruz Island, Galápagos Islands. In preparing the proposal, I recommended that John W. Wells (doyen of coral systematics), and Chuck (Charles) Birkeland (designated official photographer, Fig. 2.5) be included as members of the expedition with Jerry as our local guide and co-investigator. Happily, all parties were satisfied with this arrangement and accepted their respective assignments.

John traveled to Panama from Ithaca, New York and was our house guest before departing with Chuck and me for Ecuador. He regaled us with many intriguing stories relating to coral reef science. For example, he told us how he rescued rare coral books during the German retreat across Europe in World War II. John was an intelligence officer, a member of the OSS (Office of Strategic Services). One of his chief duties was to search for information on German troop strength and movements in vacated headquarters, often in library and museum buildings. It was necessary to move quickly because the retreating enemy was employing the so-called "scorched earth" policy. John also told us how he demonstrated the slowing of Planet Earth's

Fig. 2.5 Chuck Birkeland giving a presentation at the University of Guam's Marine Laboratory, commemorating its 50th Anniversary (May 7, 2022). Chuck was Director of the laboratory from 1979 to 1982 during his time at the University of Guam (1975–2000). Photograph courtesy of Laurie Raymundo

rotation since the Devonian period, from about 400 days per year to 365 days now, by counting the epithecal daily growth ridges of Middle Devonian fossil corals (Wells 1963). This insightful study no doubt contributed to John's election to the U.S. National Academy of Sciences. John's abiding interest in the architectural design of cathedrals led us to Panama Viejo and the ruins of Panama's original cathedral that was sacked by Henry Morgan in 1671. John delighted in stepping off distances to measure floor spaces, which he compared with various European cathedrals.

After arriving in Guayaquil, Ecuador, on our flight from Panama, John, Chuck, and I located a hotel, a quaint turn-of-the-nineteenth century inn, and we began unpacking in our rooms. Suddenly, there was a loud screech from John's room down the hallway. Chuck and I went running only to discover that John's scream was one of delight. His water closet was equipped with a Crapper toilet, named after Thomas Crapper, a London plumber. The toilet's water tank was high, near the ceiling, with a long lift chain for flushing. John was gleefully flushing the toilet and reveling in the inevitable spray from the water tank. This was the first time in decades that he had seen an operational Crapper toilet. We learned later that John collected toilet seats and other similar and bizarre paraphernalia. The next day we rented a car and

Fig. 2.6 A contemplative moment in *Beagle III*'s dinette, Peter Glynn, John Wells, and Jerry Wellington, left to right, January 1975. While cruising between known island coral sites we discussed visiting unknown areas of interest

drove along coastal areas not far from Guayaquil. Sandy and muddy beaches were the rule, not conducive for coral development.

We met Jerry at the CDRS (Charles Darwin Research Station), Santa Cruz Island, the next day after arriving on our flight from Guayaquil. He briefed us on his coral survey findings and suggested a cruise itinerary, aboard the M/Y *Beagle III* with Bernhard Schreyer as Master. He proposed visiting coral rich areas that he knew as well as sites of unknown coral development (Fig. 2.6). We then visited the CDRS natural history research museum where John examined the scleractinian corals collected by Jerry over the course of his surveys. The coral collection included seven azooxanthellate and 12 zooxanthellate species. The latter included four genera that commonly contribute to reef building in the eastern Pacific, namely *Pocillopora*, *Porites*, *Pavona*, and *Gardineroseris*.

Our research cruise was planned for two weeks (10–25 January 1975) and would include time for surveys at southern, central, and the northern-most islands of Wenman (Wolf) and Darwin (Culpepper) in the Galápagos Archipelago. The present narrative is necessarily a brief overview of our investigations in 1975 and also includes research results from cruises in 1976. Jerry and I were able to extend our surveys to unexplored sites and complete projects in 1976 aboard local boat charters and the schooner M/Y *Palawan*. The schooner leg of our itinerary was made possible by Thomas J. Watson, Jr.[6] It involved only four working days, but this amounted to a savings of about US $1000 in charter fees. A more detailed treatment of investigations in the mid-1970s is available in our book, *Corals and Coral Reefs of the Galápagos Islands* (Glynn and Wellington 1983). For the identification and systematic status of corals during this time, the reader is referred to John Wells' appendix in our book.

Our first surveys along the northwestern coast of San Cristóbal Island, during the *Beagle III* cruise, were disappointing in that pocilloporid coral surface cover amounted to only 10–20% maximum, with no evident framework development. Coral development was more pronounced at Punta Pitt, the eastern-most end of San Cristóbal. Here we found incipient pocilloporid reef construction at some sites, amounting to 30–40% live coral cover. In addition, we discovered a linear array of large colonies of *Porites lobata* extending seaward, perpendicular to the shoreline. Most of these colonies measured 1 to 2 m in height—a few were 4 m high — suggesting a growth record of several centuries. The growth history, age estimates, and mortality of these massive colonies were noted with respect to the severity of the 1982–83 El Niño event (Glynn 1990). Several of the *Porites* and *Pavona* colonies that bleached and died during this recent El Niño disturbance, at five island sites, were 100, 200, and 300+ years in age. In addition, the ages of *Pocillopora* patch reefs that bleached and died in 1982–83, at seven Galápagos sites, were 100 and 200 years old; one of the oldest dead reefs revealed an estimated age of ~500 years. These data suggest that Galápagos reef-building corals did not experience an El Niño event equal to the intensity of the 1982–83 warming disturbance for several hundreds of years.

[6]Tom Watson, a munificent entrepeneur and philanthropist, offered us the use of his schooner M/Y *Palawan* in 1976 to continue with research on the newly discovered coral formations in the Galápagos Islands. Tom served as a World War II aviator, was the second company president of IBM, and national president of the Boy Scouts of America. During his air force service, he flew President Franklin D. Roosevelt to the Yalta Conference in 1945. When Tom assumed leadership of IBM, he launched a research and development initiative, which resulted in the creation and production of a transistorized supercomputer. He became the U. S. Ambassador to the Soviet Union in 1979. Tom regaled us with many fascinating personal stories. At a farewell banquet at the CDRS, in appreciation of the time spent in the Galápagos Islands, Tom gave an unrestricted gift of US $10,000 to the Darwin research station. After departing the Galápagos Islands, M/Y *Palawan* sailed for Rapa Nui (Easter Island), the first of several Polynesian island stops on her voyage across the Pacific Ocean. An engaging account of sailing adventures was published in Tom Watson's book *Pacific Passage* (1993). I am disappointed, however, that Tom failed to mention in his book the recent coral discoveries that Jerry and I were documenting in the Galápagos Islands. It seems that Jerry and I failed to rouse Tom's awareness of the beauty of the coral world and the discoveries in progress. Corals were scarcely mentioned in Tom's book except as navigational hazards. In introducing Jerry and I in his book, when we first met at Hood (Española) Island, Tom mentioned that I directed the Smithsonian Panama Station (i.e., STRI) for eight years. For the record, I was never the director of the Smithsonian Tropical Research Institute. Martin H. Moynihan, a renowned animal behaviorist, was the founding and sole director of STRI during my tenure there (1967–1983).

Fig. 2.7 Devil's Crown (Onslow Island), a remnant crater rim located off the northeast corner of Floreana Island, Galápagos Islands (8 June 2007). Exposed rock is dark colored basalt with white patches of guano. Photograph courtesy of J. S. Feingold

The northeastern sector of Floreana Island was our second survey site and this area proved to be more coral-rich than San Cristóbal Island. A small donut-shaped island with the ominous local name "La Corona del Diablo" (Devil's Crown) lies just offshore of Floreana and was the major focus of our investigations at this location (Figs. 2.7 and 2.8).[7] The centrally located shallow lagoon, the "donut hole",

[7] Devil's Crown (English name Onslow Island) is a long extinct and heavily eroded volcanic crater; it probably emerged around 1.5 to 2.3 million years ago when Floreana Island formed (Geist et al. 2014). The high points of the crown or crest are now jagged spires circling a shallow lagoon. This island has two major passes that allow for vigorous circulation by the Peru Oceanic Current, from southeast to west by northwest. Sometimes this current is so strong that it is nearly impossible to maintain one's position inside the crown. On one occasion we measured the current velocity, which ranged from 0.58 to 1.17 knots. Divers and others partaking of water activities should always be aware of strong currents and one's inability to swim against such currents for any extended period. It is worth relating two incidents underlining the hazards of strong currents. The first involves a diver rescued at Darwin Island, the northern most of the Galápagos islands. Upon surfacing from a dive at Darwin Island, one member of our team heard a faint, distant cry. We began to visually scan the ocean's surface and caught sight of a reflection far to the west of the island. We pulled anchor and headed to the unidentified light source. Upon approach, we became aware that it was a diver in distress, far from the island and his support boat. He told us that he became separated from his diving partner and when he surfaced, he could not swim against the current to return to his dive boat. When we returned him to his main live-aboard boat, the members of his diving party had no idea of his whereabouts. The nearest landfall from Darwin Island, fol-

Fig. 2.8 Drone (DJI Mavic Air 2) image of Onslow Island, Corona del Diablo, at an altitude of 396 m. Light-colored north-west oriented sediment banks consist dominantly of biogenic calcareous skeletons. The north-south distance across the emergent crown is ~200 meters. Photographed on 6 December 2021 by Andrés Cruz, courtesy of María José Guarderas

was nearly totally covered with a vibrant pocilloporid patch reef and several associated massive corals (*Porites lobata*, *Pavona clavus*). Live coral cover was high on the reef, ranging from maximum values of 60% to 70%. This patch reef was well known by local *Galapageños*, mapped in detail by Jerry, and certainly an iconic natural attraction that needed protection. Exploring Devil's Crown revealed some

---

lowing the Peru Current, was the Marquesas Islands, about 5500 km to the west! This was a stark reminder of the importance of exercising all precautions when diving in open ocean waters. From this near-tragic outcome, we began attaching scuba safety tube sausages to our buoyancy compensators. These tubes can be quickly inflated from one's regulator or mouth, are about 1.2 m in length, and brightly colored for easy visibility and location. The second unsettling experience involving strong currents occurred to me while deep diving on the northeast side of Devil's Crown. I signaled to my "dive buddy" that my air supply was getting low. I began my ascent by following the anchor line to our inflatable dive boat. The current was strong, probably in the range of 0.8 to 1.0 knots. My hands were full of camera equipment, making it difficult to hold onto the inflatable's ropes. When I realized I was losing my grip, I dropped the camera and accessories into the boat and was immediately swept away by the current. I could see that my present trajectory was carrying me about 50 meters to the north and west of Devil's Crown with no landfall in sight. I noticed a strong eddy on the north side of the Crown, so I began to swim south toward the eddy, not against the current but at a right angle to the current. This route brought me to the eddy within which I became entrapped, moving in a counterclockwise direction. I inflated my orange sausage and waited for the divers to surface and spot my position, and predicament, for pickup. Thankfully, I was aboard the inflatable and back with my team in short order, and not on an impossible westerly course to the Marquesas Archipelago.

surprises and unexpected scenes. Sea lions were omnipresent inside the crown, frolicking over the reef and on occasion mouthing and attempting to swim away with our sampling gear and camera equipment. Chuck Birkeland was ecstatic; he witnessed and photographed an exceptional scene, a penguin (*Spheniscus mendiculus*) swimming over a coral reef. He repeatedly remarked on the rarity of such a scene—corals and penguins, emblematic of tropical and frigid waters respectively. This endemic and endangered penguin is the only one known to occur north of the equator (Fig. 2.9).

Another unusual and conspicuous sight was the large numbers of sea urchins (*Eucidaris galapagensis*) feeding on live and dead corals (Fig. 2.10). Quadrat counts revealed densities as high as 50 and 70 individuals per m^2. Initially we assumed these large and robust sea urchins were a variant of *Eucidaris thouarsii*, which is common at eastern Pacific mainland sites. Haris (Harilaos) Lessios et al. (1999) noted the long-established identity of *E. galapagensis* as a distinct species and its presence at Cocos Island. Speculating on how reef growth could occur under such an onslaught of destructive grazing, we compared coral growth rates with losses due to sea urchin erosion and concluded that *Eucidaris* at high densities could limit reef growth in the Galápagos Islands (Glynn et al. 1979).

Just outside of Devil's Crown, upstream to the east and in deeper water (15–30 m depth), Jerry escorted us to an extensive community of mushroom (fungiid) corals, thousands of closely packed individuals spread across approximately 1 hectare of a sandy-rubble substrate. A small nodular coral (*Psammocora stellata*) was also

Fig. 2.9 Endemic Galápagos penguins, *Spheniscus mendiculus*, at rest on volcanic rocky shore near Las Tintoreras, south of Puerto Villamil, Isabela Island, Galápagos Islands (22 May 2007). Photograph courtesy of J. S. Feingold

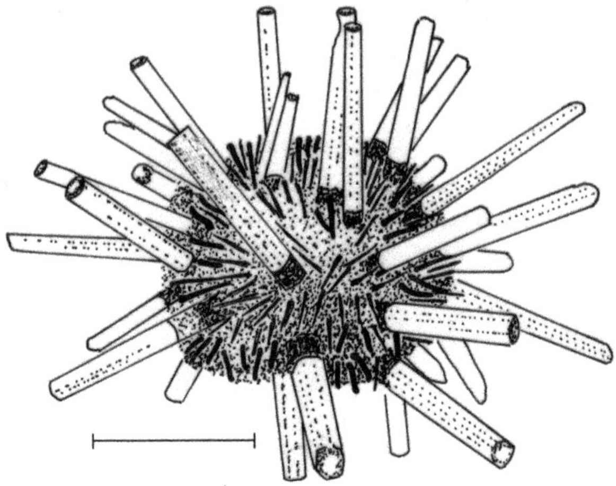

Fig. 2.10 *Eucidaris galapagensis*, the slate pencil sea urchin, scale bar = 3 cm

present intermingled with two fungiid species (*Cycloseris distorta* and *Cycloseris curvata*). Joshua Feingold, one of my early doctoral students, has monitored this community for 30+ years, since 1990, documenting marked variations in abundance in relation to ENSO bleaching events and predation (Feingold and Brulé 2020). Aggregations of a gastropod mollusk, *Babelomurex hindsii*, are occasionally seen feeding on *C. curvata*. Mortality from predation, however, was minor compared with El Niño bleaching disturbances. Sediment cores have revealed the existence of this fungiid coral community for nearly 8000 years (Feingold et al. 2020). While fungiid coral fossils occur in the Galápagos and elsewhere in the equatorial eastern Pacific, it is puzzling why this is the only known living and abundant aggregation south of the Gulf of California (Reyes-Bonilla et al. 1997).

Before leaving the northeastern sector of Floreana Island, Jerry suggested we visit nearby Champion Island, about 16 km southeast of Corona del Diablo. Seaward from the shoreline, on the western side of the island, was an impressive buildup of massive corals, principally *Pavona clavus* and *Pavona gigantea*. The maximum vertical framework height of this formation ranged between 4 and 5 m; radiocarbon dating performed in a later study by Ian Macintyre and colleagues yielded an age of 420 ± 60 yr. BP, reflecting a mean vertical growth rate of about 1 cm per year (Macintyre et al. 1992).

As our surveys of coral reef development progressed across the Galápagos Islands, we continued to encounter incipient pocilloporid fringing and patch reefs, commonly about one meter high, rarely three meters in height. Coral buildups and reefs constructed of massive corals (*Porites lobata*, *Pavona clavus*) were less common, but generally larger, having attained 3 to 5 m in height. Both reef types were usually present in bays and along shores protected from strong wave action. Much of Isabela's west coast was subject to upwelling, the shoaling of the Equatorial Undercurrent (Cromwell Current) with cool and nutrient-rich waters unfavorable for coral growth. Three sites in the Bolívar-Elizabeth Canal (formerly Bolívar Canal), in the lee of Fernandina Island and upwelling, were of considerable interest regarding development of zooxanthellate and azooxanthellate coral communities. Jerry was intimately familiar with this area and recommended three stops in our itinerary: (a) the tectonically uplifted Urvina Bay shelf, (b) Tagus Cove, and (c) the basalt coral pools at Punta Espinosa. Based on the unique presence of endemic fishes, corals, and other invertebrate species, this relatively small area is now regarded as a distinct faunal zone, namely the Elizabeth Bioregion (Edgar et al. 2002).

In 1954, following the eruption of the Alcedo volcano, a large expanse of the Urvina Bay shelf was suddenly raised over 6 meters above sea level due to an upward movement of magma (Colgan and Hollander 1987). Rich coral development, including several patch reefs and numerous large, massive coral colonies were present (and recently dead) on the newly uplifted shelf (Colgan 1990). The mix of coral formations and arid plant communities was surreal but convenient for study. A gargantuan colony of *Pavona clavus* measured 12 m in diameter, qualifying as the largest known coral colony in the eastern Pacific region (Fig. 1.7). We speculated that reef-building corals were able to flourish along this stretch of the Isabela coast thanks to the sheltering effect of Fernandina Island, which prevented the influx of cool, nutrient rich upwelling waters. With his wife, Sain, and his field assistants, Mitchell Colgan established a field base in the mid-1980s in the remote and harsh Urvina Bay environment. Mitchell and his dedicated team spent three field seasons at this site, totaling 9 months, mapping coral formations and relating their growth histories to past El Niño events.

Had Jerry not directed us to Tagus Cove we might well have by-passed it due to its steep shoreline cliffs and partially hidden entrance. Jerry warned us that we would encounter few if any zooxanthellate corals at the cove, but the site would nonetheless excite our interest. Tagus Cove is a breached volcanic crater surrounded by steep cliffs that descend to relatively deep water and a silt/sandy bottom, an ideal site for ships to anchor, shelter, and remain out of sight. There is also a source of drinking water at the landing site. The rock walls are painted with the names of sailing ships from a bygone era, ships engaged in tortoise and fur seal hunting, whalers, and brigands. The absence of direct sunlight beneath the overhanging cliffs and low underwater visibility of 3 to 4 m at most, did not make diving at this site particularly inviting.

This quickly changed upon viewing the azooxanthellate coral communities—the vertical rock walls supported a startling profusion of brilliant yellow/orange, red/purple and black colored coral colonies at 2–5 m depth. We quantified the surface cover of four coral species to a depth of 21 m, and found that these corals together occupied up to 30% of the substrate. In deeper water, suspended from the ceiling of a ledge, we encountered numerous white/pale-pink colonies of *Madrepora galapagensis*, which covered in excess of 40% of the available substrate. Tagus Cove was truly the cornucopia of azooxanthellate scleractinian corals. John Wells was ecstatic with the profusion of corals at this site and described three of the species as new to science (Wells 1982). One of the new species, *Tubastraea tagusensis,* has been recently and accidentally introduced in Brazil and is seriously disrupting native benthic populations there (De Paula and Creed 2004).

Our third reconnaissance site was Punta Espinosa, Fernandina Island, which lies across the Bolívar-Elizabeth Canal and about 8 km due west of Tagus Cove. Jerry again aroused our curiosity with another unusual community awaiting our arrival at this location—incipient coral frameworks filling lava rock ponds. This is a highly productive area as evidenced by cool, nutrient rich waters that support luxurious marine algal communities. These rich plant communities in turn support enormous numbers of marine iguanas that dive and graze on the algal carpeted rock substrates (Figs. 8.2 and 8.3). Less than a kilometer from this wave swept shore are several 1 and 2 m deep basalt rock ponds that narrowly connect to and are tidally flushed by coastal waters.

Our first views of the basalt rock ponds were surreal; most of them were nearly completely filled with live pocilloporid coral buildups. Could these 1–2 m high coral patches have formed after the violent eruption of the Fernandina (Narborough) Volcano in 1825, described by the eye-witness account of Captain Benjamin Morrell Jr. (1832). Fernandina is the most active volcano in the Galápagos Archipelago, having erupted and producing lava flows reaching the sea several times during the past two centuries (Smithsonian Institution 2023). If pocilloporid corals were present in the basalt ponds during the 1825 eruption, it is possible they did not survive due to the extreme heat produced. Captain Morrell reported stifling local heating and a sea temperature of 66 °C several kilometers from the affected coastline. Considering the mean vertical branch growth rate of *Pocillopora* in the Galápagos Islands, 2.4 cm per year, it is likely that the 1–2 m high frameworks formed after 1825, during the past 50–100 years.[8]

Next on our itinerary were the northern-most islands in the archipelago, Wenman and Darwin, which bordered and were often influenced by the Equatorial Front and warm tropical waters. Wenman Island revealed rich coral development along its eastern side (Shark Bay, named for the constantly present school of Hammerhead Sharks, Figs. 1.6, 8.1). The insular shelf is narrow and quickly descends to deep water and depths below zooxanthellate coral development in the eastern Pacific. Zooxanthellate coral presence essentially ended at 31 m in our surveys at this site.

Our first dive at Darwin Island, near Darwin's Arch, was unsettling and short-lived. Just as Chuck, Jerry and I settled on the bottom to begin with the coral survey we were surrounded by fast-swimming Galápagos sharks (*Carcharhinus galapagensis*). I estimated the school size at 10 with each shark 2 to 2.5 m in length. The sharks began darting around and between us at a furious pace. Then they began bumping into us. One swam between my legs and nearly knocked me over. I glanced toward Chuck and saw that he was heading toward the surface for our support skiff.

[8] Rapid colonization of *Pocillopora* spp. on lava flows have occurred on the island of Hawai'i (Grigg and Maragos 1974) and in Indonesia (Tomascik et al. 1996). Five years after the Indonesian volcanic eruption coral colonization resulted in high live cover, colony abundance, and species richness. Contributing to this rapid coral development were the stability of the basalt substrates and high nutrient concentration (from effusion of geothermal fluids) promoting coral growth. These conditions would also likely boost coral growth at Punta Espinosa. In addition, solar warming of the tidally flushed ponds and the local productive environment would be expected to promote coral growth.

I looked at Jerry and he motioned that all was well, his hand signal indicating no need to surface. After experiencing another bump, I decided not to heed Jerry's advice and ascended quickly for our skiff. I was greeted by the boatman, "What's wrong, did the sharks chase you here?". A few minutes later Jerry flew out of the water and slid quickly into the skiff. Galápagos Sharks have a nasty reputation, and we decided not to risk being bitten so far from the clinic at Academy Bay on Santa Cruz Island.

This incident caused us to be extra cautious on the remainder of our dives at Darwin Island. In our continuing reconnaissance we came across a 1 m high framework of *Pocillopora* but missed finding the largest known structural coral reef in the Galápagos Islands, which was discovered, mapped and dated 40 years later (Glynn et al. 2009, 2015, 2017). This reef is about 800 m long, constructed chiefly of massive *Porites* corals forming 2 to 4 m high towers at 12 to 16 m depths. Since this is a subsurface reef, present several meters below the sea surface, it is best regarded as a bank reef. It is the only one of this type presently known in the Galápagos Islands. We have named it the Wellington Reef, to honor Jerry Wellington's pioneering coral studies in the Galápagos Islands. Viktor Brandtneris and Tyler Smith enjoyed a post diving break near the Wellington Reef with Darwin's Arch visible in the background (Fig. 2.11).

In addition to the Galápagos coral book (Glynn and Wellington 1983), other sources of information on the early state of corals are available in a biological resource report prepared by Jerry for the Ecuadorian government (Wellington 1975) and in a book chapter of key Galápagos environments edited by R. Perry (Wellington

Fig. 2.11 Viktor Brandtneris (left) and Tyler Smith relaxing after diving on the Wellington reef at Darwin Island, Galápagos Islands (4 June 2012). Darwin's Arch, a rock formation in the background, still intact before its collapse in May 2021. Photograph courtesy of Tyler Smith

1984). In summary, abundant reef-building coral communities, incipient reef development, and small patch and fringing coral reefs were observed from south to north across the Galápagos Archipelago. Branching *Pocillopora* species and massive *Porites lobata* and *Pavona clavus* colonies contributed most commonly to the reef framework formations. Joubin's red chart markings indicated the occurrence of coral reefs, however, his locations were frequently off the mark, often present on exposed rocky shores and in environments generally unfavorable for coral development.

In the 1970s, arriving by plane to and especially departing from the Galápagos Islands, was sometimes problematic. At that time the resident human population in the archipelago was small, numbering only a few thousand inhabitants. A single once-daily flight, TAME airlines, was the sole means of flying in and out of the islands. The only landing field was on Baltra Island, an abandoned World War II U.S. air base. The population today is more than 30,000 souls with two airline companies servicing the islands, LATAM and Avianca. Due to our numerous research objectives and formal course responsibilities at the University of Miami (UM), we needed to follow a tight travel schedule with no free days to spare. Thus, maintaining our planned flights and connections was always of great concern. For example, once after being "bumped" (denied boarding) from a Baltra-to-Quito flight I arrived in Miami one day late and needed to take a taxi from the airport directly to UM to deliver a lecture. I arrived at the lecture hall and dais just in time for a marine science class of about 80 students! A near disaster avoided, a skipped class by its professor.

An incident I experienced with Jerry resulted in frightfully bad but necessary "travel- survival" behavior on our part. Jerry and I were "bumped" from our booked and confirmed flight when attempting to leave the islands after a research cruise. Because of missing this flight, it was necessary to retrace our steps to the Charles Darwin Research Station on Santa Cruz Island and re-book the flight for the next day. We requested permission to remain and sleep overnight at the Baltra Island airport but were denied this plan because the island is an "off limits" military base. After stowing our gear and samples on the bus's rooftop we entered the crowded bus and stood for the ride to the Baltra Canal, and then a boat ride to the north shore of Santa Cruz Island. In transferring our gear from one ride to the next, we were usually too late to claim a seat. After the bus ride to Puerto Ayora, on the south side of Santa Cruz Island, it was then necessary to find transportation back to the Darwin station. This roundtrip journey consumed most of the day. After making the trip to Baltra the next morning—truck, bus, boat, bus, plus "schlepping" all our gear—we were promised seats on the flight to the mainland by the local airline manager. I handed him a $20 bill in the hopes of securing our passage. To our consternation, and to several other hopeful ticket holders waiting in line, numerous passengers filed onto the plane, the boarding line closed, the plane's entrance door was shut, and the plane began to taxi down the runway. With engines whining, the plane suddenly stopped, a door opened, and the manager pointed in our direction signaling that two additional passengers could be accommodated. With that gesture all ticket holders, about 20 individuals (including us), began running to the plane juggling baggage and trying to stay in the lead. Jerry was quite athletic, and he alone was able to jump up and hold onto the open doorway. I tossed up our gear, Jerry extended his arm to me for a lift, but then a complication occurred. Some ticket holders caught on to our scheme; one man grabbed hold of my legs and another my torso and then began climbing up my body to reach the open door. I shouted to Jerry, "What should I do?" Without hesitation, Jerry instructed me to pummel and kick at the interlopers so that he could hoist me aboard the departing flight. I followed his instructions, shred the body climbers and entered the plane. The manager pointed to two available seats. The door was slammed shut and secured, and soon after we were airborne heading for the mainland. Jerry and I, in the throes of an adrenalin rush, exchanged disbelieving glances. On future flights to the Galápagos Islands, I wondered if I might be recognized by any of the men that I so shamefully exfoliated from my body that day.

On our return to mainland Ecuador, we planned a three-day layover in Guayaquil to further explore coastal areas to the north. The first coral reef we discovered on the continental shores of Ecuador was on 30 January 1975, during an overflight in a chartered fixed wing airplane. This reef was located a short distance offshore near the hamlet of Machalilla. We use the verb 'discover' in the sense of its first recognition by coral reef researchers vis-à-vis its structural character and biotic composition. The reef was certainly known earlier by fishers, mariners, and indigenous people for decades and probably centuries. John Wells, Chuck Birkeland and I rented a car in Guayaquil and travelled up the coast the next day to examine the reef. When we arrived at Machalilla it was high tide, overcast, and sea conditions were unsettled. The outline of the reef was barely visible from the shore. I approached a group of local fishers with *cayucos* (dugout canoes) to negotiate renting one to reach and survey the reef. Ha!, they were ready for me; their price was an exorbitant 1200 S (Ecuadorian sucres), equivalent at that time to about $50 US dollars. "Respectfully good sirs," in my most courteous Spanish, "I do not wish to purchase a *cayuco*, just rent one for two hours". After extensive pleading by me for a lower and fair price, the fishers stood firm at $40 dollars. So, Chuck and I donned our surface diving gear, leaped into the water, and began swimming toward the reef. About halfway to the reef, we noticed that a *cayuco* was a short distance behind us. The *cayuco* rental price was now 600 S, and as we reached the reef it was lowered to 300 S, a reasonable rate. Turns out — long story short — the fisher efficiently and happily supported our survey. He led us to rich coral patches, was very accommodating in following our directions, and in the end friendly. I noticed that during our survey he was eyeing my Timex dive watch. Before leaving Machalilla I gave him the watch and my swim fins, items easily replaced in Panama.

Returning to Guayaquil, we were ecstatic over the successful outcome of this brief survey. Surface observations of the coral formations and zonation, specifically eroded craters, and reef edge escarpments, suggested that this was a true structural reef, a limestone edifice at least a few meters thick. Seven zooxanthellate coral species, in the genera *Pocillopora*, *Porites*, and *Pavona*, made up the limestone reef frameworks, with three azooxanthellate coral species also present as well as diverse invertebrate taxa and fishes that are associated with reefs in Costa Rica, Colombia, and Panama. Thus, this biotically constructed formation, located just shy of 1°30'S, qualified as the southern-most coral reef yet known on continental eastern Pacific shores. Before this record can be accepted, it will be necessary to survey Santa Clara Island, Gulf of Guayaquil, where coral buildups were reported at a site below 3°S by René Espinosa. René was a skilled diver and one of our volunteer assistants for several years, familiar with coral community and reef development in the Galápagos Islands.

Continuing surveys along the Ecuadorian coast in the 1980s and more recently have revealed extensive coral development on the leeward, northeastern shores of Isla de La Plata and several mainland sites to 2° S latitude (Glynn 2003; Rivera and Martínez 2011; Glynn et al. 2017). Massive *Pavona clavus* colonies 3 to 5 m in diameter were discovered at Isla de La Plata, ranking them among the Methuselah colonies present in the Galápagos Islands.

Chile

Postcards circulating in the 1980s with underwater views of coral communities at Rapa Nui (Isla de Pascua, Easter Island) caught our attention and interest. The coral communities in the pictures appeared to be paucispecific, i.e., consisting of few species belonging to the genera *Porites* and *Pocillopora*, the principal reef-building corals in our eastern Pacific study areas. Fortunately, Jerry Wellington was awarded a research grant from the U. S. National Geographic Society to investigate the Rapa Nui coral formations and their responses to environmental disturbances such as ENSO stressors. Jerry invited me to participate in this study as well as Chilean marine ecologists (Evie A. Wieters and Sergio A. Navarrete, Catholic University in Santiago). An initial result of our surveys in 1999, focused on the identification of corals and descriptions of community structure, was published in Jorge Cortés' book on Latin American Coral Reefs (Glynn et al. 2003). Widespread coral bleaching, coincident with an El Niño warming event, was documented on a return research trip in 2000 (Wellington et al. 2001). These first studies were followed by a more detailed biogeographic analysis that explored the relationship of the Rapa Nui coral fauna with that of the Central Polynesian and Hawaiian faunal groups.

Although Rapa Nui is relatively near the Pitcairn Islands in the South Pacific Ocean (Fig. 3.6) and has been grouped with this biogeographic province (Veron 1995), we found that its coral fauna is most closely related to the eastern Pacific region (Glynn et al. 2007). We hypothesized that this condition is a result of the east to west surface current flow. This flow direction is due to the interaction between the mesoscale eddy field and the bottom topography, thus blocking eastward flow carrying potential coral propagules from the Polynesian Faunal group to the eastern Pacific region (Glynn et al. 2007). In accordance with this isolation from the west, the coral faunal affinity of Rapa Nui is most similar to the oceanic islands of Malpelo and Clipperton in the eastern Pacific.

Fig. 2.12 Rapa Nui investigators, Evie Wieters (with son Julien in arms), Bernhard Riegl, Alba Medrano, and Hagen Buck-Wiese, left to right (December 2012). The Ahu Tongariki ceremonial platform supporting Moai statues is visible in the background. The coral eyes are missing; one statue is adorned with a hat (pukao)

Fig. 2.13 Data scribing and coral marking in a Rapa Nui coral community. Evie Wieters, diver in foreground, and Alba Medrano, diver barely visible in background (April 2012)

Evie Wieters and co-workers are continuing with coral community functional ecological research at Rapa Nui (Figs. 2.12 and 2.13; Wieters et al. 2014; and see Glynn et al. 2017) and Jimmy (James) Herlan is investigating the spatial relationship between coral community composition and wave exposure gradients around the island.[9] These studies demonstrate the strong influence of ENSO temperature stress effects and wave assault on coral communities, especially at shallow depths. The recent availability of remotely operated vehicles (ROVs) has allowed exploration of mesophotic or deep-water coral communities. Our studies utilizing compressed air were limited to ~60 m. More recently, ROV technology has permitted observations to 350 m depth. Deep water surveys have revealed extensive mushroom coral (*Cycloseris vaughani*) communities (Hoeksema et al. 2019), and incipient framework development of platy *Leptoseris* sp. colonies (Sellanes et al. 2021). The discovery of mesophotic coral communities is not surprising considering the high light penetration through the water column and the deep thermocline structure at Rapa Nui. Unfortunately, the *Leptoseris* sp. community off Hanga Roa, the main village on the island, was mostly dead and covered by filamentous cyanobacteria-like mats and filamentous algae. Healthy *Leptoseris* communitries occurred off the least populated northern and southeastern sides of the island. Javier Sellanes and co-workers attributed the coral death to the growing human population and resulting eutrophication (Sellanes et al. 2021).

One afternoon while recharging scuba tanks near Hanga Roa we caught sight of a mixed gender band of students racing to the beach after their school classes ended. They lifted a large outrigger canoe resting on the beach and deftly launched it into the surf. They paddled out to the offshore breaking surf zone and began catching and riding waves. Their skill at maneuvering the canoe was executed with finesse. It was obvious that they had enjoyed this sport on many earlier occasions. At the end of their surfing amusement and now resting on the beach, melodious strains from guitars and singing began drifting across the cove, reminding us of the fundamental connection of the Rapa Nui community with the South Pacific Polynesian culture. This connection was re-enforced by our accommodations at Rapa Nui—we arranged to stay at a hostel managed by an indigenous family, the home of Doña Ana Paoa Rangitopa and Don Jorge E. Rapahango. In the evenings after dinner Don Jorge regaled us with stories of the early Rapa Nui people and their challenges. We were especially surprised to learn of Don Jorge's numerous travels to French Polynesia to visit with his distant relatives and to inspect his land holdings there.

A long-standing question was whether the enormous statues of Rapa Nui (Moai) were originally fitted with inlaid eyes. Thor Heyerdahl suggested in 1976 that this was the case (Heyerdahl 1989). A few years later Sonia Haoa, an archaeology

[9] I was surprised and delighted to learn of Jimmy's involvement in coral research at Rapa Nui. Jimmy was enrolled in the Marine Biology core course I taught at the University of Miami in 2002. He joined ESMOI (Millennium Nucleus for Ecology and Sustainable Management of Oceanic Islands) based at the Universidad Católica del Norte in Coquimbo, Chile. This government supported research center is dedicated to the management and biodiversity conservation of Rapa Nui, Sala y Gómez and other of Chile's oceanic islands. Jimmy is presently engaged in graduate studies in the Ecology, Evolutionary Biology and Behavior program at The City University of New York with plans to continue coral studies at Rapa Nui.

student and native of Rapa Nui, unearthed a complete eye which perfectly fit the eye socket of one of the statues at Anakena, near the shore of one of our dive sites. This remarkable discovery was reported by Sergio Rapu. The eye consisted of a central disc of dark lava rock serving as the pupil and iris and was surrounded by an almond-shaped substance mimicking the white sclera. Only one kind of material seemed to satisfy this need, the limestone skeleton of the coral species *Porites lobata*. This was the most abundant coral present in most of our surveys, from 10 to 40 m depth (Glynn et al. 2003). Large pieces of this coral can be found on beaches and when it ages and weathers it turns bone white. The skeleton is a relatively soft limestone and easy to sculpt. The nature of the Moai's eyes was of great interest. Skeptics had argued that inlaid eyes in stone images would not be found at Rapa Nui because they were unknown in Polynesia. The presence of inlaid eyes in stone and wooden statues in some advanced ancient cultures in western continental America could be evidence supporting Heyerdahl's hypothesis of east to west migration. As a noteworthy aside, a 10-meter diameter colony of *P. lobata*, hundreds of years old and located off Anakena, is the largest known poritid colony in the eastern Pacific region (Hubbard and Garcia 2003).

Most of the coral reef discoveries discussed in this chapter were made in the 1970s in the equatorial eastern Pacific region, namely Panama, Costa Rica, Colombia, and Ecuador. Other little-known eastern Pacific localities supporting coral growth at higher latitudes were surveyed and studied by my colleagues, students, and me in the 1990s. These areas included Clipperton Atoll, an overseas state possession of France, Huatulco, a coral reef tract in the southern Mexican state of Oaxaca, and Easter Island (Rapa Nui), far west of the Chilean coast. Research performed at these localities is discussed, where relevant, in later chapters. It is reassuring to recognize that discoveries continue to be made by a new generation of rapidly growing Latin American coral reef researchers. In the golden 50-year issue of the Atoll Research Bulletin (2001), I noted that several eastern Pacific areas were still unexplored or little studied (see Fig. 1.1 in Glynn 2001b). As of this writing, 20 years later, most of these areas have been surveyed and have revealed thriving coral communities and reefs. The extent of coral development in southern Ecuador and northwestern Peru, however, is still largely unknown. A late Pleistocene, ~13,300 cal BP, archaeological specimen of *Pocillopora damicornis* was discovered at Huaca Prieta, a prehistoric Peruvian settlement (see Chap. 12, *Pocillopora* occurrences in Peru). It is likely, however, that this occurrence was due to early overland transport by humans (Glynn et al. 2022).

References

Colgan MW (1990) El Niño and the history of eastern Pacific reef building. In: Glynn PW (ed) Global ecological consequences of the 1982–83 El Niño-Southern Oscillation. Elsevier oceanography series 52. Elsevier, Amsterdam, pp 183–229

Colgan MW, Hollander D (1987) Sedimentologic succession of uplifted coral community, Urvina Bay, Isabela Island, Galapagos archipelago, Ecuador. Am Assoc Petrol Geol 71:5. Annual convention, Los Angeles, 71

Cortés J (1990a) Coral reef decline in Golfo Dulce, Costa Rica, eastern Pacific: anthropogenic and natural disturbances. Ph.D. dissertation. University of Miami, Coral Gables, Florida

Cortés J (1990b) The coral reefs of Golfo Dulce, Costa Rica: distribution and community structure. Atoll Res Bull 344:1–37

Cortés J (1991) Los arrecifes coralinos de Golfo Dulce, Costa Rica: aspectos geológicos. Revista Geología América Central 13:15–24

Cortés J, Jiménez C (2003) Corals and coral reefs of the Pacific coast of Costa Rica: history, research and status. In: Cortés J (ed) Latin American coral reefs. Elsevier Science, Amsterdam, pp 361–385

Cortés J, Macintyre IG, Glynn PW (1994) Holocene growth history of an eastern Pacific fringing reef, Punta Islotes, Costa Rica. Coral Reefs 13:65–73

Crossland C (1927) Marine ecology and coral formations in the Panama region, Galapagos, and Marquesas Islands, and the atoll of Napuka. R Soc Edinb Trans 55(2):531–554

Dana JD (1890) Corals and coral islands, 3rd edn. Dodd, Mead, New York, p 440

Darwin CR (1842) The structure and distribution of coral reefs. Being the first part of the geology of the voyage of the *Beagle*, under the command of Capt. Fitzroy, R. N. during the years 1832 to 1836, vol 12. Smith Elder, London, p 115

De Paula AF, Creed JC (2004) Two species of the coral *Tubastraea* (Cnidaria, Scleractinia) in Brazil: a case of accidental introduction. Bull Mar Sci 74:175–183

Durham JW, Barnard JL (1952) Stony corals of the eastern Pacific collected by the Valero III and Valero IV. Allan Hancock Pacific Exp 16(1):1–110

Edgar GJ, Fariña JM, Calvopiña M, Martínez C, Banks S (2002) Comunidades submareales rocosas II: Peces y macroinvertebrados móviles. In: Danulat E, Edgar GJ (eds) Reserva Marina de Galápagos. Linea Base de la Biodiversidad. Fundación Charles Darwin y Servicio Parque Nacional Galápagos. Santa Cruz, Galápagos, Ecuador, pp 68–97

Feingold JS, Brulé B (2020) Population fluctuations of the fungiid coral *Cycloseris curvata*, Galápagos Islands. Ecuador Adv Mar Biol 87:141–166. https://doi.org/10.1016/bs.amb.2020.08.005

Feingold JS, Riegl B, Hendrickson K, Toth LT, Cheng H, Edwards RL, Aronson RB (2020) 7700-year persistence of an isolated, free-living coral assemblage in the Galápagos Islands: a model for coral refugia. Coral Reefs 39:639–647. https://doi.org/10.1007/s00338-020-01935-5

Geist DJ, Snell H, Goddard C, Kurz MD (2014) A paleogeographic model of the Galápagos Islands and biogeographical and evolutionary implications. No. 8. In: Harpp KS, Mittlestaedt E, d'Ozouville N, Graham DW (eds) The Galápagos: a natural laboratory for the earth sciences, Geophysical monograph 204, 1st edn. American Geophysical Union, Wiley, New York, pp 145–166

Glynn PW (1972) Observations on the ecology of the Caribbean and Pacific coasts of Panamá. In: Jones ML (ed) The Panamic biota: some observations prior to a sea-level canal. Bull Biol Soc Wash 2:13–30

Glynn PW (1983) Final report on the effects of the sea star *Acanthaster* on Omani coral reefs with some recommendations for further study. In: Report to Ministry of Agriculture and Fisheries and the Omani-American joint Commission for Economic and Technical Cooperation, Muscat, Oman

Glynn PW (1990) Coral mortality and disturbances to coral reefs in the tropical eastern Pacific. In: Glynn PW (ed) Global ecological consequences of the 1982–83 El Niño-Southern Oscillation. Elsevier oceanography series 52. Elsevier, Amsterdam, pp 55–126

Glynn PW (1993) Monsoonal upwelling and episodic *Acanthaster* predation as probable controls of coral reef distribution and community structure in Oman, Indian Ocean. Atoll Res Bull 379:1–66. https://doi.org/10.5479/si.00775630.379.1

Glynn PW (2001a) Henry von Prahl (1948-1989). Bull Mar Sci 69(1):5–6

Glynn PW (2001b) Eastern Pacific coral reefs: new revelations in the twentieth century. Atoll Res Bull 494:118–158

Glynn PW (2003) Coral communities and coral reefs of Ecuador. In: Cortés J (ed) Latin American coral reefs. Elsevier, Amsterdam, pp 449–472

Glynn PW, Colley SB (2001) A collection of studies on the effects of the 1997-98 El Niño-Southern Oscillation event on corals and coral reefs in the eastern tropical Pacific. Bull Mar Sci 69(1):1–288

Glynn PW, Macintyre IG (1977) Growth rate and age of coral reefs on the Pacific coast of Panama. In: Proceedings of the 3rd international coral reef symposium, Miami, vol 2, pp 251–259

Glynn PW, Stewart RH (1973) Distribution of coral reefs in the Pearl Islands (Gulf of Panama) in relation to thermal conditions. Limnol Oceanogr 18:367–379

Glynn PW, Wellington GM (1983) Corals and coral reefs of the Galápagos Islands. With an annotated list of the scleractinian corals of the Galápagos by John W. Wells. University of California Press, Berkeley

Glynn PW, Stewart RH, McCosker JE (1972) Pacific coral reefs of Panama: structure, distribution and predators. Geol Rundsch 61:483–519

Glynn PW, Wellington GM, Birkeland C (1979) Coral reef growth in the Galápagos: limitation by sea urchins. Science 203:47–49

Glynn PW, von Prahl H, Guhl F (1982) Coral reefs of Gorgona Island, Colombia, with special reference to corallivores and their influence on community structure and reef development. Anales del Instituto de Investigaciones Marinas de Punta de Betín 12:185–214

Glynn PW, Druffel EM, Dunbar RB (1983) A dead central American coral reef tract: possible link with the Little Ice Age. J Mar Res 41:605–637

Glynn PW, Dillehay TD, Netherly PJ, Riegl BM (2022) Prehistoric reef-building coral occurrence in North Peru. J Island Coastal Archeol 19:420–432. https://doi.org/10.1080/15564894.2022.2072026

Glynn PW, Riegl B, Correa AMS, Baums IB (2009) Rapid recovery of a coral reef at Darwin Island, Galápagos Islands. Galapagos Res 66:6–13

Glynn PW, Wellington GM, Wieters EA, Navarrete SA (2003) Reef-building coral communities of Easter Island (Rapa Nui), Chile. In: Cortés J (ed) Latin American coral reefs. Elsevier, Amsterdam, pp 473–494

Glynn PW, Riegl B, Purkis S, Kerr JM, Smith TB (2015) Coral reef recovery in the Galapagos Islands: the northern-most islands (Darwin, Wenman). Coral Reefs 34:421–436. https://doi.org/10.1007/s00338-015-1280-4

Glynn PW, Wellington GM, Riegl B, Olson DB, Borneman E, Wieters EA (2007) Diversity and biogeography of the scleractinian coral fauna of Easter Island (Rapa Nui). Pac Sci 61:67–90

Glynn PW, Alvarado JJ, Banks S, Feingold JS, Jiménez C, Maragos JE, Martínez P, Maté JL, Moanga DA, Navarrete S, Reyes-Bonilla H, Riegl B, Rivera F, Vargas-Ángel B, Wieters EA, Zapata FA (2017) In: Glynn PW, Manzello DP, Enochs IC (eds) Coral reefs of the eastern tropical Pacific: persistence and loss in a dynamic environment. Coral reefs of the world 8. Springer, Dordrecht, pp 107–176

Grigg RW, Maragos JE (1974) Recolonization of hermatypic corals on submerged lava flows in Hawaii. Ecology 55:387–395

Guzmán HM (1986) Estructura de la comunidad arrecifal de la Isla del Caño, Costa Rica, y el efecto de perturbaciones naturales severas. Universidad de Costa Rica, San Pedro, Costa Rica, M.Sc. thesis, 179

Guzmán HM, Cortés J (1989) Coral reef community structure at Caño Island, Pacific Costa Rica. Pubbl Staz Zool Napoli (I: Marine Ecology) 10:23–41

Guzmán HM, Cortés J (2001) Changes in reef community structure after fifteen years of natural disturbances in the eastern Pacific (Costa Rica). Bull Mar Sci 69:133–149

Heyerdahl T (1989) Easter Island—the mystery solved. Random House, New York, p 255

Hoeksema BW, Sellanes J, Easton EE (2019) A high-latitude mesophotic *Cycloseris* field at 85 m depth off Rapa Nui (Easter Island). Bull Mar Sci 95(1):101–102. https://doi.org/10.5343/bms.2018.0053

Hubbard D, Garcia M (2003) The corals and coral reefs of Easter Island—a preliminary look. In: Loret J, Tanacredi JT (eds) Easter Island: scientific exploration into the world's environmental problems in microcosm. Kluwer Academic, New York, pp 53–77

Jiménez C, Bassey G, Segura A, Cortés J (2010) Characterization of the coral communities and reefs of two previously undescribed locations in the upwelling region of Gulf of Papagayo (Costa Rica). Rev Mar Cost 2:95–108

Joubin LI (1912) Carte des banc et récifs de coraux. Inst Océanogr Monaco Ann 4:1–7. 5 maps

Lessios HA, Kessing BD, Robertson DR, Paulay G (1999) Phylogeography of the pantropical sea urchin *Eucidaris* in relation to land barriers and ocean currents. Evolution 53:806–817

Macintyre IG, Glynn PW, Cortés J (1992) Holocene reef history in the eastern Pacific: mainland Costa Rica, Caño Island, Cocos Island, and Galápagos Islands. In: Proceedings of the seventh international coral reef symposium, Guam, vol 2, pp 1174–1184

Morrell B Jr (1832) A narrative of four voyages to the South Sea, North and South Pacific Ocean, Chinese Sea, Ethiopic and Southern Atlantic Ocean, Indian and Antarctic Ocean from the year 1822 to 1831. J. and J. Harper, New York, p 492

Perrin C, Bosellini FR (2013) The late Miocene cold spot of z-coral diversity in the Mediterranean: pattern and causes. C R Palevol 12(5):245–255

Porter JW (1972) Ecology and species diversity of coral reefs on opposite sides of the Isthmus of Panama. In: Jones ML (ed) The Panamic biota: some observations prior to a sea-level canal. Bull Biol Soc Wash 2:89–116

Reyes-Bonilla H, Riosmena-Rodríguez R, Foster MS (1997) Hermatypic corals associated with rhodolith beds in the Gulf of California. Pac Sci 51:328–337

Rivera F, Martínez P (2011) Guia fotográfica de corales y octocorales Parque Nacional Machalilla y Reserva de Producción Faunística Marino Costera Puntilla de Santa Elena. With the contribution of Odalisca Breedy. NAZCA and Conservación Internacional, Ecuador, p 78

Ross CL, Falter JL, Schoepf V, McCulloch MT (2015) Perennial growth of hermatypic corals at Rottnest Island, Western Australia (32°S). PeerJ 3:e781. https://doi.org/10.7717/peerj

Salm RV (1993) Coral reefs of the Sultanate of Oman. Atoll Res Bull 380:1–85

Schmidt-Roach S, Miller KJ, Lundgren P, Andreakis N (2014) With eyes wide open: a revision of species within and closely related to the *Pocillopora damicornis* species complex (Scleractinia; Pocilloporidae) using morphology and genetics. Zool J Lin Soc 170(1):1–33. https://doi.org/10.1111/zoj.12092

Sellanes J, Gorny M, Zapata-Hernández G, Alvarez G, Muñoz P, Tala F (2021) A new threat to local marine biodiversity: filamentous mats proliferating at mesophotic depths off Rapa Nui. PeerJ 9:e12052. https://doi.org/10.7717/peerj.12052

Sheppard CRC, Salm RV (1988) Reef and coral communities of Oman, with a description of a new coral species (order Scleractinia, genus *Acanthastrea*). J Nat Hist 22:263–279

Smithsonian Institution (2023) The volcanoes of the world database. Global Volcanism Program. National Museum of Natural History, Washington, DC

Tomascik T, van Woesik R, Mah AJ (1996) Rapid coral colonization of a recent lava flow following a volcanic eruption, Banda Islands, Indonesia. Coral Reefs 15:169–175

Vargas-Ángel B (1988) Contribución al conocimiento de las formaciones coralinas del litoral Pacífico colombiano. Mem. VI Sem. Nal. Ciencias del Mar, Bogotá, pp 562–570

Vargas-Ángel B (1989) Contribución al conocimiento de la dinámica y distribución de las formaciones coralinas de la Ensenada de Utría, Chocó, Colombia. B.Sc. thesis. Department of Biology, University of del Valle, Cali, Colombia

Vargas-Ángel B (1996) Distribution and community structure of the reef corals of Ensenada de Utría, Pacific coast of Colombia. Rev Biol Trop 44:443–651

Vargas-Ángel B (2001) Corals and coral reefs of the Pacific coast of Colombia with special reference to spatial and temporal patterns of environmental disturbances. PhD dissertation. University of Miami, Coral Gables, Florida

Vaughan TW (1906) Reports on the scientific results of the expedition to the eastern tropical Pacific … 6. Madreporaria. Bull Mus Comp Zool Harv Coll 50(3):61–72

Veron JEN (1995) Corals in space and time. The biogeography and evolution of the Scleractinia. Comstock, Cornell

Veron JEN (2000) Corals of the world, vol 2. Australian Institute of Marine Science, Townsville, p 429

Veron JEN, Marsh L (1988) Hermatypic corals of Western Australia. Records and annotated species list. Records of the Western Australian Mus Suppl 29:1–136

Verrill AE (1864) List of the polyps and corals sent by the Museum of Comparative Zoology to other institutions in exchange, with annotations. Bull Mus Comp Zool Harv Coll 1(3):29–60

Verrill AE (1866) On the polyps and corals of Panama with descriptions of new species. Proc Boston Soc Nat Hist 10:323–357

von Prahl H, Guhl F, Grögl M (1979) Gorgona. Futuro Grupo Editorial, Bogotá, p 279

Watson TJ Jr (1993) Pacific passage. Mystic Seaport Museum, Mystic, Connecticut, p 179

Wieters EA, Medrano A, Pérez-Matus A (2014) Functional community structure of shallow hard bottom communities at Rapa Nui, Easter Island. Lat Am J Aquat Res 42:827–844

Wellington GM (1975) The Galápagos coastal marine environments: a resource report to the Department of National Parks and Wildlife, Quito, Ecuador, p 357. [Available as a digital file upon request from the library at RSMAES, University of Miami]

Wellington GM (1984) Marine environment and protection. In: Perry R (ed) Key environments: Galapagos. Pergamon, Oxford, pp 247–263

Wellington GM, Glynn PW, Strong AE, Navarrete SA, Wieters E, Hubbard D (2001) Crisis on coral reefs linked to climate change. Eos Trans AGU 82(1):5–5

Wells JW (1957) Coral reefs. In: Hedgpeth JW (ed) Treatise on marine ecology and paleoecology 1: 609-631. The Geological Society of America, memoir 67. Waverly, Baltimore, Maryland

Wells JW (1963) Coral growth and geochronometry. Nature 197:948–950

Wells JW (1982) Notes on Indo-Pacific scleractinian corals, part 9. New corals from the Galápagos Islands. Pac Sci 36:211–219

Zapata FA, Vargas-Ángel B (2003) Corals and coral reefs of the Pacific coast of Colombia. In: Cortés J (ed) Latin American coral reefs. Elsevier, Amsterdam, pp 419–447

Chapter 3
Origin of the Eastern Pacific Coral Reef Province

Abstract Today the coral faunas, reef development, and reef morphologies differ substantially on the Caribbean and Pacific coasts of the Central American isthmus. The contemporary eastern Pacific reef-building coral fauna consists of ~45 species and the Caribbean ~75 species with *Porites* the only genus common to the two regions. Zooxanthellate coral richness reached 55 species during the early Oligocene epoch, ~30 Ma, and then steadily declined until Holocene time, 11–12,000 years ago. Species of the fire coral genus *Millepora* have been recently discovered in the eastern Pacific, all of Indo-Pacific identities and with restricted distributions (Gulf of Chiriquí, Panama, and Clipperton Atoll). Reef development is robust and widespread in the Caribbean and less strong (particularly among *Pocillopora* reefs), and often confined to sheltered areas in the eastern Pacific. Western Atlantic and eastern Pacific coral faunas were continuous, the same, during most of Cenozoic time but became separated with the emergence of the Central American land bridge about 3 Ma. Morphometric and molecular evidence support a strong genetic affinity between eastern Pacific and Caribbean *Porites* species, indicating local speciation (vicariance premise) and a frequent influx of Indo-Pacific species via long distance larval transport during strong El Niño events (dispersal premise). The tropical "eastern Pacific barrier", the wide expanse of deep, open ocean waters preventing the dispersal of shallow marine organisms, is porous and best regarded as a filter bridge. Nearly all recent eastern Pacific coral species resulting from dispersal (88%) greatly outnumber endemic species (12%).

Introduction

Although separated today by a narrow land bridge only some 70 km wide, coral communities and coral reefs on opposite sides of the Isthmus of Panama are very different. They share few species and structural similarities. Reconnaissance flights over the northeastern Caribbean coast of Panama reveal prominent fringing reefs, coral cays, patch and bank reefs, and barrier reefs that are especially prominent in the San Blas Archipelago. Many of the approximately 50 coral cays in this area

© The Author(s), under exclusive license to Springer Nature Switzerland AG 2025
P. W. Glynn, *Coral Reef Resilience in the Anthropocene*,
https://doi.org/10.1007/978-3-031-81421-1_3

support villages of the Native American Kuna Yala people, also known as the Gunas. A barrier reef in this area, the Holandés Keys (Islas Maqui), possesses a prominent algal ridge,[1] and spurs and grooves, reminiscent of the windward reef fronts of mid-Pacific atolls and barrier reefs.

In 1969, I was fortunate to share a reconnaissance overflight of the San Blas Islands with Tom (Thomas F.) Goreau, a renowned coral reef scientist from the University of the West Indies, Jamaica. The helicopter flight was part of an information gathering survey connected with a proposal to construct a sea level canal directly linking western Atlantic with eastern Pacific waters (Challinor 1972). Tom and I marveled at the remarkable coral development and varied reef formations in this area. A few years later I teamed up with Ian G. Macintyre, a notable carbonate sedimentologist from the Smithsonian Institution to core drill a north coast Panamanian fringing reef at Galeta Point (near Colón) and the Holandés reef (Macintyre and Glynn 1976; Macintyre et al. 2001; Fig. 3.1). Sometimes when cruising along the north coast of Panama, and with sea conditions permitting, we would ask Jack Bryan, the skipper of our research vessel, the RV *Tethys*, to activate the ship's sonar to locate deep reef formations. We often found submerged reefs at depths of 20–30 m, with diverse biotic communities, including branching, dome, and platy-shaped corals. Recent studies have since shown that these relatively deep reefs, then referred to as "drowned reefs", are widespread in the Caribbean Sea (Hubbard et al. 2008). Deep reefs, now termed mesophotic reefs, occur at intermediate light level depths. Light-dependent corals and algae could not keep pace with the rapid rise of sea level, beginning about 20,000 years BP (before present), at the end of the last Ice Age. Sea level rise and coastal flooding also increased shoreline erosion and sedimentation, which likely increased turbidity and also interfered with reef accretion. The Caribbean north coast reef scape contrasts markedly from the Pacific south coast of Panama, which reveals only small fringing and patch reefs, best seen during midday, extreme low tides when the tops of corals are briefly aerially exposed.

"The study of the geographic distribution of life and the reasons for it" (Veron 1995) is one of many definitions of biogeography. A key word in this definition, "life", in this case meaning "species", is crucial. Biogeographic studies require that species are correctly identified. Similarities among species in assemblages of living organisms are a measure of their affinity or structural and functional ecologies. For example, a strong similarity of species in different assemblages or areas suggests the potential for closely interacting ecological and evolutionary processes, a weak similarity the opposite effect. Biogeographers define a major geographic region, consisting of closely related species associations or communities, a province. One of my first tasks in Panama was to identify and compare the coral species on both

[1] The elevated seaward buttresses, encrusted with crustose coralline algae, superficially appear to be true algal ridges constructed primarily of calcareous algae. Core-drilled samples, however, revealed that the ridges consist of eroded storm deposits of reef-building corals overlain with a thin veneer of coralline algae (Macintyre et al. 2001). Thus, functionally similar, this feature is not structurally identical to Indo-Pacific algal ridges.

Fig. 3.1 Ian G. Macintyre cataloging coral core samples from the Punta Islotes reef, Golfo Dulce, Costa Rica (February 1989). Photograph courtesy of Jorge Cortés

the Caribbean and Pacific coasts to assess the relationships of the faunas in these two separated regions.

How closely related were Caribbean and Pacific coral communities in the past? Did some present day genera and species coexist at some time when the two oceans were joined, and are they still around today as descendants of Miocene (23–5.3 mya, million years ago) and later faunas? Two hypotheses have been proposed to explain the origin of the eastern Pacific faunal region: vicariance and dispersal. Vicariance occurs when the range of a species is fragmented as a barrier develops, such as the Panama land bridge that separates the eastern Pacific from the Caribbean Sea; with the passage of time the separated disjunct species populations can potentially evolve into two or more distinct species. In the dispersal hypothesis, individuals, larvae, or seeds are carried to new areas from donor populations, such as a west to east journey across the East Pacific Barrier (EPB)—the deep oceanic expanse separating the Central and east Pacific regions—and establish new populations that can potentially evolve into new species. Two camps of scientists energetically disputed these hypotheses in the 1970s. Tom Dana (1975) favored the dispersal hypothesis, and

McCoy and Heck (1976) and Heck and McCoy (1978) advocated for vicariance. Because modern Eastern Pacific corals exhibited the closest affinities with Indo-western Pacific species, the dispersal hypothesis gained strong support among early researchers (Vaughan and Wells 1943; Squires 1959; Durham 1966). The dispersal hypothesis was also supported by the occurrence of many Indo-west Pacific species of mollusks, crustaceans, echinoderms, and fishes that apparently succeeded in crossing the EPB and became established on oceanic eastern Pacific islands (summarized in Briggs 1995). Because some Caribbean coral reef genera are still present today in the eastern Pacific, a handful of researchers maintain that a vicariance origin for this fauna cannot be dismissed (Budd 1989; Budd et al. 1994a, b).

Tom (Thomas F.) Dana, then a doctoral student at the Scripps Institution of Oceanography, University of California San Diego, proposed that a critical step in the development of the eastern Pacific coral fauna involved the northwestward movement of the northern Line Islands via sea floor spreading during the Pliocene Epoch (5.3–2.6 mya). This plate repositioning brought coral-rich islands into the path of the east-flowing North Equatorial Counter Current (NECC) and established important steppingstones in the dispersal of planktonic coral larvae across the Central and East Pacific regions (Fig. 3.2). Today, the northern Line Islands include part of the Republic of Kiribati plus Kingman Reef, Palmyra Atoll, and Jarvis Island, which are administered by the U.S. Pacific Remote Islands Marine National Monument.

Tom's publication immediately revived interest in the origin of the eastern Pacific coral reef region. When I visited the Biology Department at Florida State University, soon after publication of Tom's study, I was approached by two enthusiastic and well-informed graduate students, Ken (Kenneth L. Jr.) Heck and Earl D. McCoy, who asked if I agreed with Tom's dispersal hypothesis. "Yes!", I responded, "I believe Tom's sequence of events leading to the post-Pliocene migration of corals to tropical western America is based on sound reasoning and credible evidence." We were soon engaged in a heated but civil discussion of the evidence supporting each hypothesis. There were some major difficulties, for example, the eastern Pacific region has few fossil sites for establishing the past identity of coral species and direct evidence of coral dispersal was lacking. I encouraged Ken and Earl to publish

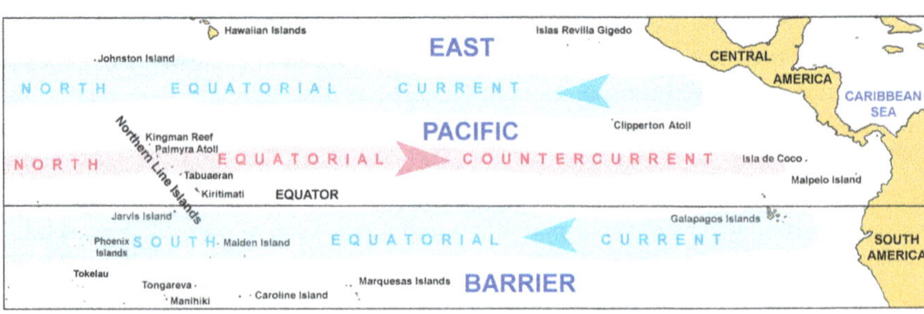

Fig. 3.2 The principal equatorial currents flowing to the west, towards the Central and South Pacific Islands, and to the East Pacific Coral Province (after Fiedler and Lavin 2017)

their opposing views. What better way to advance progress on a scientific question? As it turns out, the origin of eastern Pacific and Caribbean coral faunas is complex. It has involved isolation, massive extinctions, speciation, and immigration processes. Recent documentation of long-distance dispersal and molecular genetic evidence of speciation have shed light on the ancestry of coral lineages as revealed later in this chapter. Spoiler alert: both isolation and dispersal have played roles in the formation of the eastern Pacific coral reef region (Budd 1989; Glynn and Ault 2000; Glynn et al. 2007; Veron et al. 2015; López-Pérez 2017).

The Coral Species

First and foremost, how different or similar are the species of modern eastern Pacific and Caribbean coral faunas? Jim (James W.) Porter, a graduate student at Yale University, and I began to collect and identify Caribbean and Pacific corals in the early 1970s. I initially concentrated study on Caribbean corals because I was already familiar with the coral species of Puerto Rico, which are mostly present on the Caribbean coast of Panama. We both published tentative species lists in 1972 (Glynn 1972; Glynn et al. 1972; Porter 1972). John W. Wells at Cornell University, the doyen of coral systematics, was especially helpful in confirming and identifying Caribbean species. He established a comprehensive and well-organized coral reference collection at the Smithsonian Tropical Research Institute (STRI) in Panama. Many of the earliest-collected Pacific corals were identified by J. Wyatt Durham, University of California Berkeley, and later by John Wells, following his Galápagos surveys with us in 1975. I had the great privilege to study corals with Prof. Durham in the paleontological collections at UC Berkeley and with Prof. Wells at the museum of the Charles Darwin Research Station in the Galápagos Islands.[2]

In addition to scleractinian or stony corals, the distant cnidarian relatives commonly called "fire corals", in the genus *Millepora* (Class Hydrozoa, Order Anthoathecata), also secrete limestone skeletons and contribute importantly to reef

[2] I have continued to consult with other systematists and colleagues to confirm the identity of eastern Pacific corals as new collections and problematic specimens (unidentified and misidentified) have required. I am grateful to the following for their invaluable assistance in resolving species identifications: Francesca Benzoni, Stephen D. Cairns, Jorge Cortés, Michel Pichon, Richard H. "Dick" Randall, Héctor Reyes-Bonilla, and J. E. N. "Charlie" Veron. I have also examined corals in numerous research museums and heartily thank the curators for access to the collections and other courtesies: University of California (Berkeley), Museum of Paleontology; United States National Museum of Natural History (Department of Invertebrate Zoology); American Museum of Natural History, Dry Coral Collection; Museo de Zoología, Departamento de Zoología, Universidad de Concepción, Chile; Museo de Zoología, Universidad de Costa Rica, San Pedro, Costa Rica; California Academy of Sciences, research museum; Santa Barbara Natural History Museum, research collections; Charles Darwin Research Station, coral collection; Smithsonian Tropical Research Institute, coral collection; University of Guam Marine Laboratory, coral reference collection.

building in many areas. Fire corals are named according to the intense burning pain caused by contact with their potent nematocysts or stinging cells. Once while demonstrating the structure of *Millepora* to a group of students, my teaching assistant removed a fire coral from a drying oven set at a temperature of 103 °C. A student asked, "Are the stinging cells still potent after exposure to extreme heat?" The teaching assistant answered emphatically, "No", and made a near-fatal mistake by quickly placing the coral in his mouth, too quickly for me to intervene. Severe pain immediately followed with a swelling of the tongue and near blockage of airways, which required emergency medical care.

Millepora species possess symbiotic zooxanthellae and secrete massive skeletons, sometimes forming large clones that can dominate shallow reef zones. Jim Porter listed four *Millepora* species present on the Caribbean reefs of Panama. None was reported from the eastern Pacific region until our discovery of three species of *Millepora* at Uva Island, Gulf of Chiriquí, in March 1970. This was an exciting time, finding three unreported species at one coral reef site. Fire corals were unknown from the eastern tropical Pacific region. I immediately contacted Prof. Dr. Hilbrand Boschma of the Rijksmuseum of Natural History in Leiden, The Netherlands, requesting help in the identification of these newly discovered corals. He informed me that two of the hydrocorals belonged to well-known Indo-Pacific species (*Millepora intricata* and *Millepora platyphylla*), and one was likely a new, un-named species. He agreed to name the new species but sadly passed before he could engage in this task. During the 1982–83 El Niño warming event all known colonies of this new and unnamed hydrocoral species bleached (due to the loss of pigmented zooxanthellae) and died. We searched for surviving colonies on the Uva Island reef, and at other similar habitats, but none was found. This was concerning; is it possible that the unnamed hydrocoral species became extinct before being scientifically documented? Was this possibly a "centinelan extinction" as described by E. O. Wilson (1992)?[3] With this possibility in mind, I contacted W. H. "Wally" de Weerdt, a Dutch systematist (also from Leiden) who was familiar with the Caribbean species of *Millepora*. When Wally and I began to describe this new hydrocoral species, first recognized by Prof. Boschma, we decided that it would be appropriate to name the species in his honor. Thus, it received the official name *Millepora boschmai* Weerdt and Glynn, 1991 (Fig. 3.3). The three eastern Pacific species of *Millepora* are all morphologically different from the Caribbean species. The fate of the Pacific *Millepora* species following the 1982–83 El Niño event, a decidedly

[3]A centinelan extinction, as described by E. O. Wilson (1992), is an extinction of a species unknown to science and hence unrecorded and likely forever forgotten. Field biologists Alwyn Gentry and Calawey Dodson of the Missouri Botanical Garden, St. Louis, observed this type of extinction on a small ridge in the Andean foothills of Ecuador in 1978. There, they discovered several unnamed herbaceous plant species with uncharacteristic black leaves. When this site was revisited in 1986 it was completely changed, clear-cut by local farmers and planted in cacao and other crops. An unknown number of the unnamed species were lost. This kind of destruction and biodiversity loss are occurring across the Americas and worldwide with the increasing demand for agricultural expansion and development (Wilson 1992).

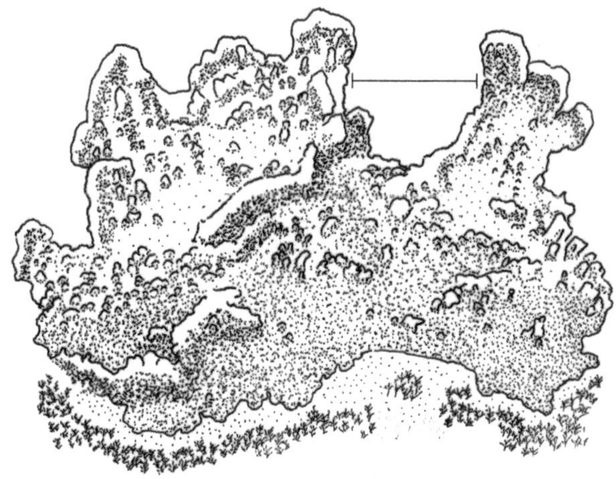

Fig. 3.3 *Millepora boschmai*, a now regionally extinct fire coral, scale bar = 10 cm

concerning question, and status of other more recent fire coral discoveries in the eastern Pacific are discussed in Chaps. 5 and 10.

In 2003, Jorge Cortés noted that the modern zooxanthellate coral faunas of the Caribbean and eastern Pacific sides of the Isthmus of Panama shared four genera—*Siderastrea*, *Leptoseris*, *Acropora*, and *Porites*—but not a single species. The scleractinian coral genus *Madracis* is also present on both sides of the Central American isthmus, but Pacific species are without zooxanthellae and generally occur in deep, off reef waters. More recent studies indicate that only one, or possibly two coral genera at most, occur in both Caribbean and eastern Pacific waters (Veron et al. 2015). The exclusion of three genera, formerly considered amphi-American in distribution, has occurred for very different reasons.

In 1992, Héctor Guzmán discovered five colonies of *Siderastrea* at Urabá Island (Taboga Islands group) on the Pacific side of Panama. *Siderastrea* is a well-known Caribbean genus and also includes species that occur in all major coral reef provinces worldwide. Before Héctor's discovery, *Siderastrea* was not known to occur in the eastern Pacific later than the Pliocene Epoch (Durham 1966; López-Pérez and Budd 2009). Based on morphometric analysis, this eastern Pacific coral was described as a new species and named after yours truly as *Siderastrea glynni* by Budd and Guzmán (1994), for which honor I am grateful. Unfortunately, extensive searching in the Taboga Islands archipelago turned up no additional colonies. Then, a molecular study by Forsman et al. (2005) reported that *S. glynni* and the common Caribbean *S. siderea* shared nearly identical genetic sequences (ITS haplotypes), which suggested that the two corals were one and the same species. Further, LaJeunesse et al. (2016) showed that the zooxanthella species in *S. glynni* and *S. siderea* were the same. Surprise! This represented the first zooxanthellate coral with identical species populations present in both Caribbean and eastern Pacific waters. Alas, these studies showed that the species named after me should be assigned to "junior synonymy" status—i. e., when a name is no longer valid because the first described name takes precedence. Science marches on.

This rare transisthmian occurrence of *S. siderea* raised the question of the source of those few colonies at Urabá Island. Had they somehow recently found their way into the eastern Pacific from the Caribbean or were they a relict population, survivors from the Pliocene? The latter possibility did not seem feasible because of the extensive research diving at this location and lack of previous sightings. Other possibilities considered by Forsman et al. (2005) involved the breaching of the Central American Isthmus approximately 2 mya or a contemporary introduction by a ship vector. The planula larvae could be transported across the isthmus in ballast water or as fouling growth on a ship's hull. These questions remained with me for several years until Benjamin Grassian,[4] working in my laboratory on a bioerosion project, called my attention to a study by Karl Kleemann (1990, 2013). Kleemann's study in the early 1980s involved the transplantation of fragments of *S. siderea* on plates, presumably dead, from the Caribbean side of Panama to the same site where Héctor discovered the *Siderastrea* colonies 10 years later. I contacted Karl about the possibility that live tissues might have remained on his transplanted corals. Indeed, his photographic records showed coral regrowth on the plates after 6 months, demonstrating survivorship after transplantation to the Pacific waters of Panama. I invited Karl to join Benjamin, Juan Maté and me to study and document the identity of the supposed *S. glynni* (Glynn et al. 2016). The mystery was quickly solved, and we removed the transplanted corals from the Pacific waters where they did not occur naturally.

We now better understand the status of the three other corals that were regarded as amphi-American genera, namely *Leptoseris*, *Acropora*, and *Porites*. *Leptoseris* was formerly considered to have a single species in the Caribbean, *L. cuculatta*. This species has been subject to a name change, however, and is now classified as *Helioseris cuculatta* (see Kitahara et al. 2012, and WoRMS, World Register of Marine Species). Nevertheless, according to Michel Pichon (pers. comm.) the assignment of species to the two genera *Leptoseris* and *Helioseris*, based on morphology, is still unresolved. Molecular methods will now be necessary to shed light on their genetic similarities and possible ancestral relationships in order to decide which species belongs in which genus.

Von Prahl and Mejía (1985) reported the occurrence of colonies of *Acropora valida*, an Indo-western Pacific species, at Gorgona Island, Colombia, in 1983. It has not been observed since in eastern Pacific waters, nor has any other *Acropora* species.[5] Therefore, *Acropora*, although a prominent coral genus in the western

[4] Benjamin served as an undergraduate marine science volunteer in my laboratory, assisting in various coral reef studies. He is presently completing studies for the Ph.D. degree at the Graduate School of Oceanography, University of Rhode Island. His doctoral research has involved the development of new acoustic methods in the study of deep, zooplankton scattering layers.

[5] The tragic death of Henry von Prahl following this remarkable discovery was not the end of misfortune. In an attempt to obtain more information on the discovery of *Acropora*, one of our colleagues visited Ángela Mejía, the co-discoverer and co-author of the 1983 publication. It was too late, Ángela had suffered a near fatal motorbike accident and was not able to communicate. This unpleasant episode became known as "the curse of *Acropora*." I overheard some of our colleagues

Atlantic and Indo-western Pacific provinces, cannot now be considered a member of the eastern Pacific coral fauna.[6]

As currently understood, *Porites* is the only zooxanthellate coral genus shared by the modern Caribbean and eastern Pacific coral faunas. A molecular study, again from Forsman's laboratory, demonstrated that the current Caribbean *Porites* complex of six species and the eastern Pacific group of nine species are closely related genetically (Forsman et al. 2006, 2010). Caribbean and Pacific waters, however, share no *Porites* species today. *Porites sverdrupi* from the Gulf of California and *Porites panamensis*, found throughout the eastern Pacific except in the Galápagos Islands, are both descended from a common pan-Tethys ancestral stock that gave rise to the Caribbean species.

Holocene Coral Revival

To help shed light on the complex history of eastern Pacific coral reefs, I propose here a broad outline of the development of the coral fauna and the emergence of modern reef ecosystems during the last 65 million years, in the Cenozoic Era (Fig. 3.4). The recent comprehensive review by Andrés López-Pérez (2017) has served as an especially valuable source of information (Fig. 3.5). Additional information on the historical development of the eastern Pacific coral fauna was obtained from several relevant studies. A caveat emphasized by Andrés is the emphasis on Mexican coral fossil records supporting this overview. The fossil assemblages of Baja California Sur (northern Mexico) may not be representative of all regional reef

vow not to report the presence of *Acropora*, even if re-discovered in the eastern Pacific. Just let it be.

[6]There are three additional coral species that have been noted as belonging to the eastern Pacific coral fauna, but their occurrence in this region remains elusive. Two species in the family Acroporidae, described in the nineteenth century, have not been found since, in the eastern Pacific or elsewhere. In the original description of *Acropora crassa*, the locality was indicated as unknown. A later examination of a specimen of *A. crassa* in the Paris Museum revealed a label indicating collection from the Galápagos Islands. No specimens of *Acropora* have yet been discovered in the Galápagos in the twentieth or twenty-first centuries (Glynn and Wellington 1983; Wells 1983; Glynn et al. 2017). The acroporid, *Montipora fragosa*, noted as coming from La Paz in the Gulf of California was, according to Squires (1959), probably mislabeled. Don (Donald F.) Squires and the scientific team of the Puritan Expedition searched unsuccessfully for this species in the Gulf, occupying over 30 stations from San Lucas Bay to San Felipe, a distance of about 1200 km. Photographs of the holotype of this species are included in Glynn et al. (2017, page 26, Fig. 1.20); this is the specimen described by A. E, Verrill (1870), from the Peabody Museum, Yale University. A few small colonies of *Porites rus*, a prominent west-central Pacific species, were collected in Costa Rica following the 1982–83 El Niño event (Cortés and Murillo 1985). This species has not been observed again after extensive searching at its original site of discovery. Why is there this fascination with small, isolated, and enigmatic purported species occurrences? Perhaps some of these species are colonists that have successfully arrived from afar and settled but have not established "permanent" populations.

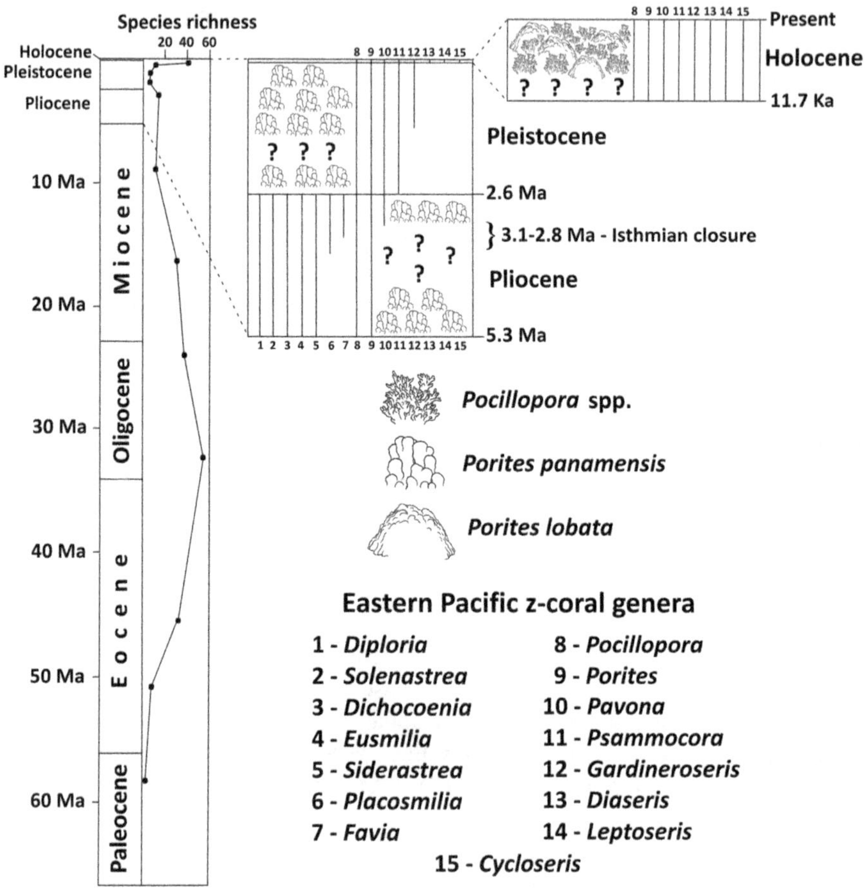

Fig. 3.4 Development of the eastern Pacific scleractinian coral fauna and coral reefs during Cenozoic time, ~60 Ma to present. Species richness or number of species reported in geologic time intervals from the Paleocene to the present. Occurrence of reef coral genera and coral reef formations from Pliocene (5.3–2.6 Ma) through Pleistocene times. Post Ice Age, or Holocene occurrence of eastern Pacific reef coral genera and structural reef formations constructed by *Pocillopora* spp. and *Porites lobata* corals. Question marks denote the absence of coral reef frameworks in the Pliocene, Pleistocene, and Holocene time intervals. Coral icons denote approximate temporal occurrence and identity of reef frameworks. The Pliocene is marked by the Isthmian closure event between 3.1-2.8 million years ago. Z-coral: zooxanthellate taxa. Records from López-Pérez (2017)

development, especially at equatorial latitudes. Coral deposits are uncommon along the Central American Pacific coast owing to the subduction of active margin deposits and a lack of well-developed sedimentary basins (Coates and Obando 1996). Other factors related to the scarcity of fossil reefs in the eastern Pacific, as summarized by Jorge Cortés (1993, 2003), are (a) the poor structural integrity of reef frameworks, due in part to a low abundance of binding crustose coralline algae; (b)

Fig. 3.5 Andrés López Pérez at an in situ Late Pleistocene terrace, El Saladito, Baja California Sur (24°30′56″; 110°43′57″). The terrace contained *Porites panamensis* corals <20 cm high and fossil gastropods and echinoderms. Photograph courtesy of A. L. Pérez (28 May 2022)

relative absence of submarine cementation, and (c) high rates of bioerosion.[7] I am hopeful this précis will stimulate research on this subject and deepen our understanding of the provenance, development, and nature of contemporary eastern Pacific coral reefs.

[7] Jorge Cortés and I can attest to the rarity of fossil eastern Pacific coral deposits. During our several decades of coral reef research, which have involved mostly underwater activities, we have also engaged in surveys of shoreline sedimentary rock outcrops for signs of fossil corals. Jorge has reported on only a single site bearing *Pocillopora* fragments. He found the fossil corals in tectonically uplifted beach rock at Puerto Escondido on the central Pacific coast of Costa Rica. The corals showed no signs of abrasion or transport and were assumed to have originated on-site. Their ages ranged from 1272 ± 60 to 2010 ± 60 years BP (Denyer and Cortés 2001). Bob (Robert N.) Ginsburg, a renowned carbonate sedimentologist, colleague, and longtime friend, and I, on separate occasions, observed a modest coral reef framework near the western shoreline of Baltra Island in the Galápagos Islands. Here, the corals appeared to be in growth position and 0.5–1.0 m in vertical height. This was a happenstance finding and unfortunately neither one of us had obtained a permit to study and sample this formation. Bob collected specimens for study but was obliged to return them. In the early 1930s, Leo Hertlein investigated a fossil-bearing bed, about 3 m thick, at the southwestern end of Baltra. This deposit contained two reef-building coral species, *Pavona gigantea* and *Psammocora* sp., both of late Pliocene-early Pleistocene age (Hertlein 1972). These coral species were also recorded in the late Pliocene-early Pleistocene of Seymour Island, Galápagos Islands (Hertlein 1972), and the late Pleistocene of Islas Coronados (Durham 1947, 1950), Punta Baja, Las Ánimas, and La Ventana, Gulf of California (López-Pérez 2012). Hertlein also recovered the modern sea urchin, *Eucidaris galapagensis*, a corallivore and potent bioeroder, from this fossil assemblage. Evidently past environmental conditions at Baltra favored coral growth, deposition and preservation. We are hopeful that these fossil coral sites receive renewed attention regarding their growth history and possible clues to past ecological conditions and interactions.

Ancient Reef Formations

Coral reefs constructed by zooxanthellate scleractinian corals and coralline algae first appeared in the early Mesozoic era (Hubbard et al. 2001; Stanley 2001). Many kinds of organisms have formed diverse types of reef structures over the past 3.5 billion years of Earth's history. The earliest reefs were constructed by cyanobacteria or blue-green algae that formed layered, cemented, sediment mounds (stromatolites). They first appeared 3.45 billion years ago and dominated shallow tropical shores for more than 2 billion-years. They still construct modest structures in a few limited environmental settings today, including western Australia and the Bahamas Islands. During more recent geologic time, reefs were constructed by sponges, bryozoans, foraminifera, ancient corals (tabulate and rugosan taxa), calcareous algae, and bivalve (rudistid) mollusks.[8] The ancestors of modern scleractinian corals have been building reefs since Triassic time (~200 mya). Such reefs were widespread along the ancient Tethys Sea,[9] a broad seaway that circled the tropics during much of the Mesozoic Era.

Scleractinian coral fossil deposits, including isolated or scattered corals as well as those building reefs, have been found in western North and South America from Mid-Triassic to Early Cretaceous time, a span of some 100+ my (Flügel and Senowbari-Daryan 2001; Leinfelder 2001; Wood 2011). The earliest of these reef-building corals were small and contributed little to reef growth and bulk volume but increased in importance through time. Calcified sponges, calcareous algae, rudistid mollusks and other skeleton-forming organisms contributed to reef buildups on the shallow shelves of the former Tethys Seaway, the antecedent basin of the present-day Mediterranean Sea. Globally, Scleractinia first appeared in the fossil record in the middle Triassic (240 mya) and then contributed extensively to reef framework construction throughout much of the Mesozoic Era. Rudist mollusks replaced scleractinian corals as shallow water reef builders of the warming waters of the late Cretaceous period and were responsible for reef development until their sudden demise in the early Cenozoic Era. Scleractinian corals then returned, steadily

[8] Following a pathway around Magueyes Island in southwestern Puerto Rico, I marveled at the rudist fossils embedded in Cretaceous rocks, only a few meters distant from living scleractinian corals. Rudists displaced corals on reefs 100 mya but became extinct around the same time as the dinosaurs. These heavy bivalve fossils made excellent door stops at the University of Puerto Rico's Magueyes marine laboratory, where I reveled in my first faculty position after graduation.

[9] The circumtropical Tethys Sea, an east-west trending equatorial seaway that was present during much of the Mesozoic era (251–65.5 mya) was an important cradle for the evolution and diversification of warm water marine biotas and reef development in Earth's history (Stanley 2001; Hou and Li 2018). This seaway formed during and after the fragmentation of the supercontinent Pangea, located between the large global land masses of Laurasia in the north and Gondwana in the south. As the Tethys Seaway gradually broke apart during Triassic and Pliocene time, due to shifting continents, its Atlantic, Pacific, and Indian Ocean masses became the regions of today's five coral reef biogeographic provinces.

diversified to ~225 genera[10] to Recent times (Veron 1995), and with coralline algae dominated reef building in the Cenozoic era. Throughout Cenozoic time, generic diversity fluctuated due to stress events such as volcanic eruptions, anoxia, impact of bolides, changes in sea level, sea water chemistry, and global climate change caused extinctions.

Ancient (Late Paleozoic and Early Mesozoic) fossil corals are known from high and low latitudes in Alaska, British Columbia, Idaho, Oregon, California, Mexico, Peru, and Chile (Blodgett and Stanley 2008). These coral faunas belonged to the ancestral Tethys stock that gave rise to modern reef-building scleractinian corals of the Cenozoic Era. Due to the mobile dynamics of plate tectonics, however, most of the coral assemblages presently found at higher latitudes in the western Cordillera (from Alaska to the Andean Mountain ranges), did not originate there but likely existed at tropical to subtropical latitudes in far-flung regions of the ancient Pacific Ocean (Panthalassa), considerable distances from the ancient American continents. A truly amazing discovery relating to the provenance of these corals, posits the migration of displaced fragments (terranes) in the Late Paleozoic and Early Mesozoic Panthalassa Ocean (Coney et al. 1980; Soja 1996). Evidence suggests that corals and reef formations originated on volcanic islands or land masses in the ancient Pacific. Sea floor spreading, propelled the various terranes, many carrying cargos of fossilized reefs and corals, eastward and northward where eventually they were accreted to the western side of North America. Post-accretion tectonics are responsible for the gradual shifting of many of these allochthonous terranes along transform faults, like the San Andreas fault in California, to latitudes where they are found today.

As noted above, reef coral assemblages were greatly reduced in numbers and area during the Cretaceous period and were severely affected by a global mass extinction, known as the K/P boundary extinction event that marked the beginning of the Cenozoic Era about 66 mya.[11] In geo-speak, the K stands for Cretaceous and the P for Paleocene—the first Epoch of the Cenozoic Era. Only two coral genera (*Pocillopora*, *Porites*) and five species (*Pocillopora damicornis*, *P. meandrina*, *P. verrucosa*, *Porites lobata*, *P. sverdrupi*) are known from eastern Pacific fossil deposits over a period of about 10 million-years following the extinction event (López-Pérez 2017). In 1980, the Alvarez's, father and son, proposed an audacious hypothesis to explain the cause of the extreme Cretaceous-Tertiary (K-T) extinction event that exterminated three-quarters of all plant and animal species on Earth (Alvarez et al. 1980). They marshaled evidence suggesting that, about 66 mya, an extraterrestrial bolide, the ~10 km diameter Chicxulub asteroid, crashed into the Gulf of Mexico off the Yucatán Peninsula. This powerful impact ejected large quantities of dust, ash, sulfur and other aerosols that entered and circulated in the Earth's

[10] The identification of higher taxonomic categories, such as genera, is a more reliable estimate of diversity than species due to the generally poor preservation of skeletal characters in fossil specimens.

[11] The K/P extinction event is also known as the K-T (Cretaceous-Tertiary) extinction event that was responsible for the disappearance of approximately three-quarters of Earth's living biota.

atmosphere. After intense heating of the atmosphere, the blanketing cloud cover screened sunlight and triggered global cooling. Much of terrestrial life, as well as shallow-living marine organisms such as reef-building corals and calcareous algae that depended upon sunlight, was extinguished.

The bolide impact hypothesis has been subject to vigorous scrutiny; a widely supported alternative hypothesis involves massive volcanism. The Deccan Traps, a large igneous province that spans much of India and the adjacent deep sea, released ~500,000 km^3 of lava, sulfurous gasses, and CO_2 around the same time as the K/P event. This large-scale outgassing could have led to lethal hothouse conditions and ocean acidification. Chiarenza et al. (2020) summarized the presumed effects of both extinction scenarios, asserting chief support for the bolide impact hypothesis, but suggesting a role for volcanism as a joint driver of K/P extinctions.

As environmental conditions improved in the Cenozoic era, coral species richness (the number of species recorded at a given locality and time, whether living or fossil) gradually increased in the eastern Pacific from two to more than fifty species over an approximately 25 million-year span, from Late Paleocene time to the Early Oligocene (Fig. 3.4). Species richness then declined over the next 20 million years, from the Late Oligocene to the Pliocene, and remained extremely low through Pleistocene time (2.58 mya to 11,650 years BP) due to climate cooling and sea level fluctuations associated with the series of glacial events commonly known as the Ice Ages. The emergence of the Isthmus of Panama, between 3.1 and 2.8 mya, gradually closed the Central American Seaway—the section of the Tethys Sea that separated North and South America—dividing the previously continuous, pan-Tethyan coral assemblages into separate tropical western Atlantic and East Pacific regions. Fewer than five coral genera are known from eastern Pacific deposits in the mid-Pleistocene (Fig. 3.4). Coral species richness rebounded rapidly, however, following the final closing of the Central American Seaway. Species richness increased from the Pleistocene to the present, from five to 41 species over a ~2 my period (Fig. 3.4). But this was not just a simple increase in species. Several Caribbean holdovers became extinct during the late Pliocene, and other species arrived via dispersal from the West-Central Pacific, producing a dramatic turnover in coral genera. This amounted to a sudden faunal transformation. Despite the woefully incomplete fossil record, a few small patch reefs of *Porites panamensis* have been reported from Pliocene and Pleistocene-age deposits (from 5 mya to present; López-Pérez 2017). Reefs composed of *Pocillopora* species and *Porites lobata*, often together on any given reef, made an appearance about 6000 years ago. *Porites lobata*, a Central Pacific species, initiated reef development in Costa Rica about the same time (Cortés et al. 1994).[12] Currently, the oldest known reefs in Panama and the Galápagos

[12] A recent genetic study by Michael Hellberg et al. (2016) demonstrated that the colonization and establishment of Central Pacific *Porites* species on American shores is more complicated than previously understood. They suggested that after *Porites evermanni* became extinct in the eastern Pacific when the Central American Seaway closed it recolonized via immigration from the Central Pacific. Michael's team hypothesized that another species, *Porites lobata*, arrived more recently and was less hardy than *P. evermanni*, which is more suited to the marginal conditions of the east-

Islands each formed ~7000 years ago (Toth et al. 2017). In summary, in less than 3 my, eastern Pacific coral species richness increased from five to 41 species.[13] By contrast, it took between 20 and 30 million years during the Eocene and Oligocene Epochs following the K/P extinction event for a similarly rich coral fauna of 55 species to develop. The rapid development of the eastern Pacific coral fauna occurred ostensibly by the influx of species from West-Central Pacific source populations and not from accelerated evolution of autochthonous species.

Following the Mesozoic era, which ended some 66 Ma, the western Atlantic and eastern Pacific marine regions remained connected via the Tethyan Central American seaway for millions of years and supported a continuous and similar marine biota, including corals that extended around the world in an unobstructed, warm tropical sea (Fagerstrom 1987; Duque-Caro 1990; Veron 1995). With the formation of barriers leading to the fragmentation of the Tethys Sea, and isolation of once continuously distributed biotas, the pace of evolution increased and gradually produced five modern coral provinces: the East Pacific, West Atlantic, East Atlantic, Mediterranean Sea, and Indo-West Pacific (Fig. 3.6). The breakup really began as Africa and Eurasia collided to close the connection between the Indian Ocean and Mediterranean Sea about 24–21 Ma. Then, 16–15 Ma uplift associated with the collision of Central and South America, and increased volcanic activity, began to form islands in the Central American Sea (Coates and Obando 1996; Collins 1996). The Central

---

ern Pacific. (Two different kinds of studies are needed, however, to verify the relative hardiness of these two species. Examining [a] the distributions of these two corals in the eastern Pacific should reveal whether *P. evermanni* occurs more commonly under stressful conditions, and [b] experimental studies testing the effects of extreme temperatures and sedimentation would allow direct assessments of hardiness.) The first post-extinction arrival, the supposedly more resilient *P. evermanni*, could have introduced genes into *P. lobata* by means of two processes: hybridization and introgression. The latter term refers to the incorporation of genes from one population or species into the genome of another with which it interbreeds (= hybridizes). Introgression, however, is a double-edged sword. As Todesco et al. (2016) report, it can also cause one of the hybridizing species, commonly the rarer of the two, to go extinct. Still, Hellberg and colleagues argue that such newly acquired genes could offer relatively rapid genetic variation for adaptive change to cope with harsh environmental conditions. Thus, the earlier colonizing resilient *P. evermanni* could have facilitated the later colonization of the more sensitive *P. lobata*. Different coral species, however, reproduce in different ways. In this case, *P. evermanni* reproduces chiefly by the asexual fragmentation of colonies whereas *P. lobata* is a broadcast spawner; it releases eggs and sperm into the water for fertilization (Boulay et al. 2014). To what extent these different reproductive modes would impede hybridization is another factor that needs study. A caveat: I was recently informed by Andrés López-Pérez (Department of Hydrobiology, National Autonomous University of Mexico, Iztapalapa, Mexico) that he recorded the presence of *Porites lobata* in the Pleistocene of Cabo Pulmo, Baja California (a single colony in the collections of the Los Angeles County Museum of Natural History) identified by traditional morphometry (López-Pérez 2012). This record will likely require revision of the Hellberg team's hypothesis.

[13] Differences in coral species numbers (species richness) reported in the literature may be an annoyance to some readers. For example, here I report 41 currently known eastern Pacific zooxanthellate coral species, and as recently as 2017 Andrés reported 45 species (López-Pérez 2017). These differences are expected and result from on-going taxonomic studies (morphological and molecular) and field surveys documenting the deletion or addition of new coral records.

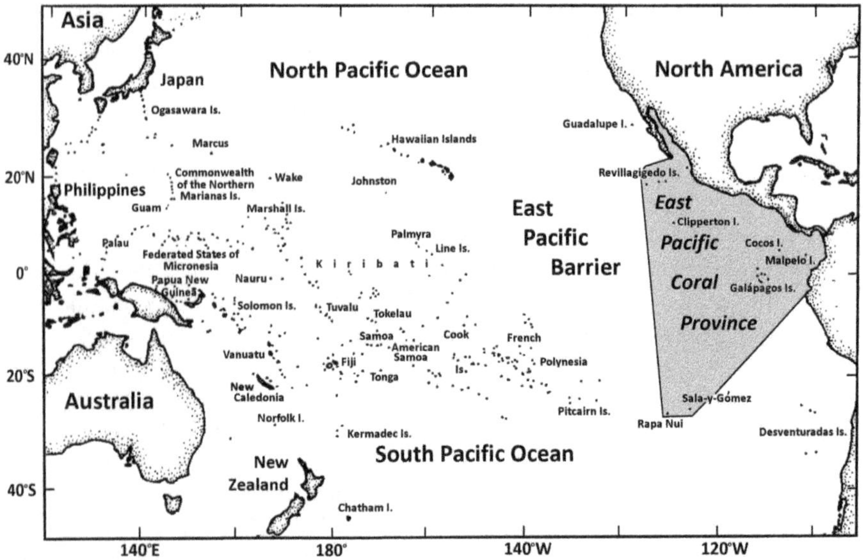

Fig. 3.6 Pacific Ocean showing location of the East Pacific Coral Province (shaded triangle), now recognized to extend south to Rapa Nui (Easter) and Sala-y-Gómez Islands (Glynn et al. 2007). Map adapted and edited by Charles G. Messing

American seaway continued to narrow through Pliocene time due to increasing tectonic subduction and uplift, with the Isthmus of Panama first completely closing around 3.5–3.1 Ma (Coates and Obando 1996). The Pacific and Indian oceans closed their connection with the northward movement of Australia.

The emergence of the Central American Isthmus had profound effects on both Caribbean and Pacific marine biotas; each was now isolated and subject to very different environmental conditions and evolutionary pathways (Jackson and Budd 1996). Extinctions exacted a heavy toll on Caribbean and Pacific coral faunas due to episodes of Pliocene-Pleistocene cooling, sea level fluctuations, and changes in the circulation of ocean currents and nutrient availability (Cortés 1986; Budd 1989; Budd et al. 1994a, b; López-Pérez and Budd 2009).[14] Although it is known that these potentially stressful conditions can affect reef-building corals, the specific factor, or combination of factors, responsible for the Late Pliocene-Pleistocene

[14]The sea level lowering during the most recent glacial advance, 20,000 years ago, was about 120 m below present-day sea level. At that time, a terrestrial ecosystem occurred on the exposed continental shelf between mainland Panama and the Pearl Islands. From time to time, fossil mastodon teeth appear in shrimp trawler nets in the Gulf of Panama, to the great excitement of boat crews. A shrimp boat captain once came to my laboratory to show me a mysterious, dark-colored, multifaceted object that I immediately recognized as a mastodon tooth. He thanked me for the identification, but I do not think he believed me. How could there be huge elephantine animals browsing in the Gulf of Panama. I offered the captain a fair price for the fossil tooth; he declined to sell because he believed it would bring him good luck.

extinctions remain unresolved. After closure of the Central American Seaway, the western Atlantic coral fauna experienced a period of sudden extinctions and rapid speciation. The extreme impoverishment of the eastern Pacific coral fauna following the isthmian emergence also experienced a rapid enrichment over a geologically brief period (<1 Ma, Late Pleistocene to the beginning of Holocene time, Fig. 3.4). As noted earlier, this sudden increase in reef-building corals occurred despite the presence of the Eastern Pacific Barrier, the vast expanse of deep water separating the Central and eastern Pacific regions (~4000–7000 km, depending on source and end immigration routes).

The Porous Eastern Pacific Barrier

The Eastern Pacific Barrier (EPB), notably effective in limiting the dispersal of shallow shelf marine organisms, was considered an "impassable barrier" by Charles Darwin, and is often regarded as the most effective marine barrier in the world today (Grigg and Hey 1992). The EPB is exceedingly effective, yet numerous diverse marine species with planktonic larval development have successfully surmounted this barrier.[15] Plankton samples examined from this oceanic barrier have revealed a great variety of invertebrate larvae belonging to six major groups (Cnidaria, Polychaeta, Sipuncula, Mollusca, Echinodermata, and Hemichordata) (Scheltema 1988; Richmond 1990). Genetic evidence also supports the long distance dispersal of reef fishes across the EPB, from west to east and east to west (Lessios and Robertson 2006). In addition to drifting freely in open waters, Paul Jokiel (1990) demonstrated that coral larvae can settle and grow on flotsam such as net floats and pumice, a process known as rafting, which may greatly extend the distance of dispersal. Although the EPB is certainly an impediment to the long-distance dispersal of marine life, because numerous species have successfully crossed this lengthy deepwater stretch, perhaps it is more accurate to regard it as a filter rather than a barrier. Following Rudi (Rudolf S.) Sheltema's lead, Jerry Ault and I prefer the sobriquet "East Pacific Filter Bridge" (EPFB) (Glynn and Ault 2000). Several lines of evidence (ecological and genetic) support the immigration and colonization of the eastern tropical Pacific by Indo-west Pacific species. Many of these immigrant species are coral reef associates, iconic and functionally important members of eastern Pacific coral reefs, for example the reef-building corals themselves, *Trapezia* spp. (coral guard crabs), *Acanthaster* (the coral-consuming crown-of-thorns sea star), Harlequin shrimp (a predator of *Acanthaster*), and Guineafowl pufferfish (a predator of corals). Although these migrants represent only a small fraction of the species-rich East Pacific shelf fauna (Lessios and Baums 2017), without them

[15] Most coral reef benthic invertebrates and fishes cannot cross this barrier because of the great distance involved, the absence of stepping-stones (shallow settlement surfaces along the dispersal path), and the relatively short larval lives of the dispersing source populations.

pivotal species interactions and functional processes of eastern Pacific coral reefs would be very different or absent.

Ocean surface currents are the principal means by which organisms cross the EPFB, and the tropical currents of relevance for coral reefs include the east-to-west moving North and South Equatorial Currents, and the west-to-east flowing North Equatorial Counter Current (Fiedler and Lavin 2017) (Fig. 3.2). Westward current flow, and presumably dispersal, was preeminent in the tropics during the Cretaceous period, about 100 mya (Grigg and Hey 1992). Westward dispersal has been documented recently for *Porites lobata* from Clipperton Atoll to the Hawaiian Islands (Baums et al. 2012). Wood et al. (2016) offered additional support for east to west dispersal based on a biophysical coral larval dispersal model. Briefly defined, the biophysical dispersal model traces the paths of particles (coral larvae in the present case) using ocean surface current data over varying time periods and larval mortality rates. It is called biophysical, because it integrates a model of physical oceanography with a model of larval biology, such as swimming ability, time of settlement, depth of settlement and other parameters. It is thus possible to realistically estimate the behavior of viable and dispersing larvae embedded in their physical environment. The bulk of studies over the years, however, have offered abundant evidence that the eastward-flowing NECC is the principal means of dispersal across the EPFB (Dana 1975; Grigg and Hey 1992; Veron 1995; Lessios et al. 1998; Glynn and Ault 2000; Robertson et al. 2004; Cortés 2011; Lessios and Baums 2017). The location, velocity and regularity of the NECC have been considered favorable for supporting west to east dispersal. The Northern Line Islands in the Republic of Kiribati support a rich coral reef biota and have been regarded as a prime source of larvae and potential donor locality for the East Pacific (Dana 1975; Grigg and Hey 1992).[16] For more recent accounts of the coral reefs of Kiribati see the engaging and informative books by Rohwer (2010), and Stone and Obura (2013). Still, Wood et al. (2016) challenged the traditional view that the NECC has played a primary role in the eastward dispersal of Indo west-Pacific species into the eastern Pacific.[17]

---

[16] The results of the modeling study by Wood and colleagues were surprising considering the wealth of evidence supporting the NECC as a west-to-east dispersal route. Their overall conclusion that the NECC is ineffective as a means of coral larval dispersal across the EPFB was based on a modeling analysis of the dispersal of virtual larvae during El Niño and neutral El Niño events. Most eastern Pacific reef-building corals, 36 of 41 species (87.8%), also occur in Central and South Pacific localities (Glynn et al. 2017). These species are assumed to have originated from western occurring source populations whose propagules immigrated to, colonized, and successfully integrated into the eastern Pacific coral reef biota. This colonization period would have occurred following the late Pliocene and Pleistocene isthmian coral extinctions. The many factors involved in the origin of the eastern Pacific coral reef region—geological, oceanographic, and biological—make this a complex subject. I recommend the following publications for gaining insight into the processes and contested issues surrounding this question: Baums et al. (2012), Lessios et al. (2012); Wood et al. (2016); Lessios and Baums (2017).

[17] Jim (James) Maragos, an accomplished coral systematist with many years of experience in the Indo-Pacific region, sent me the following anecdote about the potential efficacy of west-to-east transport of coral larvae by mid-Pacific equatorial currents. Jim's observations concern four (of 11) atolls or islets of the Line Islands that lie in a line (hence the name) extending from southeast to

Grigg and Hey (1992) noted that Christmas (Kiritimati) Island and Clipperton Atoll (Fig. 3.7), both within the path of the NECC and separated by 5400 km, represented the shortest distance between the Central Pacific and eastern Pacific regions (Figs. 3.2, 3.6, and 3.8). At maximum current flow (150 cm/s), and assuming a

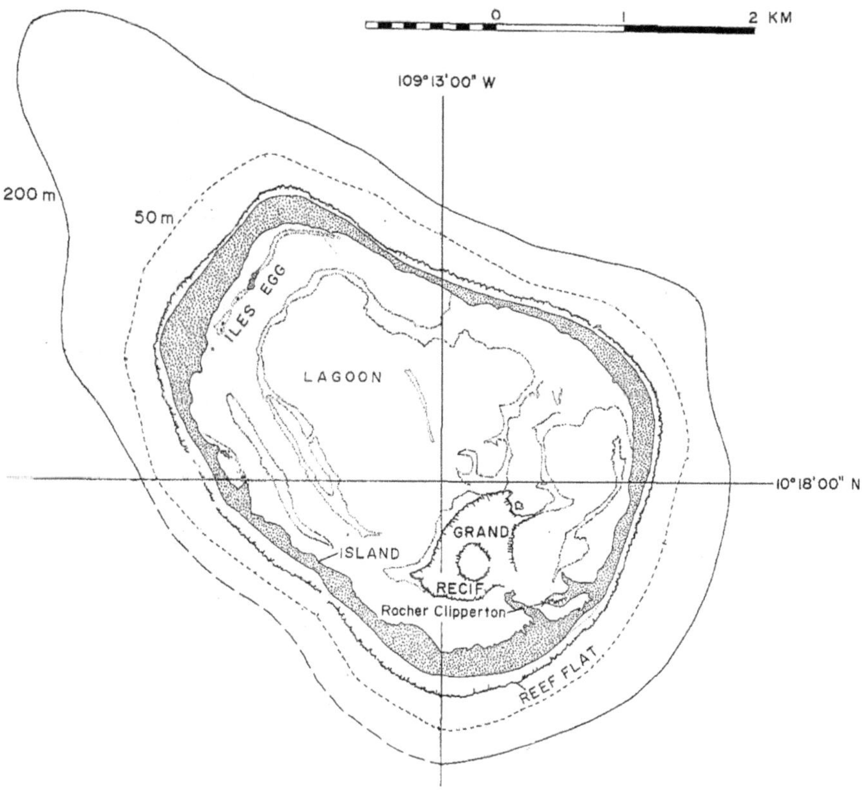

Fig. 3.7 Clipperton Atoll, map adapted and edited from Glynn et al. (1996) by Diana A. Moanga. The emergent Clipperton rock (Rocher Clipperton) is located in the almost atoll's southeast quadrant

---

northwest and centered at 160° W longitude, just north of the equator. These islands lie nearly perpendicular to the west-flowing North Pacific Equatorial Current (NPEC) and east flowing North Equatorial Counter Current (NECC). From north to south, the islands are Kingman Reef (6.4°N), Palmyra Atoll (5.9°N), Washington (Teraina) (4.7°N), and Fanning (Tabuaeran) (3.5°N). In considering one of these islands as a source of coral propagules for dispersal to the eastern Pacific, Jim related, paraphrasing: "I recommend Palmyra. I spent many months there from 1979 to 2010 and have an excellent list of corals from there. Palmyra is indeed right in the middle of the ITCZ (Inter Tropical Convergence Zone) and at times I have even seen the NECC come right off the northern side of the atoll. It was an exciting opportunity seeing large numbers of fish and marine mammals and the two currents, the NPEC and NECC, pass by one another, with the northern buoys heaving toward the west while the southern buoys were tilting eastward under substantial tension. Unforgettable."

Fig. 3.8 A shore party departing the *Royal Star* for landing on Clipperton Atoll (April 1994). At the inflatable's stern, from left, Peter Glynn and Charlie Veron waving, and Tim Ekstrom manning the outboard motor. Photograph courtesy of Henry Chaney

straight path trajectory, they calculated that a larva could cover this distance in 155 days. Closer to the equator but still within the NECC, the travel time between Kiribati and the Galápagos Islands, at maximum current velocities of 125–150 cm/s, was projected to require 70 and 55 days, respectively. These faster current flow rates, however, were for the eastbound equatorial undercurrent, also called the Cromwell Current, a subsurface current with lower temperatures and reduced light, not ideal for coral larvae. In addition, trajectories of satellite tracked drifting buoys suggest that travel times between the Northern Line Islands and Clipperton Atoll (and the Galápagos Islands) are normally much longer because the easterly bound current does not flow in a uniformly straight line. Two of my colleagues, Charlie (J. E. N. Veron) and Jerry (G. W. Wellington), and I (see Fig. 4A–C in Glynn et al. 1996) found that the tracks of 11 buoys deviated substantially from straight line trajectories and displayed multiple loops, reversed directions, and meanders of 5–10° of latitude. Such variability in these equatorial currents make it likely that travel times may reach 200 days or more. The proposed rapid dispersal times of 55–70 days (Grigg and Hey 1992) and 60–120 days (Glynn et al. 1996) probably need to be significantly lengthened to account for the circuitous routes we observed.

For any lengthy dispersal journey, planktonic coral larvae must remain competent for long periods, that is, they must maintain the ability to settle on a suitable substrate and to metamorphose (a process referred to as competency). A larva may become competent even before it is released from its maternal polyp, or it may be triggered by some external factor such as temperature or contact with a favorable substrate, e. g. calcareous algae. For reef-building corals, competency involves leaving the water column for the seafloor, selection of a suitable substrate, firm attachment, and initiation of calcification (formation of a skeleton), followed by development into a polyp that will give rise to a colony through asexual reproduction. One of Bob (Robert) Richmond's post-doctoral studies at STRI involved the question of competency in *Pocillopora damicornis*: "Was the life history larval phase of this pan-Pacific coral capable of long-distance dispersal from the Central Pacific to the eastern Pacific?". Based on calculations of larval energy reserves and daily metabolic expenditures, Bob found that planula larvae could survive approximately 100 days and still settle successfully (Richmond 1987, 1988). More recently, Graham et al. (2008) reported maximum larval life spans of 195–244 days in five Indo-Pacific corals.[18] They found that larval energy reserves did not reach critically low levels until approximately 100 days after spawning, which supported Bob Richmond's findings of the potential for coral dispersal across the EPFB.

Even under favorable sea current conditions, in terms of direction and velocity, larvae still face numerous biological constraints and challenges. First, pelagic larvae with a protracted planktonic life (teleplanic larvae) would have the best odds of completing long distance dispersal. Such coral larvae typically possess photosynthetic zooxanthellae that can augment nutritional needs during their planktonic phase. Indeed, my colleague Jerry (Jerald S.) Ault and I and others (Glynn and Ault 2000; Baird et al. 2009), found that the majority of eastern tropical Pacific transpacific coral species produce larvae that already have zooxanthellae before they are released from the maternal polyp. The timing of coral gamete spawning or larval release by source populations is critical because this should occur when transporting currents are present and favorable. If larvae manage to reach insular or continental shelves, they must be able to find suitable substrates for settling and then undergo metamorphosis. Even given successful metamorphosis, however, the developing young corals will be subject to an onslaught of competitors and predators probably unknown in their communities of origin. Little is known about these early life history impediments and challenges, and how they are surmounted.

The very strong El Niño events of 1982–83 and 1997–98, the most extreme of the twentieth century, caused severe coral bleaching and mortality in the eastern Pacific. During and following these events, marine biologists reported an influx of Indo-western Pacific species into equatorial eastern Pacific waters, including corals (Cortés and Murillo 1985; Guzmán and Cortés 2007), mollusks (Emerson and Chaney 1995), echinoderms (Lessios et al. 1996), and fishes (Lessios et al. 1998; Victor et al. 2001). Under normal circumstances Trade Winds in the eastern Pacific drive both the westbound North and South Equatorial Currents. During El Niño events, Trade Winds relax and the NECC increases in velocity and outspread, driving a warm pool of western Pacific water to the east (McPhaden and Picaut 1990; Picaut et al. 1996), conditions that should favor eastward larval dispersal. During very strong El Niño events the flow time of the NECC across the Pacific greatly accelerates, from ~100–155 days to ~50–80 days. This more rapid transit time

[18] The broadcast spawning corals in this study were *Acropora latistella*, *Favia pallida*, *Pectinia paeonia*, *Goniastrea aspera*, and *Montastraea magnistellata*. None of these species is known to occur in the eastern Pacific region.

occurs within the known competency periods of many teleplanic larvae (Richmond 1990). Although the assumed large-scale influx of Indo-West Pacific species during strong El Niño events seemed reasonable to several researchers, including myself, it is not direct evidence of immigration. In fact, elevated sea water temperature stress and low productivity, characteristic of El Niño conditions, would not be conducive to larval survival. Some investigators have challenged the validity of the heightened El Niño-immigration hypothesis (Robertson et al. 2004; Wood et al. 2016). This is another subject worthy of further study.[19]

In summary, the origin and evolution of the eastern Pacific coral reef province can be visualized as a three-step process. Step one involved the creation of the Isthmus of Panama—the Central American land barrier that separates and isolates American Pacific corals from those of the western Caribbean. These formerly connected regions shared a common Tethyan coral fauna. The second stage involved coral extinctions and speciation, which at first left impoverished coral faunas in both eastern Pacific and Caribbean waters. In the third and final (and still ongoing) stage eastern Pacific coral speciation has produced a few endemic species (~12%), but the majority (~88%) are Indo-West Pacific immigrants arriving via long-distance dispersal from Central Pacific source populations. This west-to-east migration across the porous EPFB may have occurred during rare ENSO pulses, at intervals of hundreds to thousands of years.

Many generations of humankind have been fortunate to be living in the Late Cenozoic Era, during the stable Holocene period, a time of diverse, flourishing, and widespread ecosystem development, including coral reefs. This period, following the Ice Ages, about 12,000 years BP, marks the beginning of the Neolithic (Agricultural) Revolution. We have recently entered a new phase of Earth's history, defined by the imposing influence of humans on our planet's unique biosphere. Although the onset of this so-called Anthropocene Epoch remains contested—did it begin with the Industrial Revolution?, the Nuclear Age? or with widespread plastic pollution?—Planet Earth's human populations have not treated our world kindly or wisely. Reef-building corals and their symbiotic algae are very sensitive to environmental change, especially rising thermal conditions, increasing nutrient concentrations, sedimentation, decreasing alkalinity, and rapid sea level rise, all potentially stressful conditions, and all intensifying due to humankind's ever-increasing numbers and resource demands (Riegl and Glynn 2020).

[19] Information on the recent appearance of Central and Indo-Pacific fishes in the Galápagos Islands, following the strong 2023–24 El Niño event, was reported by Billy (William T.) Bensted-Smith (personal communication). These fishes have not been observed previously in the Galápagos Islands, probably representing new records: surgeon fishes (*Acanthurus*, two species; *Naso*, one species); butterflyfishes (*Chaetodon*, *Ctenochaetus*, two and one species, respectively). Billy also noted a tentative new coral record (a species of *Pavona*). Of 14 described species of *Pavona* recognized by Veron (2000), six occur in both the Indo-Pacific and eastern Pacific regions. One species is an eastern Pacific endemic and is therefore present only in the eastern Pacific. These records possibly represent additional evidence of El Niño assisted, long distance west to east dispersal across the eastern Pacific filter bridge. The details and implications of Billy's findings are presently under review for publication.

References

Alvarez LW, Alvarez W, Asaro F, Michel HV (1980) Extraterrestrial cause for the Cretaceous-Tertiary extinction. Science 208:1095–1108

Baird AH, Guest JR, Willis BL (2009) Systematic and biogeographical patterns in the reproductive biology of scleractinian corals. Annu Rev Ecol Evol Syst 40:551–571

Baums IB, Boulay JN, Polato NR, Hellberg MB (2012) No gene flow across the eastern Pacific barrier in the reef-building coral *Porites lobata*. Mol Ecol. https://doi.org/10.1111/j.1365-294X.2012.05733.x

Blodgett RB, Stanley GD Jr. (2008) The terrane puzzle: new perspectives on paleontology and stratigraphy from the North American Cordillera. Special paper 442, The Geological Society of America, Boulder, 326 p

Boulay JN, Hellberg ME, Cortés J, Baums IB (2014) Unrecognized coral species diversity masks differences in functional ecology. Proc R Soc B 281:20131580. https://doi.org/10.1098/rspb.2013.1580

Briggs JC (1995) Global biogeography. Elsevier, Amsterdam. 452 p

Budd AF (1989) Biogeography of Neogene Caribbean reef corals and its implications for the ancestry of eastern Pacific reef corals. Mem Ass Australas Palaeontols 8:219–230. ISSN 0810-8889

Budd AF, Guzmán HM (1994) *Siderastrea glynni*, a new species of scleractinian coral (Cnidaria: Anthozoa) from the eastern Pacific. Proc Biol Soc Wash 107(4):591–599

Budd AF, Johnson KG, Stemann TA (1994a) Plio-Pleistocene extinctions and the origin of the Modern Caribbean reef-coral fauna. In: Ginsburg RN (ed) Proceedings of the colloquium on global aspects of coral reefs: health, hazards and history, 1993. Rosenstiel School of Marine and Atmospheric Science, University of Miami, Miami, pp 7–13

Budd AF, Stemann TA, Johnson KG (1994b) Stratigraphic distributions of genera and species of Neogene to recent Caribbean reef corals. J Paleontol 68(5):951–977

Challinor D (1972) Background for a new, sea-level, Panama Canal. In: Jones ML (ed) The Panamic biota: some observations prior to a sea-level canal. Bull Biol Soc Wash 2:7–12

Chiarenza AA, Farnsworth A, Mannion PD, Lunt DJ, Valdez PJ, Morgan JV, Allison PA (2020) Asteroid impact, not volcanism, caused the end-Cretaceous dinosaur extinction. Proc Natl Acad Sci USA 117(29):17084–17093. https://doi.org/10.1073/pnas.2006087117

Coates AG, Obando JA (1996) The geologic evolution of the Central American Isthmus. In: Jackson JBC, Budd AF, Coates AG (eds) Evolution and environment in tropical America. University of Chicago Press, Chicago, pp 21–56

Collins LS (1996) Environmental changes in Caribbean shallow waters relative to the closing tropical American seaway. In: Jackson JBC, Budd AF, Coates AG (eds) Evolution and environment in tropical America. University of Chicago Press, Chicago, pp 130–167

Coney PJ, Jones DL, Monger JWH (1980) Cordilleran suspect terranes. Nature 288:329–333

Cortés J (1986) Biogeografía de corales hermatípicos: el istmo centroamericano. Anales del Instituto de Ciencias del Mar y Limnología, Universidad Autónoma de México 13:297–304

Cortés J (1993) Comparison between Caribbean and eastern Pacific coral reefs. Rev Biol Trop 41(Supl 1):19–21

Cortés J (2003) Coral reefs of the Americas: an introduction to Latin American coral reefs. In: Cortés J (ed) Latin American coral reefs. Elsevier, Amsterdam, pp 1–7

Cortés J (2011) Eastern tropical Pacific coral reefs. In: Hopley D (ed) The encyclopedia of modern coral reefs: structure, form and process. Springer, Berlin, pp 351–358

Cortés J, Murillo MM (1985) Comunidades coralinas y arrecifes del Pacífico de Costa Rica. Rev Biol Trop 33:197–202

Cortés J, Macintyre IG, Glynn PW (1994) Holocene growth history of an eastern Pacific fringing reef, Punta Islotes, Costa Rica. Coral Reefs 13:65–73

Dana TF (1975) Development of contemporary eastern Pacific coral reefs. Mar Biol 33:355–374

Denyer P, Cortés J (2001) Levantamiento tectónico de la costa del Pacífico central de Costa Rica. Rev Biol Trop 49(Suppl 2):179–183 de Weerdt WH, Glynn PW (1991) A new and presumably now extinct species of *Millepora* (Hydrozoa) in the eastern Pacific. Zoologische Mededelingen Leiden 65(20):267–276

Duque-Caro H (1990) Neogene stratigraphy, paleoceanography, and paleobiology in northwest South America and the evolution of the Panama Seaway. Palaeogeogr Palaeoclimatol Palaeoecol 77:203–234

Durham JW (1947) Corals from the Gulf of California and the north Pacific coast of America. Geol Soc Am Mem 20:1–68

Durham JW (1950) 1940 E. W. Scripps cruise to the Gulf of California. Part II. Megascopic paleontology and marine stratigraphy. Geol Soc Am Mem 43:1–216

Durham JW (1966) Coelenterates, especially stony corals, from the Galápagos and Cocos Islands. In: Bowman RI (ed) The Galápagos, Proceedings of the symposium of the Galápagos International Scientific Project. University of California Press, Berkeley, pp 123–135.

Emerson WK, Chaney HW (1995) A zoogeographic review of the Cypraeidae (Mollusca: Gastropoda) occurring in the eastern Pacific Ocean. Veliger 38:8–21

Fagerstrom JA (1987) The evolution of reef communities. Wiley, New York, 600 p

Fiedler PC, Lavin MF (2017) Oceanographic conditions of the eastern tropical Pacific. In: Glynn PW, Manzello DP, Enochs IC (eds) Coral reefs of the Eastern Tropical Pacific. Coral reefs of the world 8. Springer, Dordrecht, pp 59–83

Flügel E, Senowbari-Daryan B (2001) Triassic reefs of the Tethys. In: Stanley GD Jr (ed) The history and sedimentology of ancient reef systems. Vol. 17, Topics in geobiology. Kluwer Academic/Plenum, New York, pp 217–249

Forsman ZH, Guzmán HM, Chen CA, Fox GE, Wellington GM (2005) An ITS region phylogeny of *Siderastrea* (Cnidaria, Anthozoa): is *S. glynni* endangered or introduced? Coral Reefs 24:343–347

Forsman ZH, Hunter CL, Fox GE, Wellington GM (2006) Is the ITS region the solution of the "species problem" in corals? Intragenomic variation and alignment permutation in *Porites*, *Siderastrea* and outgroup taxa. In: Proceedings of the 10th international coral reef symposium, Okinawa, vol 1, pp 14–23

Forsman ZH, Barshis DJ, Hunter CL, Toonen RJ (2010) Shape-shifting corals: molecular markers show morphology is evolutionarily plastic in *Porites*. BMC Evol Biol 9:45. https://doi.org/10.1186/1471-2148-9-45

Glynn PW (1972) Observations on the ecology of the Caribbean and Pacific coasts of Panamá. In: Jones ML (ed) The Panamic biota: some observations prior to a sea-level canal. Bull Biol Soc Wash 2:13–30

Glynn PW, Ault JS (2000) A biogeographic analysis and review of the far eastern Pacific coral reef region. Coral Reefs 19:1–23

Glynn PW, Wellington GM (1983) Corals and coral reefs of the Galápagos Islands. With an annotated list of the scleractinian corals of the Galápagos by John W. Wells. University of California Press, Berkeley, 330 p

Glynn PW, Stewart RH, McCosker JE (1972) Pacific coral reefs of Panama: structure, distribution and predators. Geol Rundsch 61(2):483–519

Glynn PW, Veron JEN, Wellington GM (1996) Clipperton Atoll (eastern Pacific): oceanography, geomorphology, reef-building coral ecology and biogeography. Coral Reefs 15:71–99

Glynn PW, Wellington GM, Riegl B, Olson DB, Borneman E, Wieters EA (2007) Diversity and biogeography of the scleractinian coral fauna of Easter Island (Rapa Nui). Pac Sci 61:67–90

Glynn PW, Grassian B, Kleemann KH, Maté JL (2016) The true identity of *Siderastrea glynni* Budd & Guzmán, 1994, a highly endangered eastern Pacific scleractinian coral. Coral Reefs 35:1399–1404. https://doi.org/10.1007/s00338-016-1470-8

Glynn PW, Alvarado JJ, Banks S, Cortés J, Feingold JS, Jiménez C et al (2017) Eastern Pacific coral reef provinces, coral community structure and composition: an overview. In: Glynn PW, Manzello DP, Enochs IC (eds) Coral reefs of the Eastern Tropical Pacific. Coral reefs of the world 8. Springer, Dordrecht, pp 107–176

Graham EM, Baird AH, Connolly SR (2008) Survival dynamics of scleractinian coral larvae and implications for dispersal. Coral Reefs 27:529–539

Grigg RW, Hey R (1992) Paleoceanography of the tropical eastern Pacific Ocean. Science 255:172–178

Guzmán HM, Cortés J (2007) Reef recovery 20 years after the 1982-83 El Niño massive mortality. Mar Biol 151:401–411

Heck KL Jr, McCoy ED (1978) Long-distance dispersal and the reef-building corals of the eastern Pacific. Mar Biol 48:349–356

Hellberg ME, Prada C, Tan MH, Forsman ZH, Baums IB (2016) Getting a grip at the edge: recolonization and introgression in eastern Pacific *Porites* corals. J Biogeogr 43:2147–2159. http://wileyonlinelibrary.com/journal/jbi. https://doi.org/10.1111/jbi.12792

Hertlein LG (1972) Pliocene fossils from Baltra (South Seymour) Island, Galápagos Islands. Proc Calif Acad Sci 4th Ser 39(3):25–46

Hou Z, Li S (2018) Tethyan changes shaped aquatic diversification. Biol Rev Camb Philos Soc 93(2):874–896. https://doi.org/10.1111/brv.12376

Hubbard DK, Gill IP, Burke RB (2001) The role of framework in modern reefs and its application to ancient systems. In: Stanley GD Jr (ed) The history and sedimentology of ancient reef systems. Topics in geobiology, vol 17. Kluwer Academic/Plenum, New York, pp 351–386

Hubbard DK, Burke RB, Gill IP, Ramirez WR, Sherman C (2008) Coral-reef geology: Puerto Rico and the US Virgin Islands. In: Riegl BM, Dodge RE (eds) Coral reefs of the USA. Coral reefs of the world 1. Springer, Dordrecht, pp 263–302

Jackson JBC, Budd AF (1996) Evolution and environment: introduction and overview. In: Jackson JBC, Budd AF, Coates AG (eds) Evolution and environment in tropical America. University of Chicago Press, Chicago, pp 1–20

Jokiel PL (1990) Long-distance dispersal by rafting: reemergence of an old hypothesis. Endeav New Ser 14(2):66–73

Kitahara MV, Stolarski J, Cairns SD, Benzoni F, Stake JL, Miller DJ (2012) The first modern solitary Agariciidae (Anthozoa, Scleractinia) revealed by molecular and microstructural analysis. Invertebr Syst 26(3):303–315. https://doi.org/10.1071/is11053

Kleemann K (1990) Boring and growth in chemically boring bivalves from the Caribbean, eastern Pacific and Australia's Great Barrier Reef. Senckenberg Marit 21:101–154

Kleemann K (2013) Fast and massive settlement of boring bivalves on coral slabs at Taboga Islands, eastern Pacific, Panama. Bollettino Malacologico 49:104–113

LaJeunesse TC, Forsman ZH, Wham DC (2016) An Indo-West Pacific 'zooxanthella' invasive to the western Atlantic finds its way to the eastern Pacific via an introduced Caribbean coral. Coral Reefs 35:577–582. https://doi.org/10.1007/s00338-015-1386-6

Leinfelder RR (2001) Jurassic reef ecosystems. In: Stanley GD Jr (ed) The history and sedimentology of ancient reef systems. Vol. 17, Topics in geobiology. Kluwer Academic/Plenum, New York, pp 251–309

Lessios HA, Baums IB (2017) Gene flow in coral reef organisms of the tropical eastern Pacific. In: Glynn PW, Manzello DP, Enochs IC (eds) Coral reefs of the Eastern Tropical Pacific: persistence and loss in a dynamic environment. Coral reefs of the world 8. Springer, Dordrecht, pp 477–499

Lessios HA, Robertson DR (2006) Crossing the impassable genetic connections in 20 reef fishes across the eastern Pacific barrier. Proc R Soc B Biol Sci 273:2201–2208

Lessios HA, Kessing BD, Wellington GM, Graybeal A (1996) Indo-Pacific echinoids in the tropical eastern Pacific. Coral Reefs 15:133–142

Lessios HA, Kessing BD, Robertson DR (1998) Massive gene flow across the world's most potent marine biogeographic barrier. Proc R Soc Lond B 265:583–588

Lessios HA, Lockhart S, Collin R, Sotil G, Sanchez-Jerez P, Zigler KS, Perez AF, Garrido MJ, Geyer LB, Bernardi G, Vacquier VD (2012) Phylogeography and bindin evolution in *Arbacia*, a sea urchin genus with an unusual distribution. Mol Ecol 21(1):130–144

López-Pérez RA (2012) Late Miocene to Pleistocene reef corals in the Gulf of California. Bull Am Paleontol 383:1–78

López-Pérez A (2017) Revisiting the Cenozoic history and the origin of the eastern Pacific coral fauna. In: Glynn PW, Manzello DP, Enochs IC (eds) Coral reefs of the eastern tropical Pacific: persistence and loss in a dynamic environment. Coral reefs of the world 8. Springer, Dordrecht, pp 39–57

López-Pérez RA, Budd AF (2009) Coral diversification in the Gulf of California during Late Miocene to Pleistocene. In: Johnson ME, Ledesma-Vasquez J (eds) Atlas of coastal ecosystems in the Gulf of California: past and present. University of Arizona Press, Tucson, pp 58–71

Macintyre IG, Glynn PW (1976) Evolution of modern Caribbean fringing reef, Galeta Point, Panamá. Am Assoc Pet Geol Bull 60(7):1054–1072

Macintyre IG, Glynn PW, Steneck RS (2001) A classic Caribbean algal ridge, Holandés Cays, Panamá: an algal coated storm deposit. Coral Reefs 20:95–105

McCoy ED, Heck KL Jr (1976) Biogeography of corals, seagrasses and mangroves: an alternative to the center of origin concept. Syst Biol 25:201–210

McPhaden MJ, Picaut J (1990) El Niño-Southern Oscillation displacements of the western equatorial Pacific warm pool. Science 250:1385–1388

Picaut J, Ioualalen M, Menkes C, Delcroix T, McPhaden MJ (1996) Mechanism of the zonal displacements of the Pacific warm pool: implications for ENSO. Science 274:1486–1489

Porter JW (1972) Ecology and species diversity of coral reefs on opposite sides of the Isthmus of Panama. In: Jones ML (ed) The Panamic biota: some observations prior to a sea-level canal. Bull Biol Soc Wash 2:89–116

Richmond RH (1987) Energetics, competence, and long-distance dispersal of planula larvae of the coral *Pocillopora damicornis*. Mar Biol 93:527–533

Richmond RH (1988) Competency and dispersal potential of planula larvae of a spawning versus a brooding coral. In: Proceedings of the sixth international coral reef symposium, Townsville, 2:827–831

Richmond RH (1990) The effects of the El Niño/Southern Oscillation on the dispersal of corals and other marine organisms. In: Glynn PW (ed) Global ecological consequences of the 1982-83 El Niño-Southern Oscillation. Elsevier oceanography series 52. Elsevier, Amsterdam, pp 127–140

Riegl BM, Glynn PW (2020) Population dynamics of the reef crisis: consequences of the growing human population. Adv Mar Biol 87(1):1–30

Robertson DR, Grove JS, McCosker JE (2004) Tropical transpacific shore fishes. Pac Sci 58(4):507–565

Rohwer F, Youle M (2010) Coral reefs in the microbial seas. Plaid, Granada Hills, 201 p

Scheltema RS (1988) Initial evidence for the transport of teleplanic larvae of benthic invertebrates across the east Pacific barrier. Biol Bull 174:145–152

Soja CM (1996) Island arc carbonates: characterization and recognition in the ancient geologic record. Earth-Sci Rev 41:31–65

Squires DF (1959) Reports of the Puritan-American Museum of Natural History expedition to Western Mexico, 7. Corals and coral reefs in the Gulf of California. Bull Am Mus Nat Hist 118(art. 7):367–432

Stanley GD Jr (2001) Introduction to reef ecosystems and their evolution. In: Stanley GD Jr (ed) The history and sedimentology of ancient reef systems. Topics in geobiology, vol 17. Kluwer Academic/Plenum, New York, pp 1–39

Stone GS, Obura D (eds) (2013) Underwater Eden. The University of Chicago Press, Chicago, 170 p

Todesco M, Pascual MA, Owens GL, Ostevik KL, Moyers BT et al (2016) Hybridization and extinction. Evol Appl 9(7):892–908. https://doi.org/10.1111/eva.12367

Toth LT, Macintyre IG, Aronson RB (2017) Holocene reef development in the eastern tropical Pacific. In: Glynn PW, Manzello DP, Enochs IC (eds) Coral reefs of the eastern tropical Pacific. Coral reefs of the world 8. Springer, Dordrecht, pp 177–201. https://doi.org/10.1007/978-94-017-7499-4_6

Vaughan TW, Wells JW (1943) Revision of the suborders, families, and genera of the Scleractinia, vol 44. Geological Society of America, Boulder

Veron JEN (1995) Corals in space and time, the biogeography & evolution of the Scleractinia. Comstock/Cornell, Ithaca, 321 p

Veron JEN (2000) Corals of the world, vol 2. Australian Institute of Marine Science, Townsville, 429 p

Veron JEN, Stafford-Smith M, DeVantier L, Turak E (2015) Overview of distribution patterns of zooxanthellate Scleractinia. Front Mar Sci 1(81):2–19. https://doi.org/10.3389/fmars.2014.00081

Verrill AE (1870) Notes on Radiata. Part 7. On the geographical distribution of the polyps of the west coast of America. Trans Conn Acad Arts Sci 1:558–570

Victor BC, Wellington GM, Robertson DR, Ruttenberg BI (2001) The effect of the El Niño Southern Oscillation event on the distribution of reef-associated labrid fishes in the eastern Pacific Ocean. Bull Mar Sci 69:279–288

von Prahl H, Mejía A (1985) Primer informe de un coral acropórido, *Acropora valida* (Dana, 1846) (Scleractinia: Astrocoeniida: Acroporidae) para el Pacífico americano. Rev Biol Trop 33(1):39–43

Wells JW (1983) Annotated list of the scleractinian corals of the Galápagos. In: Glynn PW, Wellington GM (eds) Corals and coral reefs of the Galápagos Islands. University of California Press, Berkeley, pp 212–295, 330 p

Wilson EO (1992) The diversity of life. W. W. Norton, New York. 424 p

Wood R (2011) General evolution of carbonate reefs. In: Hopley D (ed) Encyclopedia of modern coral reefs: structure, form and process. Springer, Amsterdam, pp 452–469

Wood S, Baums IB, Paris CB, Ridgwell A, Kessler WS, Hendy EJ (2016) El Niño and coral larval dispersal across the eastern Pacific marine barrier. Nat Commun 7:12571. https://doi.org/10.1038/ncomms12571

Chapter 4
Physical Environmental Challenges

Abstract Many environmental factors can stress and cause the mortality of reef-building corals with elevated thermal conditions predominant. The historically early explanation for the absence of coral reefs in the eastern Pacific region—"cold ocean currents and upwelling"—does account for some degree of coral bleaching and mortality but is superseded by region-wide mass mortality caused by El Niño warming events that in recent years are superimposed on increasing global warming. Major eastern Pacific reef-building coral species are highly tolerant of prolonged low salinity conditions and have even demonstrated modest reef development in an estuarine environment. Extreme midday low tidal events in the dry season and during La Niña events result in shallow reef zone coral exposure, bleaching, and mortality. Recovery from these disturbances occurs in <10 years. Tropical cyclones take place at Mexico's latitude, and seismic events in the equatorial eastern Pacific, generally causing only local and minor damage to coral reefs. Ocean acidification in the Galápagos Islands has minimized internal cementation and lowered coral skeletal development. A high incidence of lesions and skeletal abnormalities is observed in corals in some areas, but no confirmed disease outbreaks have been reported.

Introduction

Always cognizant of the long-held tenet of the cause of poor coral development in the eastern Pacific—"cold ocean currents and upwelling"—, the occurrence of low thermal stress events were constantly on the top of our minds. In the 1970s there was even the specter of the Earth heading into an imminent ice age, a scenario supported by a respected cadre of climate scientists (Peterson et al. 2008). At that time, an opposing view of climate change, based on increasing atmospheric CO_2 and supported by modeling studies, predicted the occurrence of future global warming. As climate change progresses, extreme sea temperature conditions have indeed been observed to be a chief cause of coral decline, however, at the upper thermal tolerance limits of marine biota, i. e. extreme stressful high temperature conditions. Corals are sensitive to many other factors besides temperature, and these are

© The Author(s), under exclusive license to Springer Nature Switzerland AG 2025

P. W. Glynn, *Coral Reef Resilience in the Anthropocene*,
https://doi.org/10.1007/978-3-031-81421-1_4

examined to further increase our understanding of reef development and persistence in the eastern Pacific region.

Physical conditions limiting reef development, often associated with marginal reef environments, were examined by Joanie (Joan Ann) Kleypas et al. (1999) to gain insight into the effects of current and projected climate change scenarios. In addition to temperature, key environmental variables—salinity, light, nutrients, and aragonite saturation state—are also considered here in relation to eastern Pacific reef condition. Overviews of natural disturbances and anthropogenic stressors in eastern Pacific coral areas in the book, *Latin American Coral Reefs*, edited by Jorge Cortés (2003), have been especially useful in recognizing site-specific disturbances such as coral tidal exposure events, storms, and seismic activity. A brief discussion of the increasing occurrence of coral diseases, a biotic disturbance closely correlated with rising global sea temperature (Harvell et al. 1999), concludes this chapter.

From the earliest speculations on the absence of coral reefs along eastern Pacific tropical shores, "cold ocean currents and upwelling" were considered primarily responsible for this condition (Couthouy 1842; Darwin 1842; Dana 1843; Verrill 1870). Cyril Crossland (1927) emphasized the adverse effects of mud bottoms in limiting the downward growth of reefs in the Gulf of Panama. Our in situ surveys of reef substrates, including reef base zones, revealed primarily coarse sediments, not supporting Crossland's view. As recently as 1963 the eminent marine biologist, C. M. Yonge (Sir Charles Maurice Yonge), concluded that "…conditions, largely of temperature, prevent the formation of reefs." in the subtropical and tropical eastern Pacific region. This supposed thermal limitation was puzzling because long term sea surface temperature trends from the lower half of the Gulf of California to southern Ecuador, were generally within the thermal limits supporting coral growth. Yearly low winter temperature records ranged from 20 to 21 °C and high summer temperatures from 26 to 29 °C, well within the optimum thermal limits of coral reef development.

The supposed absence of coral reefs in the Gulf of Panama puzzled Charles Darwin because Alexander von Humboldt's early American surveys reported this area as one of the hottest pelagic districts in the world. Synoptic charts illustrating regional sea surface isotherms (contour lines of equal temperature) were available in the early literature (e.g., Murphy 1936; Hubbs and Roden 1964; Wooster and Hedgpeth 1966; Forsbergh 1969) and indicated that most eastern Pacific areas could support reef growth, at least in terms of favorable thermal conditions. There are three major coastal upwelling centers, however—in southern Mexico, in an area straddling Nicaragua and Costa Rica, and in Panama—that experience seasonal upwelling when surface temperatures can drop temporarily to 18 °C and lower. These upwelling areas, including offshore islands (e.g., the western sector of the Galápagos Islands), are all located within thermal environments that support coral reef growth (Kleypas et al. 1999). Eastern Pacific oceanic islands (e.g., Clipperton Atoll, the Revillagigedo Islands, and Cocos Island) are all located within tropical surface waters. Why then could this region not support coral reef growth? Actually, J. Wyatt Durham (1966) was an early investigator who recognized the occurrence of coral reef development ("in the broad sense") in the eastern tropical Pacific region, e.g., at Clipperton Island (Allison 1959) and in the Gulf of California (Squires 1959).

Temperature is undeniably an important factor in influencing coral growth, however, other environmental conditions such as salinity, nutrients, sedimentation, illumination, and aragonite saturation state can play important limiting roles as well. This chapter will focus on physical conditions influencing the spatial occurrence of eastern Pacific coral reefs. A brief consideration of coral infectious agents (diseases and parasitism) concludes this chapter. Biological factors, such as competition, predation, bioerosion, reproduction and genetic connectivity, will be considered in later chapters. It will become evident that corals are affected by numerous physical and biological processes, in addition to the time-honored impacts of "cold ocean currents and upwelling".

Sea Temperature

Extreme temperatures, both high and low, can negatively impact zooxanthellate corals, leading to bleaching and partial or total colony mortality. Bleaching refers to whole coral colonies or parts of colonies that have turned stark white, a stress response caused by the loss of zooxanthellae and their colored photosynthetic and accessory pigments. Healthy coral tissues display varying shades of brown, golden-brown, green, yellow-orange and other colors due primarily to their endosymbiotic zooxanthellae and accessory pigments. With the loss of zooxanthellae, the white calcareous coral skeleton becomes visible, hence the term bleaching. Bleaching can also result from other stressors such as high light levels, low tidal exposure of corals to the atmosphere, and diseases. Coral tissue loss due to predation, for example by the crown-of-thorns sea star *Acanthaster*, can be confused with bleaching. In this case, however, a predator removes the tissue, leaving the exposed white coral skeleton.

Temperatures stressful to corals in the eastern Pacific region are associated with (1) eastern boundary currents, (2) upwelling centers, (3) shoaling of internal waves, and (4) El Niño Southern Oscillation (ENSO) events. Large and rapid changes in sea surface temperature occur over narrow latitudinal distances from Baja California (~23°N) to immediately south of the Gulf of Guayaquil (~5°S) due to the influence of the cool, southerly flowing California Current and the northerly flowing Peru Coastal Current (Humboldt Current), respectively. In the coolest seasons, suboptimal 16–20 °C isotherms commonly occur along these coastal areas (Fiedler and Lavín 2017).

Eastern Boundary Currents Eastern boundary currents are relatively slow-moving cold currents that occur off the western coasts of continents. They are the eastern arms of the major current gyres of the world's oceans, delivering cold water from high to low latitudes. The latitudinal (north-south) difference in distance between eastern and western boundary currents greatly influences the geographic extent of coral reef development in the Pacific Ocean. For example, the cold California and Peru currents, equatorward bound, limit coral reef development to 22° of latitude or ~2400 km along eastern Pacific shores. As the eastern boundary currents move westward at equatorial latitudes, they warm and enter the western Pacific basin over a broad area. The resulting geographic swath of warm, tropical waters encompasses 60° of latitude or ~6700 km. Augmenting the area of potential reef development in the West Pacific are the countless South and Central Pacific islands and archipelagoes. The influence of the California Current limits the occurrence of zooxanthellate corals on the western side of the Baja California Peninsula (Reyes-Bonilla and López Pérez 1998). *Porites panamensis*, a small nodular colony, is the only zooxanthellate coral species reported from the Pacific shores of Baja California (Reyes-Bonilla et al. 2005). Less is known of the influence of the cold Peru Current on zooxanthellate corals in southern Ecuador and their possible occurrence in northern Peru.

Although I have never surveyed coastal Peruvian waters, I have accepted requests to identify corals from that area. Corals from northern Peru were of exceptional interest in terms of their possible occurrence in a suboptimal thermal environment and potentially from an immigrant population in the process of a southerly range extension. Unfortunately, a definitive answer as to whether or not any of the corals were actually alive in Peruvian waters could not be firmly established,

The first coral specimen was brought to my RSMAES office by a Peruvian gentleman. After exchanging greetings and pleasantries, the Peruvian's demeanor turned serious, his eyes sparkled with excitement and he whispered to me that he was carrying a marvelous item in his satchel. He carefully unwrapped the mysterious item, which he claimed originated from outer space; it had mysteriously appeared (landed) on an unidentified beach in northern Peru. I was told that the specimen belonged to a high-ranking Peruvian military officer and my visitor was charged with depositing it in a secure bank vault in Miami for safe keeping. I confidently identified the coral as *Pocillopora damicornis*: it was a clean and perfectly preserved fist-sized colony. I told my visitor that this coral was a common species in the eastern tropical Pacific and Indo-Pacific waters. I asked if he would consider donating a fragment of the colony to our Marine Invertebrate Museum for study. I received a polite but firm no. He appeared agitated and did not accept my identification or that this well-known coral species inhabited the marine waters of Planet Earth. I failed to photograph the specimen and did not obtain other information, but not for want of trying. It was obvious that my Peruvian visitor was disappointed, and he quickly departed. This incident resulted in a tantalizing but undocumented occurrence of a reef-building coral from Peru. To my mind, it was much more likely the coral was from Peru and not from outer space.

The second request for Peruvian coral identifications involved two species of *Pocillopora*. This request began with e-mail messages and photographs sent to me by Dr. Patricia J. Netherly, an anthropologist at Vanderbilt University. One of the corals was a small fragment recovered from Pleistocene deposits at the Huaca Prieta study site in northern Peru. This archeological site is located near the mouth of the Chicama River, about 650 km south of coastal areas in Ecuador with known living populations of *Pocillopora damicornis*. The coral was associated with hearths, stone tools, and other artifacts, and was of late Pleistocene age (Dillehay 2017; Glynn et al. 2022). The other species of *Pocillopora* was collected at Máncora, presumably alive. Máncora is slightly more than 400 km north of Huaca Prieta, located at the southern boundary of the Gulf of Guayaquil. Additional information on the occurrence of the *Pocillopora* species in this transitional zone is presented in Chap. 12.

Coral communities and modest reef building have been reported as far south as Santa Clara Island in the Gulf of Guayaquil, an area of exceedingly high freshwater discharge (Glynn and Ault 2000; Glynn 2003; Alvarado et al. 2017). An incipient reef formation at this island was described to me by René Espinosa, a trusted field assistant in Ecuador in the 1980s. The structure and composition of these coral communities, located immediately north of the westward bend of the cold Coastal Peru Current, need detailed documentation.

Upwelling Centers I was puzzled but delighted to observe several coral reefs in the upwelling environment of the northern Pearl Islands, Gulf of Panama, during our earliest surveys in the eastern Pacific (Glynn and Stewart 1973). Pocilloporid reefs were present at Saboga and Contadora Islands and, as surveys progressed, were found at other islands in the Pearl Islands Archipelago (Glynn and Macintyre 1977). According to the surveys of early researchers, coral reefs were not observed in upwelling centers and therefore considered to be absent from upwelling environments.

The principal upwelling centers along the tropical coasts of the eastern Pacific are the Gulf of Tehuantepec (southern Mexico), the Gulf of Papagayo (straddling the border of Nicaragua and Costa Rica), and the Panama Bight, including the Gulf of Panama (Panama). These are all areas of low topographic relief, allowing seasonal, northeastern Trade Wind jet streams to cross from the western Atlantic to Pacific shores. Wind-induced friction and Ekman transport (wind stress and the earth's rotation) of Pacific surface waters result in their offshore movement with deeper and cold, nutrient-rich waters upwelling to the surface (Smayda 1966). Some parts of Central America with high mountains, for example the Cordillera de Talamanca with a peak height of 3819 m, impede Trade Wind flow and thus prevent upwelling.[1] The Talamanca mountain chain extends from western Panama to southern/central Costa Rica, and thus shelters leeward Pacific localities from northeastern winds and upwelling such as the Gulf of Chiriquí in Panama, and Golfo Dulce and Caño Island in Costa Rica.

[1] My son, Peter J. Glynn and friends, when in high school in Panama, decided to explore the Talamanca mountains by hiking and bivouacking along the summits of the mountain range. When they reached the extinct crater of Volcán de Barú at 3474 m, they decided to encamp there but experienced a restless night with little sleep. They were chilled to the bone by frigid, cross-isthmian wind jet streams. The next morning the water in their canteens was frozen. I did mention to Peter to be prepared for chilly conditions but was not forceful enough in convincing him of this likely discomfort. Lower temperatures associated with increasing elevation, rain squalls, and upwelling can quickly lead to exposure and the loss of body heat; this can bring on chills when one is acclimated to a warm, tropical environment. Our first research scuba diving in upwelling environments in the eastern Pacific were performed in brief bathing suits with protective long-sleeved cotton shirts. This apparel was soon replaced with wet suits that allowed for more comfortable diving and multiple dives per day. *Chocolate caliente* (hot chocolate) drinks were always welcome after our bone-chilling, research dives in upwelling waters.

Since upwelling driven by Trade Winds occurs during the dry season, there is generally high light penetration at such times that benefits primary production. Once phytoplankton production is in full swing, however, generating thick water blooms (sometimes "red tides"), these can result in opaque conditions, reducing water column visibility to near zero. I once experienced such conditions at Señora Islet in the Pearl Islands that led to an unsettling and rare shark encounter. I was on scuba in a thick, green algal bloom in search of a study site when suddenly a shark and I, on a collision course, literally swam into each other, bumping heads. The shark, I believe a white tip reef shark, scurried off in one direction and I in another (Fig. 4.1). It seemed we were both equally startled. I decided to return to our study site later when underwater visibility improved.

Besides the direct effect of extreme low temperatures causing coral bleaching and mortality, several other components of coral community ecology are affected by upwelling. Tyler B. Smith,[2] one of my doctoral students, investigated the effects of upwelling on coral community herbivory and competitive interactions in the Gulf of Panama. One aspect of this research involved the effects of temperature on the feeding rates of fish herbivores and how low temperatures might interfere with feeding and thus allow for increased algal growth (Smith 2008). Direct observations and field experiments of the grazing rates of a prominent parrotfish herbivore (*Scarus ghobban*) demonstrated significant declines in feeding at temperatures below 24 °C.

During intense upwelling in February 2001 temperatures at the Saboga Island reef fell below 18 °C for 16 days with hours-long drops below 15 °C and a minimum recorded temperature of 14.7 °C (Smith 2006; Glynn and Fong 2006). At extreme low upwelling temperatures (15–20 °C), this *Scarus* species and other reef fish herbivores would often enter a state of torpor, stop feeding and seek shelter in reef recesses. Also, numerous porcupinefish and puffers washed across the reef, having suffered high mortality.

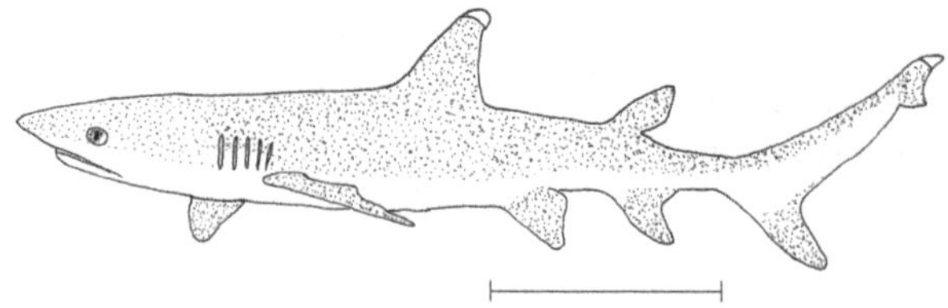

Fig. 4.1 *Triaenodon obesus*, the whitetip reef shark, is ubiquitous in eastern Pacific waters. scale bar = 0.5 m

[2] Tyler is presently a Research Professor of Marine Biology in the Center for Marine and Environmental Studies (CMES), University of the Virgin Islands (UVI), St. Thomas, U. S. Virgin Islands. His research interests are in coral reef ecology, with primary focus on reef resilience to disturbances, reef refuges and refugia, and mesophotic coral ecosystems. Coral research opportunities in the Virgin Islands have allured other UM marine science graduates. Tyler's wife, Marilyn E. Brandt, also earned her doctoral degree at RSMAES and is a Research Professor at CMES, UVI. Her research is directed toward marine diseases and coral restoration ecology. Victor Brandtneris, another UM marine science graduate, earned an M.Sc. degree at UVI under Tyler's supervision. His research centered on the trophic ecology of a mesophotic coral species (Brandtneris et al. 2016).

To evaluate experimentally the effects of upwelling nutrification on benthic algal growth (primary production) and herbivore grazing, Tyler performed caging experiments. This involved caging reef algae to exclude herbivores, thus minimizing grazing, and comparing algal production in caged and uncaged treatments. A gain in caged algal production was observed, and this was nearly two times greater than that observed in uncaged plots (Smith 2006). Clearly, algal growth benefited from upwelling nutrification. When water temperatures rise during non-upwelling periods for 9–10 months annually, fish herbivore activity increases and uncaged reef substrates are not overgrown by algae. These results suggest that reef herbivores, under normal upwelling conditions, are capable of limiting increases in benthic algal biomass. Chuck Birkeland (1977) also demonstrated the important role of fish herbivore grazing in benefiting the survival of coral recruits during upwelling in the Gulf of Panama.

Coastal upwelling occurs in other parts of Mexico besides the Gulf of Tehuantepec, for example off Cabo Pulmo (Baja California) and in the Cabo Corrientes/Banderas Bay area (Nayarit and Jalisco states) along the central Pacific coast. Long-shore northern winds contribute to upwelling in these areas, especially in the spring season (López-Sandoval et al. 2009). Internal wave propagation in coastal canyons also causes localized upwelling (Plata-Rosas and Filinov 2007), and mesoscale (10–1000 km in horizontal extent) mixing of the California Current and Gulf of California waters (Torres-Orozco et al. 2005) contribute to the complexity of the thermal dynamics in this area. Coral communities and a few small coral patch reefs are present in these coastal upwelling localities. Coral bleaching during El Niño warming events is not so severe in these areas as in equatorial localities, likely a result of upwelling and possibly the influence of the cool California Current (Carriquiry and Reyes-Bonilla 1997; Reyes-Bonilla 2001). The seasonal and annual current patterns along the central Mexican coast are in need of further study. El Pulmo Reef, and the captivating description of its rich biota and complexity by John Steinbeck in the *Sea of Cortez* (Steinbeck and Ricketts 1941)—the only coral reef[3] encountered during their epic voyage on the *Western Flyer*—is typically subject to intense upwelling in the summer months.

A major oceanic upwelling center, the Galápagos Cold Pool (GCP), is located on the western sides of Isabela and Fernandina Islands. Upwelling at this location is due to the Equatorial Undercurrent (EUC), a subsurface (~100 m deep), rapid (up to 1.5 m/s), east-moving current (see Fig. 1.6). The EUC crosses most of the Pacific Ocean and transports an enormous volume of water, about 1000 times the volume of the Mississippi River. A recent study of the EUC pathway, measured directly by employing autonomous underwater gliders, reported that this subsurface current splits into north and south branches at 93°W, about 220 km west of the Galápagos

[3] An in-depth study of the Cabo Pulmo coral formation revealed significant carbonate production by reef-associated biota, but also significant off-shelf transport resulting in little to no sediment retention in reef frameworks (Riegl et al. 2007). Thus, while El Pulmo reef functions sedimentologically like a coral reef, in the strict sense it is not a true actively accreting coral reef.

Archipelago (Jakoboski et al. 2020). The north and south branches continue flowing eastward around Isabela Island as cold, nutrient-rich filaments that support primary phytoplankton production. Where the EUC directly contacts western island shores, these are highly productive areas devoid of zooxanthellate corals but supporting abundant macroalgal communities and diverse vertebrate consumers such as whale sharks, marine iguanas, the Galápagos penguin and Galápagos fur seal. Coral communities and incipient reef growth do occur in this zone, however, in areas protected from direct upwelling, such as lava pools and shallow habitats in the lee of Fernandina Island (Glynn and Wellington 1983; Colgan 1990). A recent modeling study of the GCP locality indicates that the strength of EUC upwelling is governed by local northward winds and not, as previously understood, by remotely forced or random changes in the current's intensity or path (Forryan et al. 2021). A full understanding of Galápagos upwelling processes is important vis-à-vis recent major reductions in GCP phytoplankton production (often coinciding with El Niño events) and cascading trophic impacts on threatened endemic species.

Internal Waves Internal wave-forcing moving cool, mixed layer thermocline waters into shallow depths can cause coral bleaching and mortality. This sort of effect was documented in the northern Galápagos Islands over a 6-month period in 2006–2007 (Banks et al. 2009). Temperatures were observed to fall from 26 to 28 °C, during a moderate El Niño warm event, to 16 °C over a 6-day period, initiating a strong La Niña cold event. Temperature records to 2008 showed additional "cold shock" bleaching of *Porites lobata* and *Pocillopora* spp. across the archipelago.

Jerry Wellington and I, in one of our early 1975 surveys in the southern Galápagos Islands, experienced a shoaling internal wave at Champion Island. We were scuba diving at the deep edge (~8 m) of a coral buildup when suddenly a murky, cold-water wave pushed us rapidly upwards. Numerous fishes were caught up in this swirling maelstrom and their swimming movements appeared confused, as were ours. Rapidly shoaling cold water filaments have also been observed and recorded by our team during the dry season at the Uva Island reef, Gulf of Chiriquí, Panama.

Temperature drops caused by internal waves can be quite dramatic (especially for a thinly clad diver not expecting such cold water) and have a beneficial effect on corals by rescuing them from otherwise lethally high temperatures. During the 2015–16 very strong El Niño event, corals in the northern Galápagos islands of Wolf (Wenman) and Darwin (Culpepper) began bleaching due to a pool of heated, shallow surface waters that formed. They were, however, exposed to a series of sharp temperature drops that were presumably caused by internal waves (Riegl et al. 2019a). Temperature recordings and geomorphological assessments at both Wolf and Darwin suggest that the abundant coral populations at Wolf and reef-building on Darwin may be related to the occurrence of these surfacing internal waves. They appear to be most frequent during the warm season (Riegl et al. 2019a, b), thus offering hope that reef corals may persist at these northernmost islands. This may have relevance regionally in the equatorial eastern Pacific because coral larvae from Darwin and Wolf can potentially disperse to Costa Rica and Colombia (Riegl et al. 2019b).

El Niño Southern Oscillation (ENSO)

Following is a brief introduction to the El Niño climate pattern to facilitate understanding of its development and timing related to coral reef impacts. Events and evidence linking ENSO to coral bleaching and mortality are introduced here. ENSO, the acronym for El Niño Southern Oscillation, refers to two processes that are functionally closely interrelated. El Niño, the Christ child in the Spanish language, was the name given to the sporadically occurring warm sea conditions off Peru and northern Chile, usually near the Christmas season or Epiphany. Southern Oscillation refers to Pacific-wide inter-annual variations in atmospheric pressure centers that affect ocean currents. During normal weather conditions the southeast Trade Winds drive strong equatorial flow from the eastern to the western Pacific. Approximately every 3–7 years the southeast Trade Winds weaken or cease, and westerly winds form in the Central Pacific. This sets in motion an easterly flow of Pacific warm pool waters, the largest area of warm water in the world, thus generating an El Niño event. Two kinds of El Niño events are now recognized, (a) a Central Pacific El Niño centered near the International Date Line (180°W), and (b) the eastern Pacific El Niño in the far eastern Pacific equatorial region (Wang et al. 2017). The eastern Pacific El Niño is our principal focus. When atmospheric and sea conditions are reversed, soon or a few seasons after an El Niño event, a sustained cooling phase may occur, which has been named La Niña (the girl child). The neutral phase of ENSO, the years of normal weather conditions, includes about half the years in the climate cycle. I will use the name La Niña since it is now universally adopted. I agree with some of my Latin American colleagues, however, who find this epithet a poor choice, a nonsensical attempt to describe an opposite physical state. When vigorous La Niña conditions prevail, this often results in frequent upwelling and stressful low sea temperature conditions, high nutrient concentrations, algal blooms, and low sea level stands leading to coral aerial exposure and mortality. El Niño-related impacts on climate conditions extend far and wide globally and often result in drought conditions in the western Pacific, the Indian subcontinent, South Africa, and northern South America. At times, devastating wildfires occur in these regions. In the central and western Pacific marine environment, the thermocline depth usually shallows, and sea surface height drops, thus exposing corals present in near-surface habitats.

Our research cruise to the Gulf of Chiriquí was scheduled for 17–23 March 1983. The local maritime weather report was favorable, no unusual conditions expected. We had prepared for advancing routine coral investigations with no hint of the dreadful surprise awaiting our arrival. A few months before the Chiriquí cruise I had accepted an academic position with RSMAES, University of Miami, and decided we should conduct possibly final coral reef monitoring surveys and predator-prey behavioral observations (initiated in 1970) before departing Panama for Miami in November. After dropping anchor and entering the water at our Uva Island study site, I was astonished to see a stark white reef scape with interspersed green-brown patches. "HORRORS" is the first entry in my 1983 field record book. This was the first massive coral bleaching event I had ever witnessed, and what a shocking and depressing sight it was. A truly baffling scene, what could cause such a tragic mortality event? Here I will describe the initial responses of the stressed and moribund corals and discuss the avenues of research that were followed to uncover the causes of this extraordinary event. In Chap. 5, "El Niño Southern Oscillation", I will summarize immediate and longer-term species interactions and effects on community structure following the severe disturbance. My prior research agenda would be fundamentally changed by the course of events set in motion by the 1982–83 El Niño disruption.

Upon closer examination of the bleached corals—nine species of zooxanthellate scleractinian corals (in the genera *Pocillopora*, *Porites*, *Pavona*, *Gardineroseris*, and *Psammocora*) and three species of hydrocorals (*Millepora*)—live tissues were present with normally extended polyps, but all without color. The green-brown patches were dead corals, recently overgrown by filamentous algae. These dead coral patches were possibly the remnants of the unconfirmed bleaching reports by sports divers in the Gulf of Chiriquí in February 1983.

What could be the cause of this massive, reef-wide coral mortality? I was aware of Tom Goreau Sr's. report of massive coral bleaching in Jamaica, the result of low salinity stress from heavy rains during Hurricane Flora (Goreau 1964). Low rainfall and minimal river streamflow had occurred, however, during the coral bleaching event in Panama (Glynn 1983). Since agricultural pesticides were in wide use in the Chiriquí province, perhaps they could enter gulf waters and be a factor in the bleaching event? We speculated that herbicides, designed to adversely affect the physiology of plant growth, could negatively impact the coral's symbiotic algae (zooxanthellae). We collected tissue samples from unbleached and moribund corals, which we delivered to Gene (Eugene) Corcoran's laboratory at RSMAES, University of Miami. Gene was a professor at RSMAES with a laboratory active in the analysis and study of marine trace contaminants. Gene was very supportive of the coral reef work; I teamed up with him immediately and Alina Szmant later to investigate trace element contaminant loads in corals in southeastern Florida (e.g., Glynn et al. 1989). As for the Panama corals, relatively high levels of herbicide residues were detected in the tissues of bleached corals. Accordingly, pesticide contaminants could not be dismissed as a causative agent of coral bleaching.

While the coral bleaching disturbance was still in progress in Panama, in the Gulf of Panama following the upwelling, cool water season, I was invited to participate as a co-instructor in the graduate Coral Reef Population Biology course at Coconut Island, Hawaii Institute of Marine Biology, University of Hawaii. There, I met Steve (L. S.) Howard and briefed him on the coral bleaching observations and the presence of herbicides in coral tissues. Steve was a student in the coral course with prior experience in chemical pollutant studies in the marine environment. We decided it would be helpful to conduct controlled experiments of herbicide effects on corals. Our proposal to perform these experiments was approved as a course project. We selected *Pocillopora damicornis* as the study species, the same species affected in Panama; herbicide concentrations equivalent to those detected in bleached coral tissues from field samples were tested. Our experiments demonstrated that herbicides cause bleaching and mortality at relatively low concentrations and for brief exposure periods (Glynn et al. 1984). In presenting these laboratory results and the recent field observations in Panama to the coral class, collectively we were not convinced that agricultural contaminants alone were responsible for this large-scale coral mortality event. Troubling inconsistencies included (a) bleaching of corals on reefs without tissue pesticide residues, (b) no bleaching in corals with pesticide loads, (c) coral bleaching in localities without pesticide use, and (d) the confounding effects of El Niño high stressful temperatures.

There was another concern, could coral epizootics or contagious diseases be involved? I had attended a presentation by Arnfried Antonius at the Third International Coral Reef Symposium in Miami in 1977 and wondered if the Panamanian mortality event was similar to the coral diseases and "shut down" reaction described by Arnfried in the south Florida area. To pursue this question, I enlisted the help of Esther Peters and Len (Leonard) Muscatine, experts in coral histopathology and zooxanthella physiology, respectively. During the bleaching event in Panama we grafted normal and affected coral branches from donor colonies to normal and affected recipient colonies to test for possible infectious reactions. The grafts were fitted snugly into the branches of host colonies and their condition was observed 1, 3, and 7 months later. Branches were grafted to different positions of the same colony (isografts), from donor to different recipient colonies of the same species (allografts), and from donor to recipient colonies of different species (xenografts). None of the recipient corals that received bleached branches showed signs of infection. All colonies retained their normal pigmentation, and histopathological examination revealed high zooxanthella concentrations without signs of infectious agents (Glynn et al. 1985).

By mid-summer 1983, it became generally recognized that El Niño was active and causing unprecedented—a once in 100 years event—elevated sea temperature anomalies in the equatorial eastern Pacific region, including the coral reef areas of Panama. Paul Jokiel, a University of Hawaii professor and co-instructor in the coral course, began to make a convincing argument that high temperatures were probably contributing importantly to coral stress and bleaching. The earlier studies of Paul and Steve Coles on thermally induced bleaching demonstrated that zooxanthellate corals were especially sensitive to small increases in temperature (Jokiel and Coles 1974, 1977). This work showed that corals are living close to their upper thermal tolerance limits. Long-term (weekly) increases of 1–2 °C above summer season high temperatures or short-term (daily) increases of 3–4 °C cause coral bleaching and mortality. From these discussions I decided that two subjects needed investigation, (a) the spatial pattern of coral bleaching in relation to El Niño positive temperature anomalies,[4] and (b) experimental evidence that small increases in temperature would match the observed field bleaching/mortality response.

[4]Temperature anomalies, i.e., the increasing (positive) or decreasing (negative) changes in temperature, compared with the long-term thermal conditions at a given site, are preferable to simple, one time temperature measurements. This is because organisms are adapted to, and perform best physiologically, to the long term thermal conditions of their native environments. Corals living in a high temperature environment, for example at 28 °C, would likely bleach due to a sudden temperature increase to 30 or 31 °C (a 2–3 positive temperature anomaly). Corals living in a cool thermal environment, however, say at 25 °C, would probably bleach if temperatures suddenly increased to 27–28 °C (a 2–3 positive temperature anomaly). Coles et al. (1976) reported significant thermal tolerance differences for corals from tropical versus subtropical environments.

The first basin-wide sea surface temperature (SST)[5] records available for 1982–83 were monthly averaged ship observations compiled by the Inter-American Tropical Tuna Commission. In hand-plotting these SSTs, a strong spatial and temporal correspondence between extreme high temperatures and coral bleaching/mortality was apparent (Glynn 1984). Additional oceanographic observations documented a significant warming trend in the Galápagos Islands and over large parts of the equatorial eastern Pacific (e.g., Halpern et al. 1983; Rasmusson and Hall 1983). Coral bleaching coincident with elevated SSTs was also reported in Costa Rica (Cortés et al. 1984), Colombia (von Prahl 1983), and the Galápagos Islands (Robinson 1985). I vividly remember the alarm expressed by Henry von Prahl and the flurry of "snail mail" correspondence between the two of us regarding these massive coral mortality events in Colombia and Panama. The vibrant Gorgona Island coral reefs that we jointly investigated just a few years earlier were experiencing bleaching of 85% or more of the coral substrate cover (Glynn et al. 1982).

As data became available on coral mortality at different eastern Pacific localities, it was apparent that mortality increased significantly from Caño Island, Costa Rica (XG = 51%, XG is the grand mean value, i.e., the mean or average of all sample means) to the Gulf of Chiriquí, Panama (XG = 75%) and the Gulf of Panama, Panama (XG = 85%), to the Galápagos Islands (XG = 97%). This equator-ward increase in mortality corresponded with increasing SST anomalies, longer duration of warming, and increased rates of warming (Glynn et al. 1988).

Remarkably, in mid- to late-1983 we began to receive reports of coral bleaching in the Caribbean of Costa Rica (Cortés et al. 1984) and Panama (Lasker et al. 1984), the Florida Keys (Jaap 1985), Venezuela (Lessios et al. 1983), and in the Western and South Pacific (Glynn 1983; Brown 1987; Coffroth et al. 1990). Many of these bleaching events occurred during periods of sea warming, but some outside of the equatorial eastern Pacific occurred during sudden sea water cooling or extreme sea level lowering and aerial exposure events. In general, the eastern Pacific field evidence supported the co-occurrence of coral bleaching and positive sea temperature anomalies. Now it was time to determine experimentally the responses of corals to small increases in temperature.

In order to assess the effects of small temperature increases on corals, in the range of observed elevated SSTs during the El Niño event, mesocosm experiments were planned for 1985 at the STRI Naos Island marine laboratory. I discussed the relevance of this work with Luis D'Croz, Director of the marine laboratory of the University of Panama (Centro de Ciencias del Mar y Limnología) and was delighted when he enthusiastically accepted my invitation to collaborate on this project.

[5] SST is an acronym for sea surface temperature. This is the sea water temperature measured in the upper-most layer of the ocean's surface, for example remotely by satellite, ship hull sensors, moored buoys or manually by hand-held thermometers. These upper-most layer temperature measurements, at a range of depths (e.g., mm to cm), are often higher than measurements in deeper coral habitats. Due to the sensitivity of zooxanthellate organisms to small temperature differences, it is important to report the exact method and microhabitat of measurement.

Temperature tolerance experiments were conducted over a 10-week period employing *Pocillopora damicornis* as the test species from populations in the gulfs of Chiriquí and Panama. Temperature treatments ranged from ambient (normal) to stepwise (2 °C) increases to 30 and 32 °C. Corals demonstrated negative responses at higher temperatures with increasing exposure time, namely: (a) loss of zooxanthellae, (b) lower concentrations of chlorophyll *a* and coral protein, and (c) higher incidence of histopathological abnormalities. The fundamental conclusion of this experimental study was that bleaching and mortality mirrored the field responses of corals to the elevated sea temperatures observed in 1983.

Our 1985 experimental study was published in a special issue ("Coral bleaching") of the journal *Coral Reefs* in 1990, edited by Barbara Brown. Barbara, a professor at the University of Newcastle Upon Tyne, was engaged extensively in coral bleaching research in Indonesia and Thailand. Most of the studies in this special journal issue supported a close correspondence between elevated sea temperatures and coral bleaching and mortality across the tropical Pacific Ocean in 1982–83, and the Indo-Pacific and Caribbean regions during a following El Niño event in 1986–87. In addition to elevated thermal stress, some studies offered evidence that high irradiance and UV radiation could also contribute to and exacerbate coral bleaching (Brown 1987; Gleason and Wellington 1993).

Josh (Joshua) Feingold, one of my first doctoral students at RSMAES, who is now a professor in the Department of Marine and Environmental Sciences at Nova Southeast University (Fort Lauderdale-Davie, Florida) performed temperature tolerance experiments on Galápagos corals, which mimicked the responses of the corals in our experiments in Panama. In addition, Josh demonstrated that deep-living Galápagos mushroom corals (*Diaseris*) likely survived the 1982–83 El Niño warming event by exposure to periodic cooling conditions at 15 m depth (Feingold 1995).

Thermal Indices Because of the importance of exposure time to stressful elevated temperatures, a heat index has been introduced by investigators to assist with predicting coral bleaching events. Two of these indicators are referred to as "degree-heating-weeks" (Gleesen and Strong 1995) and "ocean hot-spots" (Goreau and Hayes 1994). Guillermo Podestá, a RSMAES colleague in the Division of Meteorology and Physical Oceanography, and I proposed a degree-heating-days index incorporating the summed temperatures above the long-term mean of site-specific seasonal patterns. This index was developed from long-term in situ temperature observations and observed natural coral bleaching events in the Galápagos Islands and Panama. The performance of our degree-heating-days index (spanning our 1970–1998 study period) was satisfying in that it closely tracked the duration of stressful, threshold temperatures during the four El Niño bleaching events in the Galápagos Islands—1982–83, 1986–87, 1990–92, and 1997–98—and the single bleaching event documented in the Gulf of Panama—1982–83 (Podestá and Glynn 1997, 2001; Glynn et al. 2001). A coral bleaching index was also developed by Ray Berkelmans with data from several reefs on Australia's Great Barrier Reef. Ray proposed the use of "bleaching threshold curves", which incorporated non-linear coral responses to sea warming. In this case, "non-linear" simply means that corals in different locations do not respond proportionately to a given temperature increase. This approach predicted coral bleaching with high reliability (Berkelmans 2002).

Global Thermal Extremes The first widespread coral bleaching/mortality reports occurred mainly during El Niño warming events (e.g., 1982–83, 1987–88, 1997–98). Bert (Ernest) Williams and Lucy Bunkley-Williams, both professors at the University of Puerto Rico Mayagüez, kindly shared a manuscript with me in which they linked the 1982–83 and 1987–88 events with global warming (Williams and Bunkley-Williams 1990) (Fig. 4.2). They listed the morbidity and mortality of numerous reef-associated organisms (e.g., coralline algae, sponges, bryozoans, mollusks, echinoderms, and tunicates) during these mass bleaching events, in addition to zooxanthellate corals; they proposed the term "coral reef bleaching" to more accurately describe the broad scope of this disturbance. Their in-depth study also revealed a strong link with elevated sea temperatures as the causative factor of coral reef bleaching, which prompted them to conclude: "We believe the world-wide coral reef bleaching complex "cycle" is caused by increased global temperatures of the 1980s.". The conceptual model proposed by Bert and Lucy showed a global warming trend coupled with decreasing resilience of coral reef organisms (due to anthropogenic stressors), yearly seasonal temperature increases, with the latter becoming more stressful as they are nudged upwards by increasing global warming (Fig. 4.2). Global warming is now unyielding, often exacerbating El Niño warming stress events, and also contributing to the occurrence of marine hot spots and heat waves that are responsible for mass coral reef bleaching across all tropical seascapes (e.g., Goreau and Hayes 1994; Hoegh-Guldberg 1999; Eakin et al. 2010; Fordyce et al. 2019). Catastrophic coral mortality in Florida and globally during the record-breaking northern hemispheric warming in 2023, coincident with strong El Niño development, is discussed further in Chap. 13.

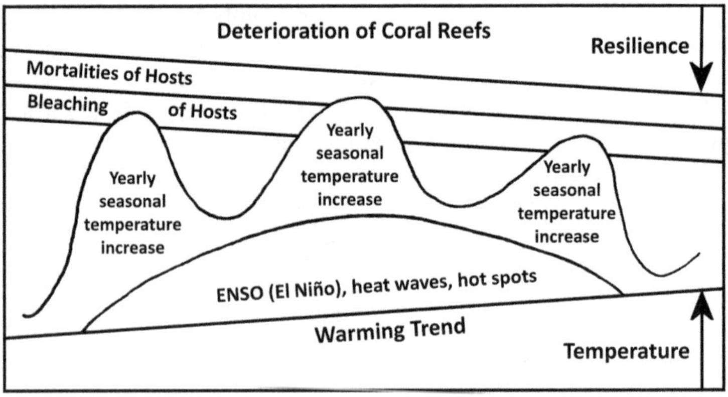

Fig. 4.2 Conceptual model of increasing thermal stress adversely impacting global coral reefs. Modified after Williams and Bunkley-Williams (1990)

Salinity, Sedimentation, Turbidity, and Nutrients

At least two, and often four, of the factors herein considered vary concurrently during upwelling and high riverine input. For example, in coastal areas influenced by river discharge, heavy rainfall resulting in a sudden influx of river water could quickly lower salinity and increase turbidity, suspended sediment loads, and nutrient levels. Such conditions frequently occur in the wet season in the eastern Pacific equatorial zone and sporadically in western Mexico (from 12 to 25°N) during the summer tropical cyclone season. Remarkably, modest coral reef development occurs in the coastal Chocó area and at Gorgona Island, Colombia, a region of exceptionally high rainfall (>7 m annually) and freshwater discharge (maximum rate of 17,650 m^3/s (Glynn and Ault 2000). Two of the largest Colombian coastal reefs are La Chola (10.5 ha) and Tebada (4.5 ha), constructed predominantly by *Pocillopora* corals but with other eastern Pacific species present, e.g., *Pavona* spp., *Porites* spp., and *Psammocora* spp. (von Prahl and Erhardt 1985; von Prahl 1986). More recent studies have documented declines of non-pocilloporid corals, likely due to disturbances associated with the 1982–83 El Niño event (Vargas-Ángel 1996; Vargas-Ángel et al. 2001). Probing of the Tebada reef by Bernado Vargas-Ángel indicated a vertical pocilloporid framework thickness of 4 m, which could represent 2000 years of accretion.[6] Reef formations at Gorgona Island, located approximately 35 km from the Colombian mainland, are more developed than the coastal reefs (Zapata and Vargas-Ángel 2003). All of the Gorgona Island reefs are also constructed of *Pocillopora* corals, and two of the largest reefs (La Azufrada and Playa Blanca) are mature fringing reefs, each extending about 1 km along the coast. The maximum measured vertical frameworks ranged from 6.0 to 8.3 m, indicating possibly 3000–4000 years of accretion (Fig. 1.9). If salinity, sedimentation and other coastal disturbances have not appreciably influenced insular reef growth, it is possible these reefs are younger than those on the mainland.

Two observations in Panama are compelling regarding the high tolerance of eastern Pacific corals to poor water quality conditions. The first of these involved corals that were held in flow-through sea water tanks at the STRI Naos marine laboratory. The sea water supplied to the tanks was pumped from Panama Bay; sediments were removed in settling tanks and by passing the water through cloth filter bags. Four coral genera (*Pocillopora*, *Porites*, *Pavona*, and *Psammocora*), with multiple species of *Pocillopora* and *Pavona*, were held in the tanks for several months as prey for the sea star predator *Acanthaster*. We were conducting prey-choice experiments and needed a steady supply of various coral species. Daily measurements of the salinity, beginning 1 September 1973 (in the wet season), continued until 30

[6]This estimate of vertical reef growth is based on an accretion rate of 2 mm/year determined for a *Pocillopora* reef in Panama (Glynn and Macintyre 1977). Vertical accretion estimates in Panama ranged from 1.6 to 7.5 mm/year. A low-end rate was selected for this Colombian coastal reef due to inevitable disturbances and interrupted reef growth on the Chocó coast (Zapata and Vargas-Ángel 2003).

November. The initial salinity was 26 PSU, which dropped to 20 PSU over the 3-month period (Glynn 1974). Two notably low daily measurements, one in October and one in November, were both 19 PSU. All coral species held continuously in low salinity conditions for an additional 2 months (5 months total) remained healthy, showing no pigment loss or other signs of stress. Colonies of *Pavona gigantea*, among the corals under observation, gained 6.1% in wet mass over 5 months. Kleypas et al. (1999) observed that the mean salinity reported for global reefs was 34–35 PSU. Eastern Pacific corals were remarkably tolerant of low salinity conditions.

The second observation of coral reef development under extreme conditions was serendipitous, an unexpected discovery of coral reefs (small patch reefs) where they should not be. Sometimes our scientific crew would travel by automobile from Panama City to meet our research vessel in the Gulf of Chiriquí. Our rendezvous location was in a major estuary just inside Punta Entrada. On one occasion, before pulling anchor and motoring to our offshore island study reefs, we noticed an odd brown patch near the shoreline close to our anchorage; it was an extreme low tide on this day. Our ship's captain (David West) gave us permission to swim to the structure and investigate. Eureka!, a living pocilloporid patch reef. All colonies were a vibrant brown to golden-brown in color with no signs of bleaching. Further, several species associated with pocilloporid reefs on islands offshore, such as guinea fowl puffers and diverse symbiotic coral crustaceans, were present on this reef. The water clarity was very poor, about 1–2 m horizontal visibility. We estimated that the coral frameworks were 1–2 m in height. Other exposed pocilloporid patches were visible along the estuarine shoreline, but we did not have an opportunity to verify their identity and map their locations. We needed to abide by the ship's schedule. The seaward-flowing surface waters were brackish to fresh (salt-free) to the taste. We suspect that these patch reefs had developed in bottom, high salinity water, i.e. under salt-wedge estuarine conditions. Unfortunately, our plans to return to Punta Entrada to study this unusual occurrence of coral development were never realized.

My RSMAES students, colleagues, and I were especially excited with the upcoming Eighth International Coral Reef Symposium, scheduled to be held in Panama City, Republic of Panama, during June 24–29, 1996. This would give us an opportunity to showcase our research findings to an international group of experts, documenting the occurrence of coral reefs in the equatorial eastern Pacific region (Glynn et al. 1972); and presenting detailed information on their biology and geology. In addition, we would be able to offer post-symposium field trips to our prime research sites in the Gulf of Chiriquí and the Gulf of Panama. Juan Maté and I had written a detailed field guide to the coral reefs of Pacific Panama with information on the environmental setting and several photographs of reefs; coral species and other organisms, 40 figures in all, were illustrated in the guide (Glynn and Maté 1997). Several coral reef researchers signed up to participate in the field excursions. During the symposium in Panamá City it began to rain; it rained every day, all day, then we began to receive reports of flooding in the province of Chiriquí. We began to worry about the flooding and how this might affect water clarity at our planned visitation sites. Several large rivers draining the Talamanca Mountain Range enter the deltas south of the cities of David and Pedregal, and the Punta Entrada coastal area. Our worst fears were realized. Underwater visibility at two prime reef sites in the Contreras and Secas Islands, 20–30 km offshore of the mainland, was 0.5–1 m! Due to this extremely reduced visibility, I asked all divers to remain close to one another. It was too close for Peggy Fong, one of our field guides, who was kicked smartly on the head by an unidentified diver. She still reminds me of this mishap 25 years after the incident. Regrettably, the extreme load of suspended sediments did not allow us to proceed with our eastern Pacific coral tutorial. Although coral bleaching, tissue necrosis, and mortality have been reported in the eastern Pacific, due to excessive sedimentation, these adverse effects often resulted from humankind's activities (Cortés 1990; Zapata and Vargas-Ángel 2003). Water clarity returned to normal at our two reef study sites and throughout the Gulf of Chiriquí after 2–3 weeks with no signs of coral bleaching or mortality.

Increasing nutrient concentration often accompanies high sediment loading from terrestrial sources, and this can benefit plant growth in benthic and plankton communities. Nutrients are also brought to sunlit surface waters during upwelling, which in high concentration can stimulate phytoplankton blooms and "red tides". In a few instances, red tides dominated by toxic dinoflagellates have been reported in equatorial eastern Pacific waters with impacts to coral reefs. In the second half of 1985, a mass mortality event occurred at Caño Island, Costa Rica, affecting fishes (e.g., parrotfishes, triggerfishes, and pufferfishes) and diverse invertebrates, especially pocillopord corals with up to 100% mortality to 3 m depth (Guzmán et al. 1990). Héctor Guzmán, one of Jorge Cortés' graduate students, first observed this event in Costa Rica, alerting us to check for similar red tide effects in Panama. The water surrounding Uva Reef in Panama was red-brown and replete with viscous foam, likely produced by the dense concentrations of living and decaying dinoflagellates. Reef fishes were not affected on the Uva reef, and live pocilloporid coral cover was reduced by only 12.8%. *Pocillopora* spp. bleaching was observed at La Chola Reef in Colombia in late 1989, during a dinoflagellate bloom, with recovery of normal pigmentation a week later (Vargas-Ángel 1996). These algal blooms occurred during the intense 1984–85 and 1988–89 La Niña events, which respectively followed closely after the 1982–83 and 1986–87 El Niño disturbances (Glynn et al. 2017). Coral reef recovery in Costa Rica experienced a setback due to these secondary water bloom mortality events. As if El Niño elevated and lethal thermal effects were not enough, rebounding La Niña occurrences of extreme low temperatures, plankton blooms, and aerial exposures also contribute to ENSO coral mortality.

An example of an indirect effect of elevated nutrients, stimulating algal overgrowth of corals, is under investigation at Rapa Nui (Easter Island). A recent study of deep coral communities has revealed filamentous mats of cyanobacteria and green and brown algae overgrowing dense aggregations of *Leptoseris* corals (Sellanes et al. 2021). This recent discovery employed a remotely operated vehicle (ROV) equipped with a video camera that recorded incipient reefs of *Leptoseris* spp. between 70 and 117 m depths. Our surveys at Rapa Nui, conducted with scuba, were restricted to 60–65 m depth and revealed only dispersed colonies of *Leptoseris*

(Glynn et al. 2003). The most serious algal overgrowth effects were observed in the ROV surveys at 70 and 95 m off Hanga Roa, the main settlement on Rapa Nui. Sellanes and co-workers hypothesized that this condition was the result of anthropogenic eutrophication, the direct introduction to coral communities of organic matter, nutrients, and contaminants by runoff and percolation through volcanic rock aquifers. Since the 1990s, the human population of Hanga Roa has increased from 2770 to 7750 souls, and visitors from 20,000 to 150,000 per annum, the latter number reported just before the Covid-19 pandemic.

In terms of these four water quality conditions it appears that individually they, or their combined effects, have not prevented at least moderate construction of coral frameworks in the eastern Pacific region. Impacts on reefs are reported more frequently in recent years, likely a result of elevated contamination of natural water quality due to humankind's relentless coastal development, deforestation, nutrient inputs, and other disturbances.

Nutrients need not always be harmful, however. As shown by Joerg Wiedenmann's research team at the University of Southampton, moderately elevated levels of dissolved inorganic nitrogen and phosphorus can increase the fitness of zooxanthellae and their proliferation rates, which can improve coral survivorship and recovery from bleaching (D'Angelo and Wiedenmann 2014). Of key importance is that the nutrients are available in a favorable stoichiometry (Buckingham et al. 2022). Moreover, by feeding on and utilizing the excess metabolites of their expanding symbiont populations, the coral hosts can tap into the pool of dissolved inorganic nutrients that is otherwise not accessible; corals use the vital N and P to sustain their own growth (Wiedenmann et al. 2023). Such a situation likely occurred during the 2015–16 El Niño event in the northern Galápagos Islands when nutrient-rich internal waves ascended into shallow coral habitats. Surviving zooxanthellae from bleaching were now exposed to cool, nutrient-rich waters. This effect could kick their metabolic machinery into high gear, preventing further bleaching and allowing the corals to regenerate (Riegl et al. 2019a). The Wiedenmann team's most recent study demonstrated that coral hosts can consume (eat) zooxanthella cells in addition to benefiting from the translocation of nutrients (transfer of fixed carbon) from zooxanthellae.[7]

Tidal Aerial Exposures

Although not considered by Kleypas et al. (1999) as a factor affecting global reef development, extreme low tidal exposures can result in massive and widespread coral bleaching and mortality. Coral mortality occurs in the vertical dimension and has been reported frequently in the equatorial eastern Pacific, namely in Costa Rica,

[7] Robert van Woesik in a NOAA coral list message acknowledged the important advances made by Wiedenmann's group but noted that early workers also found that corals hosting zooxanthellae could consume these endosymbionts (e.g., Boschma 1925, page 429 and references therein.)

Panama, Colombia, and Ecuador (Cortés 2003). The equatorial region experiences the greatest tidal amplitude in the eastern Pacific, ranging from 2.50 to 3.25 m (Fiedler and Lavín 2017). In the Gulf of Panama, the maximum tidal range is 6 m, and 5 m in the Gulf of Chiriquí. Tide-related coral mortality in the equatorial eastern Pacific can be likened to a "perfect storm" with stressful conditions occurring during astronomical extreme low tidal exposures in the dry season, on cloudless, midday periods, coincident with La Niña low sealevel events. During these approximately 2-h exposures, which typically recur over a 3–4-day period, corals are subject to desiccation, extreme heating, and intense solar radiation. Reef flat coral mortality in Panama during subaerial exposures ranged from 61 to 70% in 1974–75 (Glynn 1976), and 97% in 1989 (Eakin and Glynn 1996). Nocturnal aerial exposures can also result in coral mortality. Juan Maté reported a midnight exposure event during heavy rains in May 1999 that resulted in massive coral bleaching and mortality (Maté 2003).

Numerous reef organisms besides corals are affected by extreme aerial exposure events (Eakin and Glynn 1996; Zapata and Vargas-Ángel 2003). Like corals, crustose coralline algae also bleach and die after prolonged exposure. Due to the substrate irregularity of reef flats, receding tides often leave pools with restricted water exchange that quickly heat up and become oxygen depleted. Animals trapped in these pools, such as coral crustacean symbionts and fishes, succumb in response to the stagnant conditions.[8]

I was stunned and disheartened when I first witnessed these widespread mass mortality events. Their magnitude was immense as one could follow the tidally-exposed band of white coral skeletons along the shoreline as far as the eye could see. As dreadful as this mass mortality event seemed, I came to realize that it was transitory, characteristic of coral-dominated reef flats as they accrete vertically and approach sealevel (Thornborough and Davies 2011). Joshua Feingold and I set up a temporary platform to measure irradiance levels during an extreme tidal exposure event on the Uva Island coral reef (Fig. 4.3). Adversely affected reef flat coral communities resulting from tidal exposure events are subject to physical and biological erosive processes, but often recover. Surviving coral branches (e.g., from basal

[8] In the February 1989 exposure event at Uva Island (Gulf of Chiriquí), I decided to observe conditions in some of the affected pools during the following rising tide. With the increasing water level and wave agitation, coral mucus strings and tissues began lifting from moribund and dead pocilloporid corals. Within a few minutes numerous fishes began swimming onto the reef flat. They immediately began consuming mucus and loose tissues, and even nipping at damaged coral colonies. Interestingly, fishes biting coral colonies selected only branches with damaged tissues; healthy tissues were not targeted. The following species engaged in this scavenging mode of feeding: the Spotted Sharpnose Puffer (*Canthigaster punctatissima*), the damselfish Beaubrummel Gregory (*Stegastes flavilatus*), the King Angelfish (*Holacanthus passer*), the Cortez Rainbow Wrasse (*Thalassoma lucasanum*), and the butterflyfish Barberfish (*Johnrandallia nigrirostris*). After 2–3 weeks, the formerly stark-white bleached corals began to turn green, a result of algal colonization of the dead skeletons. These algal encrusted corals were now subject to the grazing activities of fish herbivores: the Convict Surgeonfish (*Acanthurus triostegus*), the Bluechin Parrotfish (*Scarus ghobban*), and damselfishes, the Panamanian Sergeant (*Abudefduf troschelii*) and Acapulco Gregory (*Stegastes acapulcoensis*).

colony tissues) continue growing asexually and sexual larval recruitment augment recovery. Absent further disturbances, over a period of about 10 years, corals continue to grow vertically to sea level, thus restoring/revitalizing the reef flat zone.[9] Repeated cycles of mortality, erosion, transport, and re-growth of pocilloporid corals must generate enormous amounts of carbonate sediments to replenish and add to the expansion of reef sediment habitats.

Tropical Storms

Among the most frequent physical disturbances in the eastern Pacific are tropical cyclones that occur annually off western Mexico, originating over the eastern Pacific warm pool (10–12°N) from May through November (Fiedler and Lavín 2017). Coral communities and reefs in Oaxaca, Nayarit, and the Gulf of California are usually impacted less than once annually, however, the offshore, oceanic Revillagigedo Islands are subject to frequent cyclonic disturbances, with as many as three storms per annum (Reyes-Bonilla 2003).

Just 18 months after Gerardo Leyte Morales and I completed initial surveys of the Huatulco coral reef tract (Glynn and Leyte Morales 1997), three major cyclones impacted this coastal locality consecutively within a 6-week period. These cyclones were associated with disturbances that occurred during the very strong 1997–98 El Niño event. Gerardo, a professor at the Universidad del Mar in Puerto Ángel (Oaxaca), joined our studies at Huatulco and offered essential support in describing reefs and documenting storm damage. The most severe storm, Hurricane Pauline, was a category 4 storm with maximum sustained winds of 213 km/h. Remarkably, damage to the predominantly pocilloporid reefs in this area was relatively minor.

[9] Comparing eastern tropical Pacific sea level rise of 0.16 cm/year for the period 1950–2000 (Bindoff et al. 2007, Intergovernmental Panel on Climate Change, IPCC) with *Pocillopora damicornis* average vertical branch extension of 2.6 cm/year (Glynn 1977), it is probable that coral recovery could occur within a decade. In 10 years, projected sea level rise would be 1.6 cm higher than present and coral growth 26 cm higher, greatly outpacing sea level rise. This projection assumes high coral cover and continuing rapid growth and recovery following tidal exposure mortality events. It is quite certain that sea level rise will accelerate in the coming century. If the high end of the IPCC projected rate of ~0.4 cm/year is realized a sea level rise of 4 cm per 10 years would still lag substantially behind upward coral growth. A recent study of the carbonate budget and accretion rate of a pocilloporid reef in southern Mexico also concluded that vertical reef growth could keep pace with projected sea level rise (Cabral-Tena et al. 2023). These projections of near linear sea level rise are, unfortunately, probably too low considering the continuing high fossil fuel emissions occurring this century. Two modeling studies, based on paleoclimate data and increasing greenhouse gas emissions, have predicted accelerating Antarctic and Greenland ice melt and eventual shutdown of the AMOC (Atlantic meridional overturning circulation), increasing superstorms, and sea level rise of several meters over a timescale of 50–150 years (Hansen et al. 2016; Ditlevsen and Ditlevsen 2023). This is deeply concerning and would result in a "give-up" reef growth mode with global reefs quickly becoming drowned reefs subject to intensified storm-related erosion (Neumann and Macintyre 1985).

Fig. 4.3 Joshua Feingold and Peter Glynn conducting underwater irradiance measurements at the Uva Island coral reef (20–21 February 1992). Joshua (foreground) on the tidally exposed reef flat is recording data and Peter (in water) is positioning the sensor over corals at various depths. Photograph courtesy of J. S. Feingold

We had anticipated significant destruction to the delicately branching pocilloporid colonies and reef frameworks at this locality.

Damage to corals along the southern Mexican coast was patchy in severity and spatial distribution as documented by Diego Lirman and team (Lirman et al. 2001). Diego, one of my colleagues at RSMAES, led our post-storm surveys at Huatulco. The fragmentation of coral branches and dislodgment of framework blocks were the most common types of damage observed. Prior sea urchin and sponge bioerosion likely enhanced the dislodgment and toppling of large 1–3 m pocilloporid blocks and massive corals, such as species of *Porites* and *Pavona*. Some reefs exhibited minimal damage, particularly those situated in bays and thus partially protected from storm-generated waves. Of six reefs quantitatively sampled, three lost significant amounts of coral and three reefs actually gained live coral cover. The increases were small and due to live, broken pocilloporid branches that were swept over and settled onto reef surfaces. Re-surveys of the Huatulco storm-damaged reefs would be worthwhile for determining the extent of recovery and manner of coral regrowth. Perhaps such knowledge could be used to recognize traces and frequency of past storm damage.[10]

[10] Such a study may not now be possible in light of the severe 2023 El Niño coral mortality event across the Huatulco reef tract (López-Pérez et al. 2023, 2024).

Clipperton Atoll lies close to the southern limits of eastern Pacific cyclonic activity. When I visited the atoll in 1994 there were abundant signs of physical damage to corals, for example uprooted and toppled massive colonies to 10–20 m depth, and large displaced colonies on the reef flat and strewn along the strand line (Glynn et al. 1996). This was especially evident in the northwestern sector of the atoll. Since so little is known of the storm history of this remote and uninhabited atoll, it is difficult to determine if this damage was due to passing nearby storms or to distantly generated swells. Also, the extent bioerosion contributed to the dislodgement of corals is unknown.

Gary Robinson endured the 1982–83 El Niño event in the Galápagos Islands and made eyewitness observations on coral mortality resulting from warm water bleaching and physical wave forces (Robinson 1985). When normal south and southeastern swell directions were temporarily reversed to the northwest, Gary observed that large colonies of massive corals (*Porites* and *Pavona*) were dislodged from shallow subtidal areas and deposited above the high tide line. During our 1975 Galápagos surveys, John Wells remarked on the numerous, long-dead massive corals present on beaches above the high tide line. He wondered how Charles Darwin missed seeing these large, displaced coral colonies. Alexander Agassiz (1892) also remarked on the abundance of reef rock high and dry on most Galápagos beaches. Several years later when I was sampling the population densities of the sea urchin *Eucidaris galapagensis* it was necessary to tilt large, massive corals on their sides to expose sea urchins seeking shelter under and within the colonies. To my amazement, I found that most large coral colonies were not firmly attached to the substrate and could be easily tilted to one side or the other. The colonies were relatively light and mere shells with large interior cavities. I soon stopped tilting the colonies realizing how vulnerable they were to movement and transport by a reversal of wave direction. A question immediately arose: Why were these large coral skeletons so hollowed out and light? In addition to the bioerosive effects of boring sponges, date mussels, and sea urchins, was there another erosive effect at play? Could "ocean acidification" provide a clue? Read on.

Summarizing, exacerbating storm damage are two potentially interacting effects that could play important roles: (a) reversal of wave and swell direction, and (b) the weakening of coral frameworks by bioeroders (and ocean acidification?) prior to storm events. Storm damage is generally local and a usual part of reef growth off western Mexico and in the equatorial eastern Pacific where El Niño impacts occur most frequently.

Seismic Events

Our first experience with a notably unusual consequence of a seismic event was at Urvina Bay on the western shoreline of Isabela Island in the Galápagos Islands (Figs. 1.7 and 1.8). This was in 1975, about 20 years after the uplift along the coast of the Alcedo volcano. The emplacement of magma beneath Urvina Bay uplifted a portion of the bay nearly 7 m above sea level and moved the shoreline about 1.2 km seaward. An entire living coral community was suddenly exposed to the scorching tropical sun, which resulted in a massive, putrefying die-off. Walking through the white, ghostlike corals in the heat of the day, avoiding thick patches of shrubs and catching glimpses of land iguanas, was a surreal experience. The coral skeletons were in notably good condition and the different species could be readily identified (Fig. 4.4). In addition, sea urchin and mollusk species engaged in bioerosion before the uplift event were still present and their effects readily apparent (Colgan 1990). We were hopeful that the Galápagos National Park would endeavor to protect this unique stranded coral community.[11] In 2012 we were fortunate to be able to re-survey subtidal sites along the Urvina Bay shallow shelf with M/Y *Golden Shadow* ship support, courtesy of the Khaled bin Sultan Living Oceans Foundation.[12] Live coral cover was sparse with no signs of recovery approaching the abundance and diversity of the uplifted coral community prior to the 1956 seismic event.

Strong earthquakes have occurred relatively frequently in the Tumaco area, about 200 km southwest by south of the coral reefs of Gorgona Island. The 1906 temblor, with an estimated magnitude of 8.4–8.8, did not result in any apparent structural damage when we surveyed the Gorgona Island reefs in 1979. A second strong earthquake occurred near Tumaco in December 1979 (magnitude 7.9–8.1), which resulted in nearby sea floor subsidence of 0.8 m. Neither subsidence nor the accompanying tsunami caused reef damage at Gorgona Island (Zapata and Vargas-Ángel 2003).

The 1906 "Miracle of Tumaco" again spared the fringing reefs of Gorgona Island, and human lives (see Chap. 2, Colombia). Not so at Aceh Province and other coastal areas of the Indian Ocean in 2004. Hundreds of square km of shallow reef areas were destroyed and tragically a devastating toll of human death and suffering occurred, with an estimated 232,000 lives lost and 1.2 million people displaced (Foster et al. 2006).

The Galápagos (Urvina) and Indonesian (Aceh) examples of tectonic uplift resulted in coral mortality over large areas. What about earthquakes that cause subsidence or the sinking of reefs? A study of a tectonic subsidence event in the Solomon Islands actually demonstrated a positive effect in the shallow reef flat zone

[11] Inti Keith, Director of the Charles Darwin Research Station, informed me (on 6 November 2023) that the Galápagos National Park now requires permission to undertake scientific studies on the Urvina Bay uplifted shelf. Tourists may observe exposed subfossil corals from designated coastal trails, but only under the supervision of Galápagos naturalist guides.

[12] The Khaled bin Sultan Living Oceans Foundation, founded by His Royal Highness Prince Khaled bin Sultan of Saudi Arabia, is a non-profit research organization established to help preserve, protect, and restore the health of the world's oceans. Our research in the Galápagos Islands was conducted with support of the 67 m research vessel M/Y *Golden Shadow*. Thanks are extended to the captain and crew and the ship's research team for facilitating our studies, especially A. Bruckner, P. G. Renaud, and B. Beck. In addition to our research goals, the Galápagos mission was one of several circumtropical surveys conducted over a 10-year period to evaluate the condition of coral reefs following recent physical and biotic disturbance events. Lamentably, after global reef surveys and mapping, including ~20% of the world's reefs, it was concluded that coral reefs and reef fish populations are being lost or degraded at a frightfully high rate (Carlton et al. 2021).

Fig. 4.4 Examining corals at an Urvina Bay shelf uplift site, Isabela Island, Galápagos Islands, P. W. Glynn, Gerard M. Wellington, and John W. Wells (left to right), 15 January 1975. Photograph courtesy of Charles Birkeland

(Saunders et al. 2015). The sudden sinking of the reef resulted in a relative sea level rise of 0.6 m, which offered new space for corals (*Acropora* spp.) to expand vertically and horizontally, thus increasing coral cover. This study suggested that coral reef accretion could keep pace with sea level rise to the year 2100 based on a projected sea level rise of 0.6 m. It was rightly noted, however, that the rapidly deteriorating environmental conditions associated with global climate change and coral mortality would interfere with reef growth and prevent this scenario in the future.

Ocean Acidification

Ocean acidification (OA) refers to increasing greenhouse gas emissions, primarily CO_2, which is absorbed into the world's oceans and results in a corresponding increase in acidity (lower pH). This condition is due primarily to the accelerating use of fossil fuels and the destruction of forested lands. OA ultimately interferes with the calcification or skeletogenesis of marine organisms. Upwelling regions, such as the eastern tropical Pacific, experience lower pH values than non-upwelling environments. This is a result of the surfacing of CO_2-saturated waters from below the thermocline (caused by the decomposition of plankton and other organic matter). During upwelling this CO_2-enriched water lowers the pH of surface waters and depresses calcification of shallow-occurring coral reef organisms.[13,14] The aragonite saturation state Ω_{arag} is a measure of the availability of the carbonate ion (CO_3^{-2}) for marine organisms. The carbonate ion is an essential component of calcareous ($CaCO_3$) skeletons in corals and many other calcareous reef organisms. Thus, the oceanic uptake of anthropogenic CO_2 and the subsequent decline in Ω_{arag} reflects the amount of increasing acidification and OA's potential impact on calcareous organisms.

During the *Golden Shadow* cruise, a high priority research objective was to investigate the integrity of coral communities in the northern and southern Galápagos Islands. A key question centered around the condition and recovery potential of coral communities and coral reefs impacted during the 1982–83 El Niño event. Before retiring one evening, I found Derek on the stern deck of the *Golden Shadow* measuring the pH of water samples collected from over coral communities in the southern Galápagos Islands. He was incredulous of the low values he was recording.

---

[13] The absorption of CO_2 by sea water reacts with H_2O to form aqueous carbonic acid (reaction 1) which dissociates to bicarbonate and hydrogen ions (reaction 2), and in turn to carbonate and hydrogen ions (reaction 3). As the acidity increases with the release of hydrogen ions (H^+), the concentration of CO_3^{-2} quickly decreases. It is apparent from the basic reaction of calcification (reaction 4) that a lowering of the carbonate ion concentration would interfere with the precipitation of $CaCO_3$ and hence coral skeletogenesis. Calcium ions are in relatively high concentration in sea water and not limiting.

1. $CO_2 + H_2O \leftrightarrow H_2CO_3$
2. $H_2CO_3 \leftrightarrow HCO_3^- + H^+$
3. $HCO_3^- \leftrightarrow CO_3^{-2} + H^+$
4. $Ca^{+2} + CO_3^{-2} \leftrightarrow CaCO_3 \downarrow$

Aragonite saturation state (Ω_{arag}) is determined by the equation $\Omega_{arag} = [Ca^{+2}]\,[CO_3^{-2}] / K_{sp}$ where the product of the concentrations of calcium and carbonate ions is divided by the solubility product of the aragonite mineral. The subscript "arag" refers to the special crystalline form of calcium carbonate characteristic of coral skeletons. The degree of aragonite saturation varies with latitude, e.g., $\Omega_{arag} \approx 3.3$–4.1 at tropical coral reef locations with robust reef development, and $\Omega_{arag} \approx 1.5$ at the earth's polar latitudes. This difference is due primarily to the effect of higher temperatures in equatorial waters. The threshold concentration of Ω_{arag} supporting coral communities is as low as 3.1 (Kleypas et al. 1999). The Ω_{arag} equation indicates that a lowering of the concentration of CO_3^{-2} (all other things being equal) would therefore decrease the rate of skeletogenesis.

[14] A meta-analysis by Chan and Connolly (2013) found that, on average, the calcification of corals declines by ~15% per unit decrease in aragonite saturation. The lower temperatures and increased nutrient levels of upwelled waters impact calcification rates in addition to the pH drop. Teasing out the individual impacts of these co-occurring stressors is quite difficult. Predicting their combined outcome on coral growth is even more difficult. Cumulative and synergistic stress responses abound, and even those relationships may be influenced by exposure history, or even species/genotype specific tolerances.

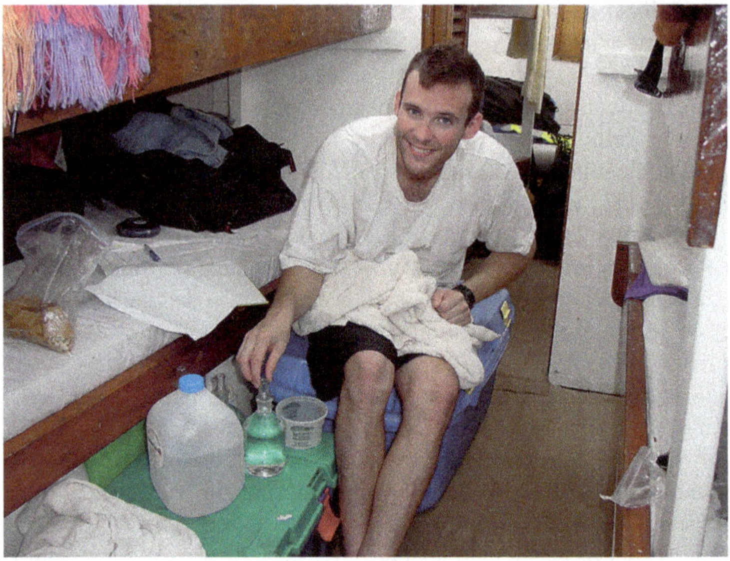

Fig. 4.5 Derek Manzello, in close quarters in the *Pirata* cabin fixing water samples for analysis, Santa Fe Island anchorage (13 May 2003). Photograph courtesy of J. S. Feingold

Mean pH and calculated values of Ω_{arag} ranged from 7.90 to 7.98 and 2.4 to 3.0 respectively (Manzello et al. 2014). The Ω_{arag} levels were considerably lower than the levels thought to be critical for coral reef development ($\Omega_{arag} > 3$). In earlier studies, Derek found that the intraskeletal pore spaces of remnant coral frameworks were devoid of inorganic cementation (Manzello et al. 2008; Manzello 2010). In the northern-most Galápagos Islands, the Wellington Reef at Darwin Island was in full recovery in 2014 at mean pH and Ω_{arag} values of 8.02 and 3.3 respectively (Glynn et al. 2015). Perhaps the unattached corals and the shell-like skeletons of massive corals observed at Devil's Crown (Floreana Island) were a result of the more corrosive conditions observed at this site? The fact that calcified corals were present at Devil's Crown suggests that OA will not likely cause mortality, but the effects could be subtle and sublethal, impacting the maintenance and integrity of reef frameworks. Derek's working space was much restricted aboard our usual support boat, the M/V *Pirata* in the Galápagos Islands (Figs. 4.5 and 4.6).

The consequences of diminished calcification resulting from OA are manifold. Weakening of a protective skeleton could render corals more vulnerable to predation. A reduction of the skeletal mass and density could increase the susceptibility of coral colonies to bioerosion and storm damage. In addition, reduced coral cementation would destabilize the colonies and associated crustose coralline algae that assist in the binding of calcareous reef frameworks. Based on current CO_2 trajectories, some investigators have predicted that most coral reefs will experience a functional collapse from warming and acidification by mid-century (van Hooidonk et al. 2014). Klein et al. (2022) are in support of the negative consequences of OA, but

Fig. 4.6 The M/V *Pirata* at anchorage off Punta Cormorant, Floreana Island, Galápagos Islands (11 June 2007). Left to right, Captain Lenín Cruz, Ángel Cedeño (cook), Diego Ruiz (research assistant) and Jorge Velez (crewman) (behind boom), Gina Shure and Amber Little (in front of boom). Photograph courtesy of Joshua S. Feingold

conclude its effects are primarily additive and emphasize the dominant role of intensifying marine heat waves.

Coral Infectious Agents

Extreme physical conditions weakening biota and compromising species' immune systems can render affected populations vulnerable to infectious diseases and elevated parasite loading (Peters 2015). Pathogen and parasite responses to climate change vary depending on the temperature sensitivity of the interacting species. Changing environmental conditions, perhaps including salinity, OA and other factors besides temperature, may be detrimental to both an infectious agent and a potential host species (Byers 2021). Increasing attention has been directed to marine diseases, particularly to those associated with and following widespread coral bleaching and mortality during the El Niño 1982–83 and 1997–98 extreme sea warming episodes (Harvell et al. 1999). With firsthand knowledge of the frightful variety of coral and octocoral diseases in the wider Caribbean, and the widespread coral loss they have caused since the 1970s, the subject of climate change and the potential threat of diseases affecting eastern Pacific coral reefs is very real. The great variety of coral diseases and reef organisms affected, and some of their major ecological ramifications are presented in a compelling review of the subject by Esther Peters (2015). There is overwhelming evidence that increasing sea water temperature is one of the principal drivers of coral diseases globally (e.g., Rosenberg and Loya 2004; Weil and Rogers 2011).

In contrast to the prevalence of coral diseases observed on reefs in the western Atlantic and Indo-Pacific regions, corals in the eastern Pacific were relatively unblemished and disease-free until about the turn of the century. Then increasingly, lesions and coral tissue discoloration were reported during stressful El Niño elevated temperatures and the occurrence of La Niña low temperature events. Two of my colleagues, Mariana Vera and Stuart Banks, formerly of the Marine Ecosystem Research Programme at the Charles Darwin Research Station (Galápagos Islands), drew my attention to the high frequency and variety of gross coral anomalies, potentially indicative of disease or parasitic infections (Work et al. 2016).

Mariana and Stuart concentrated their surveys of coral health anomalies at three of the northern-most Galápagos Islands (Marchena, Wolf, and Darwin), far removed from human habitation and considered to be close to pristine condition in terms of health. Of the six coral species evaluated in eight coral communities over three monitoring cruises from 2005 to 2007, the overall prevalence of abnormalities (colonies presenting disease signs) was 23.9% (Vera and Banks 2009). Affected corals with abnormalities were present in all surveyed coral communities. Some of the abnormal conditions recorded were "discoloration tissue thinning" in *Porites*, *Porites* trematodiasis, *Porites* abnormal overgrowth, *Pocillopora* white band, and *Pavona* white spot. *Porites* trematodiasis was the most common health anomaly with 32% overall colony prevalence. This is a parasitic disease caused by the larval stage of a digenetic trematode. The larvae encyst in the tissues of *Porites* spp. and are visible on the coral's surface as "pink pimples." While the observed signs match this disease, Mariana and Stuart acknowledged that laboratory study is necessary for confirmation. Our team has observed the characteristic pink spots on *Porites* colonies in the southern Galápagos Islands, coastal Ecuador, and Panama. Also, in 2018 a high trematodiasis infection rate (27%, 31% incidence) was reported in massive *Porites* and *Pavona* colonies at remote Clipperton Atoll (Pogoreutz et al. 2021). Interestingly, the tissue infections of this disease alter the growth of the coral skeleton preventing the normal, full retraction of coral polyps. Studies in Hawaii have demonstrated that this abnormality renders exposed polyps susceptible to predation by butterflyfish (Aeby 2002). The butterflyfish are not harmed by ingesting the parasites but serve as vectors for spreading the disease to uninfected colonies. We have observed butterfly-fish feeding on coral polyps in the eastern Pacific, but not targeting the pink spots of *Porites*. Several other fishes from different families, however, for example parrotfishes, pufferfishes, and damselfishes, frequently bite and remove the skeleton and soft tissues of a variety of corals, including those with pink spots (Enochs and Glynn 2017). These fish corallivores also could act as vectors of infectious pathogen transmission.

Many of the lesions and abnormalities observed on corals are not necessarily disease-related but a result of dead tissue sloughing from thermal stress, pollution, partial predation, and other sources of injury. In a focused study of tissue loss in seven species of *Pocillopora* along the Pacific coast of Mexico, 84% of over 1000 cases observed were associated with predation by gastropods, the crown-of-thorns sea star, and fishes (Rodríguez-Villalobos et al. 2015). The remainder of tissue loss (15%) was unexplained and tentatively classified as the white syndrome disease (WS). Histological examination failed to reveal infectious agents associated with these lesions. High proportions of coral-associated biota, corallivores and algal and metazoan overgrowths, were also responsible for many of the health abnormalities reported in the Galápagos Islands (Vera and Banks 2009). It is possible that many of the small patches of tissue loss are a result of partial predation by fishes and invertebrate micro-predators. It is often difficult to distinguish between partial predation by micro-corallivores and tissue loss due to parasites (Glynn 1988). Nonetheless, the variety of corallivores and the high frequency of coral predation could indirectly promote infections by creating open tissue wounds and physiological stress. In an historical review of coral disease studies in the eastern Pacific, Rodríguez-Villalobos and Reyes-Bonilla (2019) have concluded that to date research efforts on this topic are inadequate. I echo the concerns of these investigators and support their call for increased use of histopathological, microbial, and molecular methods to help identify suspected infectious agents.

An impressive array of environmental factors that can potentially cause coral bleaching and mortality and interfere with coral reef development in the eastern tropical Pacific region has been briefly examined in this chapter. Thermal conditions, especially sea temperature extremes associated with ENSO activity are recurring disturbances that have strong effects in this region. As firmly believed over a century ago, "cold ocean currents and upwellings" do have a limiting effect on coral reef development, but it is now abundantly clear that extreme high temperatures are more impactful than low temperatures on coral reef ecosystems in the eastern Pacific. Several newly recognized disturbances, such as extreme tidal exposures, eutrophication, aragonite saturation deficits, and coral health anomalies, can aggravate and accentuate temperature stress effects. In Chap. 5, the wide-ranging short- and long-term effects of ENSO on coral reefs and other eastern Pacific marine ecosystems are examined.

References

Aeby GS (2002) Trade-offs for the butterflyfish, *Chaetodon multicinctus*, when feeding on coral prey infected with trematode metacercariae. Behav Ecol Sociobiol 52:158–165

Agassiz A (1892) General sketch of the expedition of the "Albatross", from February to May, 1891. The Galapagos Islands. Bull Mus Comp Zool 23:56–74. Harvard College

Allison EC (1959) Distribution of *Conus* on Clipperton Island. Veliger 1:32–34

Alvarado JJ, Aburto-Oropeza O, Abad R, Barraza E, Brandt M, Cantera J, Estrada F, Gaymer CF, Guzmán-Mora AG, Herlan JJ, Maté JL (2017) Coral reef conservation in the eastern tropical Pacific. In: Glynn PW, Manzello DP, Enochs IC (eds) Coral reefs of the eastern tropical Pacific: persistence and loss in a dynamic environment. Coral reefs of the world 8. Springer, Dordrecht, pp 565–591

Antonius A (1977) Coral mortality in reefs: a problem for science and management. In: Proceedings of the third international coral reef symposium, Miami, 2:617–623

Banks S, Vera M, Chiriboga A (2009) Establishing reference points to assess long-term change in zooxanthellate coral communities of the northern Galapagos coral reefs. Galapagos Res 66:43–64

Berkelmans R (2002) Time-integrated thermal bleaching thresholds of reefs and their variation on the Great Barrier Reef. Mar Ecol Prog Ser 229:73–82

Bindoff NL, Willebrand J, Artale V, Cazenave A, Gregory J, Gulev S, Hanawa K, Le Quéré C, Levitus S, Nojiri Y, Shum CK, Talley LD, Unnikrishnan A (2007) Observations: oceanic climate change and sea level. In: Solomon S, Qin D, Manning M, Chen Z, Marquis M, Averyt KB, Tignor M, Miller HL (eds) Climate change 2007: the physical science basis. Contribution of Working Group I to the fourth assessment report of the Intergovernmental Panel on Climate Change. Cambridge University Press, Cambridge, UK

Birkeland C (1977) The importance of rate of biomass accumulation in early successional stages of benthic communities to the survival of coral recruits. In: Proceedings of the third international coral reef symposium, Miami, 1:15–21

Boschma H (1925) On the feeding reactions and digestion in the coral polyp *Astrangia danae*, with notes on its symbiosis with zooxanthellae. Biol Bull (Woods Hole) 49(6):407–439

Brandtneris VW, Brandt ME, Glynn PW, Gyory J, Smith TB (2016) Seasonal variability in energy content is greater in mesophotic corals. PLoS One 11:e0151953

Brown BE (1987) Worldwide death of corals – natural cyclical events or man-made pollution? Mar Pollution Bulletin 18:9–13

Buckingham MC, D'Angelo C, Chalk TB, Foster GL, Johnson KG (2022) Impact of nitrogen (N) and phosphorus (P) enrichment and skewed N:P stoichiometry on the skeletal formation and microstructure of symbiotic reef corals. Coral Reefs 41(4):1147–1159

Byers JE (2021) Marine parasites and disease in the era of global climate change. Annu Rev Mar Sci 13:3.1–3.24. https://doi.org/10.1146/annurev-marine-031920-100429

Cabral-Tena RA, Medellín-Maldonado F, Granja-Fernández R, González-Mendosa TM, López-Pérez A (2023) Presupesto de carbonato en el parche de arrecife de la Entrega, Oaxaca, sur del Pacífico Mexicano. Hidrobiológica 33(2):201–210

Carlton R, Dempsey A, Thompson L, Heemsoth A, Lubarsky K, Faisal M, Purkis S (2021) Global reef expedition final report, vol 15. Khaled bin Sultan Living Oceans Foundation, Annapolis

Carriquiry JD, Reyes-Bonilla H (1997) Community structure and geographic distribution of the coral reefs of Nayarit, Mexican Pacific. Ciencias Marinas 23(2):227–248

Chan NC, Connolly SR (2013) Sensitivity of coral calcification to ocean acidification: a meta-analysis. Glob Chang Biol 19(1):282–290. https://doi.org/10.1111/gcb.12011. Epub 18 Oct 2012. PMID: 23504739

Coffroth MA, Lasker HR, Oliver JK (1990) Coral mortality outside of the eastern Pacific during 1982-1983: relationship to El Niño. In: Glynn PW (ed) Global ecological consequences of the 1982-83 El Niño Southern Oscillation. Elsevier oceanography series 52. Elsevier, Amsterdam, pp 141–182

Coles SL, Jokiel PL, Lewis CR (1976) Thermal tolerance in tropical versus subtropical Pacific reef corals. Pac Sci 30:159–166

Colgan MW (1990) El Niño and the history of eastern Pacific reef building. In: Glynn PW (ed) Global ecological consequences of the 1982-83 El Niño-Southern Oscillation. Elsevier oceanography series 52. Elsevier, Amsterdam, pp 183–232

Cortés J (1990) The coral reefs of Golfo Dulce, Costa Rica: distribution and community structure. Atoll Res Bull 344:1–37

Cortés J (2003) Coral reefs of the Americas: an introduction to Latin American coral reefs. In: Cortés J (ed) Latin American coral reefs. Elsevier, Amsterdam, pp 1–7

Cortés J, Murillo MM, Guzmán HM, Acuña J (1984) Pérdida de zooxantelas y muerte de corales y otros organismos arrecifales en el Caribe y Pacífico de Costa Rica. Rev Biol Trop 32:227–232

Couthouy JP (1842) Remarks upon coral formations in the Pacific; with suggestions as to the causes of their absence in the same parallels of latitude on the coast of South America. Boston J Nat Hist 4(66–105):137–162

Crossland C (1927) The expedition to the South Pacific of the SY 'St. George'. Marine ecology and coral formations in the Panama region, the Galapagos and Marquesas Islands, and the Atoll of Napuka. Trans R Soc Edinb 55:531–554

Dana JD (1843) On the temperature limiting the distribution of corals. Am J Sci 45:130–131

D'Angelo C, Wiedenmann J (2014) Impacts of nutrient enrichment on coral reefs: new perspectives and implications. Curr Opin Environ Sustain 7:82–93

Darwin CR (1842) The structure and distribution of coral reefs. Being the first part of the geology of the voyage of the *Beagle*, under the command of Capt. Fitzroy, R. N. during the years 1832 to 1836. Smith Elder, London

Dillehay TD (2017) Where the land meets the sea: fourteen millennia of human history at Huaca Prieta, Peru, 1st edn. University of Texas Press, Austin

Ditlevsen P, Ditlevsen S (2023) Warning of a forthcoming collapse of the Atlantic meridional overturning circulation. Nat Commun. https://doi.org/10.1038/s41467-023-39810-w

Durham JW (1966) Coelenterates, especially stony corals, from the Galápagos and Cocos Islands. In: Bowman RI (ed) The Galápagos, Proceedings of the symposium of the Galápagos International Scientific Project. Univ. California Press, Berkeley, pp 123–135. 318 p

Eakin CM, Glynn PW (1996) Low tidal exposures and reef mortalities in the eastern Pacific. Coral Reefs 15:120

Eakin CM, Morgan JA, Heron SF, Smith TB, Liu G, Alvarez-Filip L, Baca B, Bartels E, Bastidas C, Bouchon C, Brandt M, Bruckner AW, Bunkley-Williams L et al (2010) Caribbean corals in crisis: record thermal stress, bleaching and mortality in 2005. PLoS One 5(11):e13969. https://doi.org/10.1371/journal.pone.0013969

Enochs IC, Glynn PW (2017) Corallivory in the eastern Pacific. In: Glynn PW, Manzello DP, Enochs IC (eds) Coral reefs of the eastern tropical Pacific: persistence and loss in a dynamic environment. Coral reefs of the world 8. Springer, Dordrecht, pp 315–337

Feingold JS (1995) Effects of elevated water temperature on coral bleaching and survival during El Niño disturbance events. PhD. Dissertation, University of Miami, Coral Gables. 236 p

Fiedler PC, Lavín MF (2017) Oceanographic conditions of the eastern tropical Pacific. In: Glynn PW, Manzello DP, Enochs IC (eds) Coral reefs of the eastern tropical Pacific: persistence and loss in a dynamic environment. Coral reefs of the world 8. Springer, Dordrecht, pp 59–83

Fordyce AJ, Ainsworth TD, Heron SF, Leggat W (2019) Marine heatwave hotspots in coral reef environments: physical drivers, ecophysiological outcomes, and impacts upon structural complexity. Front Mar Sci. https://doi.org/10.3389/fmars.2019.00498

Forryan A, Naveira Garabato AC, Vic C, George Nurser AJ, Hearn AR (2021) Galápagos upwelling driven by localized wind-front interactions. Sci Rep 11:1277. https://doi.org/10.1038/s41598-020-80609-2

Forsbergh ED (1969) On the climatology, oceanography and fisheries of the Panama Bight. Inter-Am Trop Tuna Comm Bull 14(2):49–385

Foster R, Hagan A, Perera N, Gunawan CA, Silaban I, Yaha Y, Manuputty Y, Hazam I, Hodgson G (2006) Tsunami and earthquake damage to coral reefs of Aceh, Indonesia. Reef Check Foundation, Pacific Palisades. 33 p

Gleason DF, Wellington GM (1993) Ultraviolet radiation and coral bleaching. Nature 365:836–838

Gleesen MW, Strong AE (1995) Applying MCSST to coral reef bleaching. Adv Space Res 16:151–154

Glynn PW (1974) The impact of *Acanthaster* on corals and coral reefs in the eastern Pacific. Environ Conserv 1(4):295–304

Glynn PW (1976) Some physical and biological determinants of coral community structure in the eastern Pacific. Ecol Monogr 46(4):431–456

Glynn PW (1977) Coral growth in upwelling and nonupwelling areas off the Pacific coast of Panama. J Mar Res 35:567–585

Glynn PW (1983) Extensive "bleaching" and death of reef corals on the Pacific coast of Panama. Environ Conserv 10(2):149–154

Glynn PW (1984) Widespread coral mortality and the 1982-83 El Niño warming event. Environ Conserv 11(2):133–146

Glynn PW (1988) Predation on coral reefs: some key processes, concepts and research directions. In: Proceedings of the sixth international coral reef symposium, Townsville, 1:51–62

Glynn PW (2003) Coral communities and coral reefs in Ecuador. In: Cortés J (ed) Latin American coral reefs. Elsevier Science, Amsterdam, pp 449–472

Glynn PW, Ault JS (2000) A biogeographic analysis and review of the far eastern Pacific coral reef region. Coral Reefs 19:1–23

Glynn PW, Fong P (2006) Patterns of reef coral recovery by the regrowth of surviving tissues following the 1997-98 El Niño warming and 2000, 2001 upwelling cool events in Panamá, eastern Pacific. In: Proceedings of the 10th international coral reef symposium, Okinawa, 2:624–630

Glynn PW, Leyte Morales GE (1997) Coral reefs of Huatulco, West Mexico: reef development in upwelling Gulf of Tehuantepec. Rev Biol Trop 45(3):1033–1047

Glynn PW, Macintyre IG (1977) Growth rate and age of coral reefs on the Pacific coast of Panama. In: Proceedings of the third international coral reef symposium, Miami, 2:251–259

Glynn PW, Maté JL (1997) Field guide to the eastern Pacific coral reefs of Panama. In: Proceedings of the eighth international coral reef symposium, Panama, 1:145–166

Glynn PW, Stewart RH (1973) Distribution of coral reefs in the Pearl Islands (Gulf of Panama) in relation to thermal conditions. Limnol Oceanogr 18:367–379

Glynn PW, Wellington GM (1983) Corals and coral reefs of the Galápagos Islands. With an annotated list of the scleractinian corals of the Galápagos by John W. Wells. University of California Press, Berkeley. 330 p

Glynn PW, Stewart RH, McCosker JE (1972) Pacific coral reefs of Panama: structure, distribution and predators. Geol Rundsch 61:483–519

Glynn PW, von Prahl H, Guhl F (1982) Coral reefs of Gorgona Island, Colombia, with special reference to corallivores and their influence on community structure and reef development. Anales del Instituto de Investigaciones Marinas de Punta de Betín 12:185–214

Glynn PW, Howard LS, Corcoran E, Freay AD (1984) The occurrence and toxicity of herbicides in reef building corals. Mar Pollut Bull 15:370–374

Glynn PW, Peters EC, Muscatine L (1985) Coral tissue microstructure and necrosis: relation to catastrophic coral mortality in Panama. Dis Aquat Org 1:29–37

Glynn PW, Cortès J, Guzmán HM, Richmond RH (1988) El Niño (1982-83) associated coral mortality and relationship to sea surface temperature deviations in the tropical eastern Pacific. In: Proceedings of the sixth international coral reef symposium, Townsville, 3:237–243

Glynn PW, Szmant AM, Corcoran EF, Cofer-Shabica SV (1989) Condition of coral reef cnidarians from the northern Florida reef tract: pesticides, heavy metals, and histopathological examination. Mar Pollut Bull 20(11):568–576

Glynn PW, Veron JEN, Wellington GM (1996) Clipperton Atoll (eastern Pacific): oceanography, geomorphology, reef building coral ecology and biogeography. Coral Reefs 15:71–99. https://doi.org/10.1126/science.253.5015.69

Glynn PW, Maté JL, Baker AC, Calderón MO (2001) Coral bleaching and mortality in Panama and Ecuador during the 1997-1998 El Niño-Southern Oscillation event: spatial/temporal patterns and comparisons with the 1982-1983 event. Bull Mar Sci 69(1):79–109

Glynn PW, Wellington GM, Wieters EA, Navarrete SA (2003) Reef-building coral communities of Easter Island (Rapa Nui), Chile. In: Cortés J (ed) Latin American coral reefs. Elsevier, Amsterdam, pp 473–494

Glynn PW, Riegl B, Purkis S, Kerr JM, Smith TB (2015) Coral reef recovery in the Galápagos Islands: the northern-most islands (Darwin, Wenman). Coral Reefs 34:421–436. https://doi.org/10.1007/s00338-015-1280-4

Glynn PW, Alvarado JJ, Banks S, Cortés J, Feingold JS, Jiménez C et al (2017) Eastern Pacific coral reef provinces, coral community structure and composition: an overview. In: Glynn PW,

Manzello DP, Enochs IC (eds) Coral reefs of the eastern tropical Pacific: persistence and loss in a dynamic environment. Coral reefs of the world 8. Springer, Dordrecht, pp 107–176

Glynn PW, Dillehay TD, Netherly PJ, Riegl BM (2022) Prehistoric reef-building coral occurrence in north Peru. J Island Coast Archeol. 19(2):420–432. https://doi.org/10.1080/15564894.2022.2072026

Goreau TF (1964) Mass expulsion of zooxanthellae from Jamaican reef communities after Hurricane Flora. Science 145:383–386

Goreau TJ, Hayes RL (1994) Coral bleaching and ocean "hot spots". Ambio 23(3):176–180

Guzmán HM, Cortés J, Glynn PW, Richmond RH (1990) Coral mortality associated with dinoflagellate blooms in the eastern Pacific (Costa Rica and Panama). Mar Ecol Prog Ser 60:299–303

Halpern D, Hayes SP, Leetmaa A, Hansen DV, Philander SGH (1983) Oceanographic observations of the 1982 warming of the tropical eastern Pacific. Science 221:1173–1175

Hansen J, Sato M, Hearty P, Ruedy R, Kelley M, Masson-Delmotte V, Russel G, Tselioudis G, Cap J, Rignot E et al (2016) Ice melt, sea level rise and superstorms: evidence from paleoclimate data, climate modeling, and modern observations that 2 °C global warming could be dangerous. Atmos Chem Phys 16:3761–3812. https://doi.org/10.5194/acp-16-3761-2016

Harvell CD, Kim K, Burkholder JM, Colwell RR, Epstein PR, Grimes DJ, Hofmann EE, Lipp EK, Osterhaus ADME, Overstreet RM, Porter JW, Smith GW, Vasta GR (1999) Emerging marine diseases—climate links and anthropogenic factors. Science 285:1505–1510

Hoegh-Guldberg O (1999) Climate change, coral bleaching and the future of the world's coral reefs. Mar Freshw Res 50(8):839–866. https://doi.org/10.1071/MF99078

Hubbs CL, Roden GI (1964) Oceanography and marine life along the Pacific coast of Middle America. In: Hubbs CL, Roden GI (eds) Natural environment and early cultures, Vol. 1, Handbook of Middle American Indians. University of Texas Press, Austin, pp 143–186

Jaap WC (1985) An epidemic zooxanthellae expulsion during 1983 in the lower Florida Keys coral reefs: hyperthermic etiology. In: Proceedings of the fifth international coral reef congress, Tahiti, 6:143–148

Jakoboski J, Todd RE, Owens WB, Karnauskas KB, Rudnick DL (2020) Bifurcation and upwelling of the Equatorial Undercurrent west of the Galápagos Archipelago. J Phys Oceanogr 50:887–904

Jokiel PL, Coles SL (1974) Effects of heated effluent on hermatypic corals at Kahe Point, Oahu. Pac Sci 28:1–18

Jokiel PL, Coles SL (1977) Effects of temperature on the mortality and growth of Hawaiian reef corals. Mar Biol 43:201–208

Klein SG, Geraldi NR, Anton A, Schmidt-Roach S, Ziegler M, Cziesielski MJ, Martin C, Rädecker N, Frölicher TL, Mumby PJ, Pandolfi JM, Suggett DJ, Voolstra CR, Aranda M, Duarte CM (2022) Projecting coral responses to intensifying marine heatwaves under ocean acidification. Glob Chang Biol 00:1–13. https://doi.org/10.1111/gcb.15818

Kleypas JA, McManus JW, Meñez LAB (1999) Environmental limits to coral reef development: where do we draw the line? Am Zool 39:146–159

Lasker HR, Peters EC, Coffroth MA (1984) Bleaching of reef coelenterates in the San Blas Islands, Panama. Coral Reefs 3:183–190

Lessios HA, Glynn PW, Robertson DR (1983) Mass mortalities of coral reef organisms. Science 222(4625):715

Lirman D, Glynn PW, Baker AC, Leyte Morales GE (2001) Combined effects of three sequential storms on the Huatulco coral reef tract, Mexico. Bull Mar Sci 69(1):267–278

López-Pérez A, Ramirez E, Granja Fernández R, Rodriguez Zaragoza FA, Valéncia-Méndez O, Antonio Perez V (2023) El Niño 2023 event in the Southern Mexican Pacific: a preliminary report of an imminent disaster. Reef Encounter 38(2):64–66

López-Pérez A, Granja-Fernández R, Ramírez-Chávez E, Valencia-Méndez O, Rodríguez-Zaragoza FA, González-Mendoza T, Martínez-Castro A (2024) Widespread coral bleaching and mass mortality of reef-building corals in southern Mexican Pacific reefs due to 2023 El Niño warming. Oceans 5(2):196–209. MDPI

López-Sandoval DC, Lara-Lara JR, Lavín MF, Álvarez-Borrego S, Gaxiola-Castro G (2009) Primary productivity in the eastern tropical Pacific off Cabo Corrientes, Mexico. Ciencias Marinas 35(2):169–182

Manzello DP (2010) Coral growth with thermal stress and ocean acidification: lessons from the eastern tropical Pacific. Coral Reefs 29:749–758

Manzello DP, Kleypas JA, Budd DA, Eakin CM, Glynn PW, Langdon C (2008) Poorly cemented coral reefs of the eastern tropical Pacific: possible insights into reef development in a high-CO_2 world. Proc Natl Acad Sci USA 105(30):10450–10455. www.pnas.org/cgi/doi/10.1073/pnas.0712167105

Manzello DP, Enochs IC, Bruckner A, Renaud PG, Kolodziej G, Budd DA, Carlton R, Glynn PW (2014) Galápagos coral reef persistence after ENSO warming across an acidification gradient. Geophys Res Lett 41. https://doi.org/10.1002/2014GL062501

Maté JL (2003) Corals and coral reefs of the Pacific coast of Panamá. In: Cortés J (ed) Latin American coral reefs. Elsevier Science, Amsterdam, pp 387–417

Murphy RC (1936) Oceanic birds of South America, vol 1. Macmillan, New York. 640 pages, 38 plates

Neumann AC, Macintyre IG (1985) Reef response to sea-level rise: keep-up, catch-up or give-up. In: Proceedings of the 5th international coral reef congress, Tahiti, vol 3, pp 205–210

Peters EC (2015) Diseases of coral reef organisms. In: Birkeland C (ed) Coral reefs in the Anthropocene. Springer, Dordrecht, pp 147–178

Peterson TC, Connelley WM, Fleck J (2008) The myth of the 1970s global cooling scientific consensus. Bull Am Meteorol Soc 2370:1325–1337. https://doi.org/10.1175/2008BAMS2370

Plata-Rosas L, Filinov A (2007) Internal tide in the northwestern part of Banderas Bay, Mexico. Ciencias Marinas 33(2):197–215. https://cienciasmarinas.com.mx/index.php/cmarinas/article/view/1013

Podestá GP, Glynn PW (1997) Sea surface temperature variability in Panama and Galápagos: extreme temperatures causing coral bleaching. J Geophys Res 102(C7):15749–15759

Podestá GP, Glynn PW (2001) The 1997-98 El Niño event in Panama and Galápagos: an update of thermal stress indices relative to coral bleaching. Bull Mar Sci 69(1):43–59

Pogoreutz C, Clua EE, Langarica JDJAT (2021) Current ecological status of remote coral reefs of Clipperton Island, tropical eastern Pacific. Research Square

Rasmusson EM, Hall JM (1983) El Niño, the great equatorial Pacific Ocean warming event of 1982-1983. Weatherwise 36:166–175

Reyes-Bonilla H (2001) Effects of the 1997-1998 El Niño-Southern Oscillation on coral communities of the Gulf of California, Mexico. Bull Mar Sci 69(1):251–266

Reyes-Bonilla H (2003) Coral reefs of the Pacific coast of Mexico. In: Cortés J (ed) Latin American coral reefs. Elsevier, Amsterdam, pp 331–349

Reyes-Bonilla H, López Pérez A (1998) Biogeography of scleractinian corals of the Pacific coast of México. Ciencias Marinas 24:211–224

Reyes-Bonilla H, Calderón Aguilera LE, Cruz Piñón G, Medina Rosas P, López Pérez RA, Herrero Pérezrul MD, Leyte Morales GE, Cupul Magaña AL, Carriquiry Beltrán JD (2005) Atlas de corales pétreos (Anthozoa: Scleractinia) del Pacífico mexicano. México. http://www.somac.org.mx/publicaciones

Riegl BM, Halfar J, Purkis SJ, Godinez-Orta L (2007) Sedimentary facies of the eastern Pacific's northernmost reef-like setting (Cabo Pulmo, Mexico). Mar Geol 236(1–2):61–77

Riegl B, Glynn PW, Banks S, Keith I, Rivera F, Vera-Zambrano M, d'Angelo C, Wiedenmann J (2019a) Heat attenuation and nutrient delivery by localized upwelling avoided coral bleaching mortality in northern Galapagos during 2015/16 ENSO. Coral Reefs 38:773–785. https://doi.org/10.1007/s00338-019-01787-8

Riegl B, Johnston M, Glynn PW, Keith I, Rivera F, Vera-Zambrano M, Banks S, Feingold J, Glynn PJ (2019b) Some environmental and biological determinants of coral richness, resilience and reef building in Galapagos (Ecuador). Sci Rep 10322. www.nature.com/articles/s41598-019-46607-9

Robinson G (1985) Influence of the 1982-83 El Niño on Galápagos marine life. In: Robinson G, del Pino EM (eds) El Niño in the Galápagos Islands: the 1982-1983 event. Charles Darwin Foundation for the Galápagos Islands, Quito, Ecuador, pp 153–190

Rodríguez-Villalobos JC, Reyes-Bonilla H (2019) History of perspectives on the study of coral disease in the eastern tropical Pacific. Dis Aquat Org 136:243–253. https://doi.org/10.3354/dao03411

Rodríguez-Villalobos JC, Work TM, Calderon-Aguilera LE, Reyes-Bonilla H, Hernández L (2015) Explained and unexplained tissue loss in corals from the tropical eastern Pacific. Dis Aquat Org 116:121–131. https://doi.org/10.3354/dao02914

Rosenberg E, Loya Y (eds) (2004) Coral health and disease. Springer, Berlin

Saunders MI, Albert S, Roelfsema CM, Leon JX, Woodroffe CD, Phinn SR, Mumby PJ (2015) Tectonic subsidence provides insight into possible coral reef futures under rapid sea-level rise. Coral Reefs. https://doi.org/10.1007/s00338-015-1365-0

Sellanes J, Gorny M, Zapata-Hernández G, Alvarez G, Muñoz P, Tala F (2021) A new threat to local marine biodiversity: filamentous mats proliferating at mesophotic depths off Rapa Nui. PeerJ 9:e12052. https://doi.org/10.7717/peerj.12052

Smayda TJ (1966) A quantitative analysis of the phytoplankton of the Gulf of Panama. III. General ecological conditions, and the phytoplankton dynamics at 8° 45′ N, 79° 23′ W from November 1954 to May 1957. Inter-Am Trop Tuna Comm Bull 11(5):355–612

Smith TB (2006) Benthic algal production in an upwelling coral reef environment (Gulf of Panamá): evidence for control by fish herbivores. In: Proceedings of the 10th international coral reef symposium, Okinawa, vol 1, pp 854–861

Smith TB (2008) Temperature effects on herbivory for an Indo-Pacific parrotfish in Panamá: implications for coral-algal competition. Coral Reefs 27:397–405

Squires DF (1959) Reports of the Puritan-American Museum of Natural History Expedition to Western Mexico, 7. Corals and coral reefs in the Gulf of California. Bull Am Mus Nat Hist 118(art. 7):367–432

Steinbeck J, Ricketts EF (1941) Sea of Cortez, A leisurely journal of travel and research. The Viking Press, New York. 598 p

Thornborough KJ, Davies PJ (2011) Reef flats. In: Hopley D (ed) The encyclopedia of modern coral reefs: structure, form and process. Springer, Berlin, pp. 868–876

Torres-Orozco E, Trasviña A, Muhlia-Melo A, Ortega-García S (2005) Mesoscale dynamics and yellowfin tuna catches in the Mexican Pacific. Ciencias Marinas 31(4):671–683. http://www.scielo.org.mx/scielo.php?pid=S0185-38802005000500007&script=sci_abstract&tlng=en

Van Hooidonk R, Maynard J, Manzello DP, Planes S (2014) Opposite latitudinal gradients in projected ocean acidification and bleaching impacts on coral reefs. Glob Chang Biol 20:103–112. https://doi.org/10.1111/gcb.12394

Vargas-Ángel B (1996) Distribution and community structure of the reef corals of Ensenada de Utría, Pacific coast of Colombia. Rev Biol Trop 44:443–651

Vargas-Ángel B, Zapata FA, Hernández H, Jiménez JM (2001) Coral and coral reef responses to the 1997-98 El Niño event on the Pacific coast of Colombia. Bull Mar Sci 69(1):111–132

Vera M, Banks S (2009) Health status of the coral communities of the northern Galapagos Islands Darwin, Wolf and Marchena. Galapagos Res 66:65–74

Verrill AE (1870) On the geographical distribution of the polyps and corals of the west coast of America. Trans Conn Acad Arts Sci 1:559–567

von Prahl H (1983) Blanqueo masivo y muerte de corales en la Isla de Gorgona, Pacífico Colombiano. Cespedesia 12:125–129

von Prahl H (1986) Crustáceos decápodos asociados a diferente habitats en la Ensenada de Utría, Chocó, Colombia. Actualidades Biológicas 15:95–99

von Prahl H, Erhardt H (1985) Colombia: corales y arrecifes coralinos. Fondo FEN Colombia, Editorial Presencia, Bogotá, Colombia

Wang C, Deser C, Yu JY, DiNezio P, Clement A (2017) El Niño and Southern Oscillation (ENSO): a review. In: Glynn PW, Manzello DP, Enochs IC (eds) Coral reefs of the eastern tropical Pacific: persistence and loss in a dynamic environment. Coral reefs of the world 8. Springer, Dordrecht, pp 85–106

Weil E, Rogers CS (2011) Coral reef diseases in the Atlantic-Caribbean. In: Dubinsky Z, Stambler N (eds) Coral reefs: an ecosystem in transition. Springer, Dordrecht, pp 465–491

Wiedenmann J, D'Angelo C, Mardones ML, Moore S, Benkwitt CE, Graham NAJ, Hambach B, Wilson PA, Vanstone J, Eyal G, Ben-Zvi O, Loya Y, Genin A (2023) Reef-building corals farm and feed on their photosynthetic symbionts. Nature 621(7980):E44. https://doi.org/10.1038/s41586-023-06442-5

Williams EH Jr, Bunkley-Williams L (1990) The world-wide coral reef bleaching cycle and related sources of coral mortality. Atoll Res Bull 335:1–71

Wooster WS, Hedgpeth JW (1966) The oceanographic setting of the Galápagos. In: Bowman RI (ed) The Galápagos. University of California Press, Berkeley, pp 100–107

Work TM, Kaczmarski L, Peters EC (2016) Growth anomalies in corals. In: Woodley CM, Downs CA, Bruckner AW, Porter JW, Galloway SB (eds) Diseases of coral. Wiley, Hoboken, pp 291–315

Yonge CM (1963) The biology of coral reefs. Adv Mar Biol 1:209–262. Academic, New York

Zapata FA, Vargas-Ángel B (2003) Corals and coral reefs of the Pacific coast of Colombia. In: Cortés J (ed) Latin American coral reefs. Elsevier, Amsterdam, pp. 419–447. 497 p

Chapter 5
El Niño Southern Oscillation

Abstract ENSO, El Niño Southern Oscillation, generally manifests two stages in the tropical eastern Pacific, a warm, stormy, wet, high sea level phase, and a cool, calm, dry, low sea level phase. An immediate stressful period with coral bleaching and mortality occurs soon after the influx of El Niño elevated sea temperatures. Variable coral colony bleaching is due to the presence or absence of thermally tolerant dinoflagellate endosymbionts (e.g., *Durusdinium*). *Millepora* fire corals are susceptible to high temperature stress; two of three species experienced local extinctions. Symbiotic crustaceans decreased markedly in abundance with the loss of their coral habitat niche and food source. The loss of protective barriers of coral crustacean guards exposed massive corals to *Acanthaster* predation, resulting in partial mortality. Sea urchin population increases in Panama (*Diadema*) and the Galápagos Islands (*Eucidaris*), following El Niño elevated temperatures that initiated coral mortality, resulted in severe reef bioerosion that caused significant reef framework degradation. Patch reefs in the southern Galápagos Islands disappeared; extreme sea urchin bioerosion transformed coral colonies into sediments. Community wide changes in trophic resource utilization by reef fish feeding guilds (facultative corallivores, mixed diet feeders, and benthic invertivores) during post El Niño impacts are discussed in relation to coral recovery, topographic complexity, and availability of prey. Extreme and prolonged El Niño high temperature conditions depress or inhibit coral reproduction. La Niña cool conditions, with possible exceptions (depending on nutrient availability) inhibit coral reproductive activity.

Introduction

Extreme variations in sea temperature associated with ENSO are the greatest existential threat to coral reef organisms in the eastern Pacific region. From 1500 to 2017, 42 El Niño events were reported, and nine were very strong events, probably comparable to the devastating 1982–83 El Niño, with marked global effects (Colgan 1990; Glynn 1990; NOAA 2021). Further back in time, a palaeoecological study has demonstrated interrupted reef accretion in Panama between 4000 to 6000 years

© The Author(s), under exclusive license to Springer Nature Switzerland AG 2025
P. W. Glynn, *Coral Reef Resilience in the Anthropocene*,
https://doi.org/10.1007/978-3-031-81421-1_5

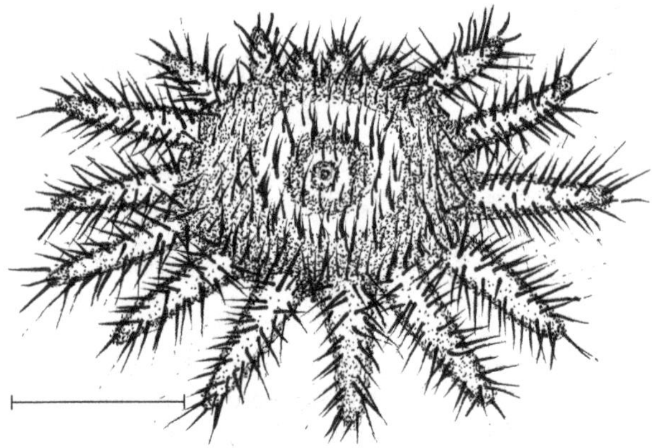

Fig. 5.1 *Acanthaster* cf. *solaris*, scale bar = 8 cm

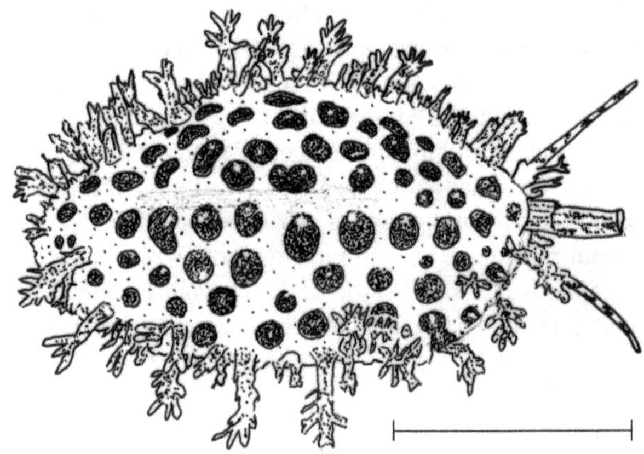

Fig. 5.2 *Jenneria pustulata*, scale bar = 1 cm

BP, attributed to ENSO variability and the upwelling phase of La Niña (Toth et al. 2012, 2017). In Chap. 4, the physical conditions initiating and driving ENSO activity were noted, and some of the biotic effects during El Niño (warm) and La Niña (cool) phases. In this chapter I focus more on biological and ecological impacts, both short term (days) and long-term (years). We begin with the initial responses of corals and their symbiotic zooxanthellae, and then discuss ENSO effects on the symbiotic crustaceans of corals, corallivores (*Acanthaster*, *Jenneria*), sea urchin bioeroders, and reef associated fishes (Figs. 5.1 and 5.2). Our team's studies revealed several species that exhibited notable effects on coral communities. Some of these species would qualify as **keystone species**, a special group of organisms having significant and disproportionate effects on community function, e.g., affecting reef accretion and decline. A recent review identified 13 ENSO biotic stress responses affecting eastern Pacific coral reefs (Glynn et al. 2017). I will introduce topics in which our team has focused their research effort and advanced understanding. Additional biotic responses to ENSO, indirect and delayed effects, are noted in other chapters (e.g., Chap. 7, corallivory, and Chap. 9, bioerosion).

Variable Coral Colony Bleaching

In viewing a bleached field of *Pocillopora* on the Uva Island reef it was apparent that all coral colonies were not bleached; some colonies still retained their normal, golden-brown pigmentation, and other colonies were variegated with a mixture of bleached and normally pigmented branches (Fig. 5.3). Laboratory studies of bleached branches showed these to be significantly depleted of zooxanthellae and chlorophyll whereas colored branches contained normal abundances of zooxanthellae and high concentrations of photosynthetic pigments. This condition raised an interesting question, "How was it that some whole colonies and some branches on individual colonies were spared from stressful elevated temperatures?". Enter Andrew C. Baker (Fig. 5.4).

Andrew was returning to England after interviewing for graduate studies at a prestigious university on the California coast. Since his itinerary included a stopover in Miami, he decided to visit the Rosenstiel marine school to learn about the on-going coral reef research at our institution. We had a pleasant and animated discussion in my office, and I thought how fortunate the west coast school was to admit Andrew to their graduate studies program. Surprise, surprise—a short while after Andrew's visit he asked if I would consider supporting his application for graduate studies at RSMAES, and to consider a research proposal to conduct studies on eastern Pacific coral reefs. In August 1993, Andrew was admitted to RSMAES as a

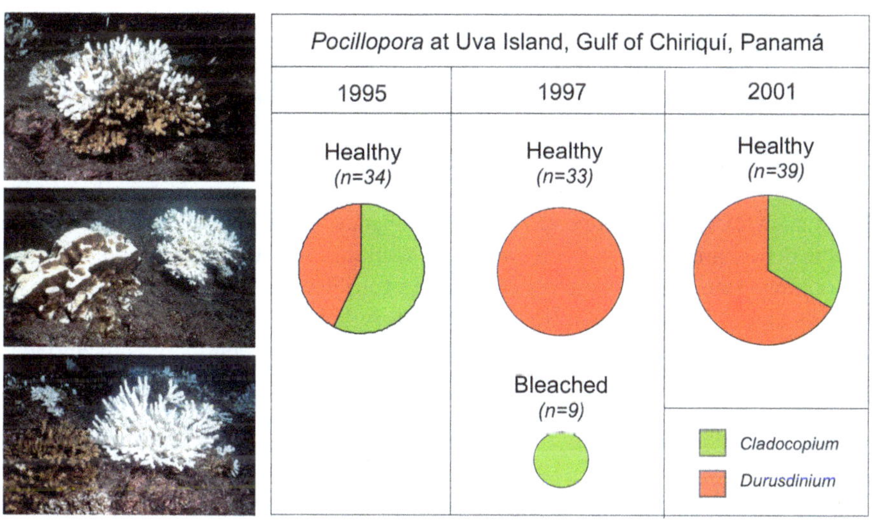

Fig. 5.3 *Pocillopora* corals containing heat-tolerant algal symbionts (*Durusdinium glynnii*) are resistant to bleaching and increase in abundance following recovery. Patchy bleaching in October 1997 at Uva Island in the Gulf of Chiriquí, Panama (see photos) revealed that colonies, or parts of colonies, hosting *D. glynnii* were very resistant to El Niño bleaching. Between 1995 and 2001 the relative abundance of *Pocillopora* hosting *D. trenchii* increased ~50% (3 × 2 contingency $c^2 = 9.71$; df = 2, $p < 0.01$). 1995 data and 1997 data and photos from Glynn et al. (2001). Data from 2001 from Baker et al. (2004). See also Palacio-Castro et al. (2023)

Fig. 5.4 Andrew Baker and Ph.D. candidate Carly (Caroline) Dennison at the UM Experimental Hatchery on Virginia Key inspecting a colony of *Eusmilia fastigiata* from the Florida Reef Tract Rescue project, May 2019. Photo credit: T. J. Levonian

Fulbright Scholar[1] from the United Kingdom. I was delighted. Soon after arriving at RSMAES, we met again and a recent publication that hypothesized coral bleaching may benefit coral hosts by allowing repopulation of symbiotic algae more resistant to high temperature stress (Buddemeier and Fautin 1993) captivated our interest. This became known as the Adaptive Bleaching Hypothesis (ABH). Admittedly, I was skeptical but intrigued by this possibility. On one of our research cruises to Panama in 1994, I introduced Andrew to Nancy Knowlton at the Smithsonian Tropical Research Institute. She was then working with Rob Rowan on a project examining intraspecific symbiont diversity in Caribbean corals. Andrew was subsequently able to spend time in Nancy's laboratory training in the molecular methods necessary for DNA extraction and determining the genetic identity of zooxanthellae inhabiting coral hosts.

[1] In the early 1970s I had the great privilege of briefing U.S.A. Senator James W. Fulbright on our coral reef initiatives and facilities at STRI's Naos Island laboratory. When walking on to our pier, I was surprised when Senator Fulbright threw his arm over my shoulder and pulled me near and asked, "What do you think of the war in Vietnam?" I immediately expressed my opposition to the war, which elicited a smile from the senator. Then, the response from the architect of the Fulbright Scholarship program, paraphrasing, "Isn't it better to direct our resources to education and research than to senseless killing and ruin?" This is still a current thought almost five decades later. The studies of three of my graduate students were supported by Fulbright Fellowships: Andrew Baker (United Kingdom of Great Britain and Northern Ireland), Jorge Cortés (Republic of Costa Rica), and Juan Maté (Republic of Panama).

The 1997–98 El Niño coral reef bleaching event offered an ideal opportunity to determine the genetic identity of algal symbionts in bleached and normal-colored *Pocillopora* corals. Andrew sampled the Uva Island *Pocillopora* population in 1995, thus providing information on the composition of the symbiont community before the El Niño bleaching event. Molecular DNA analysis of the endosymbionts from the same population of *Pocillopora* in 1995 and 1997 revealed that colonies contained two very different types of symbionts now recognized as members of two different genera, *Cladocopium* and *Durusdinium* in the family Symbiodiniaceae .[2] Bleached colonies in 1997 contained *Cladocopium* only and normally colored colonies hosted high abundances of *Durusdinium* symbionts. *Pocillopora* colonies with *Durusdinium* symbionts resisted bleaching during the warm water bleaching event (Baker 1999; Glynn et al. 2001). Andrew also found that *Pocillopora* colonies in Kenya and the Arabian Gulf that survived the 1997–98 El Niño bleaching event hosted *Durusdinium* symbionts (Baker et al. 2004). In addition, Todd LaJeunesse and co-workers reported that *Durusdinium* symbionts in *Pocillopora* in the Gulf of California resisted bleaching during periods of solar radiation stress (LaJeunesse et al. 2007) and extreme cold-water events (LaJeunesse et al. 2010), conditions often associated with ENSO activity. The *Durusdinium* symbionts are truly hardy, capable of enduring extreme physical stressors under marginal environmental conditions. These various studies provided evidence that reef-wide algal symbiont shuffling towards *Durusdinium* was one possible outcome supporting the ABH. It would be another two decades, however, before researchers could provide stronger evidence that individual coral colonies were indeed shuffling their symbionts towards thermotolerant *Durusdinium* (Palacio-Castro, personal communication; Palacio-Castro et al. 2023; Fig. 5.5)[3].

---

[2] Revisions to the taxonomy of symbiotic dinoflagellates (LaJeunesse et al. 2018) introduced the new family Symbiodiniaceae, which comprises multiple genera that correspond to the former "clades" of *Symbiodinium* that were commonly used to refer to these symbionts for almost three decades. A clade is a group of species (or other taxa) sharing a common ancestor and its descendants and closely related genetically. Most corals in the eastern Pacific are host to either *Cladocopium* or *Durusdinium* (former clades C and D, respectively, Baker and Rowan 1997), but *Breviolum* (clade B) has also been reported in some bleached *Pocillopora* in Mexico (LaJeunesse et al. 2010), and *Symbiodinium* (clade A) has been reported from milleporid fire corals in the region. Other genera occur as endosymbionts in a variety of cnidarians, foraminifers, and sponges around the world. Indeed, LaJeunesse and co-workers have suggested that the eventual number of genera (clades) may reach 15 or more as surveys continue in global coral reefs.

[3] Two new species of *Durusdinium*, *Durusdinium trenchii* and *Durusdinium glynnii*, were recently introduced by Todd LaJeunesse and co-workers (2014; Wham et al. 2017). The generic name *Durusdinium* is derived from the Latin roots "tough" and "whirling", with reference to their stress resistance and the presence of flagella in their motile stage. I feel twice honored, for the "*glynnii*" epithet and again for this hardy symbiont species to be genealogically close to my esteemed Belizean colleague and friend Bob (Robert) K. Trench. Bob passed in 2021. His considerable impact in the field of symbiosis biology is gracefully summarized in his obituary written by former students and colleagues (LaJeunesse et al. 2021). To Todd and colleagues, I send a well-deserved congratulations on their *tour de force* revision of the Symbiodiniaceae with fascinating information on the diversity, ecology, and genealogy of this taxon. Their analysis of molecular clock data revealed the earliest family diversification during the radiation of reef-building corals in the

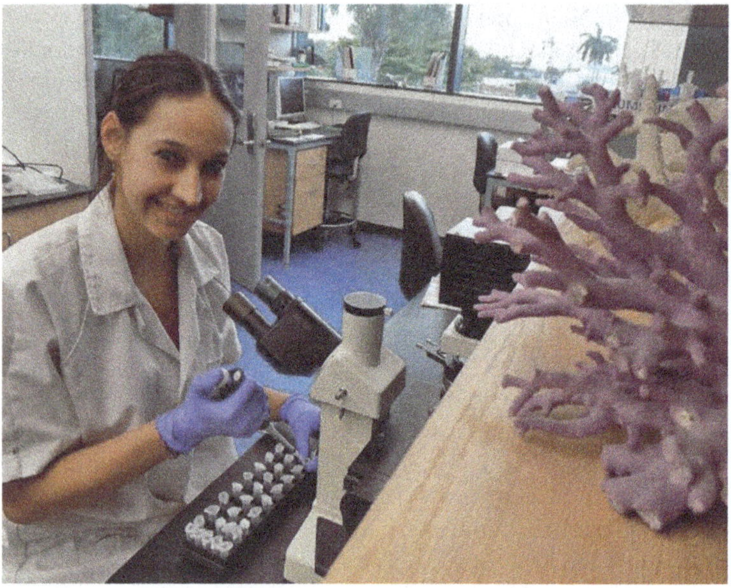

Fig. 5.5 Ana Palacio extracting DNA from coral samples in Andrew Baker's molecular laboratory at RSMAES, summer 2016. Dr. Palacio is presently an Assistant Scientist at CIMAS (Cooperative Institute for Marine & Atmospheric Studies), Miami, Florida. Ana uses a combination of field and laboratory experiments to understand the factors that shape corals' susceptibility to multiple environmental stressors, including ocean warming, ocean acidification, diseases, and nutrient pollution. Photograph courtesy of Marine-Oceane Guillermard

In addition to the bleached and dead *Pocillopora* species, closer inspection of the Uva Island coral community revealed nine other scleractinian and three hydrocoral species that were equally stressed. Several weeks after the bleaching event some species could not be found alive on the reef. Live colonies of *Porites panamensis*, a small endemic species, was not observed anywhere in the Gulf of Chiriquí for 2 years. We were fearful that this species, the only known eastern Pacific zooxanthellate coral that releases planula larvae, may be locally extinct. Thankfully, it reappeared in large numbers as sexual recruits after 3 years (Glynn 1997). Two species of mushroom corals (Family Fungiidae) experienced local extinctions at Española Island, Galápagos Islands, but survived at Floreana Island. Thirteen hardy reef building coral genera that survived severe bleaching events are generally nonbranching colonies (Glynn 2012). Species in two of these genera, *Porites* and *Pavona*, contribute significantly to reef building in the eastern Pacific.

During the 2015–16 El Niño event, we had the opportunity to track the algal symbiont communities in tagged *Pocillopora* colonies and follow the effects of

---

Jurassic Period (~160 mya). Todd received the ICRS 'Eminence in Research' award in May 2024. I thank Drew. C. Wham for first describing *Symbiodinium glynnii*, which was later reassigned to the genus *Durusdinium*.

intense heat stress at Uva Island for the third time and to compare this El Niño bleaching disturbance with the previous events. Here, tagged colonies showed shuffling away from *Cladocopium* towards *Durusdinium glynnii* following the onset of heat stress. Genetic analysis showed that the *Pocillopora* population continued to host increasing abundances of *Durusdinium* symbionts and was progressively less susceptible to heat stress over the three ENSO episodes. That was not the case, however, for other coral species. Non-*Pocillopora* corals at Uva Reef did not host high abundances of *Durusdinium* symbionts, and experienced similar bleaching and mortality across the three ENSO occurrences (Palacio-Castro et al. 2023).

Millepora Responses

The severe bleaching and mortality of the three species of *Millepora,* discovered in the Gulf of Chiriquí in 1970, was disheartening. These hydrocorals were the first ever reported, living or fossil, in the eastern Pacific region. Two of these species, *Millepora intricata* and *Millepora platyphylla*, are well known Indo-Pacific species, and *Millepora boschmai* was a new species described by Wally (Walentina) H. de Weerdt and me (de Weerdt and Glynn 1991). *Millepora* species hosting only Clade A symbionts are very susceptible to bleaching and mortality (Baker 1999, 2003). Their widespread global distributions suggest they are capable of long-distance dispersal and survival.

The exceptional occurrence of these *Millepora* species in the Gulf of Chiriquí captivated our interest. We were constantly checking on their health and numbers, and searching for new populations. In some sense, we began to feel like the guardians of eastern Pacific fire corals. In 1992, three members of our intrepid search team (Joshua Feingold, Juan Maté, and Mark Eakin) and I were inspecting a population of dead *M. boschmai* colonies, still in erect growth position, in a sheltered embayment on the north side of Uva Island. After a brief time on the bottom Mark excitedly approached me and signaled to follow him. He had discovered a living colony of *M. boschmai*, and this less than 1 year after Wally and I claimed their extinction following the 1982–83 El Niño bleaching event (Glynn and de Weerdt 1991). Further searching revealed a total of five living colonies. Viewing these living colonies evoked two thoughts, one of joy, that our newly described hydrocoral species was not extinct, and one of chagrin, contradicting our 1991 announcement in the journal *Science* of a rare coral extinction. This discovery caused some chiding from my students and colleagues. Thanks to Mark Eakin's gift of name giving, the site of rediscovery in 1992 became known as Lazarus Cove, acknowledging the resurrection of Boschma's *Millepora*.

We continued to enjoy visiting and viewing the few surviving fire corals until 2000, when again only dead colonies could be found. No live colonies of *M. boschmai* have been observed anywhere in Panama or elsewhere in the eastern Pacific at the time of this writing (November 2023). Boschma's *Millepora* was found in Indonesia in 2002, however (Razak and Hoeksema 2003); this record brings into question the endemic status of the hydrocoral in the eastern Pacific, and the mystery of how the nearly 18,000 km separation of the populations occurred. Live colonies of *M. platyphylla* have not been seen in Panama or at other continental eastern Pacific sites since 1983 (38 years), however, five colonies were discovered recently at Clipperton Atoll (Boissin et al. 2019). Small recruits of *Millepora intricata* reappeared on the Uva reef just 2 years following their disappearance in 1983. Several colonies at 12 to 18 m depth on the forereef slope of Uva Reef survived the 1982–83 and 1997–98 El Niño bleaching events. These colonies, sheltered from warm surface waters and high irradiance, likely served as a source population for larvae recruiting to shallow reef zones. Tyler Smith and colleagues referred to this scenario in proposing the thermal refuge hypothesis (Smith et al. 2014, 2017). These eclectic examples of the responses of corals to ENSO stressors, vis-à-vis (a) endosymbiont diversity and host flexibility, (b) small extinction-prone coral populations, and (c) potential refugia in habitat niche variability (different environmental conditions), offer a glimpse of the factors that need to be considered in the survival of corals to global climate change.[4]

Symbiotic Crustaceans

Like endosymbiotic zooxanthellae, several crustacean species are obligate symbionts of pocilloporid and acroporid corals. Crab (*Trapezia* spp.) and shrimp (*Alpheus lottini*) obligate symbionts live among the branches of *Pocillopora* spp. throughout the Indo-Pacific and eastern Pacific regions. These crustaceans feed on coral mucus (Knudsen 1967; Castro 1976, 1988; Patton 1976), coral "fat bodies" (Stimpson 1990), and occasionally coral tissues (Rinkevich et al. 1991).[5] In turn, they offer their coral hosts protection from predatory sea stars (*Acanthaster*, *Culcita*) and settling spores and larvae, and assist in colony cleaning activities (Glynn 1983a, b). Two common guard crabs in the Indian Ocean, *Trapezia tigrina*, inhabiting

[4] The coral holobiont is a diverse collection of organisms intimately associated with a coral animal host. The microbial community, the microbiome, represents a relatively newly recognized and vital component of the coral holobiont. Rohwer's (with Merry Youle) informative and charmingly illustrated book (2010) is recommended for an introduction to this discipline and also Thurber et al. (2017), Avila-Magaña et al. (2021), and Voolstra et al. (2021) for recent insights to the potential roles of the microbiome—(bacteria, fungi, Archaea, endolithic algae, protists), including the zooxanthellae (Symbiodiniaceae) and viruses—during periods of normal and stressful conditions.

[5] Peter Castro, Professor Emeritus, California State Polytechnic University, Pomona, has advanced our knowledge of the systematics and ecology of the symbiotic brachyuran crustaceans of *Pocillopora* in the eastern Pacific. Peter exhibited enthusiasm and outstanding performance in the first biology courses I taught at the University of Puerto Rico, Mayaguez in the early 1960s. When Peter visited my laboratory at STRI in 1974, on sabbatical leave from Pomona, we worked together on Panamanian coral reefs and shared a research experience on the Sula III Expedition at Gorgona Island, Colombia. Peter documented the frequent movements of *Trapezia* among coral colonies (Castro 1978), which was recently corroborated by Canizales-Flores et al. (2021) in Mexico. This finding would have implications for the vulnerability of the crabs to predation during disturbances that cause the crabs to leave their stressed coral hosts.

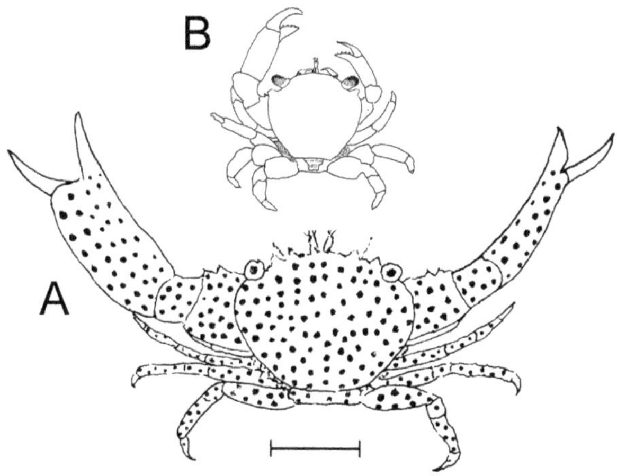

Fig. 5.6 Coral crab guards, Gulf of Oman. (**A**) *Trapezia tigrina*; (**B**) *Tetralia cavimana*. Scale bar = 1 cm

Pocillopora colonies, and *Tetralia cavimana*, a symbiont of *Acropora* species, demonstrated very different levels of protection during an *Acanthaster* outbreak in the Gulf of Oman in 1978–80. Most of the *Pocillopora* corals were effectively protected by *Trapezia*, a relatively large and aggressive guard, but sea star corallivores consumed large tracts of *Acropora*, protected by smaller and weaker *Tetralia* guards (Fig. 5.6; Glynn 1987).

During El Niño bleaching events many of the diverse crustacean species associated with *Pocillopora* became visible against their coral host's stark-white background. This is especially the case for *Trapezia* crabs and *Alpheus* shrimp whose exoskeletons are shades of orange and red. I prodded some of the larger crustaceans to determine their aggressive/escape activities and found that many were dead; those alive were lethargic and neither assumed a defensive posture nor scurried for shelter among their coral host's branches. Since these crustaceans depend on their hosts for sustenance, it is likely that the dead crabs still present on bleached and dead colonies succumbed from starvation. Mucus production is significantly diminished in moribund corals (Glynn 1983a; Glynn et al. 1985). Sandra Gilchrist, associated with my laboratory as a post-doctoral fellow at STRI,[6] observed some

[6] Sandra is presently a Professor of Biology and Marine Science, and Division Chair of Natural Sciences at New College of Florida, Sarasota. She earned a Ph.D. degree under Larry (Lawrence) Abele's guidance at Florida State University, Tallahassee. Both Sandra and Larry contributed enormously to crustacean decapod studies in Panama. Larry, also a post-doctoral fellow at STRI, was particularly interested in the factors promoting crustacean species diversity in corals and coastal habitats subject to varying degrees of disturbance (Abele 1974, 1976a, b). Recognition of biodiversity loss as early as the 1970s, especially in tropical ecosystems, pressed interest in this subject at STRI. One of our early laboratory studies of corallivores (animals that prey on the tissues of scleractinian corals), indicated that hermit crabs often scrape live coral branches causing extensive local tissue mortality and bioerosion (Glynn et al. 1972). Sandra's field observations supported this result and demonstrated that hermit crabs were significant corallivores on Pacific and Caribbean coral reefs (Gilchrist 1985). The effects of hermit crabs on bleached corals and their survivorship following El Niño bleaching events need studying.

unexpected field and laboratory feeding behaviors of *Trapezia* spp. during coral bleaching stress. On normal or non-bleached *Pocillopora*, Sandra observed that the crab symbionts stimulated the release of mucus by plucking or rubbing coral polyps with their pincers (chelipeds) and walking legs (pereopods), which is the usual way of feeding. On bleached colonies, however, the crabs engaged in cheliped fanning and pereopod waving, activities associated with plankton filter feeding. Filter feeding behavior became more active when Sandra increased the concentration of zooplankton near crabs in field and laboratory trials; in addition, gut analyses demonstrated that the crabs captured and ingested the zooplankton food source offered. To what extent this alternative feeding behavior contributed to crab survivorship during bleaching events is unknown.

Considering the high abundances of symbiotic crustacean species inhabiting live *Pocillopora* corals (Abele 1976a; Abele and Patton 1976; Glynn 1976; Gotelli et al. 1985), the loss of coral hosts must cause a heavy toll on cryptic symbiotic reef invertebrates and trophic resources for fish consumers. From numerous early studies of *Pocillopora* corals and their symbiotic crustaceans, before and following the 1982–83 El Niño bleaching event, it is possible to compare estimates of coral host mortalities and their symbiotic faunas. The pre-El Niño median density of *Pocillopora* on the Uva reef was 7.8 live colonies per m^2 (Table 5.1). This decreased to 1.8 colonies per m^2 post El Niño, resulting in a mean decline of 77%. If now the median densities of crustacean symbionts (*Trapezia* spp. and *Alpheus lottini*) per coral colony are adjusted to the number of colonies before and after the bleaching event, we can appreciate the magnitude of crustacean losses. The combined median density of crabs and shrimp before El Niño bleaching was 143.4 ind. per m^2 and 0.97 after the disturbance, resulting in a 99.3% decline. These are the largest and most conspicuous symbiotic crustaceans associated with *Pocillopora*; there were several other obligate symbiont species, such as *Fennera chacei* (8.4 ind. per colony) and *Harpiliopsis spinigerus* (2.0 ind. per colony) present on pocilloporid

Table 5.1 Estimates of *Pocillopora* spp. and crustacean symbiont (*Trapezia* spp., *Alpheus lottini*) mortalities caused by the El Niño 1982–83 bleaching event, Uva Island coral reef, Gulf of Chiriquí, Panama. Individual coral and crustacean densities (no. per m^2) are expressed as median values. The first and third quartile values (Q_1, Q_3) denote the spread of the data below and above the middle median (Md) values

Species	El Niño 1982–83		Source
	Before	After	
Pocillopora spp.[1]	7.75 (5.45–10.15) N = 12 quadrats	1.78 (1.25–2.33)[4]	Glynn (1974), Glynn et al. (1988)
Crustacean symbionts[2]	143.4 (100.8–187.8)[3] N = 56 colonies	0.97 (1.4–2.5)[3]	Glynn (1976, 1985b)

1. Predominantly *Pocillopora damicornis*, range of colony diameters 12–25 cm
2. *Trapezia* spp., *Alpheus lottini*, adult and juvenile individuals
3. Calculated from crustacean densities per colony: before El Niño Md (Q_1, Q_3) = 18.5 (14–21) ind.; after El Niño Md (Q_1, Q_3) = 0.5 (0–1) ind.
4. N = 21 affected coral reef sites sampled in the Gulf of Chiriquí; 77% mean *Pocillopora* mortality

corals before 1982–83 (Abele 1976a). These species likely experienced mass mortalities since they depend on their coral hosts for food and shelter. The mortality estimates noted are for one m^2 of the Uva Reef; the entire reef covered an area of ~2 ha or 20,000 m^2. The magnitude of coral loss, including the cryptic community of crustacean symbionts, was enormous.

Trapezia spp. and *Alpheus lottini* defend their coral hosts from sea star (*Acanthaster*) attacks. The high mortality of *Pocillopora* colonies and their crustacean guards had an unexpected indirect effect on a large patch of massive corals that have provided cover over the years for turtles, whitetip sharks, moray eels, diverse fishes, spiny lobsters, inter alia. These reef-associated animals are attracted to chamber-like shelters formed between partially eroded central coral bases and descending lateral colony skirts. This cluster of *Gardineroseris planulata*, known by our team as *Gardineroseris* City (*G* City), initially numbered 23 colonies. The largest colony had an uninterrupted vertical growth record (annual density bands) of 254 cm and a radiocarbon age of 192 years (Glynn 1985a). Before the 1982–83 El Niño bleaching event, continuous stands of live *Pocillopora* corals surrounded G City. These corals formed a protective barrier surrounding *G* City, thanks to their stinging cells (nematocysts) and crustacean guards. Before 1983, *Acanthaster* was neither observed crossing this barrier nor feeding on *G. planulata* within the refuge zone. In testing the effectiveness of the *Pocillopora* barrier, we nudged *Acanthaster* to the edges of the live coral barrier and observed immediate sea star irritation and coral guard attacks. The sea stars quickly retreated to surrounding dead coral or rubble bottom areas. Major coral bleaching and mortality had occurred during the first half of 1983 and by the end of October *Acanthaster* had crossed the dead tracts of *Pocillopora* and were preying on *Gardineroseris*. Ten percent of the surface tissues of the largest colony had been consumed by the end of October and 40% by July 1984. *Acanthaster* feeding continued for several years, but most colonies still contained patches of live tissue. Based on the uninterrupted growth of the oldest coral colony in *G* City, it is likely that the *Pocillopora* barrier has been effective for at least 192 years.[7] In a size-structured population model developed by Peggy Fong,[8] based on the *G* City coral cluster, it was found that *Acanthaster* was potentially of

---

[7] The occurrence and numbers of *Acanthaster* skeletal elements in reef sediments indicated the long-term presence (2000-3000+ years) of this sea star corallivore on the Uva Island coral reef. (See *Acanthaster* in Chap. 7).

[8] When I introduced Peggy Fong to our coral reef studies in the eastern Pacific, she was a Rosenstiel post-doctoral fellow at RSMAES (her Ph.D. studies were conducted in a joint University of California Davis and San Diego State University Program). Along with the eastern Pacific investigations, she also documented effects of Hurricane Andrew on the Florida Reef Tract and explored the relationship between nutrients and algal/coral interactions. Peggy was fascinated with our research agenda — ENSO disturbances, coral bleaching, coral community recovery, and upwelling effects — and soon joined our team. She greatly benefited our research agenda in the areas of community and ecosystem modeling and algal ecology. Peggy is now a Professor in the Department of Ecology and Evolutionary Biology, University of California Los Angeles, where she continues to study human impacts on coastal marine ecosystems, with an intent to provide solutions to environmental degradation.

overwhelming importance in causing coral mortality compared to El Niño bleaching events (Fong and Glynn 1998). Further, this simulation model was able to accurately predict the regional response of *G. planulata* to the 1997–98 ENSO disturbance event (Fong and Glynn 2000). Considering the demonstrated benefits that obligate crustaceans bestow on coral host health and survivorship, perhaps these metazoan symbionts should also be considered an essential component of the coral holobiont consortium.

Corallivore and Bioeroder Responses

Acanthaster was moderately abundant in the Gulf of Chiriquí and on the Uva Island coral reef during coral bleaching in 1983. Sea star densities ranged between 2 and 15 individuals per hectare between 1982 and 1984 (Enochs and Glynn 2017). Even this relatively low density was concerning because after bleaching the crustacean guards of *Pocillopora* corals were not effective in colony defense, and other coral species were greatly reduced in abundance, and were more exposed and accessible to *Acanthaster* predation (Glynn 1985a). Numerous surviving cryptic and shade-tolerant species present among *Pocillopora* stands, such as *Porites panamensis*, *Pavona varians*, and juvenile *Gardineroseris planulata*, were consumed by *Acanthaster* in the aftermath of the bleaching event. Predator concentration, an upsurge in the predator/prey ratio, increased on the Uva reef and resulted in higher rates of coral mortality. Surprisingly, no scleractinian species became locally extinct. Inexplicably, from 1985 to 2010, *Acanthaster* abundances declined significantly, with Uva reef sea star densities of ≤ 1 per hectare.

Another corallivore, an ovulid gastropod, *Jenneria pustulata*, experienced high mortality in 1983 with numerous empty and unbroken shells scattered around the bases of *Pocillopora* as evidence of the El Niño disturbance (Fig. 5.2). *Pocillopora* is the preferred prey of *Jenneria* and the gastropod's usual habitat. *Jenneria* is seemingly cryptic in mimicking the coral's skeletal surface and expanded polyps. The mean density of *Jenneria* on the Uva reef in 1982 ranged from 14.7 to 27.7 individuals per m^2; mean densities in 1984 ranged from 0 to 4.0 individuals per m^2. From laboratory studies of heat tolerance and coral tissue nutritional needs, the gastropod corallivore likely succumbed from a combination of high temperature stress and food deprivation (Oramas 1979; Achurra Cárdenas and Valdés Araúz 1980; Glynn 1985b).[9]

[9]As a research scientist at STRI, my primary responsibility involved coral reef research. STRI scientists were encouraged, however, to mentor upper class students in the school of Biology at the University of Panama. I greatly enjoyed this pedagogical interaction and mentored nearly twenty Panamanian students during my time at STRI. Their research projects involved marine biological investigations and mostly studies on Pacific coral reefs. Field study sites included coral reefs in the gulfs of Panama and Chiriquí, and occasionally in Costa Rica. Three of these students, Felix Oramas, Juan Achurra, and Eduardo Araúz, accomplished important advances in our understanding of the biology and ecology of *J. pustulata*. Their research involved laboratory experiments, behavioral observations, and field studies. Citations of their theses—a substantial written docu-

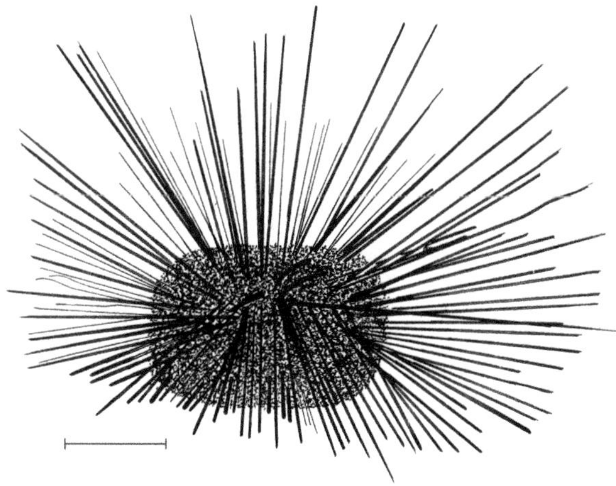

Fig. 5.7 *Diadema mexicanum*, scale bar = 2 cm

A dramatic consequence of the 1982–83 El Niño disturbance was the sudden increase in sea urchin abundances on coral reefs in Panama and the Galápagos Islands. This was of concern considering the strong eroding effects (bioerosion) of sea urchins when they graze on coral substrates. They possess a powerful feeding apparatus with five calcareous teeth (Aristotle's lantern) which they can adroitly maneuver to scrape coral surfaces. The sea urchin species in Panama is the Black Spine or Needle Urchin, *Diadema mexicanum,* and in the Galápagos Islands the Slate Pencil Sea Urchin, *Eucidaris thouarsii galapagensis.*[10] *Diadema* is primarily an herbivore that scrapes fouling algae from dead coral skeletons, and in the process erodes the underlying skeleton (Fig. 5.7). *Eucidaris* grazes on algae in a similar manner, but also preys on barnacles and live coral, primarily *Pocillopora* spp. Jerry Wellington, Chuck Birkeland, and I documented this carnivorous feeding habit in the Galápagos Islands in the mid-1970s (Glynn et al. 1979; Glynn and Wellington 1983). We observed live coral feeding behavior (carnivory) by sea urchins grazing on the edges of patch reefs. We also examined the gut contents of sea urchins and found that ingested coral tissue was digested and absorbed.

ment is a requirement for baccalaureate graduation at the University of Panamá—are listed in the "References" section at the end of this chapter. In recognition of their research accomplishments, Felix and Juan accompanied me, all expenses covered, to the meeting of the Association of Island Marine Laboratories of the Caribbean in Kingston, Jamaica in 1969. This was an exhilarating experience, an opportunity to meet Tom and Nohora Goreau, Ivan Goodbody and other researchers and students engaged in tropical marine studies.

[10] The Galápagos Slate Pencil Sea Urchin is now classified as a subspecies of *Eucidaris thouarsii* which occurs in eastern Pacific mainland waters. This sea urchin is geographically restricted to the Galápagos and Cocos Islands and is morphologically distinguishable from *E. thouarsii* (Fig. 2.10). It is larger and more robust than *E. thouarsii* and not secretive as adults, commonly observed grazing on exposed surfaces. For systematic information see Kroh and Mooi (2021). World Echinoidea Database. *Cidaris (Eucidaris) thouarsii* (Agassiz and Desor 1846). Accessed through: World Register of Marine Species at: http://www.marinespecies.org/aphia.php?p=taxdetails&id=368330 on 2021-10-22

In the 1970s, *Diadema* densities on the Uva reef and other coral reefs in Panama were low, usually ranging between 2 to 5 individuals per m^2. They began increasing in abundance in 1984, mostly on dead *Pocillopora* frameworks where they were actively grazing mostly on algae that had settled on dead coral skeletons. By 1985, their numbers had increased dramatically, attaining densities between 40–100 individuals per m^2. This density made sampling more difficult as it became necessary to carefully adjust our buoyancy to avoid sea urchin spine punctures. The reason for this population outbreak is unknown; increased dead coral habitat providing shelter and elevated productivity and food resources associated with La Niña (occurring on the heels of El Niño 1983) have been proposed as causal factors. It is also possible that pre-ENSO high live coral cover, producing mucus secretions and with nematocyst defenses, interfered with sea urchin larval settlement.

Sea urchin bioerosion greatly exceeded coral calcification for several years, until about the late 1990s when much of the framework in some reef zones was reduced to rubble and sand. One such area on the Uva reef became known as the "lunar scape" due to its barren, lifeless appearance. There was a close correspondence between the loss of coral framework, potential shelter, and the decline of sea urchins, as if they bioeroded themselves out of "house and home". In early 2000 coral recovery was in rapid progress and by 2015 had reached the extent of live cover and framework development similar to most pre-El Niño reef zones in the 1970s. The time to full recovery, with respect to live coral cover and framework construction, required a little more than 30 years at the Uva reef.

Reef Fishes

Two noteworthy reef fish responses to El Niño 1982–83 that we documented, involving a pufferfish and community-wide changes in trophic resource utilization, are briefly noted here. The Guineafowl puffer, *Arothron meleagris*, is an Indo-Pacific corallivore closely associated with eastern Pacific coral reefs (Fig. 5.8). It is commonly seen hovering over coral and occasionally nipping and ingesting 1–2 cm-long terminal branch tips of *Pocillopora* or similar-size fragments from massive corals such as *Porites* spp. and *Pavona* spp. This pufferfish possesses two striking color phases, golden-yellow and blue-black sprinkled with white spots. These color phases may represent a warning signal (aposematism) to potential predators; the pufferfish are sluggish swimmers and easily approached. They often gorge themselves with coral, typically *Pocillopora* branch tips, until their distended abdomens hang down awkwardly and appear to impede swimming. Feeding preference trials on a Costa Rican reef have shown that the Guineafowl puffer is a facultative corallivore; when *Pocillopora* is in low abundance it can switch its diet to other coral species and even non-coral prey such as coralline algae (Guzman and Robertson 1989). Before 1983, the Uva reef pufferfish population ranged from about 40 to 50 individuals, but this number declined for a few years following the catastrophic loss

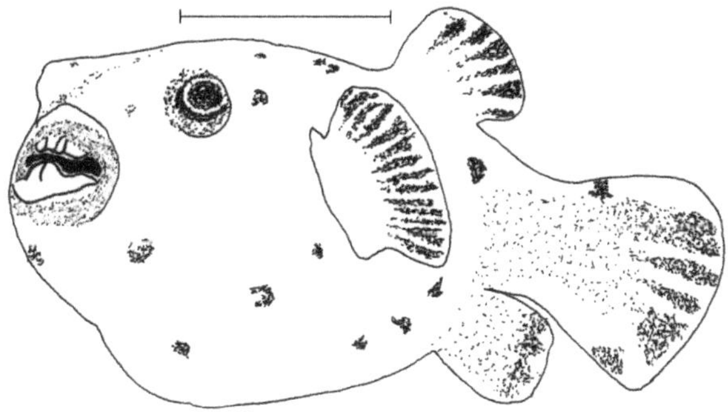

Fig. 5.8 *Arothron meleagris*, scale bar = 5 cm

of *Pocillopora*. Since *A. meleagris* was often present in off-reef habitats their mobility and flexible diets aided in their survival.

Since the early 1970s, we have studied 63 of the larger ($\geq$ 15 cm total length) fish on the Uva reef. The diet information obtained has allowed us to assign fishes to trophic groups, e.g. *facultative corallivores* (fish species consuming primarily scleractinian corals), *mixed diet feeders* (consuming fishes and invertebrates), and *benthic invertivores* (consuming primarily motile benthic invertebrates). From 1980 to 2010 we have conducted frequent quantitative fish surveys to monitor and assess fish population densities in response to disturbance events and availability of trophic resources (Glynn et al. 2014). Reef fish density (all fish densities combined) and species richness (number of species present) varied over the 30-year period but did not show a significant increasing or decreasing temporal trend. Some intriguing relationships became apparent, however, when fish trophic groups were compared with coral recovery over periods of several years. *Facultative corallivores*, such as the predominant *Arothron meleagris* pufferfish, were uncommon on Uva reef after the massive loss of *Pocillopora*. The pufferfish began to show a consistent increase in the early 1990s when recovering *Pocillopora* reached 25–30% cover. Fishes in the *mixed diet feeder* guild demonstrated a significant linear decline with increasing coral recovery. We hypothesized that as coral recovery progressed this led to an increase in topographic complexity and more sheltering sites for potential fish and invertebrate prey. *Benthic invertivores* demonstrated a significant increase in abundance over 10 years (1984–1994), up to about 20% coral cover, which corresponded closely with the recovery of *Pocillopora* and the increasing availability of coral-associated prey. As coral cover continued to increase to 30% and higher, however, *benthic invertivore* abundances began to decline. This trend was parabolic and highly significant with low invertivore abundances at low coral cover, increasing and reaching maximum abundance at moderately high coral cover, and then decreasing at full pre-disturbance cover. The above three relationships are based on correlation analysis and only suggestive cause-and-effect interactions between different fish trophic guilds and coral food and shelter resources. Additional fish and coral monitoring data are needed at Uva and other reefs in the region, as well as manipulative field experiments, to test these hypothesized functional relationships.

My colleague and close friend, Kiyoshi Yamazato,[11] invited me to attend and deliver a talk at a workshop and symposium on coral reef and mangrove conservation and management in Okinawa in 1987. This invitation arrived when I was immersed in documenting the decline of equatorial eastern Pacific coral reefs. Since Kiyoshi's invitation left open the subject of my presentation, I thought it would be an opportune time to present the alarming rapid decline of eastern Pacific reefs to attendees from the western Pacific and Indian Oceans. By this time enough data were available to quantify sea urchin bioerosion and reef framework destruction in Panama and the Galápagos Islands. When I sent my proposed presentation to Kiyoshi, I had trepidations that eastern Pacific coral bleaching and reef decline may not have been appropriate for the upcoming conference. Thankfully, Kiyoshi's response was that this subject was exactly what he had in mind. In fact, he had described a warm-water coral bleaching event in Okinawa in 1980 (Yamazato 1981) and was fully aware of the threat of global warming (Hoegh-Guldberg 1999). Also, coral bleaching and mortality had occurred on the Great Barrier Reef, in the Java Sea, French Polynesia, and the southwestern Indian Ocean during the 1982–83 period, in close association with the exceptionally strong El Niño event (Lessios et al. 1983; Coffroth et al. 1990). Indeed, soon nearly all of the world's coral reefs would be in peril and experiencing significant degradation (Glynn 1993; Wilkinson 1998). The report edited by Clive Wilkinson (1998), "Status of Coral Reefs of the World: 1998", a product of the Global Coral Reef Monitoring Network, was the first of several coral health assessments which have continued to the present (Souter et al. 2021). This is a particularly valuable resource for documenting and updating the status of coral reefs worldwide, necessary information for implementing protection, restoration, and policy decisions regarding this fragile ecosystem.

As occurred in Panama after 1983, dead coral skeletons in the Galápagos Islands were invaded by high numbers of sea urchins, in this case the Slate Pencil Sea Urchin *Eucidaris*. These elevated sea urchin densities were not due to heavy recruitment, however, but from the redistribution of existing individuals into tightly-knit aggregations. This, too, resulted in increased sea urchin grazing and the rapid bioerosion of dead coral colonies and frameworks. Our earliest studies of the grazing rates of *Eucidaris* in 1975 and 1976 revealed notably high per capita coral erosion (Glynn et al. 1979). Over a 24-hour period, a single average-sized *Eucidaris* eroded 2.0 gm dry weight of coral. The feeding rate of *Diadema* was only about 0.2 gm coral per individual per day. The high population densities of *Diadema*, however, upwards of 50 to 150 individuals per m^2, resulted in significant reef degradation (Glynn 1988;

[11] Kiyoshi and I had met in 1957 at the Hopkins Marine Station, Pacific Grove, California, when he was immersed in Don Abbott's marine invertebrate summer course. I was the TA (teaching assistant) in that class. We became steadfast friends and developed similar interests in coral reef ecology in our research pursuits on opposite sides of the Pacific Ocean. He attained full professorship rank in the Department of Biology at the University of the Ryukyus, Okinawa, and was appointed director of the Sesoko Marine Science Center.]

Eakin 1996). By the early 1990s, the erosive grazing of *Eucidaris* reduced to rubble and sand all 17 of the known coral reefs in the Galápagos Islands. Other species contributed to coral breakdown, such as sponge and mollusk internal bioeroders (Scott et al. 1988; Reaka-Kudla et al. 1996), but sea urchins were preeminently destructive.

Not only patch reefs were degraded and lost, but also stately coral colonies that were part of or on the fringes of reef communities. Before 1982–83, these Methuselah-aged colonies occurred at several islands in the southern sector of the Galápagos Islands. The majority were massive species (*Porites lobata* and *Pavona clavus*), attaining diameters or heights of 1 to 5 meters (Glynn 1990). Age estimates of these colonies ranged from 100 to 500 years, based on linear growth dimensions and annual skeletal extension rates. Radiocarbon dating of one *Pavona* colony with a 2.2 m-long growth axis revealed a date of 420 ± 60 years BP (Macintyre et al. 1992). By 1990 nearly all colonies were dead and undergoing significant erosion. From the ages of these old corals that were irreparably damaged in 1983, I hypothesized that a disturbance event comparable to the 1982–83 El Niño had not occurred in the Galápagos Islands for as long as 400–500 years. Analysis of the skeletal growth record of a 10 m diameter *Pavona clavus* colony from Isabela Island revealed no extensive breaks or discontinuities indicative of tissue mortality from 1586 to 1953 A.D., thus lending support to favorable and uninterrupted growth over the past four centuries (Dunbar et al. 1994).

In 1995, Joshua Feingold completed his doctoral research at the Devil's Crown (DC) site, also known as Onslow Island.[12] His primary research was focused on a relatively deep (15 m) coral community located about 100 m east of DC that bleached, and remained so for several weeks, but eventually recovered and survived the 1982–83 El Niño disturbance.[13] Joshua was especially interested in following the recovery of the deep coral community, but also regularly monitored the iconic shallow *Pocillopora* patch reef inside the DC extinct crater. Occasionally a few *Pocillopora* recruits would appear at the former shallow reef site, and Joshua (and our field team in general) was hopeful that this reef would re-establish and fully recover following its complete degradation (and disappearance) in the late 1980s. On the heels of completing the Ph.D. degree, Joshua initiated a monitoring program to re-visit and quantitatively survey the deep reef community and the shallow crater site. He has managed to return and sample the two sites on 14 occasions over

[12] Nearly all of the major islands and islets in the Galápagos Archipelago have both Spanish and English names. While both names are often in use, a preference is given here to the Spanish names to help avoid confusion among the Galapageños, the Galápagos residents.

[13] The corals making up this community are mushroom corals in the family Fungiidae (*Cycloseris distorta* and *Cycloseris elegans*) and a small branching species in the family Siderastreidae (*Psammocora stellata*). All of these species are unattached and rest freely on a coarse sandy bottom. A brief description of these free-living corals and their occurrence in the Galápagos Islands can be found in Feingold and Glynn (2014) and Glynn et al. (2018).

25 years (1996–2021)[14]. A steady and gradual increase of *Pocillopora* occurred at the crater site, from three colonies in 1996 to 154 colonies in 2007. This number of colonies suggested that there was a reproductive population nearby producing a source of planula larvae that settled within the DC crater. Our hopes of recovery were dashed in 2009 when the DC census revealed only 11 colonies. This abrupt decline was preceded by a severe La Niña cold shock event that persisted for 8 months, causing coral bleaching and mortality across the entire Galápagos Archipelago (Glynn et al. 2017). This mortality event was disappointing and unexpected considering the thermal stress observations on corals in Panama, which suggested that elevated temperature extremes were more detrimental to zooxanthellate corals than low temperature extremes (Glynn and D'Croz 1990). But then significant coral recruitment was again observed at DC in 2017 (151 colonies), and this trend has continued to 2019 and 2021 (J. Feingold, pers. comm.). We are cautiously optimistic that *Pocillopora* recruitment will continue but are wary of both ENSO cold shock (La Niña) and stressful warming (El Niño) disturbances in the future.

El Juego de la Gallina Enterrada

Driving north to Ayangue on the Ecuadorian coast with René Espinosa, Bob Richmind, and Fernando Rivera, early team members, we suddenly came to a roadblock at San Pablo, a small fishing village. We were on our way to survey a *Pocillopora* community at Pelado Island, which is now a marine reserve (Alvarado et al. 2017). Our automobile was suddenly surrounded by scantily clothed men in loin cloth attire, donning feathered war bonnets and brandishing spears and bows and arrows. Inebriated armed men, this was an unsettling situation. What to expect?

We were coaxed out of our vehicle and urged to join the festivities. A crowd of revelers pushed us forward to a gathering surrounding a large circle. A blind-folded

[14] In 2001, Joshua conceived of an undergraduate field course focusing on the coral ecology of the Galápagos Islands for Nova Southeastern University, Ft. Lauderdale, Florida. In the summer season he would arrive in the Galápagos with a group of students to explore island life and perform research, including the coral surveys at DC. The Galápagos experience motivated some of these students to pursue graduate studies, directing their research to the coral communities of DC. Master of Science (MS) theses resulting from these studies included: mushroom coral distributions in the eastern Pacific by Amber Little (2008); changes in the *Porites lobata* community inside DC by Nicole Paul (2012); sediment core analyses from the deep DC coral community by Katherine Hendrickson (2014); and a *Psammocora* population study at Xarifa Island by Kathryn Brown (2016). Since Katherine's research contributed importantly to a long-term analysis of the deep coral assemblage, she was awarded co-authorship on a peer-reviewed publication in the journal *Coral. Reefs* (Feingold et al. 2020). Most recently, María José Guarderas (2023) completed her thesis research on the free-living *Psammocora stellata* population in the channel of DC. These corals disappeared following the 1982–83 ENSO and have now recovered to pre-disturbance numbers (perhaps even exceeding prior densities). All five MS studies are fully referenced below because of the value of demonstrating the notable changes occurring in DC coral communities in recent decades.

and thoroughly inebriated "warrior" in the circle was violently and wildly swinging a machete. There were a few near misses as the warrior stumbled toward the ring of spectators! What was going on? We watched the spectacle in earnest to try and understand the aim of this event. Then Fernando spotted a small, feathered object in the sand wildly jerking up and down. It was a chicken's head with its entire body buried in the sand. Now we understood the object of this game, to decapitate the chicken. The successful slayer could claim the chicken as a prize to take home to cook and eat. The alcohol got the best of this warrior, and the chicken survived. Before long, several thoroughly inebriated revelers urged Fernando to take part in the sport. He first needed to be properly fortified by drinking *aguardiente* (cane liquor) and a higher-octane potion circulating among the revelers. Fernando was reluctant but urged on, not so politely. Then, into the circle he was unceremoniously pushed, blind-folded and with a machete in hand. He was then forcefully spun around several times to eliminate directional cues. Fernando's search strategy was well conceived; he gently pushed the machete flat against the sand until an obstruction was felt, and the sound of a squawking chicken. Presto!, target identified, preparing for the coup de grace chop. Fernando was immediately seized by several warriors and removed from the killing field. The chicken prize was not meant for strangers. This sport, known as "The Game of the Buried Hen" in coastal Ecuador, originated in Spain as "El Juego de Gallos" ("The Game of the Roosters"). It was brought to the New World during the early Hispanic settlement period (Díaz Viana 1982). The game was in full swing by the indigenous peoples of San Pablo when we stumbled upon the festivities in late June.

Following a few good-willed toasts (and our relief), we covertly pushed rocks out of the road and were on our way again to Ayangue.

Coral Reproduction

Coral reproduction became a great concern early in our studies of ENSO-induced coral bleaching and mortality. How did extreme fluctuations in temperature affect coral reproduction? Would coral reproduction resume in surviving colonies with the return of pre-disturbance (normal) environmental conditions? And in the final stages of reproduction how would ENSO disturbances affect larval recruitment and survivorship? The answers to these questions are of critical importance since the recovery of coral populations depends on successful reproduction, recruitment, and the stability and persistence of eastern Pacific reproductive source populations. Our research efforts focused on the seasonal development of gonads and gametogenesis (the production of gametes, i.e., eggs and sperm), observations of coral spawning in the field and in mesocosms temporarily installed aboard ship, and larval settlement and coral recruitment surveys on reefs. Sampling sites for coral reproduction were established at equatorial eastern Pacific reef localities severely affected by the 1982–83 El Niño bleaching event: Caño Island, Costa Rica; upwelling (Gulf of Panama) and non-upwelling (Gulf of Chiriquí) centers in Panama; and the southern

Galápagos Islands (Ecuador). Here attention is directed toward the responses of coral sexual reproduction to ENSO-related variations in thermal conditions. A more comprehensive treatment of coral reproduction, including the field sampling program, laboratory procedures, the attempts to observe spawning events, asexual reproduction and recruitment, and mention of the members of our team that made these investigations possible, is presented in Chap. 6, "Coral Reproduction and Recruitment".

Our studies revealed that most coral species at all sites tended to become sexually active with gamete development during seasonal sea warming periods and throughout most of the warm-wet season. Also, some species exhibited year-round sexual activity, even during cool and seasonal upwelling periods. These observations were especially welcome considering the potential for augmenting coral abundances that suffered severe declines during ENSO thermal stress. Even more encouraging were the large numbers of coral colonies exhibiting reproductive activity soon after the 1982–83 El Niño disturbance. Coral species contributing most to reef building in the eastern Pacific—*Pocillopora* spp., *Porites lobata*, and *Pavona clavus*—exhibited gonadal development and the presence of mature gametes as early as 1984, 1985, and 1986. Later in our study, and with the opportunity of sampling corals through the weak 1990–92 and strong 1997–98 El Niño bleaching events, we found that most populations remained reproductive. At the Galápagos sampling sites in 1991 the usually cool season remained warm and three species (*Porites lobata, Pavona chiriquiensis*, and *Pavona varians*) continued with gamete development for several months past the warm season. In 1997, samples revealed a greater proportion of reproductive colonies (*Pocillopora damicornis* and *Pocillopora elegans*) than in non-El Niño years and *Porites lobata* colonies exhibited all stages of gametogenesis and spawned gonads (i.e., gonads empty or with residual non-released eggs and sperm). *Pocillopora damicornis,* however, experienced a significant decline in reproductive activity in the Gulf of Chiriquí, Panamá, but demonstrated significant enhanced gametogenesis in the Galápagos Islands. These different responses were observed in other species and underscore the variability of coral reproduction during an El Niño warming event. Numerous environmental factors are at play adding to the complex and unpredictable nature of this sensitive life history process (Colley et al. 2006; Santiago-Valentín et al. 2018).

Since many coral species tend to reproduce during the warm season it is logical to hypothesize that La Niña events could depress reproduction due to abnormally lower temperatures. For corals to develop mature gonads and produce gametes it is reasonable to assume that they would require a thermal Goldilocks zone, that is temperature conditions not too hot and not too cold. Irregularly occurring La Niña cool periods affecting coral reproduction are seldom evaluated. Since La Niña cool events often follow El Nino warm conditions, separating the effects of these two stressful periods is difficult. An unusual set of observations from the Gulf of Panama, involving gametogenesis and thermal history, offers a hint of stressful cooling effects on gametogenesis. Despite the 1986–87 El Niño event, unusual sea warming did not occur in the Gulf of Panama area in 1987, but the lowest temperatures over a 40-year period were recorded 1 year later in 1988 (Glynn et al. 2017). Surprisingly, three coral species subject to this severe cooling period (*Pavona varians*, *Pavona clavus*, and *Psammocora stellata*) demonstrated high rates of gametogenesis in 1989. Mature gametes were observed during this period, ranging from >80% to 100% present in samples. These responses, at least for the three species examined, suggest that extreme low temperatures may not necessarily nullify coral reproduction. Anomalous low temperatures promote bleaching by photoinhibition and lower metabolic activity of both the coral host and symbionts; however, La Niña also promotes seasonal upwelling and increased plankton availability as a food source, which may enhance energy reserves and reproduction (Grotolli et al. 2006). In an experimental study in an upwelling area off the central Mexican Pacific coast, Paola Rodríguez-Troncoso and colleagues (Rodríguez-Troncoso et al. 2014; Fig. 5.9) reported on the physiological responses of *Pocillopora verrucosa* to cold stress conditions. This is an area subject to pronounced spring-time upwelling events and coral bleaching. These workers suggested that the stressed corals likely utilize plankton, phagocytize their algal symbionts and lastly metabolize their lipid reserves to offset the low thermal effects. The extreme thermal swings between El Niño and La Niña ENSO phases require further study. There are many other factors besides temperature, however, that influence reproductive processes, such as day length, nutrients, sea state (turbulence and calm periods), and solar and lunar irradiance. Peter Harrison's (2011) review should be consulted to gain an appreciation of the known environmental conditions that affect coral reproduction.

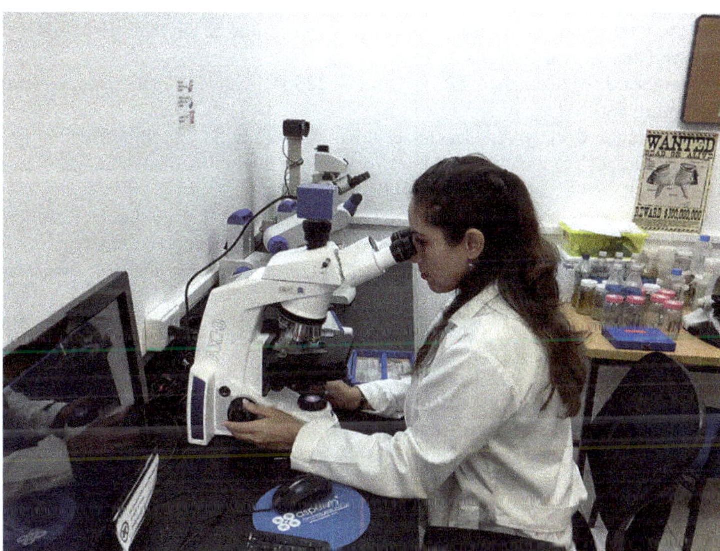

Fig. 5.9 Alma Paola Rodríguez Troncoso scanning slides for presence of gametes in a colony of *Pocillopora verrucosa* collected at Islas Marietas National Park (28 January 2021). Photograph courtesy of Leonardo Maisonnave

References

Abele LG (1974) Species diversity of decapod crustaceans in marine habitats. Ecology 55:156–161

Abele LG (1976a) Comparative species richness in fluctuating and constant environments: coral-associated decapod crustaceans. Science 192:461–463

Abele LG (1976b) Comparative species composition and relative abundance of decapod crustaceans in marine habitats of Panama. Mar Biol 38:263–278

Abele LG, Patton WK (1976) The size of coral heads and the community biology of associated decapod crustaceans. J Biogeogr 3:35–47

Achurra Cárdenas JB, Valdés Araúz E (1980) Estudios sobre los efectos predatorios de los moluscos *Coralliophila abbreviata* y *Jenneria pustulata* de ambos literales del Istmo de Panama. Tesis, Universidad de Panamá, Escuela de Biología, República de Panamá

Agassiz L, Desor PJE (1846) Catalogue raisonné des familles, des genres, et des espèces de la classe des échinodermes. Annales des Sciences Naturelles, Troisième Série, Zoologie 6:305–374

Alvarado JJ, Aburto-Oropeza O, Abad R, Barraza E, Brandt M, Cantera J, Estrada P, Gaymer CF, Guzmán-Mora AG, Herlan JJ, Maté JL (2017) Coral reef conservation in the eastern tropical Pacific. In: Glynn PW, Manzello DP, Enochs IC (eds) Coral reefs of the eastern tropical Pacific: persistence and loss in a dynamic environment. Coral reefs of the world 8. Springer, Dordrecht, pp 565–591

Avila-Magaña V, Kamel B, DeSalvo M, Gómez-Campo K, Enríquez S, Kitano H, Rohlfs RV, Iglesias-Prieto R, Mónica M (2021) Elucidating gene expression adaptation of phylogenetically divergent coral holobionts under heat stress. Nat Commun 12:5731. https://doi.org/10.1038/s41467-021-25950-4

Baker AC (1999) The symbiosis ecology of reef-building corals. Ph.D. dissertation. University of Miami, Coral Gables, p 120

Baker AC (2003) Flexibility and specificity in coral-algal symbiosis: diversity, ecology, and biogeography of *Symbiodinium*. Annu Rev Ecol Syst 34:661–689

Baker AC, Rowan R (1997) Diversity of symbiotic dinoflagellates (zooxanthellae) in scleractinian corals of the Caribbean and eastern Pacific. In: Proceedings of the 8th international coral reef symposium, Panama, vol 1, pp 1301–1306

Baker AC, Starger CJ, McClanahan TR, Glynn PW (2004) Corals' adaptive response to climate change. Nature 430:741

Boissin E, Pogoreutz C, Pey A, Gravier-Bonnet N, Planes S (2019) *Millepora platyphylla* (Cnidaria, Hydrozoa) range extended back to the eastern Pacific, thanks to a new record from Clipperton Atoll. Zootaxa 4668(4):599–600

Brown KA (2016) Population fluctuation of the nodular coral *Psammocora stellata* in the Galápagos Islands, Ecuador: an indicator of community resilience and implications for future management. MS thesis. Halmos College of Natural Sciences and Oceanography, Nova Southeastern University, p 96

Buddemeier RW, Fautin DG (1993) Coral bleaching as an adaptive mechanism. Bioscience 43:320–326

Canizales-Flores HM, Rodríguez-Troncoso AP, Rodríguez-Zaragoza FA, Cupul-Magaña AL (2021) A long-term symbiotic relationship: recruitment and fidelity of the crab *Trapezia* on its coral host *Pocillopora*. Diversity 13:450. https://doi.org/10.3390/d13090450

Castro P (1976) Brachyuran crabs symbiotic with scleractinian corals: a review of their biology. Micronesica 12:99–110

Castro P (1978) Movements between coral colonies in *Trapezia ferruginea* (Crustacea, Brachyura), an obligate symbiont of scleractinian corals. Mar Biol 46:237–245. https://doi.org/10.1007/BF00390685

Castro P (1988) Animal symbioses in coral reef communities: a review. Symbiosis 5:161–184

Coffroth MA, Lasker HR, Oliver JK (1990) Coral mortality outside of the eastern Pacific during 1982-1983: relationship to El Niño. In: Glynn PW (ed) Global ecological consequences of the

1982–83 El Niño Southern Oscillation. Elsevier oceanography series 52. Elsevier, Amsterdam, pp 141–182

Colgan MW (1990) El Niño and the history of eastern Pacific reef building. In: Glynn PW (ed) Global ecological consequences of the 1982–83 El Niño-Southern Oscillation. Elsevier oceanography series 52. Elsevier, Amsterdam, pp 183–232

Colley SB, Glynn PW, May AS, Maté JL (2006) Species-dependent reproductive responses of eastern Pacific corals to the 1997-1998 ENSO event. In: Proceedings of the 10th international coral reef symposium, Okinawa, vol 1, pp 61–70

de Weerdt WH, Glynn PW (1991) A new and presumably now extinct species of *Millepora* (Hydrozoa) in the eastern Pacific. Zoologische Mededelingen Leiden 65(20):267–276

Díaz Viana L (1982) El juego de gallos (formas, textos e interpretaciones). Rev Folk 24. https://funjdiaz.net/folklore/07ficha.php?id=208

Dunbar RB, Wellington GM, Colgan MW, Glynn PW (1994) Eastern Pacific sea surface temperature since 1600 A.D.: the $\delta^{18}O$ record of climate variability in Galápagos corals. Paleoceanography 9(2):291–315

Eakin CM (1996) Where have all the carbonates gone? A model comparison of calcium carbonate budgets before and after the 1982-1983 El Niño at Uva Island in the eastern Pacific. Coral Reefs 15:109–119

Enochs IC, Glynn PW (2017) Corallivory in the eastern Pacific. In: Glynn PW, Manzello DP, Enochs IC (eds) Coral reefs of the eastern tropical Pacific: persistence and loss in a dynamic environment. Coral reefs of the world 8. Springer, Dordrecht, pp 315–337

Feingold JS, Glynn PW (2014) Coral research in the Galápagos Islands, Ecuador. In: Denkinger J, Vinueza L (eds) The Galapagos marine reserve: a dynamic social-ecological system. Springer, Cham, pp 3–22

Feingold JS, Riegl B, Hendrickson K, Toth LT, Cheng H, Edwards RL, Aronson RB (2020) 7700-year persistence of an isolated, free-living coral assemblage in the Galápagos Islands: a model for coral refugia. Coral Reefs 39:639–647. https://doi.org/10.1007/s00338-020-01935-5

Fong P, Glynn PW (1998) A dynamic size-structured population model: does disturbance control size structure of a population of the massive coral *Gardineroseris planulata* in the eastern Pacific? Mar Biol 130:663–674

Fong P, Glynn PW (2000) A regional model to predict coral population dynamics in response to El Niño-Southern Oscillation. Ecol Appl 10(3):842–854

Gilchrist SL (1985) Hermit crab corallivore activity. In: Proceedings of the 5th international coral reef congress, Tahiti, vol 5, pp 211–214

Glynn PW (1974) The impact of *Acanthaster* on corals and coral reefs in the eastern Pacific. Environ Conserv 1(4):295–304

Glynn PW (1976) Some physical and biological determinants of coral community structure in the eastern Pacific. Ecol Monogr 46:431–456

Glynn PW (1983a) Increased survivorship in corals harboring crustacean symbionts. Mar Biol Lett 4:105–111

Glynn PW (1983b) Crustacean symbionts and the defense of corals: coevolution on the reef? In: Nitecki MH (ed) Coevolution. The University of Chicago Press, Chicago, pp 111–178

Glynn PW (1985a) El Niño-associated disturbance to coral reefs and post disturbance mortality by *Acanthaster planci*. Mar Ecol Prog Ser 26:295–300

Glynn PW (1985b) Corallivore population sizes and feeding effects following El Niño (1982-1983) associated coral mortality in Panama. In: Proceedings of the 5th international coral reef congress, Tahiti, vol 4, pp 183–188

Glynn PW (1987) Some ecological consequences of coral-crustacean guard mutualisms in the Indian and Pacific oceans. Symbiosis 4(1–3):301–324

Glynn PW (1988) El Niño warming, coral mortality and reef framework destruction by echinoid bioerosion in the eastern Pacific. Galaxea 7(2):129–160

Glynn PW (ed) (1990) Global ecological consequences of the 1982–83 El Niño-Southern Oscillation. Elsevier oceanography series 52. Elsevier, Amsterdam, p 563

Glynn PW (1993) Coral reef bleaching: ecological perspectives. Coral Reefs 12:1–17

Glynn PW (1997) Eastern Pacific reef coral biogeography and faunal flux: Durham's dilemma revisited. In: Proceedings of the 8th international coral reef symposium, Panama, vol 1, pp 371–378

Glynn PW (2012) Global warming and widespread coral mortality: evidence of first coral extinctions. In: Hannah L (ed) Saving a million species: extinction risk from climate change. Island Press, Washington, DC, pp 103–119

Glynn PW, D'Croz L (1990) Experimental evidence for high temperature stress as the cause of El Niño-coincident coral mortality. Coral Reefs 8:181–191

Glynn PW, de Weerdt WH (1991) Elimination of two reef-building hydrocorals following the 1982–1983 El Niño warming event. Science 253:69–71. https://doi.org/10.1126/science.253.5015.69

Glynn PW, Wellington GM (1983) Corals and coral reefs of the Galápagos Islands. With an annotated list of the scleractinian corals of the Galápagos by John W. Wells. University of California Press, Berkeley, p 330

Glynn PW, Stewart RH, McCosker JE (1972) Pacific coral reefs of Panamá: structure, distribution and predators. Geol Rundsch 61(2):483–519

Glynn PW, Wellington GM, Birkeland C (1979) Coral reef growth in the Galápagos: limitation by sea urchins. Science 203:47–49

Glynn PW, Perez M, Gilchrist SL (1985) Lipid decline in stressed corals and their crustacean symbionts. Biol Bull 168:276–284

Glynn PW, Cortès J, Guzmán HM, Richmond RH (1988) El Niño (1982-83) associated coral mortality and relationship to sea surface temperature deviations in the tropical eastern Pacific. In: Proceedings of the 6th international coral reef symposium, Townsville, vol 3, pp 237–243

Glynn PW, Maté JL, Baker AC, Calderón MO (2001) Coral bleaching and mortality in Panama and Ecuador during the 1997-1998 El Niño-Southern Oscillation event: spatial/temporal patterns and comparisons with the 1982-1983 event. Bull Mar Sci 69(1):79–109

Glynn PW, Enochs IC, Afflerbach JA, Brandtneris VW, Serafy JE (2014) Eastern Pacific reef fish responses to coral recovery following El Niño disturbances. Mar Ecol Prog Ser 495:233–247. https://doi.org/10.3354/meps10594

Glynn PW, Mones AB, Podestá GP, Colbert A, Colgan MW (2017) El Niño-Southern Oscillation: effects on eastern Pacific coral reefs and associated biota. In: Glynn PW, Manzello DP, Enochs IC (eds) Coral reefs of the eastern tropical Pacific: persistence and loss in a dynamic environment. Coral reefs of the world 8. Springer, Dordrecht, pp 251–314

Glynn PW, Feingold JS, Baker A, Banks S, Baums IB, Cole J, Colgan MW, Fong P, Glynn PJ, Keith I, Manzello D, Riegl B, Ruttenberg BI, Smith TB, Vera-Zambrano M (2018) State of corals and coral reefs of the Galápagos Islands (Ecuador): past, present and future. Mar Pollut Bull 133(2018):717–733

Gotelli NJ, Gilchrist SL, Abele LG (1985) Population biology of *Trapezia* spp. and other coral-associated decapods. Mar Ecol Prog Ser 21:89–98

Grotolli AG, Rodriguez LJ, Palardy JE (2006) Heterotrophic plasticity and resilience in bleached corals. Nature 440:1186–1189

Guarderas MJ (2023) *Psammocora stellata* distribution at Devil's Crown, Floreana Island, Galápagos Islands, Ecuador. MS thesis. Halmos College of Natural Sciences and Oceanography, Nova Southeastern University, p 42

Guzman HM, Robertson DR (1989) Population and feeding responses of the corallivorous pufferfish *Arothron meleagris* to coral mortality in the eastern Pacific. Mar Ecol Prog Ser 55:121–131

Harrison PL (2011) Sexual reproduction of scleractinian corals. In: Dubinsky Z, Stambler N (eds) Coral reefs: an ecosystem in transition. Springer, Cham, pp 59–85. www.Springer.com, 552 p

Hendrickson K (2014) Changes in coral community composition at Devil's Crown, Galápagos Islands, Ecuador: A 7,700 year perspective. MS thesis. Nova Southeastern University Oceanographic Center, Nova Southeastern University, p 93

Hoegh-Guldberg O (1999) Coral bleaching, climate change and the future of the world's coral reefs. Mar Freshw Res 50(8):839–866. https://doi.org/10.1071/MF99078

Knudsen JW (1967) *Trapezia* and *Tetralia* (Decapoda, Brachyura, Xanthidae) as obligate ectoparasites of pocilloporid and acroporid corals. Pac Sci 21:51–57

Kroh A, Mooi R (2021) World Echinoidea Database Tetragramma L. Agassiz, 1838

LaJeunesse TC, Reyes-Bonilla H, Warner ME (2007) Spring "bleaching" among *Pocillopora* in the Sea of Cortez, eastern Pacific. Coral Reefs 26:265–270

LaJeunesse TC, Wham D, Pettay D, Parkinson J, Keshavmurthy S, Chen C (2014) Ecologically differentiated stress tolerant endosymbionts in the dinoflagellate genus *Symbiodinium*. Clade D are different species. Phycologia 53:305–319

LaJeunesse TC, Parkinson JE, Gabrielson PW, Jeong HJ, Reimer JD, Voolstra CR, Santos SR (2018) Systematic revision of Symbiodiniaceae highlights the antiquity and diversity of coral endosymbionts. Curr Biol 28:1–11. https://doi.org/10.1016/j.cub.2018.07.008

LaJeunesse TC, Banaszak AT, Fisher CR, Shick JM, Warner ME, Porter JW, Kuris AM, Iglesias-Prieto R, Fitt WK (2021) Robert Kent Trench (1940-2021): a life devoted to symbiotic mutualisms and seeking nature's truth. Symbiosis 85:393–400

LaJeunesse TC, Smith R, Walther M, Pinzón J, Pettay DT, McGinley M, Aschaffenburg M, Medina-Rosas P, Cupul-Magaña AL, López Pérez A, Reyes-Bonilla H, Warner M, E. (2010) Host-symbiont recombination versus natural selection in the response of coral-dinoflagellate symbioses to environmental disturbance. Proc R Soc B 277:2925–2934. https://doi.org/10.1098/rspb.2010.0385

Lessios HA, Glynn PW, Robertson DR (1983) Mass mortalities of coral reef organisms. Science 222(4625):715

Little A (2008) The distribution of fungiid corals in the eastern Pacific. MS capstone. Nova Southeastern University Oceanographic Center, Nova Southeastern University, p 112

Macintyre IG, Glynn PW, Cortés J (1992) Holocene reef history in the eastern Pacific: mainland Costa Rica, Caño Island, Cocos Island, and Galápagos Islands. In: Proceedings of the 7th international coral reef symposium, Guam, vol 2, pp 1174–1184

NOAA (2021) National Oceanic and Atmospheric Administration, MEI.v2:NOAA physical sciences. Accessed 15 Sep 2021

Oramas NFA (1979) Estudio ecológico del *Jenneria pustulata* (Mollusca, Gastropoda). Tesis, Universidad de Panamá, Escuela de Biología. República de Panamá. 100 pages, 21 tables, 29 figures

Palacio-Castro AM, Smith TB, Brandtneris V, Snyder GA, van Hooidonk R, Maté JL, Manzello D, Glynn PW, Fong P, Baker AC (2023) Increased dominance of heat-tolerant symbionts creates resilient coral reefs in near-term ocean warming. Proc Nat Acad Sci 120(8):e2202388120. https://doi.org/10.1073/pnas.2202388120

Patton WK (1976) Animal associates of living reef corals. In: Jones OA, Endean R (eds) Biology and geology of coral reefs, vol 3. Academic Press, New York, pp 1–36

Paul NC (2012) Variable recovery of the massive coral, *Porites lobata*, in response to El Niño-Southern Oscillation events at Devil's Crown, Galápagos Islands, Ecuador. MS thesis. Nova Southeastern University Oceanographic Center, Nova Southeastern University, p 73

Razak TB, Hoeksema BW (2003) The hydrocoral genus *Millepora* (Hydrozoa: Capitata: Milleporidae) in Indonesia. Zoologische Verhandelingen 345:313–336

Reaka-Kudla ML, Feingold JS, Glynn PW (1996) Experimental studies of rapid bioerosion of coral reefs in the Galápagos Islands. Coral Reefs 15:101–107

Rinkevich B, Wolodarsky Z, Loya Y (1991) Coral-crab association: a compact domain of a multi level trophic system. Hydrobiologia 216(217):279–284

Rodríguez-Troncoso AP, Carpizo-Ituarte E, Pettay DT, Warner ME, Cupul-Magaña AL (2014) The effects of an abnormal decrease in temperature on the eastern Pacific reef-building coral *Pocillopora verrucosa*. Mar Biol 161:131–139

Rohwer F, Youle M (2010) Coral reefs in the microbial seas. Plaid Press, Granada Hills, California, p 201

Santiago-Valentín JD, Colley SB, Glynn PW, Cupul-Magaña AL, López-Pérez RA, Rodríguez-Zaragoza FA, Benítez-Villalobos F, Bautista-Guerrero E, Zavala-Casas DA, Rodríguez-Troncoso AP (2018) Regional and species specific sexual reproductive patterns of three zooxanthellate scleractinian corals across the eastern tropical Pacific. Mar Ecol 39. https://doi.org/10.1111/maec.12497

Scott PJB, Risk MJ, Carriquiry JD (1988) El Niño bioerosion and the survival of East Pacific reefs. In: Proceedings of the 6th international coral reef symposium, Townsville, vol 2, pp 517–520

Smith TB, Glynn PW, Maté JL, Toth LT, Gyory J (2014) A depth refuge from catastrophic coral bleaching mortality twice prevented regional extinction. Ecology 95:1663–1673

Smith TB, Maté JL, Gyory J (2017) Thermal refuges and refugia for stony corals in the eastern tropical Pacific. In: Glynn PW, Manzello DP, Enochs IC (eds) Coral reefs of the eastern tropical Pacific: persistence and loss in a dynamic environment. Coral reefs of the world 8. Springer, Dordrecht, pp 501–515

Souter D, Planes S, Wicquart J, Logan M, Obura D, Staub F (2021) Status of coral reefs of the world: 2020. Global Coral Reef Monitoring Network (GCRMN) and International Coral Reef Initiative (ICRI). https://gcrmn.net/2020-report/

Stimpson J (1990) Stimulation of fat-body production in the polyps of the coral *Pocillopora damicornis* by the presence of mutualistic crabs of the genus *Trapezia*. Mar Biol 106:211–218

Thurber R, Payet J, Thurber A et al (2017) Virus–host interactions and their roles in coral reef health and disease. Nat Rev Microbiol 15:205–216. https://doi.org/10.1038/nrmicro.2016.176

Toth LT, Macintyre IG, Aronson RB (2017) Holocene reef development in the eastern tropical Pacific. In: Glynn PW, Manzello DP, Enochs IC (eds) Coral reefs of the eastern tropical Pacific: persistence and loss in a dynamic environment. Coral reefs of the world 8. Springer, Dordrecht, pp 177–201

Toth LT, Aronson RB, Vollmer SV, Hobbs JW, Urrego DH, Cheng H, Enochs IC, Combosch DJ, van Woesik R, Macintyre IG (2012) ENSO drove 2500-year collapse of eastern Pacific coral reefs. Science 337:81–84

Voolstra CR, Suggett DJ, Peixoto RS, Parkinson JE, Quigley KM, Silveira CB, Sweet M, Muller EM, Barshis DJ, Bourne DJ, Aranda M (2021) Extending the natural adaptive capacity of coral holobionts. Nat Rev Earth Environ 2:747–762

Wham DC, Ning G, LaJeunesse TC (2017) *Symbiodinium glynnii* sp. nov., a species of stress-tolerant symbiotic dinoflagellates from pocilloporid and montiporid corals in the Pacific Ocean. Phycologia 56:396–409

Wilkinson C (1998) Status of coral reefs of the world: 1998. Global coral reef monitoring network. Australian Institute of Marine Science, Cape Ferguson, Queensland, Australia, 184

Yamazato K (1981) A note on the expulsion of zooxanthellae during summer, 1980 by the Okinawan reef-building corals. Sesoko Marine Science Laboratory Technical Report 8:9–18

Chapter 6
Coral Reproduction and Recruitment

Abstract Little was known of zooxanthellate coral reproduction in the eastern Pacific until studies were initiated following large scale coral mortality caused by the 1982–83 El Niño-sea warming disturbance. Before this El Niño event, it was generally considered that eastern Pacific corals were relatively sterile, capable of only slow recovery following disturbances; population replenishment depended on long-distance dispersal of reproductive propagules from across the eastern Pacific barrier. Three decades of research in Panama, Costa Rica, and the Galápagos Islands, employing histological methods and field observations, have demonstrated that 13 species are broadcast spawners with peak reproductive activity in the warm, wet season when equatorial areas are under the influence of the Intertropical Convergence Zone. Most coral species exhibit high fecundity and some level of reproductive activity throughout the year, with gonadal development or spawning reduced in the dry, upwelling season. Reproduction also occurs in some species during weak and mild El Niño warming. Broadcast spawning species release gametes at or a few days after the full moon, but mass (simultaneous) species spawning events were not observed. Asexual fragmentation, a result of internal (e.g., boring sponges and date mussels) and external (e.g., corallivores and benthic fish invertivores) bioerosion is prominent in the eastern Pacific. Both sexual (gamete production) and asexual reproduction (colony fragmentation) play essential roles in population recovery and reef growth. The appearance of extra-regional Indo-Pacific species (e.g., cnidarians, mollusks, crustaceans, echinoderms, and fishes) after strong El Niño events is evidence of the influx of easterly moving warm water and long-distance dispersal, but not necessary for coral reef recovery that is self-sustaining.

Introduction

Reproduction and recruitment are necessary to maintain stability and persistence of coral populations. As noted in Chap. 5, the severe reductions in coral abundances caused by the 1982–83 El Niño warming disturbance directed our attention to the

© The Author(s), under exclusive license to Springer Nature Switzerland AG 2025
P. W. Glynn, *Coral Reef Resilience in the Anthropocene*,
https://doi.org/10.1007/978-3-031-81421-1_6

reproductive capacity of affected eastern Pacific corals to repopulate depleted populations. Pre-ENSO studies that reported low rates of sexual coral recruitment in the Pacific waters of Panama were worrisome (Birkeland 1977; Highsmith 1982; Wellington 1982). Bob Richmond, in Panama on a STRI postdoctoral research fellowship, established plots on affected reefs to follow recovery after the 1982–83 El Niño mortality event. He, too, reported virtually no recruitment or sexual activity in *Pocillopora* spp., the predominant eastern Pacific reef-building coral genus (Richmond 1985, 1987a, b). Ray (Raymond C.) Highsmith (1982) demonstrated earlier that asexual fragmentation can play an important role in coral community recovery.[1] Ray's findings offered a ray of hope. Asexual fragmentation in corals depends on some level of disturbance causing the fracturing of coral colonies, for example, by storms or seismic-generated turbulence, bioerosion, and corallivores. Asexual reproduction of *Porites lobata* was reported by Héctor Guzmán (1986) in his Master's research in Costa Rica and later published by Guzmán and Cortés (1989). Is it possible there exists a latitudinal difference in coral asexual reproduction vis-à-vis physical versus biological fragmentation in the eastern Pacific? For example, storm-related coral fragmentation may be more common along the Mexican coast and bioerosion at equatorial localities (e.g., Panama and Galápagos Islands).

If sexual and asexual recruitment are infrequent then it is likely that benthic algae and non-coral invertebrates will settle on and occupy disturbed, newly opened substrates. The occupation of free space by non-coral benthos limits or prevents the settlement and growth of coral larvae in the eastern Pacific (e.g., Birkeland 1977). Such a scenario weighed heavily on our minds and was a primary motivating factor for studying reproduction, recruitment, and coral community recovery.

Collections and Data

To gain an understanding of coral reproductive responses to the seasonal migration of the ITCZ (Intertropical Convergence Zone), upwelling events, and ENSO activity in the equatorial eastern Pacific, we established sampling protocols at various sites in Costa Rica, Panama, and the Galápagos Islands. The objective was to perform sampling on a monthly schedule over several consecutive years (as travel allowed to these remote sites). This undertaking involved the combined effort of RSMAES and in-country teams to obtain the collections and properly preserve the

[1] Chuck Birkeland, Bob Richmond, Ray Highsmith and others have contributed significantly to coral reef studies in Panama thanks to pre- and post-doctoral Smithsonian Institution fellowships administered by STRI. Notable among early pre-doctoral Smithsonian awardees were Larry (Lawrence G.) Abele, Andrew C. Baker, Tom (Thomas F.) Dana, Juan L. Maté, Jim (James W.) Porter, and Jerry Wellington.

samples for laboratory analyses in Miami.[2] Sampling protocols required significant effort, collecting several cm-sized fragments from up to 11 coral species from each location. The collectors needed to be intimately familiar with the species selected for sampling. Also, care was necessary to ensure that samples were taken from the parts of colonies where reproductive tissues could develop. The coral fragments were initially fixed (to preserve tissues and cellular structure in place) in a special solution, a modified Helly's fixative, for approximately one day, then rinsed in sea water for one day, and finally preserved in 70% ethyl alcohol and labeled for transport to RSMAES. The rinsing was necessary to wash out the chemical fixative to avoid precipitate interference with the histological staining. In the 1970s to early 1990s there were no or only lightly enforced restrictions for transporting biological samples in liquid preservatives on commercial airline flights. As a precautionary safety measure, we drained the alcohol from the containers; the corals were then wrapped in diluted alcohol-soaked towels to prevent drying. Permission was occasionally required to board the flight with the collections; this was always granted after a brief description of the contents and explanation of our research objectives. One airline captain even offered to stow the coral samples as "precious cargo" beside his seat in the cockpit!

By the early 1990s, two additional permissions were required to import live or dead corals to the USA: (1) an export permit approved by the country from which a species was collected, and (2) a CITES (Convention on International Trade in Endangered Species of Wild Fauna and Flora) permit. We are grateful to our esteemed, departed colleague, Professor Emerita Nancy Voss (1929–2020), indefatigable director of the RSMAES Marine Invertebrate Museum, for obtaining CITES permission for the acquisition of corals and for her unflagging encouragement to add corals to the museum's collections.[3]

---

[2] Special thanks are due the following for help with collections and laboratory analyses: K. Attridge, A. C. Baker, R. Ball-Bailey, H. Balchowsky, O. Barrio, M. Basile, I. B. Baums, T. Baynes, K. Black, J. Borger Richardson, C. Bourbonierre, V. Brandtneris, C. Carbo, N. Chapman, S. B. Colley, A. M. S. Correa, J. Cortés, L. Cruz, D. Day, L. D'Croz, H. Dienes-Diaz, C. M. Eakin, J. M. Eakin, I. C. Enochs, E. Espinosa, R. Espinosa, J. N. Feingold, L. Ferrer, L. Flebotte, L. Flores, J. Firman, P. Fong, N. J. Gassman, J. Gomezjurado, H. M. Guzmán, D. Holstein, J. Holtvluwer Ting, C. Hueerkamp, M. Hyatt, J. Jara, C. Jiménez, C. Johnson, E. Johnson, E. Kapostasy, S. Laessig, E. Major, S. Man, P. Martínez, J. L. Maté, A. May, S. McKenna, G. Merlen, R. Menoscal, M. D. Moore, A. Moulding, A. Muentes, D. O. Obura, M. Ortiz, J. Pefiaherrera, J. Peña, D. Perrine, E. J. Prager, R. H. Richmond, F. Rivera, J. B. Del Rosario, J. Sabater, G. Saunders, D. Smith, T. B. Smith, J. H. Stout, A. Szmant, S. Thornton, L. T. Toth, B. C. Tucker, J. Valenzuela, B. Vargas Ángel, A. Velarde, N. Voss, T. Woodworth, J. A. Yager, and A. Yedid. Individuals who helped with planning, arranging for boat transport and support, and permissions are noted in the 'Acknowledgments' sections of the cited reproductive publications in this chapter.

[3] I highly recommend viewing "A Museum of Marine Life: The University of Miami," a poignant video account of Nancy's dedication to the maintenance and protection of the marine invertebrate collections (https://cephalopod.wordpress.com/nancy-voss-obit/). Nancy's association with the University of Miami began with her enrollment in a marine biology course in 1950 and later marrying her instructor, Gilbert Voss, in 1952. At that time, Gil was a graduate student teaching assistant. He later earned a Ph.D. degree at George Washington University, joined the faculty at RSMAES, and advanced academically to full professor. Gil was a staunch marine conservationist

On a flight from Quito, Ecuador to Miami in the early 1990s, an incident occurred that made me fearful of missing the flight, relinquishing our collections, and even being arrested. Shortly after boarding the plane and settling into my seat, I was called over the intercom to report to the fore door of the plane where armed guards were waiting. They briskly escorted me off the plane and to my dive bag where two German Shepard sniffer dogs were excitedly pawing at the bag. The dogs had presumably been attracted to the sea smells emanating from the bag. I was ordered to unpack the bag and spread all contents on the tarmac. The dogs smelled the dive gear and specimen containers and immediately lost interest in them. I explained our research and the importance of the coral collections to the security personnel and added that the Charles Darwin Foundation of Ecuador sanctioned our work. This information was sufficient; I was told to re-pack and re-board the flight posthaste. It turns out that the President of Ecuador, Sixto Durán Ballén, was on the Miami flight *en route* to Japan. About an hour into the flight, President Durán strolled aft to where I was sitting and began to chat with me about the importance of the Galápagos Islands to Ecuador. He expressed interest in our coral studies and encouraged us to continue with this line of research. I welcomed this unexpected and distinguished encounter and enjoyed the remainder of the flight in high spirits. Upon landing in Miami and clearing customs, a quick trip to our RSMAES laboratory was necessary to re-immerse the corals in alcohol to prevent desiccation.

Once our team was aboard a research vessel or a functional substitute, collecting corals for reproductive study required care but was not arduous. Fixing, preserving, and documenting collections on consecutive lunar days was laborious, and did require extra care. This was generally the case in Panama and the Galápagos Islands. Collecting corals in Costa Rica was another matter, demanding an uncommon effort and a flair of bravado. Jorge Cortés and Héctor Guzmán were able to successfully collect corals monthly over 2 years at Caño Island, located about 12 km off the Costa Rican mainland coast. This involved driving from the Universidad de Costa Rica (San Pedro) in San José to the Sierpe River, a seven-hour trip, and then loading a skiff with supplies to travel to the island. The road trip from San José to the Sierpe River needed to be carefully timed to coincide with a flooding or in-coming tide, which usually meant departing San José at midnight. This strict schedule is due to

and renowned scholar of cephalopod systematics. Nancy and Gil began collaborating on various research projects, including studies of planktonic larval and juvenile fishes of the Florida Current, and the marine invertebrate faunas of Soldier Key and Bimini (Bahamas Islands). In the 1960s, Nancy embarked on studies of cephalopod systematics, which led to the publication in 1969 of a seminal monograph on the squid family Histioteuthidae. This notable revision led to NSF (National Science Foundation) funding for the study of other squid families in the 1970s and 1980s. Nancy's academic appointments at RSMAES began with Research Instructor in 1957, from which she was promoted to Research Professor in 1981 and culminated with Research Professor Emerita upon her retirement in 2007. Nancy continued managing the museum collections and accommodating visiting scientists for another dozen years following her retirement. Nancy also continued to serve on doctoral and Master's student committees and supervised several undergraduate students in special museum-related research assignments. As a wife, mother, research scientist, university professor, and museum director, Nancy was a pioneer and role model for women entering the field of marine science.

the hazardous entry to the sea from the river mouth, where turbulent waters are known to capsize boats. If their boat should capsize, it would be best to occur during a rising tide so they would be pushed landward and not out onto the open sea. The ocean crossing to Caño Island was sometimes rough and wet. Jorge and Héctor needed to hurriedly offload their food and equipment for a 5 to 8 day stay when negotiating a beach landing on the island. They were offered shelter in a bungalow maintained by the Caño Island Biological Reserve, National Park Service of Costa Rica. Jorge wisely took advantage of this opportunity to invite colleagues in other disciplines—archaeology, plant ecology, and ichthyology—to engage in scientific studies on this remote, undisturbed island.

The next phase of work in our Miami laboratory, histology—the preparation of tissues for microscopical examination—required skill, patience, and a degree of artistic panache. Fortunately, we had good luck finding someone possessing these traits, Susan Colley (Fig. 6.1).[4] I invited her to join our team in 1991. Susan is an excellent teacher and several students working in my laboratory have benefited from learning the techniques of histology and coral biology from her. Alison Moulding and Bernardo Vargas-Ángel, both graduate students in my laboratory, contributed significantly to our eastern Pacific reproductive studies. After completing their doctoral degrees at the University of Miami, they joined the Nova Southeastern Oceanographic Center (Dania Beach, Florida) where they continued with studies of coral reproductive ecology.[5] Bernardo's publication on the gametogenesis and seasonality of *Acropora cervicornis* in southeastern Florida exemplifies one such model study (Vargas-Ángel et al. 2006). I briefly describe the main histological procedures below because it is necessary for one contemplating such work to understand the necessary time, labor, and expertise required to master this technique.

The first step in our laboratory sample processing involved decalcification, i.e., treating the coral fragments with acid to dissolve away the hard calcareous skeletons. The concentration of acid must be adjusted carefully so that the dissolution of

[4]As of November 2021, Susan is in graduate school in the Department of Oceanography and Coastal Sciences at Louisiana State University, continuing to research eastern Pacific coral reproductive ecology. She has been admitted to candidacy for the Ph.D. degree and her doctoral dissertation addresses the importance of diverse environmental parameters affecting gamete production at various eastern Pacific sites. Susan is enhancing our understanding of the effects of environmental drivers on coral reproduction at a critical time of rapid climate change and coral reef decline. Her present effort is an extension of the research that was completed in the eastern Pacific and the RSMAES laboratory and published in our reproductive ecology series from 1991 to 2017 (Colley et al. 2000, 2006; Glynn et al. 1991, 1994, 1996, 2000, 2008, 2011, 2012, 2017; Glynn and Colley 2008). Fast forward, Susan was awarded the doctoral degree on 11 August 2023 and is now preparing her research results for publication.

[5]Alison and Bernardo followed similar career paths after completing their doctoral studies in my laboratory at RSMAES. Both joined the Nova Southeastern University Oceanographic Center, as Research Scientists at Dania Beach, Florida. Then both continued with coral research at NOAA, the National Oceanographic and Atmospheric Administration. Alison is a coral biologist in the Protected Resources Division at the St. Petersburg laboratory in Florida and Bernardo a coral reef ecologist in the Pacific Islands Fisheries Science Center at the Honolulu laboratory in Hawaii. Unsurprisingly, their primary research focus is related to the current coral reef crisis, Alison tracking coral species' local extinctions in the Caribbean Sea and Bernardo monitoring coral bleaching/mortality and diseases in the Hawaiian Islands.

Fig. 6.1 Susan B. Colley, coral reproductive biologist, histologist, student mentor, and stalwart field and laboratory team researcher (1993–2005)

the skeleton proceeds slowly and the release of CO_2 bubbles does not disrupt the soft tissues. Decalcifying samples must be monitored carefully to maintain the integrity of tissues. Decalcified tissues were then cut into ~1 cm pieces to orient them for viewing once embedded and sliced. The acid is then washed from the tissues, which are next dehydrated with graded steps of ethanol; the ethanol is then removed with xylenes so that the tissues can be infiltrated with paraffin and embedded; care is necessary to orient the tissues as they are embedded so that properly aligned longitudinal and cross sections of the polyps can be sectioned. With the tissues stabilized in paraffin, they are ready to be sectioned into thin slices, 6–8 μm thick, with a rotary microtome cutting tool. The last stage of preparation involves staining with the Heidenhain's aniline-blue technique that preferentially stains reproductive structures and connective tissue. Our methods followed those of Alina Szmant-Froelich et al. (1980, 1985), resulting in high resolution of gonads and developing gametes.[6] Once the gonadal tissues are sectioned and stained, on slides

[6] Alina and I first crossed paths in Puerto Rico in 1964 when she was a student (a very young undergraduate freshman) in a tropical marine ecology course I taught at the Magueyes Island field station, University of Puerto Rico Mayagüez. She exhibited great enthusiasm and a flair for coral reef ecology. In 1983, we became departmental colleagues when I joined the RSMAES faculty. I then

and protected with cover slips, the microscopical examination may proceed to determine the stages of gamete development. We scored gamete development utilizing four stages of maturity, reduced from Alina's 6 (spermaries) and 5 (oocytes) stages (Szmant-Froelich et al. 1980). Changes in cell size, staining characteristics, and the appearance and location of the nucleus in oocytes, cell division and formation of tails (in spermaries) were critical criteria in defining the states of gamete development. The relative occurrence of the different developmental stages could then be related to seasons (wet and dry), lunar cycles, and time of day. In many cases we counted all spermaries and oocytes to determine fecundity.

Spawning and Release of Larvae

Observing the release of gametes into the water column in the field, spawning, is an ideal way to determine the vital maturation phase of sexual reproduction in corals.[7] Our time-consuming effort to histologically document gonadal development was

---

became her student in learning how to stage and interpret the reproductive development of corals. An example of her excellent contributions in this field is encapsulated in her model study of the reproductive ecology of Caribbean corals (Szmant 1986). Alina was regarded as the Queen of coral spawning in the Caribbean due to her studies of *Montastrea* in the 1990s. She is now a retired University of North Carolina Wilmington professor.

[7] Some corals retain their eggs until fertilized internally by spawned sperm entering polyps from colonies in surrounding areas. Fertilized eggs are brooded and develop into planula larvae which are then released; this is termed **planulation** as opposed to spawning. Species with this type of development are called **brooders** and generally possess small colonies. In addition, larval dispersal is usually short-lived, often hours to a few days, with larvae settling relatively near the maternal colonies. Only two brooding species occurred in our studies: *Porites panamensis* (zooxanthellate) and *Tubastraea coccinea* (azooxanthellate). Asexual reproduction of brooded planulae was reported in *T. coccinea* by Ayre and Resing (1986), but fertilization may also be possible in eastern Pacific populations, according to our observations (see below). Since *Pocillopora damicornis* is a brooder in Hawaii, and assumed to have colonized the eastern Pacific by long distance dispersal, Richmond (1985) held open the possibility that *P. damicornis* would likely release planulae in the eastern Pacific. The formerly recognized Hawaiian *P. damicornis*, however, based on molecular genetic evidence, is now classified as the species *Pocillopora acuta*, which is a brooder throughout its central Pacific to Indian Ocean range (Pinzón et al. 2013; Schmidt-Roach et al. 2014). *Pocillopora acuta* is not among the 12 *Pocillopora* morphospecies more recently reported in the eastern Pacific fauna (Glynn et al. 2017). The *Pocillopora damicornis* species sampled in our reproductive study (Glynn et al. 1991) was either *Pocillopora grandis* Dana, 1846 (see Hoeksema and Cairns 2024) or *Pocillopora verrucosa* (Ellis and Solander 1786), both of which are broadcast spawning species (Pinzón et al. 2013). The actual diversity of the *Pocillopora* species complex now appears to be considerably less than indicated by gross colony morphology, which demonstrates extreme phenotypic plasticity in the eastern Pacific (Pinzón and LaJeunesse 2011; Paz-García et al. 2015). Based on genetic phylogenies, population genetics, and dinoflagellate symbiont associations, the eastern Pacific *Pocillopora* lineage now stands at three species (Pinzón and LaJeunesse 2011; Johnston et al. 2017; Turnham et al. 2021): *Pocillopora grandis*, *Pocillopora ligulata*, and *Pocillopora verrucosa*. In the present discussion, I restrict the use of coral spawning to those species that release both eggs and sperm. Such species are termed **broadcast spawners**

not in vain but allowed us to establish a time interval of the days when spawning was likely to occur. From our histological slides, revealing the presence of mature gonads and spent ovaries and spermaries, devoid of or with only residual spermatozoa remaining (see Glynn et al. 1996; Colley et al. 2000), we were able to narrow our field surveys to the warm season and to nights of the full moon (most often) and to 2 days after full moon or new moon phases. This is when mature gametes were present in most species at most sampling localities. This was also the time when other investigators observed coral spawning on Indo-Pacific and Caribbean coral reefs (e.g., Szmant 1986; Harrison and Wallace 1990; Richmond and Hunter 1990). At Uva Island the full moon would usually first appear and cast light on the reef between 2000 (8 PM) and 2100 (9 PM) because the high terrain of the island blocked the early rising moon. Moonlight cues involving photosensitive cryptochrome pigments have been shown to initiate lunar spawning in some corals (Gorbunov and Falkowski 2002; Levy et al. 2007). Palpable excitement would build in our team on clear, full moon nights.

Despite many hours underwater dedicated to watching for coral spawning or planula release, we infrequently observed these key reproductive events. On weekly schedules, we dived all day, every day, and for several hours during three consecutive nights—an exhausting effort for our small field crew. We missed witnessing spawning events in two species of *Pavona* for mindless reasons. Unwittingly, our meal schedules aboard ship interfered with us observing spawning in *P. chiriquiensis* and *P. varians*. Mealtimes are difficult to change, especially with programmed crew watches and inflexible ship captains. Breakfast was served shortly after sunrise and dinner at sunset. From our histological evidence on the appearance of mature gonads, we suspected that spawning should occur on days 15–18, at and just after full moon nights (Glynn et al. 2000). For several years we watched diligently for spawning all hours of day and night during full moon, and a few days following, except during morning and evening mealtimes.

We were finally rewarded in 1997 when we decided to observe both species of *Pavona* in buckets aboard ship. Fifteen colonies of each species were apportioned into 20, 5 gallon (~ 20 liter) buckets of seawater.[8] We replenished the water overnight, approximately every two hours, which required a bucket brigade. During one

and are generally more common in coral faunas than brooding species. Brooders most likely spawn only sperm, retaining ova for internal fertilization and development. A comparison of global scleractinian corals revealed 354 broadcasting and 61 brooding species (Harrison 2011). Of these corals, 13 species exhibited both broadcasting and brooding developmental modes.

[8] For several years, we suspected that *Pavona varians* represented two distinct species. We tentatively referred to one of the morphologically distinct types as *Pavona* sp. a, which was formally described in 2001 as *Pavona chiriquiensis* (Glynn et al. 2001). The long-recognized *Pavona varians* is distinguished by prominent meandering collines (ridges) and uniformly colored polyps, whereas *Pavona chiriquiensis* bears distinct tubercles and usually colored polyp mouths. The color photographs of the two species in Hickman Jr.'s (2008) Galápagos field guide show these differences. Three additional field guides in Hickman's Galápagos Marine Life Series, Echinoderms, Mollusks, and Crustaceans, are beneficial in the identification of reef-associated species in the eastern Pacific.

of our first trials, we were assembled on the bow deck in the late afternoon, enjoying the beautiful sunset and discussing the plan for monitoring the corals that night. A sound from the sea below attracted my attention; I leaned over the railings and spotted a whale shark slowly swimming toward our ship's stern. I immediately alerted our team that set off a stampede with loud foot stomping that caused the shark to dive and disappear. Alas, a missed opportunity for most team members to see a whale shark in the wild and near a coral reef. However, with a clear view from the ship's bow of the western sea surface horizon, we were fortunate to witness the green flash during sunset.[9]

An undeserved incident, never to appear in a science publication, occurred during our initial attempts to observe coral spawning. This involved a bucket brigade, each person toting ~20 liters of sea water from the ship's stern to her bow, about 15 m distance, to replenish the water in the 20 separate sample buckets. Susan's histological slide preparations, revealing mature gametes, strongly suggested that the two species (*P. varians* and *P. chiriquiensis*) should spawn on or just after the 15th lunar day of March. We divided our team members into shifts to check and watch for spawning and performing seawater changes throughout the night. After we failed to observe spawning on the 15th and 16th lunar days, Susan vividly recalls my reaction, "…you told me they were going to spawn." This remark was unfair; it certainly was not Susan's fault that the corals did not spawn. Perhaps the treatment in setting them up for observation or the conditions in the buckets interfered with the natural spawning behavior?

We committed to another night of observations, the 17th lunar night, and at 0500 (5:00 AM) Iliana Baums awakened Susan to tell her that *P. varians* was spawning. With great merriment, everyone quickly gathered around the buckets to observe the clouds of eggs and sperm that were being released. Most colonies are sequential hermaphrodites, meaning they can develop eggs or sperm but at different times. For example, a colony may spawn eggs at full moon and sperm at full moon but a month later. Suddenly, in great excitement Iliana and Christiane Hueerkamp[10] leapt into the

---

[9] The green flash is an optical phenomenon visible for 1–2 seconds after the setting sun disappears below the horizon. We observed the green flash on 26 January,1997. It is believed to be a sign of good luck. Considering our observations of coral spawning on this cruise, it certainly did bring good luck to our investigations.

[10] Christiane became a volunteer member of our team in the late 1990s. She was then associated with the Zentrum für Marine Tropenökologie in Bremen, Germany. In addition to helping us with ENSO coral recovery research and reproductive studies she conceived of and led an experimental study on the responses of several coral species to El Niño simulated temperature elevation (Hueerkamp et al. 2001). I came to know Iliana, an unusually inquisitive international exchange student, in the Marine Ecology course I taught at RSMAES in 1996. Iliana joined our laboratory group to investigate the reproductive biology of an eastern Pacific reef coral (*Psammocora stellata*). I was privileged to chair her doctoral committee at UM; she investigated the genetic status of the threatened Caribbean coral, *Acropora palmata*, and received her Ph.D. degree in 2004. Her research resulted in understanding of the genetic structure of rapidly declining populations, important for conservation management strategies (Baums et al. 2005a, b, 2006) After graduating from UM and 2 years of post-doctoral research at RSMAES, in Alina Szmant's laboratory, she joined the faculty at Pennsylvania State University. She remained in the Biology Department there from

water with swimsuits and face masks only, still in the dark of night, to swim to the reef to confirm spawning in the field. This impulsive behavior prompted a swift reaction from David West, our boat captain. The two coral spawning enthusiasts were quickly retrieved and soundly scolded. A cardinal rule was violated, going overboard without the captain's consent. No matter, with this morning's fulfilling result, we enjoyed a late breakfast and much self-congratulatory banter.

We continued our coral bucket observations and *Pavona chiriquiensis* did not disappoint; most of the colonies spawned in a few minutes following the sunset's green flash spectacle. Some of us wondered if there was not some reality to the good luck omen of the green flash. So, predawn and sunset periods, matching the ship's breakfast and dinner mealtime calls were the spawning times of *P. varians* and *P. chiriquiensis*, respectively. We were able to confirm these spawning times on the Uva reef and elsewhere on other occasions (Fig. 6.2). Juan Maté and I witnessed an especially bountiful spawning event at Coral Collection Point, one of our study sites across the bay from the Uva reef. Juan urged me to follow him early one evening to "Big Mama," a gigantic colony of *P. chiriquiensis* encrusting 1 by 3 m of substrate. We were on scuba and hovering over "Big Mama" when it began releasing wisps then dense clouds of spawn a few minutes after sunset, which was a genuinely glorious spawning spectacle. Knowing that our coral-bucket observations matched the spawning events occurring naturally on the reef was reassuring. When corals are ready to spawn, it seems that all manner of experimental interference does not hold them back from this vital process, at least in these two species of *Pavona*.

On brief occasions outside of our overly ambitious research objectives, shipboard entertainment involved card games, usually poker, and fishing from off the side of the boat. On one of our cruises, a member of the scientific party remarked

Fig. 6.2 A still underwater image from a video of *Pavona varians* spawning at 5:30 AM (05:30), 9 March 2004, three days after the full moon, Uva Island reef. Nearly all of the coral's upper surface is covered with a mucus-gamete matrix. Barely visible are wisps of released spawn (with eggs) at the colony's summit. Courtesy of Juan Maté

2006 to 2022, having advanced to the rank of full professor with tenure. Iliana is a molecular ecologist focusing on coral conservation and management, and the responses and adaptations of corals to changing climate stressors. She is presently a professor of marine conservation at the Helmholtz Institute for Functional Marine Biodiversity, University of Oldenburg. Iliana gained Fellow status in the International Coral Reef Society and received honorable mention for the 'Mid-Career Scientist' award in May 2024.

that it was 17 March, St. Patrick's Day. We could not toast to Ireland's patron saint with a pint of Guinness, but a rum drink was the next best thing. Then traditional Irish music filled the dining mess, thanks to You Tube. This became a joyous occasion, which stirred Tyler Smith to perform Irish step dancing. His nimble leg movements were impressive and enjoyed by all—until he jumped a little too high and hit his head against the dinette's low ceiling. This resulted in only a brief pause, however, as Tyler moved onto the outdoor deck and continued dancing. His amazing finale ended by jumping onto our inflatable boat and dancing on the flexible gunwales, a serious challenge successfully executed. It seemed that we all shared a latent Irish gene or two during this festive occasion—Erin Go Bragh!.[11]

Perhaps the most unexpected and prominent coral spawning event was witnessed by Rick (Richard W.) Grigg,[12] Joshua Feingold, and me at Devil's Crown in the Galápagos Islands. Rick contacted me in 1991 about joining us on one of our Galápagos cruises. I was delighted to hear of his interest in observing eastern Pacific coral reefs in recovery mode and invited him to join us on a monitoring cruise in 1992. Our first surveys and collections were performed in the southern Galápagos Islands, with R/V Beagle[13] serving as our support vessel from March 18–25, 1992. Weak El Niño activity was in effect during this time (1990–93). We had planned a general coral collection for histological analysis to assess the influence of a mild warming event on reproductive development.

Rick, Joshua, and I were making a coral collection on scuba during the sunny afternoon of 20 March 1992, two days after the full moon. It was at peak high tide

---

[11] At times while on shore, we would find ourselves at a pool table in a cantina. Paola Rodríguez-Troncoso and Susan Colley were especially skillful at shooting pool. Once, Susan and I were challenged to a match of *billares* (billiards) at Puerto Ayora in the Galápagos Islands—a Galápagos team versus Susan and I representing the University of Miami. The Galápagos players were surprised, and slightly shocked, at Susan's skills. In the end, the UM team triumphed over the local team. Considering our opponents' frustration and growing discontent, we decided a hasty exit was necessary.

[12] I first met Rick at the Scripps Institution of Oceanography, University of California, in La Jolla when he was completing the final stages of his doctoral studies. After graduation he joined the faculty of the University of Hawaii, attained the academic rank of professor, and remained there for 36 years until his death in 2014. Rick was a pioneering big wave surfer and winner of the 1967 Duke Kahanamoku Invitational Surfing Championship (Grigg 1998). His contributions to coral reef science were many, diverse, and profound: e.g., (a) biology, ecology, and management of deep water corals (Grigg 2002); (b) paleoceanography and zoogeography (1981, 1988, and Grigg and Hey 1992), including recognition and significance of the Darwin Point, the geographic or depth limit (threshold) beyond or below which coral reefs drown (1982); (c) early demonstration of the "Intermediate Disturbance Hypothesis" (Connell 1978)—changes in species diversity during recovery—in coral communities impacted by lava flows in Hawaii (Grigg and Maragos 1974). Rick also served as Editor-in-Chief of *Coral Reefs* (1992–1997), the flagship journal of the International Society for Reef Studies.

[13] The M/Y *Beagle III* was one of the first research vessels operated by the Charles Darwin Research Station. She was sold to a Galápagos tour company and continued sailing among the islands for many years. The *Beagle IV* and *Beagle V* met untimely ends on the rocks of the Santa Cruz shoreline. It was subsequently decided to drop the numerical designation from the *Beagle* boat name. Continuing boat losses began to take on the aura of ill luck and embarrassment.

with a strong, westerly flowing current. This current is the Oceanic Peru Current that flows from east to west for most of the year. We were collecting samples from the outer rock rim in the southwestern sector of Devil's Crown. In rounding a rock islet, we encountered a patch of several colonies of *Pavona gigantea* and *Pavona clavus*. As soon as I prepared to collect a fragment from a colony of *P. gigantea* it began spawning. It exuded milky-white sheets and strings of mucus from large areas of its surface. Several other nearby *P. gigantea* colonies began spawning similarly, filling the water with milky-white clouds. Depending on the location of spawning colonies relative to microscale eddies, spawn clouds remained suspended, nearly motionless or moved slowly across nearby colonies. Downstream colonies began spawning with the passing of the released upstream spawn clouds. This spawning event lasted 25 min, from 1550 (3:50 PM) to 1615 (4:15 PM). We could not visually determine whether the spawn consisted of eggs or spermatozoa. Mature *P. gigantea* eggs are about 100 μm in diameter (0.00394 inches) and barely visible under lighting conditions. Mature Stage IV eggs are present in the histological slide in Fig. 6.3A, and spermaries in Fig. 6.3B (from samples collected earlier this spawning day). At the end of the spawning event, some colonies extruded mesenterial filaments 2 to 5 mm in length and polyp mouths were agape in other colonies. Several coral species present at this site—*Pavona clavus*, *Porites lobata*, *Psammocora profundacella*, and *Tubastraea coccinea*—did not spawn.

Our chance viewing of a coral spawning event provoked considerable discussion and speculation: were the abundances and patterns of coral distribution at this site related to microscale circulation? Could gamete retention or guidance by eddies promote fertilization? Could this local coral patch have been colonized and developed by the entrainment of larvae over successive reproductive periods? These suggested effects are simply ideas that could serve as questions for in-depth study. The impact of larval retention by mesoscale eddies have been proposed to explain the origin and maintenance of high coral cover in the Flower Garden Banks in the Gulf of Mexico (Limer et al. 2020).

After completing our work in the southern islands, we pulled anchor and began to motor north to Canal de Itabaca, a rendezvous site on the north side of Santa Cruz Island. Here, we were to meet the 32 m (105 ft) steel-hulled schooner M/V *Andando* for the second leg of our work in the northern islands of Wolf and Darwin. To save time, we planned to complete this transit from Devil's Crown to Itabaca Canal at night.[14] Unfortunately, the weather turned foul with rough seas and driving rain. The

[14] Motoring between island study sites at night can be hazardous, even under calm sea conditions. Early one dark and foggy morning, with nearly zero visibility, it was my turn to change water into tanks with live corals. Since our boat did not have a seawater pump, collecting water with a bucket and line leash was necessary, a primitive but true and tried method. A bucket is thrown forward in the direction of the moving boat and as it sinks it fills with water and is then hauled aboard. The water-filled bucket and boat's forward motion can exert considerable pull and heavy lifting. When lifting one full bucket, I began to slide on the deck toward the boat's stern, without a guard rail. I released the line just in time to avoid being pulled overboard. Slimy algae covered the aft deck. This was a narrowly avoided mishap. It is unlikely our helmsman would have heard my cries for help, man overboard! over the din of the boat's engine. I admonished our team to not repeat my recklessness and to always alert someone when going on deck.

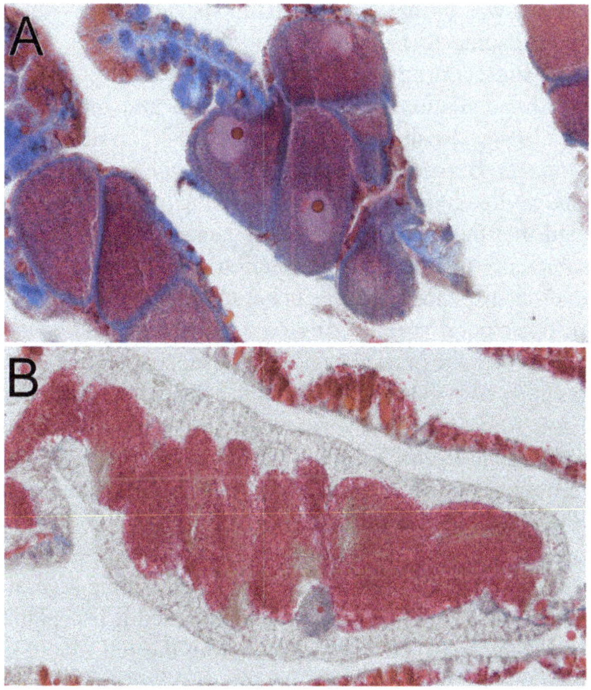

Fig. 6.3 *Pavona gigantea*, histological micrographs (200 X magnification) of gonads from hermaphroditic colonies sampled in the Galápagos Islands. Stained with Heidenhain's Aniline Blue. (**A**) Stage IV ova, 3 to 4 days before spawning, Canal de Itabaca, Santa Cruz Island, 16 March 1992. (**B**) Stage IV spermaries and a single Stage II ovum, Playa Cormoran, Floreana Island, 20 March 1992. Micrographs courtesy of Susan B. Colley

stormy weather was not a surprise considering we were amid the 1990–92 El Niño event. Our boat captain, nicknamed Bacalao (codfish), became disoriented and began heading directly for a jutting headland, Punta Robinson, instead of the entrance to Itabaca Canal. Rick politely but firmly asked Bacalao for the wheel to steer our boat away from the rocky, wave battered shore. Bacalao at first refused but then stepped away from the wheel after a swift and "friendly persuasion." Rick was able to motor the *Beagle* safely to our rendezvous port. Mutiny on the *Beagle*! We were then comfortably anchored but soon confronted with another pending disaster. In our bunks and prepared for a restful night, I began to hear an unwelcome sound, sloshing water. "Joshua, do you hear running water?" "Yes," was his instant reply. We opened the hatch to the boat's engine and spotted a steady stream of water from the propeller packing box filling the bilges! It's time to abandon ship. We gathered up all our scientific gear and personal items and rowed to shore in the *Beagle's* dingy. This was an unpleasant segment of our cruise, but tolerable considering the completed surveys, collections, and serendipitous viewing of coral spawning. We were informed later that the *Beagle* had sunk in the canal that night!

Natural spawning was primarily observed in species of *Pavona*. *Pavona clavus*, a common eastern Pacific reef coral, was observed spawning in Costa Rica and Panama soon after sunset during the full moon and a few days following. This coral is gonochoric with populations composed of male and female colonies. Broad mucus areas, enveloping sperm or eggs, typically lift off spawning colonies. One of Jorge Cortés' students, Bernadette Bezy, discovered the spawning time of *P. clavus* in Costa Rica and invited him to observe a spawning event. In recalling this experience, Jorge noted that the bottom reef waters began turning murky, but strangely without noticeable currents. On close inspection, Jorge soon found that the murky water was due not to suspended sediments but to mucus-covered spawn clouds.

Spawning in other coral species, observed immediately after the collection of samples or inferred from the presence and then disappearance of mature gametes in histological samples, also occurred commonly just after full moon (Fig. 6.4). According to Paola Rodríguez-Troncoso, of the University of Guadalajara, this was also the case in Mexican Pacific waters, inferred from coral gonadal development. After extensive observation time underwater, lamentably, Paola's team has yet to observe a natural spawning event (Fig. 6.5).

The two brooding corals present on eastern Pacific reefs, *Porites panamensis* and *Tubastraea coccinea*, released large numbers of larvae during several nights around the new moon (Fig. 6.4). Also, *P. panamensis* released a few planulae and some colonies of *T. coccinea* spawned a few days before the full moon. Could the relatively large larvae released by these species (up to 0.5–1.0 mm in length) be more susceptible to visual predators on bright moonlit nights compared to the small eggs spawned by broadcasting species?[15] As a consequence, could evolution have favored a shift from full moon to new moon planulation to reduce larval mortality in these species?[16] This is a question in need of study.

With nearly all eastern equatorial Pacific corals spawning at and a few days following full moon, we surmised that some species might spawn simultaneously, as occurs in many reef provinces. Peter Harrison (2011) noted, however, that synchronous spawning varies significantly over a broad geographic range. Many species spawn simultaneously during mass spawning events on the Great Barrier Reef, Australia, a few species overlap in spawning in the Gulf of Mexico and Caribbean, and spawning occurs independently in the Gulf of Eilat, Red Sea (Shlesinger et al. 1998).

There are potentially negative and positive effects to consider regarding multispecific spawning. On the one hand, the possibility of interspecific fertilization and

---

[15] The free-swimming planula larvae of *P. panamensis* and *T. coccinea* are visible, several times larger than spawned eggs. Obtaining a consistent length measurement is difficult, however, because the larvae often contract and expand when swimming. Estimated lengths varied from 300–500 μm (measured from histological slides) compared with 100 μm spawned egg diameters (Smith 1991; Glynn et al. 1994, 1996, 2008, 2017)

[16] It must be cautioned that our observations of brooding corals were biased towards the full moon and a few days following and focused on observing planulation. Spawning in brooding species may occur at other times of the lunar cycle.

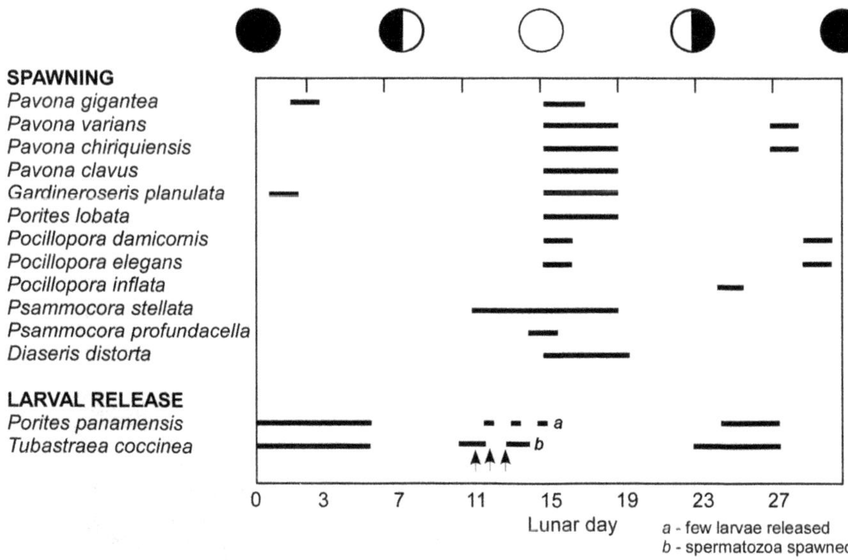

Fig. 6.4 Gamete spawning and larval release in eastern tropical Pacific scleractinian corals in relation to lunar phase. In light of recent molecular genetic studies, it is likely that the three species of *Pocillopora*, originally identified from morphological characters, belong to the single species *Pocillopora grandis* (see Pinzón and LaJeunesse 2011; Paz-García et al. 2015; and see footnote 7)

sterility could severely lower the reproductive effort of a species. On the other hand, Harrison et al. (1984) suggested that the sudden occurrence of large quantities of sexual products resulting from multi-specific spawning could increase gamete and larval survival by overwhelming potential planktonic predators. As with brooding species that release planulae, a possible benefit of spawning gametes during periods of darkness would be a reduced risk of predation by visual predators (Babcock et al. 1986). In the Caribbean, several broadcast spawners release gametes a few days or as many as 8 days after the full moon, before the moon rises, with spawning occurring in total darkness. Our observations support independent spawning of species in the eastern Pacific.

The fire coral, *Millepora intricata*, also spawns gametes but from medusae that are released from benthic polypoid colonies. The free-swimming medusae enter a brief planktonic phase of just a few hours (Soong and Cho 1998). Following successful fertilization of spawned gametes from gonochoric medusae (separate female and male medusae), the zygotes develop into planula larvae, settling and growing into calcifying colonies that often contribute to reef growth. The benthic phase of the life cycle typically grows rapidly, and in the Gulf of Chiriquí, Panama, has resulted in areas of extensive hydrocoral cover on numerous reefs. This mode of sexual-to-asexual-to sexual reproduction, characteristic of hydrozoan cnidarians, is referred to as **alternation of generations** (e.g., Hyman 1940; Lewis 2006). Fire corals disappeared from shallow reef habitats in the Gulf of Chiriquí during the 1982–1983 and 1997–1998 El Niño bleaching events but began repopulating

Fig. 6.5 Coral reef research team, Marine Ecology Laboratory, Centro Universitario de la Costa, Universidad de Guadalajara, Mexico. Left to right, Diana Elisabeth Morales de Anda, Hazal María Canizales Flores, Violeta Martínez Castillo, Alma Paola Rodríguez Troncoso, and Jeimy Denisse Santiago Valentín (3 March 2021). Photograph courtesy of Eric Bautista

shallow reef areas within 3 years (Smith et al. 2014). Evidence suggests that sexual reproduction, from deep (> 12 m) refuge populations, was responsible for the fire coral recovery. We assumed that planula larvae, resulting from the spawning of planktonic medusae from deep populations, settled and colonized shallow reef areas when seawater thermal conditions normalized. We have only indirect evidence, however, to support this hypothesis.

In one of the undergraduate courses I taught in the Marine Science Program at the University of Miami (Introduction to Marine Biology), I often encouraged students to volunteer as research assistants or enroll for research credits in my laboratory. One such student, Michael Hyatt, an undergraduate marine science major, became interested in the sexual reproduction of *Millepora* and began observing live corals and preserved specimens in my laboratory. The live corals included two species of *Millepora* from reefs in south Florida and *M. intricata* from Panama. In 1999 we transported live *Millepora intricata* colonies from Panama to Miami where we maintained them in closed (internally circulating) seawater aquaria for 1 year.[17] Alcohol-preserved specimens were collected in Panama in all seasons over a 12-year period (1985–1997). Michael meticulously monitored the live corals, which flourished but unfortunately did not produce medusae. Susan Colley and Juan Maté joined Michael in the histological examination of the preserved specimens. When *Millepora* spp. become reproductively active blister-like spherical chambers (ampullae) appear on the colony surfaces. These structures contain developing medusae that break free when gametes are mature (Lewis 1991; Soong and Cho 1998). After the release of medusae, connective tissue bands form over the openings, and calcification ensues, resulting in eventual closure. Michael's team examined the gross appearance and development of ampullae, including histological preparations, to assess the timing of reproduction. They found evidence of two seasonal reproductive periods: one in the Panamanian dry season (January–March) and the other in the wet season (July). Further, from the timing of the occurrence of ampullae and the release of medusae with mature gametes, spawning was judged to have occurred on or around both full and new moon phases. These results are like those reported for three species of *Millepora* in the Gulf of Eilat where massive medusa release has been documented (Shlesinger and Loya 2019). Because *M. intricata* is one of the first corals to colonize disturbed shallow reef communities, it would be advantageous to better understand its early life history stages, e.g., reproductive seasonality, spawning activity, capacity for dispersal, and probability of successful settlement. Hopefully, future researchers will study these processes.

The Seasonal Timing of Spawning

Another aspect of coral reproduction that interested us was seasonality or the timing of reproduction in relation to the equatorial wet and dry seasons. Our studies showed that most species were reproductively active in the warm, wet season, when equatorial regions are under the influence of the ITCZ (Intertropical Convergence Zone). However, several species, e.g., in the genera *Psammocora*, *Pavona*, *Pocillopora*, and *Porites lobata*, were also observed spawning or containing gonads with mature gametes in the dry, upwelling season. Curiously, although *Pocillopora damicornis* is a simultaneous hermaphroditic species with well-defined female and male gamete development, it has not been reported to spawn naturally in the central/west or eastern Pacific regions (Miller and Ayre 2004; Glynn et al. 2017). From histological studies and field observations, our interpretation of active reproduction was that it

[17] At first, we transported live corals in ice chests with filtered seawater and continuous aeration. This usually resulted in some coral mortality and occasionally raised the concern of seawater spillage aboard the flight. We eventually switched to an effective technique that caused no apprehension and resulted in near-zero mortality. This involved wrapping corals in a seawater-soaked light fabric (e.g., a nylon stocking) and enclosing them in polyethylene bags. The corals would survive nicely for several hours with tissue expansion occurring upon re-immersion in seawater in the laboratory.

usually occurred at non-upwelling locations or during weak upwelling years in upwelling centers. The only periods when reproductive activity was greatly curtailed were during ENSO extreme cooling (La Niña) and warming (El Niño) phases.

Since the 1982–83 El Niño disturbance, several studies in the tropical and subtropical waters of the Mexican Pacific have broadened our understanding of coral reproduction across the eastern Pacific region. Contributions from the subtropical area of La Paz, thanks to Héctor Reyes-Bonilla and co-workers (Fig. 6.6), and the tropical regions of Bahía Banderas and Oaxaca by Eugenio Carpiso-Ituarte, Paola Rodriguez and team have been incredibly informative. These investigators have shown that coral reproduction also occurs during warm periods (May–October) at tropical and subtropical latitudes in the Mexican region, with highest fecundity in July and August (Chávez-Romo and Reyes-Bonilla 2007; Chávez-Romo et al. 2013; Carpizo-Ituarte et al. 2011; Zavala-Casas 2013; Rodríguez-Troncoso et al. 2014; Santiago-Valentín et al. 2015). And like the equatorial eastern Pacific, coral reproduction is significantly compromised or ceases altogether during strong El Niño activity and upwelling (Carpizo-Ituarte et al. 2011; Rodríguez-Troncoso et al. 2011).

Fig. 6.6 Coral research team preparing for an outing off Cabo Pulmo, Universidad Autónoma de Baja California Sur, La Paz, Mexico. "Eche Palante" (forward), standing, left to right: Betsabé Luna, Magdalena Précoma, Vladimir Pérez, Violeta Martínez, María Martínez, Cecilia Mozqueda, Héctor Reyes-Bonilla, Israel Sánchez; seated, left to right, Adriana Velazco, Alexandra Álvarez (courtesy of LAVISA Laboratory, unknown photographer, summer 2011)

Coral Fecundity and Recruitment

The reproductive effort and high levels necessary for coral community recovery depend on coral fecundity, the ages of corals when they first become sexually mature, and the duration of the breeding season. Fecundity is often defined as the maximum reproductive output of an individual female; for example, the total number of eggs released by a coral colony during its lifetime. This vital life cycle function also applies to coral species that release planula larvae. We have found that the ages when corals first become reproductive vary significantly among species, from 6 months to 15–20 years. Some long-lived species, such as *Pavona gigantea*, *Pavona clavus*, *Porites lobata*, and *Gardineroseris planulata*, may spawn gametes for most of their life spans of 50 to 200 years. The death of such long-lived colonies realized across many of the Galápagos Islands in 1983 (Glynn 1990), would be expected to significantly diminish the reproductive output essential for coral community recovery and stability.

A well-defined trend was evident between the size of reproductive products and fecundity when comparing broadcast spawning with brooding species. Hundreds of large planula larvae (with maximum diameters of 800–1400 μm) are released by *Tubastraea coccinea* and *Porites panamanesis* compared with thousands of relatively small eggs (100–150 μm in diameter) that broadcast spawning corals release. Successful recruitment depends on survivorship, and it has been shown that planula larvae in some species with large energy stores settle quickly after release, thus lowering the risk of predation in the planktonic phase of the life cycle. At Uva Island, *Porites panamensis* larvae settled in planula collectors from 24 to 48 hours after release (Smith 1991).The median basal diameter of the corals increased by 17% in one week, from 750 μm to 880 μm. These large sizes are from measurements of live larvae. David Smith reported these observations from his Master's research on the Uva Island coral reef. The time spent in the water column by dispersing coral larvae, however, is highly variable. Data from numerous studies have shown that larvae released by brooding coral species may settle within hours or not until 90–100 days; larvae released by broadcast spawning species were reported by Edinger and Risk (1995) to settle within a few days and up to 28 days.[18]

A fundamental consideration regarding sexual reproduction is recruitment, i.e., the settlement and metamorphosis of planulae. Larval settlement time may be shortened or prolonged depending on suitable substrates' availability or by evolutionary dispersal strategies. As of 2017, quantitative data on coral recruitment are available for 17 scleractinian species at 25 sites in three equatorial regions (Costa Rica, Panama, and the Galápagos Islands) and three species at 16 sites in Mexico. Mean recruitment densities are highly variable within and among species, but all corals have demonstrated some degree of sexual recruitment. More recent studies have also directly or indirectly underlined the critical role of sexual reproduction in

[18] More recent estimates of larval longevity have reported maximum life spans of some broadcast spawning species of 200+ days (Graham et al. 2008).

promoting recruitment in the recovery of eastern Pacific reefs, particularly pocilloporid reefs (Romero-Torres et al. 2020; Martínez-Castillo et al. 2022; Palacio-Castro et al. 2023).

Experimental studies of the sensory behavior of coral larvae have revealed an amazing capacity to limit mortality during reentry to the reef benthos. It has been convincingly demonstrated that some coral larvae do not settle haphazardly but are influenced by reef sounds and chemical and spectral cues (Vermeij et al. 2010). Over relatively short dispersal distances, the settlement and survival of coral larvae may benefit from reef sounds, e.g., fish calls and grunts, and alpheid snapping shrimps. For long distances between regions, dispersal times and viability may be significantly increased by detecting crustose coralline algae (CCA) on flotsam, the preferred settlement substrate of some corals (Foster and Gilmour 2016). Once settlement and metamorphosis (attachment) are achieved, the juvenile corals may grow and travel on their buoyant substrates for great distances.

In summary, there now exists ample evidence of successful sexual reproduction in corals contributing to the recovery of catastrophic ENSO-driven mortality. It is unnecessary to invoke long-distance dispersal and the immigration of coral recruits from distant source populations to the eastern Pacific. Self-seeding, mainly by autochthonous sexually produced planula larvae, has greatly aided recovery. But there is another mode of reproduction, namely asexual fragmentation, that also has contributed significantly to the recovery of eastern Pacific coral reefs.

Brooding and Broadcast Spawning

How do coral species with brooding or broadcast spawning reproductive modes compare in time and space? Does one type of life history adaptation have an advantage in dispersal or survivorship over the other? The prevalence of the two reproductive strategies was examined by Edinger and Risk (1995) who reported that brooding corals comprised a greater proportion of the Caribbean Miocene fauna during stressful environmental conditions, i.e., at the onset and persistence of turbid, cold, and elevated nutrient conditions than did broadcasting species. In contrast, nearly all eastern Pacific reef-building corals today are broadcast spawning species (Glynn and Ault 2000; Glynn et al. 2017). For example, of the 14 species in Panama 12 spawn gametes and 2 brood larvae (Fig. 6.4). These species—members of the genera *Pocillopora*, *Porites*, and *Pavona*—have contributed dominantly to reef building in the marginal eastern Pacific region from hundreds to thousands of years. Is it possible that the prevalence of highly fecund broadcast spawning corals is an adaptive response to survive rapidly changing environmental conditions in the eastern Pacific region (Glynn et al. 2017)?

As noted in Chap. 3 (Origin of the Eastern Pacific Coral Reef Province), the eastern Pacific reef-building coral fauna is closely related to, and likely derived from, the central/west Pacific faunas by long distance dispersal since the closure of the Central American portal. A relatively high proportion of eastern Pacific coral species possess autotrophic larvae (with zooxanthellae) that perform photosynthesis and thus can satisfy the energetic needs of their coral hosts during dispersal. Baird et al. (2009) hypothesized that the high proportion of eastern Pacific corals with autotrophic larvae were those species successful in crossing the east Pacific filter bridge. Several species (9 of 12, 75%) are shared between central/west Pacific and eastern Pacific faunas; however, they do not have autotrophic larvae (Glynn and Ault 2000).This suggests other factors at play in effecting long distance dispersal, for example, rafting. Rafting involves the settlement of larvae on floating objects, such as pumice, vegetation or debris, and their survival and growth during drifting until reaching distant shores. Paul Jokiel (1990a, b) reported on finding many rafted coral colonies on beached pumice that was adrift in the South Pacific (e.g., Micronesian and Marshall Islands). The pumice source area in one case, identified from the distinctive elemental composition of the pumice, was 1000+ km to the east (Tonga Islands) of the Great Barrier Reef. The rafted coral species were members of the genera *Pocillopora* and *Porites*, both with autotrophic larvae that are among the dominant reef builders in the eastern Pacific region.

Asexual Reproduction

Asexual reproduction results in the clonal multiplication of new individuals without producing gametes and their subsequent fusion during fertilization. Numerous organisms are capable of asexual or vegetative reproduction. The asexual propagation of plants is well known. George Williams (1975) proposed the "strawberry coral model" to demonstrate the similarities of many plants and corals in performing and benefiting from asexual and sexual reproduction. When a strawberry plant is successfully established in a favorable environment, it will send out asexual runners to expand and fill the available space. Similarly, an established coral colony will continue producing polyp-bearing tissues asexually (by budding), expanding its habitat space. Both strawberry plants and corals also produce sexually derived propagules, seeds and larvae[19] that can be widely dispersed. Colonizing diverse, widely dispersed habitats can be advantageous when conditions worsen locally.

Corals reproduce asexually in many ways. New individuals or colonies can occur asexually by budding and fission, polyp, or polyp ball detachment, asexual planulation (as noted above), and fragmentation (Harrison and Wallace 1990; Richmond 1997). The regeneration of injured tissues also occurs by polyp budding and the re-sheeting of dead skeletal surfaces in branching and massive coral species (Glynn

---

[19] *Pocillopora damicornis* is a brooding species in the Indo Pacific region, releasing sexually produced planula larvae (Richmond and Jokiel 1984; Richmond 1985). Asexually produced planulae were discovered in *P. damicornis* in Hawaii in an electrophoretic study by James Stoddart (1983). Additional examples of asexual larvae were reported by Ayre and Resing (1986) for two azooxanthellate corals (*Tubastraea* spp.), but not for two zooxanthellate brooding species of *Acropora* and *Seriatopora*.

and Fong 2006). In eastern Pacific coral communities, colony fragmentation is the predominant mode of asexual reproduction. Two processes can lead to the breakage of coral colonies, namely bioerosion and mechanical disturbances, the latter most often by benthic fish invertivores, and water turbulence (Highsmith 1980, 1982; Glynn et al. 2017). Bioerosion by boring sponges, bivalves, sea urchins, and fishes greatly weakens coral skeletons (Scott et al. 1988; Glynn and Manzello 2015; Alvarado et al. 2017). Coral skeletons thus weakened are more susceptible to fragmentation from increased water motion, projectiles (such as corals and tree trunks), and the feeding activities of fishes. If an observer on scuba lays motionless on the bottom of an eastern Pacific reef for a reasonable amount of time, say 30 min, the variety of activities of diverse mobile reef animals is astonishing. Common reef sounds of breakage are snapping, cracking, scratching, and crunching. We are aware of many of the species producing these sounds, but not all. Corallivores feeding on coral tissues or biting off branches and fragments from colonies, fishes and turtles rooting into coral frameworks, nest-building fishes, and fishes seeking shelter in reef cavities are some examples that result in fragmentation and potentially coral recruitment through asexual reproduction. Specific examples of such activities are examined in the chapters on predation (Chap. 7, Corallivores and Reef-dwelling Predators) and bioerosion (Chap. 9, Bioerosion).

Eastern Pacific estimates of the densities of recruits resulting from asexual fragmentation are highly variable. *Pocillopora* spp. fragment densities on the Huatulco (Mexico) reef tract, 1 to 2 months after three major hurricanes, ranged from 4.8 to 26.2 fragments per m^2 (Lirman et al. 2001). Recruit densities reported 3–4 years later along the same reef tract were significantly lower, ranging from 1.3 to 13.1 colonies per m^2 (López-Pérez et al. 2007). These lower values were largely a result of the absence of hurricanes after 1997 and the low survival of a high proportion of fragments from 2001.[20] Assessing the abundances of asexual recruits is problematic because if they are not firmly attached to a suitable substrate continued movement of fragments may result in the upending of polyp surfaces or deposition onto sandy or silty sediments where they die. Recruit densities of *Pocillopora damicornis* reported at the Uva reef in Panama, ranging from 0.38 to 0.44 colonies per m^2, were notably lower than those in Mexico (Highsmith 1982). This difference may be related to Panama's low latitudinal position below the hurricane zone. The lowest reported recruit densities were for a non-branching, massive coral (*Gardineroseris planulata*) at Uva Island with 0.008 to 0.009 colonies per m^2 (Glynn et al. 2017). This coral species is seldom attacked by corallivores or fishes (probing for and extracting boring bivalves), thus resulting in infrequent colony fragmentation and asexual propagation.

[20] Andrés López-Pérez highlighted the distinction between "recruitment" and "successful recruitment," two possible outcomes that must be recognized for critical comparisons. Generally, if a coral fragment is firmly attached to the substrate, for example, by cementation, it may be classified as a successful recruit. In this case, the possibility of fragment movement and mortality should be significantly minimized. Many potential recruits may never become attached and eventually die from frequent movement or smothering in fine sediments. It is important to provide all the methodology employed when reporting coral fragment survivorship.

Pocillopora spp. patch reefs occur in marginal environmental settings in other tropical regions, e.g., Western Australia (Ward 1992) and the Gulf of Oman (Sheppard and Salm 1988; Glynn 1993). These reefs appear to be clone formations, arising from asexually produced larvae and colony fragmentation. Iliana Baums and co-workers have reported low genetic diversity of a post-El Niño *Pocillopora* patch in the Galápagos Islands, providing evidence of asexual fragmentation in its formation (Baums et al. 2014). It seems that coral fragmentation plays a vital role in developing patch reefs in the eastern tropical Pacific. Fragmentation due to hurricanes may play an essenial role at higher latitudes, for example along the Mexican coast (Aranceta-Garza et al. 2012), and corallivory and bioerosion at lower latitudes in the equatorial eastern Pacific (Costa Rica, Panama, Colombia, coastal Ecuador, and the Galápagos Islands).

What is the relative importance of sexual and asexual recruitment in the recovery and maintenance of eastern Pacific coral populations? Some data are available from study plots frequently monitored in Panama and the Galápagos Islands from 1993–1998 (Glynn et al. 2017). For massive coral species at Uva Island, Panama (*Porites lobata*, *Pavona* spp., and *Gardineroseris planulata*), sexual recruits were present in all plots sampled (n = 16), but at low densities (0.002–0.267 colonies m^{-2}). Asexual recruits were absent from 13 plots and at low densities (0.007–0.083 colonies m^{-2}) in three plots.[21] This difference of higher sexual than asexual recruitment was also observed in *Pocillopora* spp. at three sampling sites in the Galápagos Islands. Since these differences are not rates and do not consider environmental factors that could influence the two modes of reproduction, more study is necessary to gain an understanding of the relative importance of each. There are so many confounding factors to consider, as Jerry Wellington would often exclaim.

It is now clear that eastern Pacific reef recovery from ENSO events and other disturbances is highly dependent on local sexual and asexual recruitment. Locally induced recovery or self-seeding is probably the reason for the documented rapid recovery of coral cover in 10–15 years across the eastern Pacific region (Guzmán and Cortés 2001, 2007; Baker et al. 2008; Fong et al. 2017; Romero-Torres et al. 2020; Martínez-Castillo et al. 2022; Palacio-Castro et al. 2023). Caution is necessary in interpreting the extent of recovery, however, because in most studies recovery is reported as **live coral cover** before and after a disturbance event. Such measures do not consider, for example, the loss or alteration of coral framework structures and the consequent loss of critical habitat space necessary for the support/existence of associated reef species.

[21] The overall growth shape and attachment base of young colonies are usually symmetrical in sexual recruits and asymmetrical with evidence of basal attachment breaks in asexual recruits (Richmond 1985).

References

Alvarado JJ, Aburto-Oropeza O, Abad R, Barraza E, Brandt M, Cantera J, Estrada P, Gaymer CF, Guzmán-Mora AG, Herlan JJ, Maté JL (2017) Coral reef conservation in the eastern tropical Pacific. In: Glynn PW, Manzello DP, Enochs IC (eds) Coral reefs of the eastern tropical Pacific: persistence and loss in a dynamic environment. Coral reefs of the world 8. Springer, Dordrecht, pp 565–591

Aranceta-Garza F, Balart EF, Reyes-Bonilla H, Cruz-Hernández P (2012) Effect of tropical storms on sexual and asexual reproduction in coral *Pocillopora verrucosa* subpopulations in the Gulf of California. Coral Reefs 31:1157–1167. https://doi.org/10.1007/s00338-012-0941-9

Ayre DJ, Resing JM (1986) Sexual and asexual production of planulae in reef corals. Mar Biol 90:187–190

Babcock RC, Bull GD, Harrison PL, Heyward AJ, Oliver JK, Wallace CC, Willis BL (1986) Synchronous spawnings of 105 scleractinian coral species on the Great Barrier Reef. Mar Biol 90:379–394

Baker AC, Glynn PW, Riegl B (2008) Climate change and coral reef bleaching: an ecological assessment of long-term impacts, recovery trends and future outlook. Estuar Coast Shelf Sci 80:435–471

Baird AH, Guest JR, Willis BL (2009) Systematic and biogeographical patterns in the reproductive biology of scleractinian corals. Annu Rev Ecol Evol Syst 40:551–571

Baums IB, Hughes CR, Hellberg ME (2005a) Mendelian microsatellite loci for the Caribbean coral *Acropora palmata*. Mar Ecol Prog Ser 288:115–127

Baums IB, Miller MW, Hellberg ME (2005b) Regionally isolated populations of an imperiled Caribbean coral, *Acropora palmata*. Mol Ecol 14:1377–1390

Baums IB, Miller MW, Hellberg ME (2006) Geographic variation in clonal structure in a reef-building Caribbean coral, *Acropora palmata*. Ecol Monogr 76(4):503–519

Baums IB, Devlin-Durante M, Laing BAA, Feingold J, Smith T, Bruckner A, Monteiro J (2014) Marginal coral populations: the densest known aggregation of *Pocillopora* in the Galápagos archipelago is of asexual origin. Front Mar Sci 1. https://doi.org/10.3389/fmars.2014.00059

Birkeland C (1977) The importance of rate of biomass accumulation in early successional stages of benthic communities to the survival of coral recruits. In: Proceedings of the 3rd international coral reef symposium, Miami, vol 1, pp 15–21

Carpizo-Ituarte E, Vizcaíno-Ochoa V, Chi-Barragán G, Tapia-Vázquez O, Cupul-Magaña AL, Medina-Rosas P (2011) Evidence of sexual reproduction in the hermatypic corals *Pocillopora damicornis*, *Porites panamensis*, and *Pavona gigantea* in Banderas Bay, Mexican Pacific. Cienc Mar 37:97–112

Chávez-Romo HE, Reyes-Bonilla H (2007) Reproducción sexual del coral *Pocillopora damicornis* al sur del Golfo de California, México. Ciencias Marinas 33(4):495–501

Chávez-Romo HE, Paz-García DA, Correa Sandoval F, Reyes-Bonilla H, López-Pérez RA, Medina-Rosas P (2013) Diferencia en las estrategias reproductivas de los corales escleractinios (ramificado vs masivo) a lo largo de la costa occidental de México. Cienc Mar 39(4):387–400

Colley SB, Feingold JS, Peña J, Glynn PW (2000) Reproductive ecology of *Diaseris distorta* (Michelin) (Fungiidae) in the Galápagos Islands, Ecuador. In: Proceedings of the 9th international coral reef symposium, Bali, vol 1, pp 373–379

Colley SB, Glynn PW, May AS, Maté JL (2006) Species-dependent reproductive responses of eastern Pacific corals to the 1997-1998 ENSO event. In: Proceedings of the 10th international coral reef symposium, Okinawa, vol 1, pp 61–70

Connell JH (1978) Diversity in tropical rain forests and coral reefs: high diversity of trees and corals is maintained only in a nonequilibrium state. Science 199(4335):1302–1310

Edinger EN, Risk MJ (1995) Preferential survivorship of brooding corals in a regional extinction. Paleobiology 21(2):200–219

Ellis J, Solander D (1786) The natural history of many curious and uncommon zoophytes, collected from various parts of the globe. In: Systematically arranged and described by the late Daniel Solander. 4. Benjamin White, London. 1-206, pls. 1-63

Fong P, Smith TB, Muthukrishnan R (2017) Algal dynamics: alternate stable states of reefs in the eastern tropical Pacific. In: Glynn PW, Manzello DP, Enochs IC (eds) Coral reefs of the eastern tropical Pacific: persistence and loss in a dynamic environment. Coral reefs of the world 8. Springer, Dordrecht, pp 339–367

Foster T, Gilmour JP (2016) Seeing red: coral larvae are attracted to healthy-looking reefs. Mar Ecol Prog Ser 559:65–71

Glynn PW (1990) Coral mortality and disturbances to coral reefs in the tropical eastern Pacific. In: Glynn PW (ed) Global ecological consequences of the 1982–83 El Niño-Southern Oscillation. Elsevier oceanography series 52. Elsevier, Amsterdam, pp 55–126

Glynn PW (1993) Monsoonal upwelling and episodic *Acanthaster* predation as probable controls of coral reef distribution and community structure in Oman, Indian Ocean. Atoll Res Bull 379:1–66. https://doi.org/10.5479/si.00775630.379.1

Glynn PW, Ault JS (2000) A biogeographic analysis and review of the far eastern Pacific coral reef region. Coral Reefs 19:1–23

Glynn PW, Colley SB (2008) Survival of brooding and broadcasting reef corals following large scale disturbances: is there any hope for broadcasting species during global warming? In: Proceedings of the 11th international coral reef symposium, Ft. Lauderdale, Florida, vol 1, pp 368–372

Glynn PW, Fong P (2006) Patterns of reef coral recovery by the regrowth of surviving tissues following the 1997-98 El Niño warming and 2000, 2001 upwelling cool events in Panamá, eastern Pacific. In: Proceedings of the 10th international coral reef symposium, Okinawa, vol 1, pp 624–630

Glynn PW, Manzello DP (2015) Bioerosion and coral reef growth: a dynamic balance. In: Birkeland C (ed) Coral reefs in the Anthropocene. Springer, Dordrecht, pp 67–97

Glynn PW, Maté JL, Stemann TA (2001) *Pavona chiriquiensis*, a new species of zooxanthellate scleractinian coral (Cnidaria: Anthozoa: Agariciidae) from the eastern tropical Pacific. Proc Biol Soc Wash 10:210–225

Glynn PW, Gassman NJ, Eakin CM, Cortés J, Smith DB, Guzmán HM (1991) Reef coral reproduction in the eastern Pacific: Costa Rica, Panamá and Galápagos Islands (Ecuador). I. Pocilloporidae. Mar Biol 109:355–368

Glynn PW, Colley SB, Gassman NJ, Black K, Cortés J, Maté JL (1996) Reef coral reproduction in the eastern Pacific: Costa Rica, Panamá and Galápagos Islands (Ecuador). III. Agariciidae (*Pavona gigantea* and *Gardineroseris planulata*). Mar Biol 125:579–601

Glynn PW, Colley SB, Guzmán HM, Enochs IC, Cortés J, Maté JL, Feingold JS (2011) Reef coral reproduction in the eastern Pacific: Costa Rica, Panamá and Galápagos Islands (Ecuador). VI. Agariciidae, *Pavona clavus*. Mar Biol 158:1601–1617

Glynn PW, Colley SB, Maté JL, Cortés J, Guzmán H, Bailey RL, Feingold JS, Enochs IC (2008) Reproductive ecology of the azooxanthellate coral *Tubastraea coccinea* in the equatorial eastern Pacific: part V. Dendrophylliidae. Mar Biol 153:529–544

Glynn PW, Colley SB, Eakin CM, Smith DB, Cortés J, Gassman NJ, Guzmán HM, Del Rosario JB, Feingold JB (1994) Reef coral reproduction in the eastern Pacific: Costa Rica, Panamá and Galápagos Islands (Ecuador). II. Poritidae. Mar Biol 118:191–208

Glynn PW, Colley SB, Ting JH, Maté JL, Guzman HM (2000) Reef coral reproduction in the eastern Pacific: Costa Rica, Panamá and Galápagos Islands (Ecuador). IV. Agariciidae, recruitment and recovery of *Pavona varians* and *Pavona* sp. a. Mar Biol 136:785–805

Glynn PW, Colley SB, Maté JL, Baums IB, Feingold JS, Cortés J, Guzmán HM, Afflerbach JC, Brandtneris VW, Ault JS (2012) Reef coral reproduction in the equatorial eastern Pacific: Costa Rica, Panamá and Galápagos Islands (Ecuador). VII. Siderastreidae, *Psammocora stellata* and *Psammocora profundacella*. Mar Biol. https://doi.org/10.1007/s00227-012-1979-5

Glynn PW, Colley SB, Carpizo-Ituarte E, Richmond RH (2017) Coral reproduction in the eastern Pacific. In: Glynn PW, Manzello DP, Enochs IC (eds) Coral reefs of the eastern tropical Pacific: persistence and loss in a dynamic environment. Coral reefs of the world 8. Springer, Dordrecht, pp 435–476

Gorbunov MY, Falkowski PG (2002) Photoreceptors in the cnidarian hosts allow symbiotic corals to sense blue moonlight. Limnol Oceanogr 47:309–315

Graham EM, Baird AH, Connolly SR (2008) Survival dynamics of scleractinian coral larvae and implications for dispersal. Coral Reefs 27:529–539. https://doi.org/10.1007/s00338-008-0361-z

Grigg RW (1981) *Acropora* in Hawaii. Part 2. Zoogeography. Pac Sci 35(1):15–24

Grigg RW (1982) Darwin point: a threshold for atoll formation. Coral Reefs 1:29–34

Grigg RW (1988) Paleoceanography of coral reefs in the Hawaiian-Emperor Chain. Science 240(4860):1737. https://doi.org/10.1126/science.240.4860.1737

Grigg RW (1998) Big surf, deep dives, and the islands. My life in the ocean. Editions Limited, Honolulu, p 180

Grigg RW (2002) Precious corals in Hawaii: discovery of a new bed and revised management measures for existing beds. Mar Fish Rev:13–20

Grigg RW, Hey R (1992) Paleoceanography of the tropical eastern Pacific Ocean. Science 255:172–178

Grigg RW, Maragos JE (1974) Recolonization of hermatypic corals on submerged lava flows in Hawaii. Ecology 55:387–395

Guzmán HM (1986) Estructura de la comunidad arrecifal de la Isla del Caño, Costa Rica, y el efecto de perturbaciones naturales severas. M.Sc. thesis. Universidad de Costa Rica, San Pedro, Costa Rica

Guzmán HM, Cortés J (1989) Coral reef community structure at Caño Island, Pacific Costa Rica. PSZNI Mar Ecol 10:23–41

Guzmán HM, Cortés J (2001) Changes in reef community structure after 15 years of natural disturbances in the eastern Pacific (Costa Rica). Bull Mar Sci 69:133–149

Guzmán HM, Cortés J (2007) Reef recovery 20 years after the 1982-1983 El Niño massive mortality. Mar Biol 151:401–411

Harrison PL (2011) Sexual reproduction of scleractinian corals. In: Dubinsky Z, Stambler N (eds) Coral reefs: an ecosystem in transition. Springer, Cham, pp 59–85. www.Springer.com, p 552

Harrison PL, Wallace CC (1990) Coral reproduction. In: Dubinsy Z (ed) Ecosystems of the world, vol. 25 coral reefs. Elsevier Science, New York, pp 133–208

Harrison PL, Babcock RC, Bull GD, Oliver JK, Wallace CC, Willis BL (1984) Mass spawning in tropical reef corals. Science 223:1186–1189

Hickman CP Jr (2008) A field guide to corals and other radiates of Galápagos. Sugar Spring, Lexington, Virginia, p 162

Highsmith RC (1980) Passive colonization and asexual colony multiplication in the massive coral *Porites lutea* Milne Edwards & Haime. J Exp Mar Biol Ecol 47:55–67

Highsmith RC (1982) Reproduction by fragmentation in corals. Mar Ecol Prog Ser 7:207–226

Hoeksema BW, Cairns S (2024) World list of Scleractinia. Accessed at https://www.marinespecies.org/sclaractinia on 2024-09-15

Hueerkamp C, Glynn PW, D'Croz L, Maté JL, Colley SB (2001) Bleaching and recovery of five eastern Pacific corals in an El Niño-related temperature experiment. Bull Mar Sci 69:215–236

Hyman LH (1940) The invertebrates: Protozoa through Ctenophora. McGraw-Hill Book, New York, p 726

Johnston EC, Forsman ZH, Flot J-F, Schmidt-Roach S, Pinzón JH, Knapp ISS, Toonen RJ (2017) A genomic glance through the fog of plasticity and diversification in *Pocillopora*. Nat Sci Rep 7:5991. https://doi.org/10.1038/s41598-017-06085-3

Jokiel PL (1990a) Long-distance dispersal by rafting: reemergence of an old hypothesis. Endeavour, New Series 14(2):66–73

Jokiel PL (1990b) Transport of reef corals into the Great Barrier Reef. Nature 347:665–667

Levy O, Appelbaum L, Leggat W, Gothlif Y, Hayward DC, Miller DJ, Hoegh-Guldberg O (2007) Light-responsive cryptochromes from a simple multicellular animal, the coral *Acropora millepora*. Science 318:467–470

Lewis JB (1991) The ampullae and medusae of the calcareous hydrozoan *Millepora complanata*. Hydrobiologia 216(217):165–169

Lewis JB (2006) Biology and ecology of the hydrocoral *Millepora* on coral reefs. Adv Mar Biol 50:1–55

Limer BD, Bloomberg J, Holstein DM (2020) The influence of eddies on coral larval retention in the Flower Garden Banks. Front Mar Sci 7. https://doi.org/10.3389/fmars.2020.00372

Lirman D, Glynn PW, Baker AC, Leyte Morales GE (2001) Combined effects of three sequential storms on the Huatulco coral reef tract, Mexico. Bull Marine Sci 69(1):267–278

López-Pérez RA, Mora-Pérez MG, Leyte-Morales GE (2007) Coral (Anthozoa: Scleractinia) recruitment at Bahías de Huatulco, western Mexico: implications for coral community structure and dynamics. Pac Sci 61:355–369

Martínez-Castillo V, Rodríguez-Troncoso AP, Mayfield AB, Rodríguez-Zaragoza FA, Cupul-Magaña AL (2022) Coral recovery in the central Mexican Pacific 20 years after the 1997-1998 El Niño event. Oceans 3:48–59. https://doi.org/10.3390/oceans3010005

Miller KJ, Ayre DJ (2004) The role of sexual and asexual reproduction in structuring high latitude populations of the reef coral *Pocillopora damicornis*. Heredity 92:557–568

Palacio-Castro AM, Smith TB, Brandtneris V, Snyder GA, van Hooidonk R, Maté JL, Manzello D, Glynn PW, Fong P, Baker AC (2023) Increased dominance of heat-tolerant symbionts creates resilient coral reefs in near-term ocean warming. Proc Nat Acad Sci 120(8):e2202388120. https://doi.org/10.1073/pnas.2202388120

Paz-García DA, Hellberg ME, García-de-León FJ, Balart EF (2015) Switch between morphologies of *Pocillopora* corals. Am Nat 186(3):434–440. https://doi.org/10.1086/682363

Pinzón JH, LaJeunesse TC (2011) Species delimitation of common reef corals in the genus *Pocillopora* using nucleotide sequence phylogenies, population genetics and symbiosis ecology. Mol Ecol 20:311–325. https://doi.org/10.1111/j.1365-294X.2010.04939.x

Pinzón JH, Sampayo E, Cox E, Chauka LJ, Chen CA, Voolstra CR, LaJeunesse TC (2013) Blind to morphology: genetics identifies several widespread ecologically common species and few endemics among indo-Pacific cauliflower corals (*Pocillopora*, Scleractinia). J Biogeogr 40:1595–1608

Richmond RH (1985) Variations in the population biology of *Pocillopora damicornis* across the Pacific. In: Proceedings of the 5th international coral reef congress, Tahiti, vol 6, pp 101–106

Richmond RH (1987a) Energetics, competence, and long-distance dispersal of planula larvae of the coral *Pocillopora damicornis*. Mar Biol 93:527–533

Richmond RH (1987b) Energetic relationships and biogeographical differences among fecundity, growth, and reproduction in the reef coral, *Pocillopora damicornis*. Bull Mar Sci 41:595–604

Richmond RH (1997) Reproduction and recruitment in corals: critical links in the persistence of reefs. In: Birkeland C (ed) Life and death of coral reefs. Chapman & Hall, New York, pp 175–197

Richmond RH, Hunter CL (1990) Reproduction and recruitment of corals: comparisons among the Caribbean, the tropical Pacific, and the Red Sea. Mar Ecol Prog Ser 60:185–203

Richmond RH, Jokiel PL (1984) Lunar periodicities in larva release in the reef coral *Pocillopora damicornis* at Enewetak and Hawaii. Bull Mar Sci 34:280–287

Rodríguez-Troncoso AP, Carpiso-Ituarte E, Leyte-Morales GE, Chi-Barragán G, Tapia-Vázquez O (2011) Sexual reproduction of three coral species from the Mexican South Pacific. Mar Biol 158:2673–2683

Rodríguez-Troncoso AP, Carpizo-Ituarte E, Pettay DT, Warner ME, Cupul-Magaña AL (2014) The effects of an abnormal decrease in temperature on the eastern Pacific reef-building coral *Pocillopora verrucosa*. Mar Biol 161:131–139

Romero-Torres M, Acosta A, Palacio-Castro AM, Treml EA, Zapata FA, Paz-García DA, Porter JW (2020) Coral reef resilience to thermal stress in the eastern tropical Pacific. Glob Chang Biol 26:3880–3890. https://doi.org/10.1111/gcb.15126

Santiago-Valentín JD, Rodríguez-Troncoso AP, Carpizo-Ituarte E, Benítz-Villalobos F, Torres-Hernández P, López-Pérez A (2015) Reproductive pattern of the reef-building coral *Pavona gigantea* (Scleractinia: Agariciidae) off southwestern Mexico. Cienc Mar 41:233–246

Schmidt-Roach S, Miller KJ, Lundgren P, Andreakis N (2014) With eyes wide open: a revision of species within and closely related to the *Pocillopora damicornis* species complex (Scleractinia; Pocilloporidae) using morphology and genetics. Zool J Linnaean Soc 170(1):1–33. https://doi.org/10.1111/zoj.12092

Scott PJB, Risk MJ, Carriquiry JD (1988) El Niño bioerosion and the survival of East Pacific reefs. In: Proceedings of the 6th international coral reef symposium, Townsville, vol 2, pp 517–520

Sheppard CRC, Salm RV (1988) Reef and coral communities of Oman, with a description of a new coral species (order Scleractinia, genus *Acanthastrea*). J Nat Hist 22:263–279

Shlesinger T, Loya Y (2019) Mass medusae release and temporal reproductive segregation among the three Red Sea fire coral species. Ecology 100(4):e02581. https://doi.org/10.1002/ecy.2581

Shlesinger Y, Goulet TL, Loya Y (1998) Reproductive patterns of scleractinian corals in the northern Red Sea. Mar Biol 132:691–701

Smith DB (1991) The reproduction and recruitment of *Porites panamensis* Verrill at Uva Island, Pacific Panama. MS thesis. University of Miami, Coral Gables, Florida, p 64

Smith TB, Glynn PW, Maté JL, Toth LT, Gyory J (2014) A depth refugium from catastrophic coral bleaching prevents regional extinction. Ecology 95(6):1663–1673

Soong K, Cho LC (1998) Synchronized release of medusae from three species of hydrozoan fire corals. Coral Reefs 17:145–154

Stoddart JA (1983) Asexual production of planulae in the coral *Pocillopora damicornis*. Mar Biol 76:279–284

Szmant AM (1986) Reproductive ecology of Caribbean reef corals. Coral Reefs 5:43–53

Szmant-Froelich A, Reutter M, Riggs I (1985) Sexual reproduction of *Favia fragum* (Esper): lunar patterns of gametogenesis, embryogenesis and planulation in Puerto Rico. Bull Mar Sci 37:880–892

Szmant-Froelich A, Yevich P, Pilson MEQ (1980) Gametogenesis and early development of the temperate coral *Astrangia danae* (Anthozoa, Scleractinia), vol 158. Biological Bulletin of the Marine Biological Laboratory, Woods Hole, pp 257–269

Turnham KE, Wham DC, Sampayo E, LaJeunesse TC (2021) Mutualistic microalgae co-diversify with reef corals that acquire symbionts during egg development. ISME J 15(11):3271–3285

Vargas-Ángel B, Colley SB, Hoke SM, Thomas JD (2006) The reproductive seasonality and gametogenic cycle of *Acropora cervicornis* off Broward County, Florida, USA. Coral Reefs 25:110–122. https://doi.org/10.1007/s00338-005-0070-9

Vermeij MJA, Marhaver KL, Huijbers CM, Nagelkerken I, Simpson SD (2010) Coral larvae move toward reef sounds. PLoS One 5(5):e10660. https://doi.org/10.1371/journal.pone.0010660

Ward S (1992) Evidence for broadcast spawning as well as brooding in the scleractinian coral *Pocillopora damicornis*. Mar Biol 112:641–646

Wellington GM (1982) Depth zonation of corals in the Gulf of Panama: control and facilitation by resident reef fishes. Ecol Monogr 52:223–241

Williams GC (1975) Sex and evolution. Princeton University Press, Princeton, NJ, p 200

Zavala-Casas DA (2013) Reproducción sexual de *Pocillopora damicornis* (Linnaeus 1758) (Anthozoa: Scleractinia) en la costa de Oaxaca, Mexico. BSc thesis. Universidad del Mar, pp 51–497

Chapter 7
Corallivores and Reef-Dwelling Predators

Abstract The proportion of corallivore species in relation to their coral prey species is higher in the eastern Pacific than in any other global coral reef region. The 47 known corallivore species are nearly equally divided among invertebrates and fishes. Prominent invertebrate corallivores of *Pocillopora* are an ovulid gastropod, hermit crabs, a sea urchin (in the Galápagos Islands), and the Crown-of-thorns Sea star, *Acanthaster* (in Mexico and Panama). A pufferfish (*Arothron*) is a frequent predator of *Pocillopora* and triggerfishes (*Pseudobalistes*, *Sufflamen*) have an indirect impact on massive corals (*Porites*, *Pavona*) when extracting date mussels (*Lithophaga*) imbedded in coral skeletons. Predation, a vital animal function, is manifest in diverse interrelated effects in coral reef ecosystems, ranging from planktivory, corallivory (which can exercise strong effects on coral abundance and community structure), indirect trophic interactions, and cryptofaunal predation. The defensive behaviors of crab and shrimp guard symbionts in protecting their coral hosts from predation are widespread and similar in eastern Pacific and Indo-Pacific regions. Free living shrimp and worm predators relentlessly attack and subdue *Acanthaster* over periods of days. Benthic invertivores (triggerfishes and turtles) are disruptive, breaking coral colonies and reef frameworks in pursuit of prey, but also promote coral asexual reproduction. Baited traps and simulated reef frameworks revealed highly diverse and abundant cryptic predator communities. Cryptic predators (e.g., polychaete worms, crustaceans, octopuses, and several fish families) inhabiting *Pocillopora* framework habitats are numerous and prey on larvae recruiting into reef structures from the plankton. While these diverse interactions are cryptic and often fleeting, their impact on the structure and ecology of coral reefs is consequential.

Introduction

Ecologists have long recognized the fundamental role of feeding and consumer-resource interactions in modulating the structure, function, and evolution of biotic communities. This has been especially well established in coral reef studies (e.g.,

© The Author(s), under exclusive license to Springer Nature Switzerland AG 2025
P. W. Glynn, *Coral Reef Resilience in the Anthropocene*,
https://doi.org/10.1007/978-3-031-81421-1_7

Bakus 1964; Vermeij 1978; Glynn 1990; Carpenter 1997; Karlson 1999; Duffy and Hay 2001; McClanahan 2008; Hixon 2015). These trophic-related effects are often easily observed and obvious, but can be indirect and subtle, difficult to discern and corroborate. Two of my early studies addressed community-wide feeding interactions, the first in the intertidal zone of the central California coast, and the second in a reef flat coral community off southwestern Puerto Rico. In both studies, plankton and detritus trophic resources were pivotal in supporting dominant suspension feeding and detritivore guilds in an algal-barnacle association in California (Glynn 1965) and a *Porites* community in Puerto Rico (Glynn 1973a, b). A landmark study of the trophic structure and organic production of a windward coral reef community on Eniwetok Atoll by the Odum brothers, Tom (Howard) and Gene (Eugene), captivated the interest of marine ecologists in the 1950s (Odum and Odum 1955). The Odum's demonstrated the pivotal role of primary production (benthic algae, zooxanthellae, *inter alia*) in powering the trophic dynamics of an atoll coral reef ecosystem. My focus was on the secondary and higher animal groups dependent on plant production, directly or indirectly through further trophic interactions (e.g., carnivory and detritivory). Our family move from Puerto Rico (University of Puerto Rico Mayagüez) to Panama (STRI, Smithsonian Tropical Research Institute) in 1967 prevented the completion of the Caribbean *Porites* community trophic study. This move, however, heralded new research opportunities into studies of the trophic structure and predator-prey interactions of eastern Pacific coral reefs. Predation, a life-sustaining animal function, is a complex subject with numerous interspecific interactions and far-reaching ecological ramifications. Here I discuss corallivory, coral host protection by crustacean guards and damselfish, direct and indirect effects on coral community diversity, and cryptic predators.

In our first surveys of coral reefs off the Pacific coast of Panama, we were struck by the great abundance of reef-associated fishes. Dick (Richard H.) Rosenblatt and John E. McCosker, ichthyologists familiar with the eastern Pacific fish fauna, performed species identifications; they found that several prominent species belonged to the Indo-Pacific fish fauna.[1] Large schools of parrotfishes and surgeonfishes were actively grazing on benthic algae, and damselfishes were omnipresent, defending their algal "gardens." Triggerfishes and pufferfishes were also commonplace, and quickly caught our attention regarding their damaging feeding effects on corals. *Pseudobalistes naufragium*, a triggerfish reaching 60 to 80 cm in length, was observed biting off large pieces from massive coral colonies, and even excavating and destroying sizable areas of encrusting corals (Fig. 7.1). While these triggerfish were preying on boring bivalve mollusks and not feeding directly on coral tissues, their impact on the health and ecology of the coral hosts was undeniable. Other less commonly observed incidences of destructive predation to reef frameworks involved

[1]Dick was the curator of the fish collection at the Scripps Institution of Oceanography (SIO), University of California, La Jolla, and John was a STRI pre-doctoral fellow from SIO. John, later associated with the Steinhart Aquarium in San Francisco, became Chair of Aquatic Biology of the California Academy of Sciences.

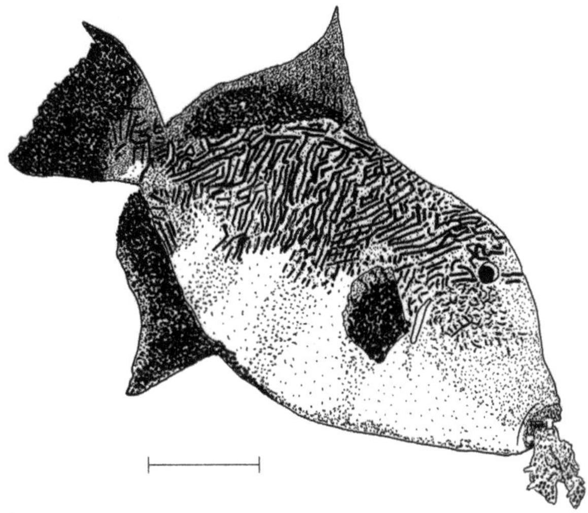

Fig. 7.1 *Pseudobalistes naufragium*, mouthing an uprooted *Pocillopora* branch. Scale bar = 6 cm

excavating sea turtles in search of sponges (Glynn 2008) and sharks in pursuit of fish prey (Jiménez 1996–1997).

The Guineafoul Pufferfish, *Arothron meleagris*, is a true predator of corals; such species that concentrate their feeding on live corals are termed **corallivores**. We observed this pufferfish biting off and ingesting the branch tips of *Pocillopora* corals. The visual effects of fishes feeding on corals were extraordinary and adding to this spectacle were various fish sounds—rasping, thumping, snapping, and clicking. This was not a "silent world" as characterized by Jacques Cousteau in his popular book, *The Silent World* (Cousteau and Dumas 1953). Diverse invertebrate corallivores were also commonly present, such as the Crown-of-Thorns sea star *Acanthaster*, an ovulid gastropod (*Jenneria*), and hermit crabs (*Trizopagurus* and *Aniculus*). Most of the invertebrate corallivores were small and secretive, however, and not easily viewed. Even *Jenneria*, with its flamboyant spotted shell and filamentous mantle extensions, closely resembled *Pocillopora* polyps (Fig. 5.2).

Broadly defined, corallivores feed on some variable amount of coral tissue, including endosymbiotic algae and metabolic products such as mucus, fat bodies, gametes, and larvae. The capture and consumption of a whole prey item, for example a crab or shrimp by a wrasse, is an example of **discrete predation**. Discrete predation is common in numerous carnivorous reef fishes. Because most reef-building corals are large colonies compared to the relatively small sizes of corallivores feeding on them, the actual consumption of coral tissues is often minimal and non-lethal, and termed **partial predation** (Jackson and Palumbi 1979; Glynn 1988a). In addition, the diets of corallivores usually include other non-coral prey.

A recent listing of eastern Pacific corallivores includes 47 species, consisting of 25 fishes and 22 invertebrates (Enochs and Glynn 2017). A portrayal of a coral-corallivore food web for the Uva Island reef consisted of 287+ species links with strong pathways between *Pocillopora* corals and diverse invertebrate and fish consumers (Glynn 2004). This large number of species feeding on corals is surprisingly high for a marginal and depauperate coral reef region. The numbers of corallivores reported by Rotjan and Lewis (2008) on species-rich reefs in the western Pacific ranged between 52 (Great Barrier Reef, Australia) and 73 (Philippines Archipelago). To help explain the high diversity of corallivores in the eastern Pacific, with the majority preying on pocilloporid coral species, Ian Enochs[2] and I hypothesized that this was due in large measure to this taxon's widespread occurrence, abundance, rapid growth rate, and high nutritional value. We also recognize that this coral resource provides a complex framework and habitat niches for numerous sheltering corallivores (Glynn 2008), and the potential for rapid recovery from disturbances. The persistence and abundance of pocilloporid corals, therefore, make these species a reliable food source and a valuable habitat.

Early studies by scientists investigating coral reefs held that corals were generally immune to predation because of their hard, mineral skeletons and stinging nematocysts (a type of defensive venomous dart). For example, in 1963 Sir Maurice Yonge, a distinguished invertebrate zoologist and leader of the Great Barrier Reef Expedition (1928–29), stated "Thanks no doubt to the presence of nematocysts, corals are little affected by predators." This pronouncement did not agree with the pervasive activity of corallivores observed on Panamanian reefs. Just 6 years later a predatory outbreak of *Acanthaster* in Guam prompted Rick (Richard) Chesher to speculate, "There is a possibility that we are witnessing the initial phases of extinction of madreporarian corals in the Pacific (Chesher 1969).[3] These seemingly contradictory views were cause for concern and heightened our attention to corallivores and their effects on eastern Pacific coral reefs. Some highlights of the discovery and investigations of eastern Pacific corallivores are presented in this chapter. Predator-prey interactions involving corals and other reef consumers are also discussed. Besides the clearly visible macropredators and macroherbivores, the feeding activities of minute, cryptic consumers and symbiotic associates of corals can also have important effects on reef community structure and organic production (Glynn 1976). Some coral consumers and common feeding pathways on a generalized coral reef are portrayed in Fig. 7.2.

[2]Atlantic Oceanographic and Meteorological Laboratory (AOML), Ocean Chemistry and Ecosystem Division (OCED), National Oceanic and Atmospheric Administration (NOAA), Miami, Florida, USA.

[3]Devastating *Acanthaster* outbreaks also were reported on the Great Barrier Reef in the 1960s (Pearson and Endean 1969; Endean 1973). In Bob (Robert) Endean's 1973 review of the history of *Acanthaster* predation, several instances of moderate to high coral mortality, and even local outbreaks resulting in near-total coral mortality, were reported for various Indo-Pacific sites before the 1960s. Whether the recent outbreaks were natural or human-induced was a hotly debated subject. A special session on the *Acanthaster* problem was held in 1973 at the Second International Coral Reef Symposium in Australia, aboard the passenger liner M. V. *Marco Polo*. A spirited barrage of questions was directed at Bob Endean when he presented hypotheses relating to the causes of outbreaks, especially his favored hypothesis of the removal of a predator by humans (the Giant Triton, a gastropod mollusk) that triggered sea star outbreaks (Endean 1974).

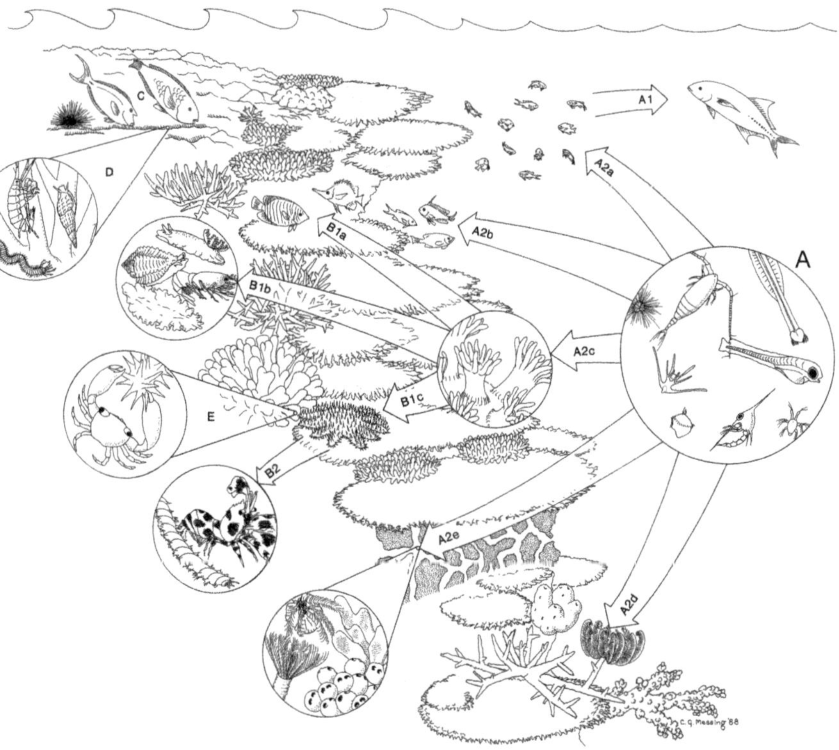

Fig. 7.2 Hypothetical coral reef front community illustrating diverse consumer organisms and trophic pathways. A, magnified view of mid-water zooplankton moving across reef front and consumed by planktivorous fishes (A2a, A2b), and suspension feeders (A2c-A2e). Many of the resident planktivores (A1) are preyed on by piscivores. Corals are consumed by corallivores (B1a-B1c). Micropredators and scavengers (B2) prey on *Acanthaster* corallivores (B2). Sea urchins and fish herbivores (C) feed on reef plants and incidentally on cryptic microfauna (D), and symbiotic crab guards gain sustenance from their coral hosts and protect them from sea star corallivore attacks (E). From Glynn (1988a, 1988b) and with permission of C. G. Messing

Diverse corallivores of *Pocillopora*, including *Acanthaster* and sea urchins (*Eucidaris galapagensis*), *Jenneria*, and *Arothron meleagris*, were commonplace in several areas of the eastern Pacific. When hermit crabs were first observed sheltering among the branches of *Pocillopora* we had no idea of their diet. Their coral feeding habit was first observed in laboratory aquaria when they became active at night. This was later substantiated during night diving observations on reefs. The hermit crabs remove coral tissues and associated carbonate skeleton by scraping and picking with their chelipeds (claw-like appendages). Sandra Gilchrist (1985) observed this feeding behavior in four species of hermit crabs, two each on Caribbean and Pacific reefs in Panama. The coral prey on Caribbean reefs were *Acropora* spp. and *Agaricia* spp.

Jenneria, Señora Islet

In order to determine the relative importance of corallivore species and to quantify their effects, I teamed up with Bob (Robert H.) Stewart and John E. McCosker, two of my colleagues in Panama.[4] We selected a *Pocillopora* community for study at Señora Islet in the Pearl Islands, Gulf of Panama. This site was relatively close to STRI, infrequently visited and undisturbed. By means of replicate quadrat and visual sampling we determined the live coral cover and biomass, as well as the population densities of four corallivore species, three invertebrates (a gastropod mollusk and two hermit crabs) and a pufferfish. About 10% of the area of the living coral community consisted of framework development. Corallivore feeding rates were determined from mesocosm observations. Because feeding behaviors can change in captivity, e.g., due to unnatural prey availability and relative abundances, laboratory and field observations of the four principal corallivores were compared. Prey selection, feeding rates, and diel activities in the field did not differ significantly from our mesocosm observations. Coral growth rates and related increases in carbonate mass were quantified and compared with overall corallivore consumption rates (Glynn et al. 1972).

Our approximation of the standing crop biomass of live *Pocillopora* coral was 105.1 metric tons per hectare with a tissue mass of 8.8 metric tons. The total coral mortality (chiefly live branches) due to the feeding of four corallivores amounted to 6.72 tons per year or 6.4% of the standing crop. Coral mortality resulting from the grazing of *Jenneria* was 5.26 tons per year or 79% of the total coral consumed. The remainder, a loss of 1.46 tons per year, resulted from hermit crabs and puffer fish feeding. The notably high coral tissue consumption by *Jenneria* was unexpected. This was due to their high abundance and efficient grazing of coral tissues. Is it possible that the high abundance of *Jenneria* benefited from the close resemblance of the corallivore to its prey (Fig. 5.2)? Due to the cryptic behavior of *Jenneria* coweries and their mostly nocturnal feeding, they are likely not easily detected by visual predators. In addition, their highly ornamented shell, variegated coloration, and mantle filaments (resembling *Pocillopora* polyps), contribute to their concealment. Mesocosm feeding observations conducted by Felix Oramas (1979)[5] demonstrated

---

[4] Bob, now deceased, was a geologist in the Engineering Division of the Panama Canal Company. I very much benefited from Bob's geological expertise, perspective, and companionship on several of our joint coral reef projects.

[5] Felix Oramas, Juan Achurra, and Antonio Montaner, conducted research on the feeding behavior of *Jenneria* and other gastropods in my laboratory to fulfill their thesis requirements for the degree of "Licenciado en Biología con Especialización en Zoología" at the University of Panama. Felix observed, by means of Y-tube choice experiments, that *Jenneria* was able to detect the presence of *Pocillopora* prey by chemosensory perception (Oramas 1979). Juan and Eduardo found that *Jenneria* had a strong feeding preference for *Pocillopora* over other coral species and died within a few days when deprived of coral food (Achurra Cárdenas and Valdés Araúz 1980). Many empty *Jenneria* shells were present on Panamanian reefs that experienced high *Pocillopora* mortality during the 1982–1983 El Niño event (Glynn 1985a). It is likely the high *Jenneria* mortality was due to the disappearance of their preferred prey. Considering the relative isolation of my laboratory I

that *Jenneria* moved to and fed on live branches at the bases of *Pocillopora* colonies when porcupinefish (*Diodon holocanthus*), one of its predators, were present. At times, *Jenneria* was easily spotted when it formed feeding aggregations, as high as 50–100 individuals per coral colony (Glynn 2004). Dense feeding aggregations of *Jenneria* were also reported to occur on colonies of *Porites panamensis* in Baja California (Paz-García et al. 2012). These dense aggregations often resulted in whole coral colony death. Since other unobserved fish and invertebrate corallivores surely contributed to coral loss, our results should be considered as minimum estimates. From this focused study in the Pearl Islands, it is evident that further study of corallivore effects on coral communities is needed.

Acanthaster, Gulf of Chiriquí

With Rick Chesher's (1969) stunning report of widespread and extreme coral destruction by *Acanthaster planci* in the western Pacific, Jim (James) W. Porter and I were concerned by the presence of sea star corallivores in the Gulf of Chiriquí (Glynn 1972; Glynn et al. 1972; Porter 1972a[6]). If Rick's dire predictions were realized, this meant that Panama's Pacific reefs could also be in jeopardy. In addition, in the 1960s and 1970s serious attention was being given to the construction of a sea-level canal (by nuclear[7] or conventional excavation) across the Isthmus of Panama (Challinor 1972). If this plan came to fruition, the possibility would have existed that *Acanthaster* could enter the Caribbean Sea via a sea-level canal if safeguards preventing an introduction were not in place. I was so excited by my first sighting of *Acanthaster* in the eastern Pacific, in April 1970, that I quickly dived to point out the sea star to others in our survey team. Naively, I extended my index finger too quickly and too near to *Acanthaster* and received a painful prick. *Acanthaster's* venomous spines are covered by a thin membrane that conceals the needle-like tips. The puncture wound caused an immediate throbbing pain that slowly subsided over a two-day period!

Initially, I mistakenly identified *Acanthaster* as *Acanthaster ellisii.* But with the help of Dave (David L.) Pawson, then Curator of Echinoderms in the Smithsonian

very much appreciated the camaraderie of these students and others from the University of Panama who shared in the joy of research and discovery.

[6] Jim was a predoctoral fellow at STRI from Yale University. His research was supported by a Grant-in-Aid from the Society of Sigma Xi, grants from the Organization for Tropical Studies, and from the Peabody Museum of Natural History, Yale University. Not always in agreement on some scientific topics, nonetheless I found our collaboration a most profitable and enjoyable experience. I was honored to serve as Jim's Best Man when he married Karen at a quaint chapel in the Panama Canal Zone. Jim continued with his research and academic pursuits, earning him a distinguished career in coral reef ecology. He is presently a Professor Emeritus, University of Georgia, Athens, and still very active in coral reef research.

[7] Nuclear excavation would have employed the use of large-scale nuclear explosions to create craters and spoil (debris) that could be removed by conventional means.

National Museum of Natural History, this error was corrected (Glynn 1974). Morphologically, the sea star belonged to the species, *Acanthaster planci*, which was experiencing population outbreaks and ravaging coral reefs in the western Pacific. *Acanthaster ellisii* was known for many years to be present in the Gulf of California (Caso 1962). A molecular analysis of *Acanthaster* populations across the Indo-Pacific region revealed that *A. planci* is a species complex consisting of at least four species (Vogler et al. 2008).[8] During the time of our observations in Panama, small *Acanthaster* populations were reported by Dana and Wolfson (1970) and Barham et al. (1973) in the Gulf of California. *Acanthaster* is still present in the Gulf and is moderately abundant in areas supporting coral growth. A feeding aggregation on El Corralito Reef reported by Jenny Carolina (Rodriguez Villalobos and Ayala-Bocos 2021) has caused high mortality in massive colonies of 100+ years in age (Fig. 7.3). *Acanthaster* aggregations in many parts of the western Pacific reached densities of thousands and tens of thousands of individuals per hectare, resulting in nearly total coral mortality on numerous large reefs (Moran 1988; Birkeland and

Fig. 7.3 Crown of thorns sea stars (*Acanthaster* cf. *ellisii*) feeding on a large, massive colony of *Pavona gigantea* at El Corralito, Espíritu Santo Archipelago, La Paz, Baja California Sur, 6 m depth, 11 August 2018. Seven feeding sea stars are visible; light-colored colony areas were stripped of coral tissues. Photograph courtesy of Jenny Carolina Rodríguez Villalobos

[8] Molecular data have shown that *A. planci* occurs in the northern sector of the Indian Ocean and a separate *Acanthaster* species occurs in the Pacific Ocean, present from the western Pacific to the equatorial far eastern Pacific region (Vogler et al. 2013; Haszprunar and Spies 2014; Haszprunar et al. 2017). Although its taxonomic status is unresolved, Haszprunar and co-workers recommended use of the name *Acanthaster solaris* for the Pacific species. Still, the status of the putative species *A. ellisii* has not yet been determined. Is this species in the Gulf of California genetically similar to *A. solaris* in the equatorial eastern Pacific or is it sufficiently isolated to have become genetically distinct? More recently Uthicke et al. (2024) recommend retention of the species name *A.* cf. *ellisii* for eastern Pacific populations pending further taxonomic study.

Lucas 1990; De'ath et al. 2012). Outbreak populations of such extraordinarily high densities have never been reported in the eastern Pacific; only localized feeding aggregations of 15 to 30 individuals per hectare are known.

Jim Porter and I were privy to an attempt of a researcher to introduce *Acanthaster* to a Panamanian Caribbean reef in the early 1970s, not long after its discovery on Panamanian Pacific reefs. This person, known by us, was curious to determine if *Acanthaster* would eat and survive on novel Caribbean coral species. There is no record of *Acanthaster* ever having lived in the Atlantic Ocean. I was astonished when informed and invited to participate in this plan. I emphatically declined and explained in detail the potential ecological and economic damage that could result from such an introduction. Sometime later I learned that a trial introduction was attempted with one individual *Acanthaster*. Fortunately for Caribbean corals the sea star, confined to a galvanized tin bucket without aeration, did not survive the trans-isthmian automobile ride and the project was abandoned. The idea of testing whether *Acanthaster* would feed and survive on novel coral prey was of interest, but the possibility of introducing an alien corallivore to the Caribbean Sea was too risky. During World War II, several swimming pool-sized tanks, each with a capacity of 115 kl (30,400 gallons) were used to store (and minimize corrosion) submarine cables employed in the security of the Panama Canal. Some years later we reasoned that these large tanks could be used to maintain and observe Caribbean reef corals and associated species in isolation. We modified the tanks to receive circulating, and filtered sea water, and the overhead opaque roof was replaced with transparent panels to allow the entry of sunlight. We then introduced corals and reef species to create mini community patches. One set of observations was designed to examine the interactions between *Acanthaster* and one of its predators, the Harlequin shrimp *Hymenocera*, which is discussed later in this chapter (Glynn 1977). A second set of observations involved coral patches with a mix of Pacific and Caribbean corals that were available to *Acanthaster* as potential prey (Glynn 1981). *Acanthaster* preyed on both Pacific and Caribbean coral species with equal alacrity. This reinforced our commitment to not allow sea star corallivores to enter and become established in western Atlantic tropical waters.

A major question centered around the ages of *Acanthaster* outbreaks. Were they cyclic, long-term natural occurrences or recent events, the result of anthropogenic disturbances? Ed (Edward) Frankel delivered a provocative report at the international coral reef meetings in Miami in 1977, offering evidence of past *Acanthaster* presence from the existence of skeletal remains in reef sediments, suggesting episodic aggregation abundances (Frankel 1977). Ed's study was conducted on Australia's Great Barrier Reef and reported the occurrence of skeletal elements on reefs from 110 to 1145 yr. BP (before present). His methodology seemed ideally suited to duplicate in Panama. I arranged for Ed to join us on a cruise to the Uva Island reef to search for evidence of past *Acanthaster* presence on an eastern Pacific coral reef.

The three basic steps in this research involve (a) collection of samples from different sediment pile depths, (b) a search for *Acanthaster* skeletal remains at different sediment depths, and (c) radiocarbon dating of skeletal elements to obtain age estimates of past sea star occurrences. We collected sediment samples with a push core, assisted by airlifting,[9] from surface and subsurface sediments to a maximum depth of about 2.5 meters. The sediments were then carefully examined microscopically to search for *Acanthaster* remains. My trusted laboratory assistant, Aníbal Velarde, pored over the sediments and began finding *Acanthaster* ossicles and spines in nearly all samples (Fig. 7.4). His son, Aníbal Eduardo Velarde, began watching the search effort with interest and was soon enlisted to help with this phase of the work. To help verify the identity of the skeletal elements, we disarticulated an *Acanthaster* specimen to compare with the remains recovered from the sediment samples. The shapes of the ossicles and spines and their mauve color were generally sufficient for identification. Margaret J. Basile, one of my first Master's (MA degree) students at RSMAES, employed scanning electron micrography (SEM) to help verify the identity of damaged and eroded skeletal elements. To avoid confusion with the occurrence of other non-*Acanthaster* echinoderm skeletal debris, for comparison Margaret disarticulated three sea star and three sea urchin species present on the Uva reef. She found that the sizes of the minute perforations in the lattice network of *Acanthaster* ossicles[10] were highly variable compared with the other species. Finally, radiometric C-14 dating allowed us to determine the approximate time of occurrence of *Acanthaster* on the Uva Island reef.[11]

The results of our study at the Uva reef were encouraging and made sense in terms of the occurrence of *Acanthaster* skeletal remains and the correspondence between their increasing age with the greater depths of sediment sampled (Enochs and Glynn 2017). Recent skeletal remains, from sea stars present on the reef during the past decades, were recovered from surface to sediment depths of about 1 meter. The ages of remains increased steadily in push core samples to maximum values of 1645 ± 300 years and 3830 ± 300 years at 2.5 meters depth in the sediment pile. No more than three skeletal elements per kg of sediment analyzed were recovered, but this convinced us of the likely long-time presence of *Acanthaster* on the Uva reef.[12] From this evidence, it seems that low to moderate abundances of *Acanthaster* have been present and preying on corals for hundreds of years on a Panamanian coral reef.

---

[9] A pushcore and airlifting device, fashioned after that employed by Gene (Eugene) A. Shinn (1968), allowed for the collection of sediment samples at specific depths.

[10] The micro-texture and ultra-structure of the ossicles reveal a lattice of minute (10–500 μm) perforations or pores, the stereom, which is unique to echinoderm taxa. Margaret observed that the highly variable pore size of the stereom was a useful distinguishing feature for old, fractured, and discolored *Acanthaster* skeletal debris.

[11] I thank Mike (Michael J.) Risk, McMaster University, for facilitating radiocarbon dating at the National Ocean Sciences Accelerator Mass Spectrometry Facility (NOSAMS), Woods Hole Oceanographic Institution, Massachusetts.

[12] This line of research was not without controversy. For example, the sediment piles on the Great Barrier Reef that suggested cyclic sea star abundances related to past outbreaks, like those of the late 1960s and later (Frankel 1978; Walbran et al. 1989a, b), were subject to disturbances that could disrupt the steady accumulation of sediments over time. Some skeptics argued that burrowing and foraging animals (agents of **bioturbation**, i. e., the mixing of sediment by excavating organisms), and storms could disrupt sedimentary records, thus resulting in biased data.

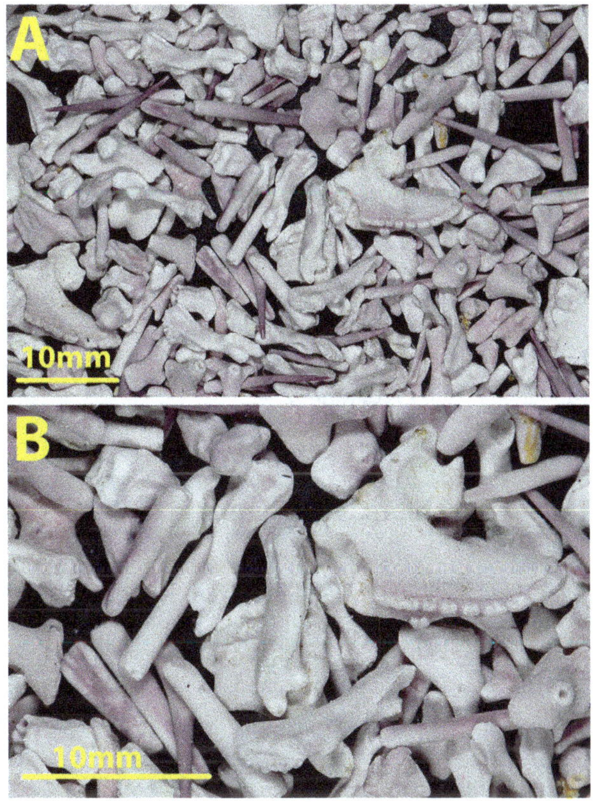

Fig. 7.4 Porous skeletal elements (ossicles) and spines (faint purple) of a disarticulated *Acanthaster* from Uva Island reef, December 1982. Amount shown in (**A**) equals about one-tenth of total from a single 30 cm diameter (arm tip to arm tip) sea star individual. (**B**), Enlarged view of A

Soon after our discovery of *Acanthaster* in Panama I planned a study to examine its feeding effects on the growth (horizontal change of live coral cover) of a *Pocillopora* coral reef. The discrete Uva Island reef, 2.5 ha in area with *Pocillopora* the predominant coral cover and with a relatively stable sea star population, was selected for study. The results of this analysis, based on the area of live *Pocillopora* (6100 m^2), potential horizontal reef expansion (21.4% per year, in the absence of corallivore mortality), the population density of *Acanthaster* (26 individuals per ha), and feeding rate (0.5 m^2 per individual sea star per month), indicated positive reef growth. The area of live *Pocillopora*, barring any major disturbances, was projected to increase more than two-fold, from 6100 m^2 to 13,700 m^2 over a 5-year period. Positive reef growth also was projected to occur at *Acanthaster* densities of 65 individuals per ha, but would experience severe coral loss at an outbreak density of 260 individuals per ha, 10 times the density of *Acanthaster* in the 1970s (Glynn 1973c).

As noted earlier, from the 1980s to the early 1990s, *Acanthaster* has undergone a dramatic and unexplained decline in abundance at the Uva Island reef. Only one or two sea star individuals were observed in censuses from 1993 to 2010, yielding extremely low densities of 0.9 individuals per ha (Enochs and Glynn 2017). At present, *Acanthaster* is not limiting reef growth at any monitored sites in the equatorial eastern Pacific. In the Gulf of California, however, a localized aggregation of *Acanthaster* has been observed killing centuries-old massive colonies of *Pavona gigantea* (Rodriguez Villalobos and Ayala-Bocos. 2021) (Fig. 7.3).

The underlying causes of *Acanthaster* outbreaks have remained elusive. Chuck Birkeland's (1982) hypothesis of terrestrial runoff as the cause of *Acanthaster* increasing abundance is a rational explanation demonstrating the complexity of factors causing sea star outbreaks. His hypothesis was supported by observations made over several years of study in the South Pacific region.[13] In qualitative modeling of outbreaks on the Great Barrier Reef, elevated nutrients and predator removal hypotheses demonstrated strong effects, however, other factors are likely involved (Babcock et al. 2016). The possible role of cryptic invertebrate predators in limiting sea star population growth in the eastern Pacific is discussed later in this chapter (under "Cryptic Predators").

Eucidaris, Galápagos Islands

It has long been known that large aggregations of sea urchins can disrupt and even denude biotic substrates. This fact did not prepare us, however, for the large numbers of Slate Pencil Sea Urchins (*Eucidaris galapagensis*)[14] observed at numerous coral sites in the Galápagos Islands in 1975. The large size of *E. galapagensis*, some individuals 16–18 cm across (spine tip to spine tip), were abundant on rock and coral substrates throughout the central and southern Galápagos Islands. Densities of 20 and as high as 50 individuals per m^2 were commonly encountered. Observations and gut content analyses revealed algae and coral (dominantly *Pocillopora*) as the principal food sources at our study sites. Since the high densities and nearly continuous grazing of *Eucidaris* on live *Pocillopora* appeared to negatively impact reef growth, we decided it would be informative to create a budget comparing carbonate gains and losses. *Eucidaris* generally concentrated its feeding on corals at the margins of reefs as did *Acanthaster* in Panama. Our analysis indicated that reef growth

[13] Chuck proposed several preconditions and interdependent events to explain *Acanthaster* outbreaks in Micronesia and Polynesia. He observed that outbreaks occur around drought-stricken or end of dry season high islands following typhoons and heavy rainfall. Nutrient runoff from the rains following dry spells stimulate plankton production, which benefits early *Acanthaster* larval development and survival. During early development post-settlement sea stars are cryptic (sheltering in reefs) and herbivorous (Moran 1986; Birkeland and Lucas 1990). Juvenile sea stars feed mostly on coralline algae, emerge onto open reef surfaces after 3 years and then feed exclusively on living corals. A recent study has demonstrated that the diet of juvenile *Acanthaster* is more flexible than previously perceived, including fleshy algae and **biofilms**, i.e., bacterial-coated surfaces (Deaker et al. 2020). Growth rates of subadult sea stars can vary depending on the availability of different food sources, which could change the timing of emergence and outbreaks.

[14] From genetic analysis, Lessios et al. (1999) distinguished between *Eucidaris galapagensis* and *Eucidaris thouarsii*, which were earlier considered to be the same species. *Eucidaris galapagensis* is restricted to the Galápagos Islands and Cocos Island whereas *E. thouarsii* is present on continental shores and islands throughout the eastern Pacific region.

(carbonate production) in the southern Galápagos Islands was robust, for example resulting in the deposition of 4 to 5 kg $CaCO_3$ per m^2 per year at 30% to 40% live coral cover (Glynn et al. 1979). We projected this level of production in the absence of grazing by *Eucidaris*. At observed *Eucidaris* densities of 30 to 40 individuals per m^2 at reef margins, however, net production was estimated to range from less than one to zero kg of calcium carbonate annually.

John Lawrence, an echinoderm specialist, became interested in our findings and questioned the nutritional benefit of coral to sea urchins. John visited my laboratory and investigated the digestive capacity of *E. thouarsii* when feeding on coral. He found that the absorption efficiency of *E. thouarsii* feeding on *Pocillopora* in Panama was 85.3% for protein and 69.3% for lipid (Lawrence and Glynn 1984). Despite this relatively high efficiency, John concluded that since the gross composition of *Pocillopora* is primarily inorganic (96.2%), large quantities of coral must be consumed to meet the sea urchin's nutritional requirements. When *Eucidaris* feeds, it rasps the branch-tips of colonies removing mostly skeletal fragments. The continuous 24-h feeding behavior of both *Eucidaris* species is probably necessary to maintain sufficient nutritive intake. This continuous grazing also contributes significantly to bioerosion, the subject of Chap. 9.

General Predation

Coming from a research background that centered on intertidal zone community ecology, I was greatly impressed with the pioneering field experiments of Joe Connell (1961) and Bob Paine (1966) who demonstrated the pivotal role of biotic interactions—competition and predation—in modulating community structural patterns. Species exercising strong community wide interactions came to be known as **keystone species.** Two definitions of a keystone species, originating from Paine's early studies: "a top predator with broad impacts on a community" and "a species with a disproportionate effect in maintaining community diversity" are commonly cited. These processes and defining properties have taken on additional flavors over the years. For example, many influential species that do not interact directly with other species, but that provide habitat space or facilitate other species have been identified. A classic demonstration by Jerry Wellington of the critical role of damselfish in restricting the entry of grazing herbivores and corallivores into their protected territories was shown to facilitate the recruitment of *Pocillopora* (Wellington 1982). With enhanced recruitment of *Pocillopora*, positive feedback results in an increase in damselfish shelter and abundance thereby promoting the development of a shallow *Pocillopora* zone. An additional example of species providing habitat, so-called **foundation species**, has been recognized and widely adopted (Bruno and Bertness 2001). The habitat space (shelter) and trophic resources provided by pocilloporid corals are cases in point. The three examples introduced earlier of strong corallivory and impacts on carbonate budgets—by a gastropod (*Jenneria*), sea star (*Acanthaster*), and sea urchin (*Eucidaris*)—clearly justify the moniker keystone species. Here I consider three areas of predator-prey interactions that my students, colleagues, and I have investigated over the years: (a) effects of corallivory on coral community diversity, (b) indirect effects of feeding, and (c) cryptic predators.

***Acanthaster* and Coral Community Diversity.** From our initial observations of the feeding behavior and prey choices by *Acanthaster*, Jim Porter and I came to different conclusions regarding the sea star's effects on coral community structure and diversity. *Pocillopora* corals are exceedingly abundant in the Gulf of Chiriquí and at other eastern Pacific sites where *Acanthaster* is present. *Acanthaster* is often observed feeding on *Pocillopora* where species in this genus and the sea star co-occur. Jim concluded that by feeding on and lowering the abundance of the coral community dominant (*Pocillopora*) that this should lead to a diversifying effect on community structure (Porter 1972a, b, 1974). I, too, observed *Acanthaster* commonly feeding on *Pocillopora* corals, but mainly on broken branches and small colonies (Glynn 1974, 1977). Undamaged and intact colonies of *Pocillopora* invariably host and shelter crab (*Trapezia* spp.) and shrimp (*Alpheus lottini*) guards that provide protection from predatory sea star attacks. The avoidance of coral colonies with crustacean guards encourages the selection of alternative unprotected coral prey, such as loose, broken coral branches.

Owing to its relatively small size and isolation from human disturbances, we have concentrated our feeding studies of *Acanthaster* on the Uva Island patch reef. Over the past several decades we have come to know this reef and its distinctive coral patches—"Stonehenge", "Richmond's 4 × 5", "lunar scape", "*Gardineroseris* city" or "G-City"—perhaps better than our own backyards. Also, to better understand the feeding behavior of *Acanthaster* we developed a non-intrusive coding system to identify individual sea star movements and coral prey selection. This system is based on the number and disposition of arms, madreporites (openings to the water-based circulatory system), anuses, and spines, which together are unique for any individual sea star (Glynn 1982; Fig. 7.5). This scheme has allowed us to quickly identify individuals, in 1 to 2 min, which we have followed around the reef for up to 2 years.

When *Acanthaster* sea stars approach and prepare to attack a *Pocillopora* colony the crustacean guards move to the threatened branches and assume defensive postures. The guards are often aroused before the sea star touches the coral. This raises the suspicion that chemical cues from *Acanthaster* may alert the guards of an approaching threat. To assess this notion, we placed *Acanthaster* in sealed transparent bags and moved the bags toward corals to mimic attacks. We also employed orange-red cloth models of *Acanthaster* of various sizes (Fig. 7.6), with white or black tube feet.[15] The bagged *Acanthaster* or the models did not alert the coral guards. When either of these treatments was infused with "*Acanthaster* water" (obtained from a bucket holding a sea star and introduced from a medical syringe),

[15] Aníbal Velarde and I wondered how *Acanthaster* size and color might play a role in eliciting the defensive behaviors of crustacean guards. Enter the Velarde family again to help with evaluating these cues. Aníbal's wife, Silvia Herrera de Velarde, kindly accepted the challenge and fabricated a variety of cloth model facsimiles of *Acanthaster*.

A

1 M/////M/M//M/M/M/M/M/M//
14 X 27 cm

2 M/////M/M/M/M//M/M/M/M//
17 X 30 cm

3 M//M///M/M//M//M/////
16 X 25 cm

4 M/////M///M/M/M/M//M///
14 X 22 cm

B

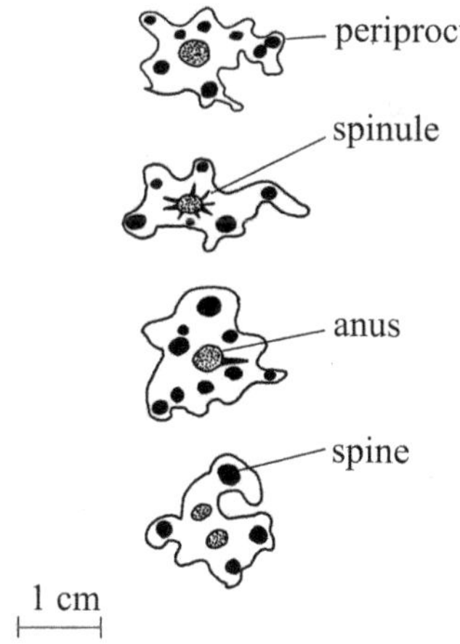

Fig. 7.5 Morphological features employed in the recognition of individual *Acanthaster*. (A) Number, and arrangement of interradial madreporites (M) and arms (/) with corresponding disc and maximum (arm tip to arm tip) sea star diameters. (**B**) Shape of smooth membranous area (periproct) surrounding the anus. Examples of four individuals whose movements were followed around the Uva Island coral reef (2 March 1982). Sea star 1 possessed 5 arms following M, 1 arm following M, 2 arms following M, and so on to M followed by 2 arms with a total of 9 M and 15 arms. The periproct contained 8 spines and 1 anus

however, strong guard defensive responses were elicited (Glynn 1980) (Fig. 7.7). The "*Acanthaster* water" alone and with the model sea stars elicited sudden and aggressive defensive responses. It is possible that organic compounds, secondary metabolites such as saponins that are present in *Acanthaster* (Kitagawa and Kobayashi 1978), might arouse the guards. Indeed, crustacean guards located in corals downstream from approaching *Acanthaster* were aroused significantly sooner than upstream guard symbionts. A likely chemical compound is still in need of identification.

The various defensive responses of the *Pocillopora* crustacean guards—*Trapezia*: (a) startle display, (b) pushing, (c) up-down jerking of *Acanthaster* (Fig. 7.8), (d) pinching and clipping of the sea star's spines and tube feet (Fig. 7.9), (e) resisting the sea star's retreat (Fig. 7.10), and *Alpheus*: (f) snapping and pinching at sea star's arm tips (Fig. 7.11)—are described in detail in Glynn (1977). A crab's response of restraining the retreat of an agitated sea star ("e" above and Fig. 7.10) was entertaining and compelling. This behavior began with a crab guard gripping a spine and maintaining a hold on the spine even though the sea star was forcefully attempting to dismount and break free from the coral colony. It appeared from the sea star's forceful attempt to retreat that the crab's chelipeds (pincers) and pereopods (legs) would separate from the body. Just before the crab appeared to experience the loss of appendages, however, it would quickly release its grip. This sudden release would invariably result in the sea star tumbling away from the coral. If the

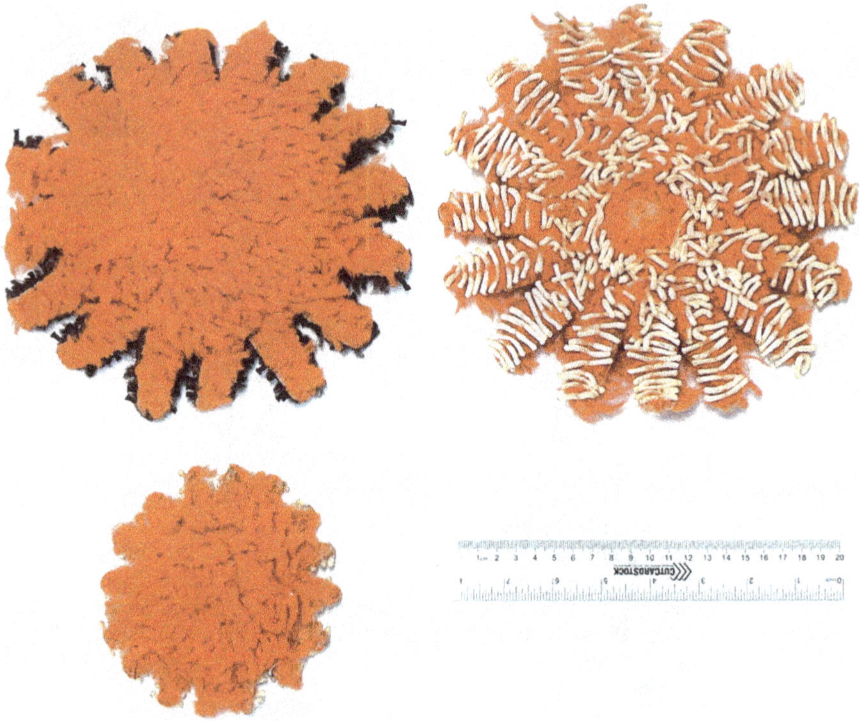

Fig. 7.6 *Acanthaster* models employed in simulated sea star attack behavior of corals. The tube feet are black and white cotton cord segments. The cloth models were filled with sand to increase their stability under turbulent conditions. A 20 cm ruler denotes size

threatened coral colony was located on an elevated surface, e.g. a vertical coral face, the disengaged sea star would flip backwards in free fall for some considerable distance. Upon landing, *Acanthaster* would usually scurry away from the area, at warp speeds of 0.5–1.0 meter per minute.

The crustacean defensive responses, those involving physical contact ("b"– "f" above), were quantified by recording their frequency over a 3-min period. This method was not without difficulties because of the uncertain deterrent effect of the different sorts of agonistic or protective engagements. Nonetheless, from trials simulating attack of *Pocillopora* by various corallivores (e.g., *Jenneria, Eucidaris, Arothron*, and hermit crabs), the defensive responses of coral guards was an order of magnitude more intense toward *Acanthaster* than for any of the other species evaluated. This could be interpreted to signify that *Acanthaster* is the most potent corallivore and that selection for the evolution of defensive behaviors has been directed toward thwarting mainly the sea star corallivore.

Acanthaster demonstrated a preference for *Pocillopora* prey, but usually for small colonies or broken branches not defended by mature crustacean guards. To understand a sea star's effect on the relative abundances of coral species it is

Fig. 7.7 Peter W. Glynn observing crustacean guards and sea star interactions on the Uva Island coral reef (14 January 1980). A model *Acanthaster* was placed over a *Pocillopora* coral to mimic a sea star's feeding posture, and water from a container with the sea star was introduced near the branches of the coral. This treatment was sufficient to elicit strong crustacean defensive responses equal to those directed toward a live attacking *Acanthaster*. A live *Acanthaster* is visible to the left. Photograph courtesy of Arcadio Rodaniche

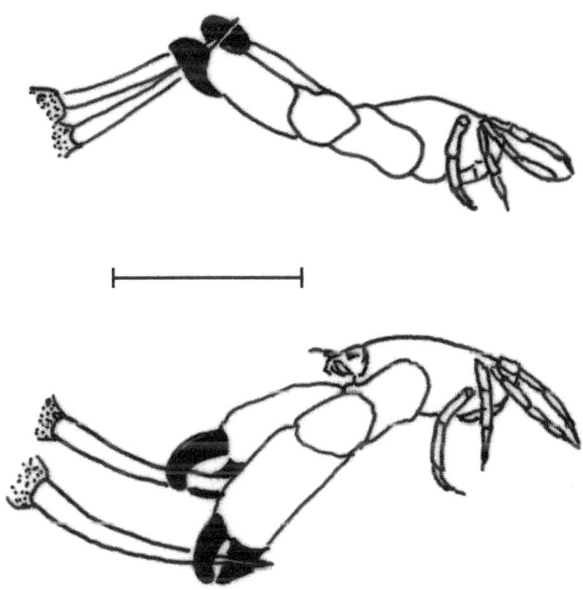

Fig. 7.8 *Trapezia ferruginea*, a *Pocillopora* crab guard grasping *Acanthaster* spines with up- and-down jerking, scale bar = 1 cm

Fig. 7.9 *Trapezia ferruginea*, a *Pocillopora* crab guard, pinching and clipping *Acanthaster* spines and tube feet, scale bar = 1 cm

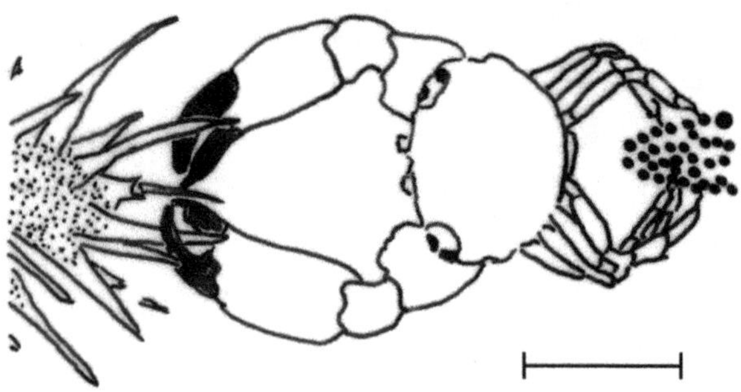

Fig. 7.10 *Trapezia ferruginea*, a *Pocillopora* crab guard resisting retreat of *Acanthaster*, scale bar = 1 cm

necessary to determine the frequency with which species are eaten in relation to their abundances and availability. Such an analysis conducted at Uva Island in 1973 demonstrated that *Pocillopora* colonies were consumed at a lower frequency than their availability (Glynn 1974). Observations of foraging *Acanthaster* showed that attempted mounting of intact *Pocillopora* colonies by sea stars usually resulted in repulsion. These colonies were regarded as non-preferred prey. Over time, prey selection of unprotected, non-pocilloporid species would have a depressing effect on coral community diversity. In our predator-prey studies at several Pacific and Indian Ocean localities (Panama, American Samoa, Guam, and Oman), we have found that *Acanthaster* predation results in the suppression of coral community diversity (Glynn 1987, 1990, 2013; Enochs and Glynn 2017).

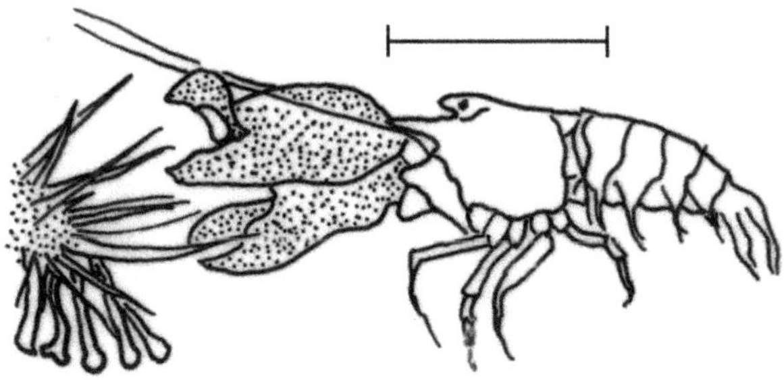

Fig. 7.11 *Alpheus lottini*, a pistol shrimp snapping at an approaching *Acanthaster* arm tip, scale bar = 1 cm

In 1982 I was invited by the Ministry of Agriculture and Fisheries (MAF), Sultanate of Oman, to study the effects of an *Acanthaster* outbreak that occurred in the Gulf of Oman during 1978–1980. This was an extraordinary predation event that caused high coral mortality and raised the alarm among local fishers who depended on coral reef habitats for their optimal fish and shellfish catches. When Aníbal and I arrived, we were told that tens of thousands of sea stars had invaded reefs in the Daymaniyat Islands area in the gulf. Unfortunately, no accurate abundance data were available. In the main Daymaniyat Archipelago we observed densities of 100–140 ind per ha (Glynn 1993). From the abundances of recently dead corals, with many sea stars still feeding, it appeared that we were witnessing the final stages of a massive *Acanthaster* outbreak event. Amazingly, nearly all the patch reefs composed mainly of *Acropora* spp. or *Montipora* spp. were recently dead and the *Pocillopora* patch reefs were still alive. While sea stars were actively foraging and feeding on remnant acroporid species, pocilloporid corals (including *Stylophora* spp.) were virtually untouched. The surviving pocilloporid colonies hosted normal complements of crustacean guard species and abundances, similar to demographic data from the eastern Pacific (Glynn 1976, 1987, 1993).

Armed with behavioral observations of crustacean guard defenses from the eastern Pacific, we were prepared to observe and compare the responses elicited by *Acanthaster* in acroporid and pocilloporid prey species. As a reminder, acroporid species do not occur in the eastern Pacific region. Crustacean guards on *Acropora* spp. performed mean simulated defensive responses of 3.8–4.8 per 3 min. compared with mean responses of 17.9 and 21.7 per 3 min. on *Stylophora pistillata* and *Pocillopora damicornis* respectively. Nearly all live *Acropora* colonies examined hosted a minute crab guard (*Tetralia cavimana*) that was generally timid and exhibited weak defensive behaviors (Fig. 5.6). The defense displayed by the crustacean guards (*Trapezia* spp. and *Alpheus lottini*) on pocilloporid colonies was intense and similar to that observed in the eastern and central Pacific regions (Glynn 1987). The levels of host colony defense by crustacean guards, ranked from most to least intense—*Pocillopora* > *Stylophora* > *Acropora*—were proportional to the resistance of these coral genera to predation and survival success following the *Acanthaster* outbreak in Oman. This pattern of defensive behavior and species-specific survivorship was similarly observed by Morgan Pratchett and co-workers on the Great Barrier Reef, Australia (Pratchett et al. 2000; Pratchett 2001). Also, McKeon et al. (2012) observed defensive synergism, i.e., elevated levels of coral host protection in coral communities with diverse guard species at Mo'orea, French Polynesia. In addition, McKeon and Moore (2014) demonstrated that for a particular coral guard species, the size of defending crabs is of critical importance. The larger the guards, the stronger the host protection. Smaller size crabs within the genus *Trapezia* (juvenile and adults), however, can play important defensive roles in thwarting attacks by relatively small corallivores.

Looking back, Jim and I were both right. *Acanthaster* could influence coral community diversity, causing increasing or decreasing species diversity depending on circumstances, e.g., abundance of foraging sea stars, relative abundances of preferred/non-preferred coral prey, protected/unprotected coral colonies, size structure and density of coral colonies, *inter alia*.

Before leaving our research activities in the Sultanate of Oman, I relate some notable experiences and express my gratitude to the Sultanate's Ministry of Agriculture and Fisheries for allowing us to work there. Aníbal and I found the Sultanate to be a most beautiful, interesting, and welcoming country. Since we were contracted to survey possible *Acanthaster* outbreaks along all of Oman's coastline, this required field work in the remote Musandam region at the Strait of Hormuz in the north and along remote coastal areas of Salalah and near Yemen in the south. Mohammad Hasan of the Omani fisheries department served as our envoy, translator, and willing field assistant. At one of our survey sites in the Musandam, we anchored our dhow off Kumzar, an isolated village nestled in a narrow mountainous valley. From our vantage point offshore, we observed footpaths winding around dwellings, no automobiles, roads, or sounds. This was a mysterious settlement, ostensibly cut off from the modern world. I asked Hasan if he would speak with Kumzar's wali (the governor of a province) for permission to visit this intriguing village—our request was quickly denied by the wali. We were granted permission, however, to conduct *Acanthaster* surveys in the inlet at the entrance to the village.

In contrast to our unwelcome experience at Kumzar, we were warmly greeted at the village of Mirbat on the south coast of Oman, and even invited into the wali's fort. This was an imposing stronghold with two turbaned guards at the front gate wearing harem or Punjabi pants and brandishing crossed lances to prevent entry. As we approached an order was called out, the lances were raised, and we were allowed entry into the fort. We were then escorted up a narrow flight of stairs to a tower with an elaborately decorated reception room. There were beautiful Persian rugs and ornate pillows for our comfort. We were served tea and flavorsome sweets. After explaining our work and requesting permission to conduct surveys along the coast—permission immediately granted—our conversation turned to the wali's memories and stories of the battle of Mirbat during the Dhofar Rebellion. This conflict (1963–1976) involved a major British intervention against the Dhofar Liberation Front, a Marxist group that aimed to create an independent communist state in the Sultanate. The wali remarked that this was the first time in a long while that he had the pleasure to speak in English with visitors. Between Mirbat and Sadh, about

60 km eastward, the coastal area is subject to seasonal (monsoon) upwelling. We did not observe *Acanthaster* in our surveys there, but abalone, kelps, and other temperate water species. Small coral colonies were present under the kelp canopies. Surprisingly, we encountered Omani women divers, covered with black garments from head to toe, harvesting abalones. I had never before witnessed such a scene and began to slowly swim toward the divers to photograph their activities. Suddenly, I was subject to a cacophony of ululating lady divers. I miscalculated the sensitivities of these Muslim women and quickly returned to our dhow, embarrassed, crestfallen and without photographs.

Indirect Interactions. It is now firmly established that crustacean guards can influence coral prey selection by *Acanthaster* across many parts of the Indo-Pacific region. Corallivores preying on corals directly, e.g., *Jenneria*, hermit crabs, pufferfish and *Acanthaster*, initiate **direct trophic effects**. Crustacean guards that influence the prey choice of *Acanthaster* (and other corallivores) by defensive behavioral interactions bring about **indirect trophic effects**. There are numerous manifestations of indirect effects, some obvious and others subtle and difficult to detect. As noted earlier in Chap. 5 under "Symbiotic Crustaceans", for many years a clone (23 colonies) of *Gardineroseris planulata* on the Uva Island reef was surrounded by a continuous stand of the live *Pocillopora*-crustacean mutualism that denied *Acanthaster* access to this aggregation of a preferred prey species. One of these corals was nearly 200 years old, suggesting decades of protection from predation. The surrounding circle of the guarded *Pocillopora* corals was serving as a refuge, thus causing an indirect positive effect (Glynn 1985b). Unfortunately, the protective *Pocillopora* barrier was broken during the 1982–83 El Niño warming period, a result of coral bleaching and mortality, allowing *Acanthaster* to enter and prey on *Gardineroseris* colonies that survived the bleaching event. Fortunately, *Acanthaster* numbers declined to 1–2 individuals reef-wide at Uva in the early 1990s, leaving G-City in a healthy state (Fong and Glynn 1998).

An intriguing pattern of uneaten preferred corals, with a halo shape, was described by Mitchell Colgan (1987) on a coral reef devastated by *Acanthaster* in Guam. Mitchell found that the living halo of corals surrounded colonies of the *Pocillopora*-crustacean mutualism. The proximity of the crustacean guards was sufficient to extend protection to nearby non-pocilloporid species. A similar effect was reported by McKeon and Moore (2014) at Mo'orea. Were there other invertebrate species besides crustacean guards that conferred some level of protection from the ravages of *Acanthaster*?

An earlier study at Mo'orea was puzzling since it reported an associational refuge from *Acanthaster* predation but made no mention of crustacean guards (Kayal et al. 2011). An **associational effect** is a form of indirect positive or negative interaction whereby one species indirectly benefits or disadvantages another by reducing or increasing consumer pressure (Bruno and Bertness 2001). When fish invertivores prey on crustacean guards, for example when crabs are migrating among coral colonies (Castro 1978), the resulting reduced numbers of guards would lower *Pocillopora* host protection and thus increase the coral's vulnerability. This is an example of an **indirect negative interaction** (INI) affecting *Pocillopora*. Field experiments by

Kayal and co-workers demonstrated that *Acropora* corals were sheltered from predation by neighboring *Pocillopora* corals that reduced detection and access by corallivores, resulting in an **indirect positive interaction** (IPI) affecting *Acropora*. In the Mo'orea study, were crustacean guards present and contributing to protection or were they absent? Were the structural refuges and density-dependent prey dilution effects sufficient for an IPI outcome? In a recent personal communication from Jim Porter, harkening back to our early 1970s studies in Panama, he stressed that we must not forget the pivotal role of crustacean guards in influencing the prey choice of *Acanthaster*.

Some possible community-based effects resulting from *Acanthaster* predation may be apparent in the abundance and age structure of potential coral prey populations. This relationship was examined from long-term monitoring in Panama, from sites with and without *Acanthaster* populations. In terms of preferred *Acanthaster* prey species, the numbers of live colonies, their larger sizes and longevity were significantly greater in the Pearl Islands, in the absence of *Acanthaster* predation, than in the Gulf of Chiriquí where the sea star is present (Glynn 1987). These results cannot be accepted as definitive cause and effect relationships, however, because of the possible confounding effects of the upwelling and non-upwelling conditions in the two gulfs. For example, seasonal upwelling in the Gulf of Panama could advantage the larger, long-living corals during stressful ENSO events (Glynn et al. 2001).

Another serious disturbance affecting corals in the Gulf of Chiriquí was the occurrence of the 1982–83 El Niño event (Chap. 5, see also Glynn 1990). Peggy Fong and I developed a simulation model to compare the effects of *Acanthaster* predation and El Niño coral bleaching/mortality on the *Gardineroseris* subpopulation (the G-City clone) at Uva Island. This analysis incorporated 25 years of field data for the size structure of *Gardineroseris* colonies and 20 years of *Acanthaster* predation on the coral clone. Model results indicated that coral mortality from persistent *Acanthaster* predation[16] demonstrated a much higher impact on *Gardineroseris* than El Niño episodic events (Fong and Glynn 1998). Predation scenarios of varying frequency (from high to low) skewed the coral size-class distributions toward smaller colonies, while simulations where populations were subjected to frequent El Niño events (e.g., every 3 years) still maintained colonies in the largest size classes. From these model results, we hypothesized that *Gardineroseris* colonies would achieve large size only under special circumstances, i.e., if protected by a biotic barrier, if present in cryptic locations, or in areas from which *Acanthaster* was absent (i.e. upwelling environments).[17]

[16] Our model results suggested that *Acanthaster* densities may not have been so abundant (10–28 ind ha^{-1}, 1976–1987) on the Uva reef during the previous 30 years (Enochs and Glynn 2017).

[17] A later regional model, modified from the earlier local version, extended the coral growth temperature relationship to include lower temperatures in areas of strong seasonal upwelling. The modified model can aid in predictions of changes in coral population dynamics in response to ENSO events of different frequency and magnitude (Fong and Glynn 2000). Finally, the very strong 1997–98 ENSO event, in many ways equal in strength to the 1982–83 event (Enfield 2001), allowed an additional modeling study to compare the *Gardineroseris* populations in the non-upwelling Gulf of Chiriquí with the upwelling Gulf of Panama environments (Fong and Glynn

A final mention of damselfishes is included here because of their aggressive behavior contributing toward indirect interactions and their widespread occurrence in eastern Pacific coral communities. Damselfishes bite off and ingest coral tissues to clear and create surfaces to cultivate "algal gardens" or "lawns". These algal patches provide a food source and are defended by damselfishes against competing herbivores and corallivores. The basic indirect interaction involves a damselfish ejecting sea urchin or *Acanthaster* corallivores from its "garden" and nearby surroundings thus providing associational protection for corals. In the Galápagos Islands, damselfishes eject *Eucidaris* (Glynn and Wellington 1983), and in Panama they eject *Diadema* (Eakin 1987). Damselfishes can significantly influence bioerosion, a subject introduced in Chap. 9. In Panama, damselfish also interfere with the entry of *Acanthaster* into their territories (Glynn 1988a, b). The damselfish ejection behavior of *Acanthaster* has been observed in American Samoa and Tahiti as well (Glynn and Colgan 1988).[18] It is intriguing that multiple crustacean and damselfish species share similar territorial defensive behaviors via INI and IPI in so widely separated regions of the world's oceans.

The agility, frequency, and persistence of damselfish in executing their repertoire of ejection activities are notable. The more common agonistic behaviors performed by large adult damselfish are the following. Damselfishes grasp sea urchin spines by their mouths and swim with them, sometimes up to one meter, and drop them outside of their territorial boundaries (Fig. 11.4). *Eucidaris*, too large to be lifted by damselfish, are dragged and rolled along the bottom until removal from the algal lawn. Spine nipping, tail beating and brushing against approaching *Diadema* have also been observed in Panama. Nipping spines and biting the tube feet of *Acanthaster* are commonly observed defensive behaviors. Forceful tail beating was usually directed at the undersides of *Acanthaster* and not at the spiny aboral surface. Smaller damselfish often engaged in non-contact displacement activities, such as biting and tail brushing against the substrate, towards approaching sea urchins or *Acanthaster*. These agonistic behaviors are illustrated in Glynn (1988b) and Glynn and Colgan (1988).

Jerry Wellington and I performed simple trials of introducing other (possibly threatening) species of echinoderms and gastropods into damselfish territories. We found that the frequency and intensity of ejections were correlated with the potential for damage to damselfish territories (Glynn and Wellington 1983). The graded responses (greatest toward *Eucidaris*) in relation to sea urchin grazing and damage to the defended territories would suggest a behavioral adaptation associated with the

2001). Model predictions and field measurements of the colony size structures were similar in the two populations. While highest coral mortality occurred in the Gulf of Panama in 1982–83, this was reversed in 1997–98 with highest mortality in the Gulf of Chiriquí. A key predictor for the mortality of *Gardineroseris* proved to be the rate of temperature increase during the El Niño warming phase.

[18] Two species of damselfishes were observed ejecting *Eucidaris* from territories in the Galápagos Islands, namely *Stegastes arcifrons* and *Stegastes beebei*. In Panama, *Stegastes acapulcoensis* defended territories from *Diadema mexicanum* and *Acanthaster*. *Acanthaster* was repelled by *Stegastes nigricans* in Tahiti, and by *Plectroglyphidodon dickii* in American Samoa.

defense and maintenance of a vital resource. The importance of protecting a contested algal food source is obvious but there are other likely benefits associated with these territories, such as (a) supplemental nutrition from corals, (b) spaces serving as shelters from predators, and (c) sheltered surfaces for depositing and protecting eggs. Perhaps defending all of these resources contributes to the fitness of territorial damselfishes.

Cryptic predators. It is a strongly held tenet that coral reefs provide shelter for a multitude of species. Coral reefs are an oasis of biodiversity, the tropical rainforests of the sea (e.g., Reaka-Kudla 1997). Bob (Robert N.) Ginsburg proffered that most of the biomass of reef metazoans is unseen and exceeds that of the visible epibenthos and nekton. These cryptic animals, both motile and sessile, belong to the **cryptofauna**, the suite of species that live within the cavities and recesses of coral reef frameworks. The numerous cavities, spaces and substrate surfaces offer an opportunity of survival for many potential prey—resident reef species as well as migrants and planktonic larval stages—to shelter from predators in relative safety. This is especially the case in the eastern Pacific where *Pocillopora* reefs, by virtue of a complex framework of interconnected skeletal branches, offer diverse protective niches. This scenario seems logical if it were not for the fact that coral reefs also provide shelter for a host of cryptic predators, scavengers, and parasites.[19] Thus, the sheltering effect of coral reefs is best viewed in relative terms.

From our earliest surveys of *Pocillopora* reefs we suspected the presence of a rich cryptofauna. Two unexpected feeding observations raised our interest in cryptic predators. The first involved the Harlequin shrimp *Hymenocera picta*, which we observed picking and cutting into the dermis of an individual *Acanthaster* that we followed around the reef for five days (Fig. 7.12). The shrimp remained on the sea star's aboral surface the whole while, riding, cutting, and consuming internal organs. This was our first sighting of the magnificently colored shrimp predator after several years of intensive study of the Uva reef. We found that *Hymenocera* could be lured out of the reef framework by baiting with various sea star species. Over a period of greater than 3 years (1975–1979) and 12 baiting trials, 10 *Hymenocera* attacks on *Acanthaster* were observed, involving a total of 30 individual shrimp (Glynn 1981). This cryptic predator, which emerges from the reef when potential sea star prey are present, is seldom seen but evidently abundant on the Uva reef.

The second observation, often made during *Hymenocera* attacks, involved the cryptic polychaete worm predator/scavenger, *Pherecardia striata*[20] (Fig. 7.13). The

[19] Preliminary data on the abundance and diversity (100+ species) of cryptic metazoans inhabiting *Porites furcata* (branching colonies) patch reefs in Puerto Rico suggest that this condition is generally common in similar coral communities in other regions (Glynn 1972, 1973d). The depletion of zooplankton biomass, 62.0% of biomass in water streaming across *Porites* reef flat communities (Glynn 1973b), was substantial. We observed nocturnal and crepuscular fish planktivores (apogonids, holocentrids, and pomacentrids) actively feeding on zooplankton, however, the amounts consumed were not quantified.

[20] The drawing of *P. striata* is from a photograph of a specimen collected in the Gulf of Chiriquí, Panama (courtesy of Leslie Harris, Natural History Museum of Los Angeles County, Los Angeles, California). According to Sergio Salazar (El Colegio de la Frontera Sur, Chetumal, México), who

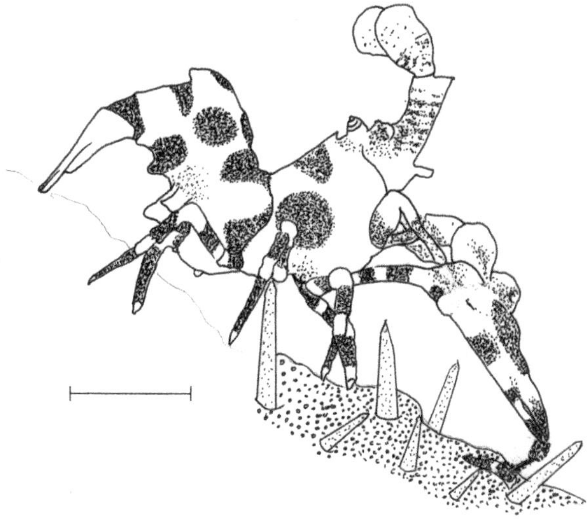

Fig. 7.12 *Hymenocera picta*, the Harlequin shrimp, removing *Acanthaster* gonads, scale bar = 1 cm

feeding habits of this worm were unknown to us until we observed *Acanthaster* being attacked by *Hymenocera* (Glynn 1984). When following and recording the behavior of *Acanthaster* under attack—sea stars usually moved quickly during the first day or two but then became torpid and immobile a day or hours before death—we observed that an individual sea star could be under attack by 2–4 shrimp and 12–18 worms feeding simultaneously. Necropsies of dead *Acanthaster* revealed extensive damage and evidence of the consumption of hepatic caeca, the pyloric stomach, gonads, and the water vascular system.[21] The opportunistic worm predator also proved to be exceedingly abundant on the Uva reef. Despite a concerted effort over several years to find juvenile *Acanthaster* in Panamanian coral communities, none has ever been seen by our team. Could it be that these early life history stages are exceedingly rare due to predation by the cryptic predators *Hymenocera* and *Pherecardia*?

In a recent study of juvenile *Acanthaster* (<4 mm) on the Great Barrier Reef, 10 of 26 cryptic invertebrate predators were reported that regularly consume the early stage sea stars (Desbiens et al. 2023). A crustacean predator, a decorator crab (*Schizophrys aspera*), consumed whole juvenile *Acanthaster* in 89% of feeding trials with a mean predation rate of 5.8 individuals day^{-1}. This high level predation rate of vulnerable juvenile *Acanthaster* could play a critical role in regulating adult corallivorous sea star abundances.

is completing a revision of the genus, there are probably two and possibly three species of *Pherecardia*. The photographed specimen from Panama probably belongs to the species *Pherecardia distincta*, whose dorsal stripes are more irregular with wider bands medially and thinner ones laterally toward the parapodia than in the other species.

[21] The feeding behaviors of the sea star and worm are unusual and a bit macabre. *Acanthaster* everts its stomach over coral prey, essentially feeding outside of its body. *Pherecardia* enters sea star prey through lessions cut by *Hymenocera*, thus feeding internally, from the inside out.

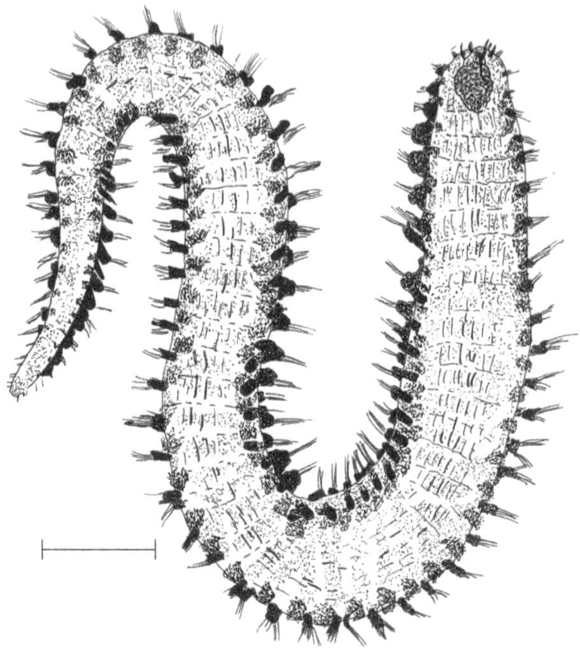

Fig. 7.13 *Pherecardia striata*, a cryptic polychaete worm predator/scavenger, scale bar = 1 cm. This drawing is probably the species, *Pherecardia distincta*, (see footnote 20)

To continue searching for cryptic predators and other trophic guilds we constructed dead *Pocillopora* rubble structures (held together by netting) that approximated reef frameworks. Dead, coral rubble plots, essentially two-dimensional habitats, were also sampled to compare with the three-dimensional simulated reef frameworks. These two treatments were deployed across the reef and sampled periodically over several years to determine and assess the abundances of motile metazoan species recruiting to the structures. Species identifications in our studies were determined from classical taxonomic procedures utilizing morphological traits. Other studies have employed DNA barcoding and metabarcoding, which have revealed formerly unrecognized biodiversity (e.g. Leray and Knowlton 2015; Ransome et al. 2017).

We also deployed scavenger traps designed by Irving Bethancourt,[22] a student volunteer. These were baited with strips of fish flesh and left overnight in reef frameworks. An inverted mesh cone allowed entry (and minimal escape) of invertebrates and fishes, up to a maximum body diameter of 2–3 cm. The trapped scavengers and predators were collected in the early morning hours, counted, and identified.

[22] Irving is a University of Panama graduate with a "Licenciado" degree in Biology and specialization in Zoology. In Panama the "Licenciado" degree normally requires 4 years of study and a thesis of original research in a student's area of specialization. This degree is approximately equivalent to the MA (Master of Arts) degree in universities in the USA. Irving volunteered as a field assistant and significantly advanced our studies of the cryptofauna, from 2003 to 2006. He recently became the Scientific Coordinator of the Naos Laboratory complex at STRI when Aníbal Velarde retired in February 2022.

On the Uva reef, a total of 46 fish species in 23 families was recovered in SRF structures compared with 20 fish species in 15 families observed over DCR substrates (Glynn 2006). Dead rubble substrates were essentially two dimensional with a thin layer, 5–10 cm thick, of mostly *Pocillopora* branches overlaying sand. Of 37 juvenile and adult cryptic fishes, 30 (81%) were carnivores and 8 (22%) were planktivores (Glynn 2013). Three of the four most abundant cryptic species were carnivores or planktivores with mean densities of about 10 to 25 ind m^{-2}. Among the fish carnivores were seven species of eels. Notable among the invertebrate predators were brachyuran crabs, octopuses, polychaete worms, stomatopods, and isopods.

The yields of species from scavenger traps embedded in the reef were no less productive than the animals captured in the SRF structures. Virtually every trap captured a variety of species. Relative abundances, in descending order, were isopods, gastropods, polychaete worms, and ostracods with mysids, brachyuran crabs, pagurid crabs, and four additional taxa (holothurians, polyclad flatworms, bivalves, and muraenid eels) (Glynn 2013). Octopuses and eels occasionally entered the traps and attacked species that were feeding on the fish bait. It is not uncommon for some species to prey on living, moribund, and dead animals at different times. In Ian's study of motile cryptofaunal communities inhabiting live and dead framework structures in the Gulf of Panama, he found a high proportion of biomass belonging to diverse carnivorous invertebrate predators (Enochs 2012). As observed in Panama, a high proportion of cryptic predators (74%) also recruited to dead coral rubble at Gorgona Island, Colombia (Mora and Zapata 2000). The high abundance of cryptic invertebrate predators (Keable 1995) and fishes (Depczynsi and Bellwood 2003) associated with reefs in the Great Barrier Reef region of Australia indicate that the predator and scavenger trophic guilds are an important component of coral reefs. Detritivory was also a major guild among fishes in Australia.

Considered as a whole, the multifaceted nature of corals within the reef ecosystem is undeniable. On the surface, they provide sustenance for a diverse assemblage of corallivorous species. From within, however, they offer shelter for a community of cryptic predators that is staggeringly complex, with orders of magnitude higher biodiversity across multiple phyla. While there is still much to learn, this peak into the cryptic reef community and observations of their trophic behavior support Ginsburg's prescient pronouncement of both the abundance and importance of animal life sheltering in the internal spaces of coral reefs. The ubiquity of carnivores and prevalence of predator-prey interactions in reef environments, however, lends some irony to their consideration as "shelters". In some ways the interstices of the reef might be likened to a "snake pit" for the many species attempting to shelter within.

References

Achurra Cárdenas JB, Valdés Araúz E (1980) Estudios sobre los efectos predatorios de los moluscos *Coralliophila abbreviata* y *Jenneria pustulata* de ambos literales del Istmo de Panama. Tesis, Universidad de Panamá, Escuela de Biología, República de Panamá. 66 pages, 24 plates, 10 figures

Babcock RC, Dambacher JM, Morello EB, Plagányi ÉE, Hayes KR, Sweatman HPA, Pratchett MS (2016) Assessing different causes of crown-of-thorns starfish outbreaks and appropriate responses for management on the Great Barrier Reef. PLoS One 11(12):e0169048. https://doi.org/10.1371/journal.pone.0169048

Bakus GJ (1964) The effects of fish grazing on invertebrate evolution in shallow tropical waters. vol 27. Allan Hancock Foundation Publications, Occasional Paper, pp 1–29

Barham EG, Gowdy RW, Wolfson FH (1973) *Acanthaster* (Echinodermata, Asteroidea) in the Gulf of California. Fish Bull 71(4):927–942

Birkeland C (1982) Terrestrial runoff as a cause of outbreaks of *Acanthaster planci* (Echinodermata: Asteroidea). Mar Biol 69:175–185

Birkeland C, Lucas JS (1990) *Acanthaster planci*: major management problem of coral reefs. CRC Press, Boca Raton, Florida, p 257

Bruno JF, Bertness MD (2001) Habitat modification and facilitation in benthic marine communities. In: Bertness MD, Gaines SD, Hay ME (eds) Marine community ecology. Sinauer Associates, Sunderland, Massachusetts, pp 201–218

Carpenter RC (1997) Invertebrate predators and grazers. In: Birkeland C (ed) Life and death of coral reefs. Chapman & Hall, New York, pp 198–229

Caso ME (1962) Estudios sobre Asteridos de México. Observaciones sobre especies Pacíficas del género *Acanthaster* y descripción de una subespecie nueva, *Acanthaster ellisii pseudoplanci*. An Inst Biol 32:313–331

Castro P (1978) Movements between coral colonies in *Trapezia ferruginea* (Crustacea Brachyura), an obligate symbiont of scleractinian corals. Mar Biol 46:237–245. https://doi.org/10.1007/BF00390685

Challinor D (1972) Background for a new, sea-level, Panama Canal. In: Jones ML (ed) The Panamic biota: some observations prior to a sea-level canal. Bull Biol Soc Wash 2:7–12

Chesher RH (1969) Destruction of Pacific corals by the sea star *Acanthaster planci*. Science 165(3890):280–283

Colgan MW (1987) Coral reef recovery on Guam (Micronesia) after catastrophic predation by *Acanthaster planci*. Ecology 68(6):1592–1605

Connell JH (1961) The influence of interspecific competition and other factors on the distribution of the barnacle *Chthamalus stellatus*. Ecology 42:710–723

Cousteau J, Dumas F (1953) The silent world: a story of undersea discovery and adventure. Harper & Brother, New York, p 266

Dana T, Wolfson A (1970) Eastern Pacific crown-of-thorns starfish populations in the lower Gulf of California. Trans San Diego Soc Nat Hist 16:83–90

Deaker DJ, Mos B, Lin HA, Lawson C, Budden C, Dworjanyn SA, Byrne M (2020) Diet flexibility and growth of the early herbivorous juvenile crown-of-thorns sea star, implications for its boom-bust population dynamics. PLoS One 15(7):e0236142

De'ath G, Fabricius Kcz E, Sweatman H, Puotinen M (2012) The 27-year decline of coral cover on the Great Barrier Reef and its causes. Proc Natl Acad Sci 109(44):17995. https://doi.org/10.1073/pnas.1208909109

Depczynsi M, Bellwood DR (2003) The role of cryptobenthic reef fishes in coral reef trophodynamics. Mar Ecol Prog Ser 256:183–191. https://doi.org/10.3354/meps256183

Desbiens AA, Mumby PJ, Dworjanyn S, Plagányi ÉE, Uthicke S, Wolfe K (2023) Novel rubble-dwelling predators of herbivorous juvenile crown-of-thorns starfish (*Acanthaster* sp.). Coral Reefs 42:579–591. https://doi.org/10.1007/s00338-023-02364-w

Duffy JE, Hay ME (2001) The ecology and evolution of marine consumer-prey interactions. In: Bertness MD, Gaines SD, Hay ME (eds) Marine community ecology. Sinauer Associates, Sunderland, Massachusetts, pp 131–157

Eakin CM (1987) Damselfishes and their algal lawns: a case of plural mutualism. Symbiosis 4:275–288

Endean R (1973) Population explosions of *Acanthaster planci* and associated destruction of hermatypic corals in the Indo-West Pacific region. In: Biology and geology of coral reefs, biology 1, 2: 389–438. Academic Press, New York

Endean R (1974) *Acanthaster planci* on the Great Barrier Reef. In: Proceedings of the second international coral reef symposium, Australia, vol 1, pp 563–576

Enfield DB (2001) Evolution and historical perspective of the 1997-1998 El Niño-Southern Oscillation event. Bull Mar Sci 69(1):7–25

Enochs IC (2012) Motile cryptofauna associated with live and dead coral substrates: implications for coral mortality and framework erosion. Mar Biol 159(4):709–722. https://doi.org/10.1007/s00227-011-1848-7

Enochs IC, Glynn PW (2017) Corallivory in the eastern Pacific. In: Glynn PW, Manzello DP, Enochs IC (eds) Coral reefs of the eastern tropical Pacific: persistence and loss in a dynamic environment. Coral reefs of the world 8. Springer, Dordrecht, pp 315–337

Fong P, Glynn PW (1998) A dynamic size-structured population model: does disturbance control size structure of a population of the massive coral *Gardineroseris planulata* in the eastern Pacific? Mar Biol 130:663–674

Fong P, Glynn PW (2000) A regional model to predict coral population dynamics in response to El Niño-Southern Oscillation. Ecol Appl 10(3):842–854

Fong P, Glynn PW (2001) Population abundance and size-structure of an eastern tropical Pacific coral after the 1997-98 ENSO: a simulation model predicts field measures. Bull Mar Sci 69(1):187–202

Frankel E (1977) Previous *Acanthaster* aggregations in the Great Barrier Reef. In: Proceedings of the 3rd international coral reef symposium, Miami, vol 1, pp 201–208

Frankel E (1978) Evidence from the Great Barrier Reef of ancient *Acanthaster* aggregations. Atoll Res Bull 220:75–93

Gilchrist SL (1985) Hermit crab corallivore activity. In: Proceedings of the 5th international coral reef congress, Tahiti, vol 5, pp 211–214

Glynn PW (1965) Community composition, structure, and interrelationships in the marine intertidal *Endocladia muricata—Balanus glandula* association in Monterey Bay, California. Beaufortia 12(148):1–198

Glynn PW (1972) Observations on the ecology of the Caribbean and Pacific coasts of Panamá. In: Jones ML (ed) The Panamic biota: some observations prior to a sea-level canal. Bull Biol Soc Wash 2:13–30

Glynn PW (1973a) Ecology of a Caribbean coral reef. The *Porites* reef-flat biotope: Part I. Meteorology and hydrography. Mar Biol 20(4):297–318

Glynn PW (1973b) Ecology of a Caribbean coral reef. The *Porites* reef flat biotope. Part II. Plankton community with evidence for depletion. Mar Biol 22:1–21

Glynn PW (1973c) *Acanthaster*: effect on coral reef growth in Panama. Science 180:504–506

Glynn PW (1973d) Aspects of the ecology of coral reefs in the western Atlantic region. In: Jones OA, Endean R (eds) Biology and geology of coral reefs, vol 2, Biol. 1. Academic Press, New York, pp 271–324

Glynn PW (1974) The impact of *Acanthaster* on corals and coral reefs in the eastern Pacific. Environ Conserv 1(4):295–304

Glynn PW (1976) Some physical and biological determinants of coral community structure in the eastern Pacific. Ecol Monogr 46(4):431–456

Glynn PW (1977) Interactions between *Acanthaster* and *Hymenocera* in the field and laboratory. In: Proceedings of the 3rd international coral reef symposium, Miami, vol 1, pp 209–215

Glynn PW (1980) Defense by symbiotic Crustacea of host corals elicited by chemical cues from predator. Oecologia 47:287–290. https://doi.org/10.1007/BF00398518

Glynn PW (1981) *Acanthaster* population regulation by a shrimp and a worm. In: Proceedings of the 4th international coral reef symposium, Manila, vol 2, pp 607–612

Glynn PW (1982) Individual recognition and phenotypic variability in *Acanthaster planci* (Echinodermata: Asteroidea). Coral Reefs 1:89–94

Glynn PW (1984) An amphinomid worm predator of the crown-of-thorns sea star and general predation on asteroids in eastern and western Pacific coral reefs. Bull Mar Sci 35:54–71

Glynn PW (1985a) Corallivore population sizes and feeding effects following El Niño (1982-1983) associated coral mortality in Panama. In: Proceedings of the 5th international coral reef congress, Tahiti, vol 4, pp 183–188

Glynn PW (1985b) El Niño-associated disturbance to coral reefs and post disturbance mortality by *Acanthaster planci*. Mar Ecol Prog Ser 26:295–300

Glynn PW (1987) Some ecological consequences of coral-crustacean guard mutualisms in the Indian and Pacific oceans. Symbiosis 4:301–324

Glynn PW (1988a) Predation on coral reefs: some key processes, concepts and research directions. In: Proceedings of the 6th International coral reef symposium, Townsville, vol 1, pp 51–62

Glynn PW (1988b) El Niño warming, coral mortality and reef framework destruction by echinoid bioerosion in the eastern Pacific. Galaxea 7(2):129–160

Glynn PW (1990) Coral mortality and disturbances to coral reefs in the tropical eastern Pacific. In: Glynn PW (ed) Global ecological consequences of the 1982–83 El Niño-Southern Oscillation. Elsevier oceanography series 52. Elsevier, Amsterdam, pp 55–126

Glynn PW (1993) Monsoonal upwelling and episodic *Acanthaster* predation as probable controls of coral reef distribution and community structure in Oman, Indian Ocean. Atoll Res Bull 379:1–66. https://doi.org/10.5479/si.00775630.379.1

Glynn PW (2004) High complexity food webs in low-diversity eastern Pacific reef-coral communities. Ecosystems 7:358–367. https://doi.org/10.1007/s10021-004-0184-x

Glynn PW (2006) Fish utilization of simulated coral reef frameworks versus eroded rubble substrates off Panamá, eastern Pacific. In: Proceedings of the 10th international coral reef symposium, Okinawa, vol 1, pp 250–256

Glynn PW (2008) Food-web structure and dynamics of eastern tropical Pacific coral reefs: Panamá and Galápagos Islands. In: McClanahan TR, Branch GM (eds) Food webs and the dynamics of marine reefs. Oxford University Press, New York, pp 185–208

Glynn PW (2013) Fine-scale interspecific interactions on coral reefs: functional roles. In: Lang MA, Marinelli RL, Roberts SJ, Taylor PR (eds) Research and discoveries: the revolution of science through scuba. Smithsonian contributions to the marine sciences, vol 39. Smithsonian Institution Scholarly Press, Washington, DC, pp 229–248

Glynn PW, Colgan MW (1988) Defense of corals and enhancement of coral diversity by territorial damselfishes. In: Proceedings of the 6th international coral reef symposium, Australia, vol 2, pp 157–163

Glynn PW, Wellington GM (1983) Corals and coral reefs of the Galápagos Islands. With an annotated list of the scleractinian corals of the Galápagos by John W. Wells. University of California Press, Berkeley, p 330

Glynn PW, Stewart RH, McCosker JE (1972) Pacific coral reefs of Panama: structure, distribution, and predators. Geol Rundsch Int J Earth Sci 61(2):483–519

Glynn PW, Wellington GM, Birkeland C (1979) Coral reef growth in the Galápagos: limitation by sea urchins. Science 203:47–49

Glynn PW, Maté JL, Baker AC, Calderón MO (2001) Coral bleaching and mortality in Panama and Ecuador during the 1997-1998 El Niño-Southern Oscillation event: spatial/temporal patterns and comparisons with the 1982-1983 event. Bull Mar Sci 69(1):79–109

Haszprunar G, Spies M (2014) An integrative approach to the taxonomy of the crown-of-thorns starfish species group (Asteroidea: *Acanthaster*): a review of names and comparison to recent molecular data. Zootaxa 3841(2):271–284. https://doi.org/10.11646/zootaxa.3841.2.6

Haszprunar G, Vogler C, Wörheide G (2017) Persistent gaps of knowledge for naming and distinguishing multiple species of crown-of-thorns-seastar in the *Acanthaster planci* species complex. Diversity 2017:9,22. https://doi.org/10.3390/d9020022

Hixon MA (2015) Reef fishes, seaweeds, and corals. In: Birkeland C (ed) Coral reefs in the Anthropocene. Springer, Dordrecht, pp 195–215

Jackson JBC, Palumbi SR (1979) Regeneration and partial predation in cryptic coral reef environments: preliminary experiments on sponges and ectoprocts. In: Levi C, Boury-Eanault N (eds) Biologie des spongaires. Colloq. Int. CNRS 291, Paris, pp 303–308

Jiménez C (1996–1997). Coral colony fragmentation by whitetip reef sharks at Coiba Island National Park, Panamá. Rev Biol Trop 44/45:698–700

Karlson RH (1999) Dynamics of coral communities. Kluwer Academic, Dordrecht, p 250

Kayal M, Lenihan HS, Pau C, Penin L, Adjeroud M (2011) Associational refuges among corals mediate impacts of a crown-of-thorns starfish *Acanthaster planci* outbreak. Coral Reefs 30:827–837

Keable SJ (1995) Structure of the marine invertebrate scavenging guild of a tropical reef ecosystem: field studies at Lizard Island, Queensland, Australia. J Nat Hist 29:27–46. https://doi.org/10.1080/00222939500770021

Kitagawa I, Kobayashi M (1978) Saponin and Sapogenol. XXVI. Steroidal saponins from the starfish *Acanthaster planci* L.(crown of thorns).(2). Structure of the major saponin thornasteroside a. Chem Pharm Bull 26(6):1864–1873

Lawrence JM, Glynn PW (1984) Absorption of nutrients from the coral *Pocillopora damicornis* (L.) by the echinoid *Eucidaris thouarsii* (Val.). Comp Biochem Physiol A Physiol 77(1):111–112. https://doi.org/10.1016/0300-9629(84)90020-3

Leray M, Knowlton N (2015) DNA barcoding and metabarcoding of standardized samples reveal patterns of marine benthic diversity. Proc Natl Acad Sci USA 112:2076–2081

Lessios HA, Kessing BD, Robertson DR, Paulay G (1999) Phylogeography of the pantropical sea urchin *Eucidaris* in relation to land barriers and ocean currents. Evolution 53:806–817. http://www.jstor.org/stable/3448876

McClanahan TR (2008) Food-web structure and dynamics of east African coral reefs. In: McClanahan TR, Branch GM (eds) Food webs and the dynamics of marine reefs. Oxford University Press, Oxford, pp 162–184

McKeon CS, Moore JM (2014) Species and size diversity in protective services offered by coral guard-crabs. PeerJ 2:e574. https://doi.org/10.7717/peerj.574

McKeon CS, Stier AC, Boyer S, Bolker BM (2012) Multiple defender effects: synergistic coral defense by mutualist crustaceans. Oecologia 169:1095–1103

Mora C, Zapata FA (2000) Effects of a predatory site-attached fish on abundance and body size of early post-settled reef fishes from Gorgona Isand, Colombia. In: Proceedings of the 9th international coral reef symposium, Bali, vol 1, pp 475–480

Moran PJ (1986) The *Acanthaster* phenomenon. Oceanogr Mar Biol Annu Rev 24379–24480

Moran PJ (1988) The *Acanthaster* phenomenon, vol 7. Australian Institute of Marine Science Monograph, p 178

Odum HT, Odum EP (1955) Trophic structure and productivity of a windward coral reef community on Eniwetak Atoll. Ecol Monogr 25(3):291–320

Oramas NFA (1979) Estudio ecológico del *Jenneria pustulata* (Mollusca, Gastropoda). Tesis, Universidad de Panamá, Escuela de Biología. República de Panamá. 100 pages, 21 tables, 29 figures

Paine RT (1966) Food web complexity and species diversity. Am Nat 100(910):65–75

Paz-García DA, Aldana-Moreno A, Cabral-Tena RA, Balart EF (2012) High predation by the corallivore sea snail *Jenneria pustulata* in a high latitude reef in the Gulf of California. Mar Biodivers Rec 5. https://doi.org/10.1017/S1755267212000826

Pearson RG, Endean R (1969) A preliminary study of the coral predator *Acanthaster planci* (L.) (Asteroidea) on the Great Barrier Reef, vol 3. Fishery Notes Department of Harbours and Marine, Queensland, pp 27–55

Porter JW (1972a) Ecology and species diversity of coral reefs on opposite sides of the Isthmus of Panama. In: Jones ML (ed) The Panamic biota: some observations prior to a sea-level canal. Bull Biol Soc Wash 2:89–116

Porter JW (1972b) Predation by *Acanthaster* and its effect on coral species diversity. Am Nat 106:487–492

Porter JW (1974) Community structure of coral reefs on opposite sides of the Isthmus of Panama. Science 186(4163):543–545

Pratchett MS (2001) Influence of coral symbionts on feeding preferences of crown-of-thorns starfish *Acanthaster planci* in the western Pacific. Mar Ecol Prog Ser 214:111–119

Pratchett MS, Vytopil E, Parks P (2000) Coral crabs influence the feeding patterns of crown-of-thorns starfish. Coral Reefs 19:36

Ransome E, Geller JB, Timmers M, Leray M, Mahardini A, Sembiring A, Collins AG, Meyer CP (2017) The importance of standardization for biodiversity comparisons: a case study using autonomous reef monitoring structures (ARMS) and metabarcoding to measure cryptic diversity on Mo'orea coral reefs, French Polynesia. PLoS One 12(4):e0175066. https://doi.org/10.1371/journal.pone.0175066

Reaka-Kudla ML (1997) The global biodiversity of coral reefs: a comparison with rainforests. In: Reaka-Kudla ML, Wilson DE, Wilson EO (eds) Biodiversity II: understanding and protecting our biological resources. Joseph Henry Press, Washington, DC, pp 83–108

Rodriguez Villalobos J-C, Ayala-Bocos. (2021) Depredación masiva por *Acanthaster planci* en el arrecife El Corralito, Golfo de California: amenaza a corto plazo. Rev Biol Trop 69(Suppl. I):272–286. https://doi.org/10.15517/rbt.69isuppl.1.46359

Rotjan RD, Lewis SM (2008) Impact of coral predators on tropical reefs. Mar Ecol Prog Ser 367:73–91

Shinn EA (1968) Burrowing in recent lime sediments of Florida and The Bahamas. J Paleontol 42:879–894

Uthicke S, Pratchett MS, Bronstein O, Alvarado JJ, Wörheide G (2024) The crown-of-thorns seastar species complex: knowledge on the biology and ecology of five corallivorous *Acanthaster* species. Mar Biol 171:32. https://doi.org/10.1007/s00227-023-04355-5

Vermeij GJ (1978) Biogeography and adaptation: patterns of marine life. Harvard University Press, Cambridge, p 332

Vogler C, Benzie J, Lessios H, Barber P, Wörheide B (2008) A threat to coral reefs multiplied? Four species of crown-of-thorns starfish. Biol Lett 4:696–699

Vogler C, Benzie JAH, Tenggardjaja K, Ambariyanto SC, Barber PH, Wörheide G (2013) Phylogeography of the crown-of-thorns starfish: genetic structure within the Pacific species. Coral Reefs 32:515–525. https://doi.org/10.1007/s00338-012-1003-z

Walbran PD, Henderson RA, Jull AJT, Head MJ (1989a) Evidence from sediments of long-term *Acanthaster planci* predation on corals of the Great Barrier Reef. Science 245:847–850

Walbran PD, Henderson RA, Faithful JW, Polach HA, Sparks RJ, Wallace G, Lowe DC (1989b) Crown-of-thorns starfish outbreaks on the Great Barrier Reef: a geological perspective based upon the sediment record. Coral Reefs 8:67–78

Wellington GM (1982) Depth zonation of corals in the Gulf of Panama: control and facilitation by resident reef fishes. Ecol Monogr 52:223–241

Yonge CM (1963) The biology of coral reefs. Adv Mar Bio 1:209–262

Chapter 8
Megafauna

Abstract Encountering large vertebrates, many of these rarely present in coral reef environments, was often a pleasant diversion, and observing their behaviors was instructive. Several of the relatively large species did not measure up to the strict definition of megafauna, exceeding 46 kg or 100 lbs. in weight. These were bony fishes, sharks and rays, sea snakes, marine iguanas, and birds. Those species fitting the definition of megafauna were some sharks and rays, sea turtles, crocodiles, sea lions, dolphins, and whales. An intriguing, and somewhat frightening encounter with a manta ray was possibly a prelude to copulation. A sea lion bull attack was a stark warning not to locate study sites near sea lion harems. One of the most beautiful birds in the world was encountered in Panama, and we managed to attract it to the Uva Reef with a dead fish. Two fortuitous encounters with whale pods of considerable size, in the Galápagos Islands (sperm whales) and off Baja California, Mexico (blue whales), brought a feeling of heightened joy to observe these species whose populations were decimated by commercial whaling during the last two centuries.

Introduction

Marine megafauna are commonly defined as large animals, including bony fishes that exceed 46 kg (100 lbs), and equally large sharks and rays (elasmobranchs), reptiles, and mammals. Here, I deviate from the usual definitions of megafauna to include birds and other relatively small vertebrates encountered in our eastern Pacific field studies. Some of these species seldom, if ever, entered coral reef environments. Our close-up encounters with them, however, revealed interesting behaviors and occasionally provoked responses worth recounting. While many of our chance meetings with relatively large vertebrates were near or in coral reef habitats, some occurred in the open sea, during travel to or from reef study sites. Most species in all these disparate groups are carnivores that prey primarily on fishes, occasionally on mammals, with fewer examples of herbivores and planktivores.

© The Author(s), under exclusive license to Springer Nature Switzerland AG 2025

P. W. Glynn, *Coral Reef Resilience in the Anthropocene*,
https://doi.org/10.1007/978-3-031-81421-1_8

Bony Fishes

Bony fishes and sharks and rays are members of the Phylum Chordata, Subphylum Vertebrata. The internal skeleton of bony fishes (Paraphylum Osteichthyes) is bone and that of sharks and rays (Paraphylum Chondrichthyes) is cartilage.[1] There are many other morphological and physiological differences separating the two groups that diverged in the Silurian Period, ~ 422 million years (Myr) ago (Zhu et al. 2009). Compared with reef-building corals the fish fauna of eastern Pacific coral reefs is diverse. A recent inventory of eastern Pacific corals lists 43 species (Glynn et al. 2017). Visual surveys of bony fishes in the Gulf of Chiriquí, Panama have reported 126 species present on coral and rocky reefs (Dominici-Arosemena and Wolff 2006). Currently, of the entire known shallow water marine fish fauna of the region, 111 species are sharks and rays, and 1193 species are bony fishes (Zapata and Robertson 2007; Robertson and Allen 2008; Cortés et al. 2017). Of these, 25 elasmobranchs and 482 bony fishes (39% of the regional fauna) have been observed on coral or rocky reefs. This amounts approximately to a 10 to 1 ratio of fish to coral species that might co-occur on an eastern Pacific coral reef. Fernando Zapata has noted that, except for six species, all the coral reef fishes in the region occur also on rocky reefs, while the reverse is not true (i.e., not all rocky reef species occur on coral reefs).[2,3]

In the early 1970s, large snappers were commonly observed on reefs in the Gulf of Chiriquí, Panama. The largest of these, the Dog Snapper (*Lutjanus novemfasciatus*), reached 1 m in length and commonly sheltered during the day in voids in large coral colonies or in rock cavities and crevices. During night dives we often saw them in open reef areas. When startled, the air bladder drumming was audible over considerable distances. In recent years (2000–2015), large dog snappers have been

---

[1]A paraphylum is a higher taxon whose member species are all descendants of a common ancestor, but that does not contain all the species descended from that ancestor. Both the Osteichthyes and Chondrichthyes taxa are paraphyla but not sharing cartilaginous or bony fish species respectively. See the following link for more information on these taxonomic categories: Monophyletic, Polyphyletic, & Paraphyletic taxa (mun.ca).

[2]As a cautionary note, Fernando Zapata observed that all six and possibly seven (a new record) of the following species are endemic to Clipperton Atoll, which is a coral environment devoid of rocky substrates: Axelrod's reef-bass, *Pseudogramma axelrodi* (Serranidae); Clipperton angelfish, *Holacanthus limbaughi* (Pomacanthidae); Baldwin's major, Clipperton Gregory, *Stegastes baldwini* (Pomacentridae); Clipperton rainbow wrasse, *Thalassoma robertsoni* (Labridae); Clipperton razorfish, Wellington's razorfish, *Xyrichtys wellingtoni* (Labridae); Clipperton fanged-blenny, *Ophioblennius clippertonensis* (Blenniidae); Clipperton cardinal soldierfish, *Myripristis gildi* (Holocentridae), a relatively new record. Since none of these species is a strict corallivore it is likely that most or possibly all could successfully inhabit rocky substrate habitats.

[3]Information on the identity and habits of eastern Pacific reef fishes can be found in Thomson et al. (1979), Merlen (1988), Humann and DeLoach (1989), Allen and Robertson (1994), and Grove and Lavenberg (1997), and Robertson and Allen (2008, 2015). In the late 1960s and early 1970s, before these sources were available, John McCosker (Steinhart Aquarium, San Francisco) and Richard Rosenblatt (Scripps Institution of Oceanography, La Jolla) kindly provided identifications of several reef fish species.

present at Uva Island, but they are extremely wary of divers. They have not been seen in our fish transect sampling over a period of 30+ years (Glynn et al. 2014). A large snapper has been observed on multiple occasions near the Uva reef at night, however, only to disappear quickly into a narrow rock grotto when approached. We are not sure if this is the same fish or different individuals. It is always heartening to know that such large, mature fish are still present. Large fish in a species population are highly fecund, producing disproportionately bigger and more eggs than smaller fish. This is critically important concerning sustainability vis-à-vis overfishing.

While transect sampling and nearly motionless on the bottom at our Secas reef study site, I occasionally saw wahoo (*Acanthocybium solandri*). These large apex predators were observed swimming along the seaward reef slope, about 2 m below the surface, as if searching for reef fish prey. I tried approaching them, but without success; they appeared to be wary of my presence. On one occasion at the Uva Island anchorage, a crew member directed our attention to a very large fish resting about 50 m from the boat. It remained there for several hours, virtually motionless. We believed it was a wahoo and estimated its length at about 2 m and its girth close to an average sized telephone poll, a truly large fish. This inactive behavior puzzled us, and we decided not to disturb it. It finally swam away in the late afternoon.

On a few rare occasions, we have observed tuna schools approaching coral reefs. On a deep dive at Clipperton Atoll, Charlie Veron and I once observed a school of Yellowfin Tuna (*Thunnus albacares*) swimming rapidly along the reef base. Audible thumping occurred as several 1.5 m long individuals rocketed by us. At Uva Island, we sometimes witnessed schools of Black Skipjack Tuna (*Euthynnus lineatus*) briefly enter the reef front. We wondered if such tuna schools could avail themselves of the abundant and diverse source of reef-fish prey? One sunny afternoon in the dry season a large school of Black Skipjack Tuna entered the Uva reef, spending about 5 min jumping and thrashing in the waters of the seaward reef zone. We observed this event at a distance from our inflatable dive boat due to the timid nature of the fish. This tuna species quickly flees if approached by a swimmer/diver. Entering the area immediately after the tuna departed, it was apparent that a feeding frenzy had just occurred on the reef. The water column sparkled with suspended and sinking fish scales. Moreover, the normally abundant schools of wrasses, damselfish, and angelfish were out of sight. Upon close inspection, numerous fishes were found seeking shelter in corals and reef framework cavities. It appeared that the quick-moving tuna school had consumed many reef fishes. How often these feeding forays occur and the amount of biomass consumed are unknown.

Of the larger fishes, several species of eels revealed a close tie to coral reefs. Owing to the abundance of shelter sites and potential fish prey, moray eels were present on virtually all reefs. The largest, and sometimes most threatening of these, were the Panamic Green Moray (*Gymnothorax castaneus*) and the Fine-Spotted Moray (*Gymnothorax dovii*). The largest individuals observed exceeded 1 m in length. One of our first encounters with Panamanian eels occurred in the Pearl Islands, Gulf of Panama. Aníbal and I were diving on a *Pocillopora* reef at Señora Islet and became aware of a large Green Moray that began following us as we moved along the reef. It began approaching us closely with mouth agape. We carried a mesh dive bag and used this to fend off the eel when it approached too closely. During one close encounter, it pressed its open mouth against my body but did not bite. It continued to follow us and returned to the reef only after we boarded our boat. Its presence was annoying and threatening. We imagined that it associated our presence with food. Perhaps the eel had encountered spear-fishing divers previously and fed on wounded or dead speared fish?

One of our team members spotted a very large specimen of the Fine-Spotted Moray on the Secas Island study reef in the Gulf of Chiriquí. The eel was swimming over the forereef and was estimated to approach 1.5 m in length, the maximum size reported for this species. I did not see this eel, but other team members did on several occasions. Its presence always evoked awe and a degree of apprehension. The greatest number of eels observed anywhere in the eastern Pacific was at Clipperton Atoll. One of the goals of our first survey at Clipperton was to create an inventory of the corals present on this remote atoll. It was necessary to collect specimens for morphological study to verify species identities. Occasionally I would find a coral colony suitable for study only to discover that it was too close to an eel's lair to be safely collected. Some eels, species unidentified, had the curious habit of swimming across the reef flat at Clipperton, even slithering over exposed algal surfaces, presumably in search of prey. Two members of the M/V *Royal Star* research party were bitten by eels, causing serious cuts to their hands that required medical attention. The ship's doctor advised us that if bitten, do not jerk arm or leg away suddenly, but remain motionless and allow the eel time to release its hold. This response would result in puncture wounds instead of tears which are much easier to treat and quicker to heal — excellent advice but requiring nerves of steel.

While discussing bites and stings, it is necessary to recall the pain caused by nematocysts, the venomous cellular products present in cnidarians. An incident involving a potent fishing tentacle of *Physalia* (Portuguese Man o' War) at Gorgona Island, Colombia, was noted in Chap. 2. Cnidarians are invertebrates with a relatively simple, two-cell layer body plan, evolutionarily distant from vertebrate taxa of the megafauna. Cnidarians do associate with and benefit the ocean sunfish (*Mola mola*), one of the heaviest bony fish. These fish giants feed on zooplankton, including diverse cnidarian jellyfish. As young and naïve surface divers in southern California, my diving companions and I would often encounter ocean sunfish in kelp beds during the summer season. These fish were 1–1.2 m in length and occurred near the surface. To assess our swimming and diving skills, we thought it would be challenging to grab hold of a resting sunfish and try to prevent it from diving. Much to our chagrin *Mola* was usually the victor, causing us to release our grip and swim for the surface to catch a breath. These were the days before wet suits; all our body parts in contact with *Mola* became inflamed and highly painful. Cnidarian nematocysts embedded in the mucus coating of *Mola* were responsible for this unpleasant experience. We surmised that this fortuitous association could protect the gentle giant *Mola* from certain predators. Some precautions regarding nematocysts include wearing wet suits, a hoodie or head cap and gloves in areas with heavy zooplankton blooms. In addition, a culinary tenderizer with the papain enzyme or fresh papaya fruit would be good antidotes to have at hand.

Cartilaginous (Class Elasmobranchii) Sharks and Rays

A flash of white from dorsal and caudal fins often heralded the presence of the inquisitive Whitetip Reef Shark (*Triaenodon obesus*). This shark is ubiquitous, commonly seen resting under massive corals or rock boulders during the day or patrolling reefs day and night in search of prey (Fig. 4.1). The Nurse Shark (*Ginglymostoma cirratum*) is another elasmobranch commonly seen on reefs where it feeds on benthic invertebrates and fishes. In terms of prey consumption and bioerosion (see Chap. 9), these two species are probably more influential on eastern Pacific reefs than are other, generally larger and more menacing, shark species.

Of the requiem sharks in the family Carcharhyinidae,[4] consisting of about 50 species, three were seen on eastern Pacific coral reefs, the Bull, Tiger, and Galápagos sharks. The Bull Shark (*Carcharhinus leucas*) frequents coastal areas and can tolerate brackish and freshwater habitats, often ascending rivers. It is present in Lake Nicaragua, and I once spotted an individual following a ship into Lake Gatun in Panama. This species occurs worldwide and is responsible for fatal attacks on humans. Aware of shark fatalities reported in Nicaragua, we remained alert in the presence of this species.

In the mid-1970s, commercial shark fishing became popular around Uva Island. When we once visited a beach landing site on the island, we found that shark fishers had used this area for butchering Bull and Tiger sharks. We frequented this beach daily when anchored at Uva Island, usually in the late afternoon, to bathe and wash off our diving gear in a nearby waterfall. While making an island landing one of our volunteers hopped into the water, waist-high, to help haul our inflatable onto the beach. As soon as he entered the water, a bull shark lunged for his legs, but thankfully it just bumped a leg instead of biting it. A quick scramble back into the boat left our volunteer stunned but grateful to have avoided a severe injury. Likely, the blood and entrails from the butchered sharks on the beach attracted sharks to this area.

Paradoxically, Tiger Sharks (*Galeocerdo cuvier*), which we had never seen while diving at Uva Island, were one of the most frequently caught species by the shark fishers in this area. We suspect that they were sometimes near us while we were diving but just beyond our visual range. It is possible that olfaction and other senses allowed tiger sharks to follow our activities unseen.

The Galápagos Shark (*Carcharhinus galapagensis*) is commonly encountered around oceanic islands. It was abundant in the Galápagos Islands and at Clipperton Atoll. We were encouraged to cut short a dive at Darwin Island, as noted in Chap. 2, due to the aggressive behavior of a school of this species. Seals and sea lions sometimes fall prey to Galápagos sharks. Galápagos sea lions are occasionally seen with

[4]The origin of the term 'requiem' is uncertain. One suggestion is that it derives from the French word for shark, *requin*. Several of the larger requiem sharks are known to have caused human fatalities, but most of the species in this group are harmless.

maimed or missing flippers or massive body scars, presumably due to shark encounters.

The Scalloped Hammerhead Shark (*Sphyrna lewini*) was restricted in distribution but when present the most abundant elasmobranch observed by our team. We most often encountered hammerheads at three sites in the Galápagos Islands, at Devil's Crown (Floreana Island), Wolf (Wenman) Island, and at Darwin Island (Fig. 8.1). Large schools of hammerheads, 2–3 m in length, and numbering up to 30–40 individuals, were nearly always present in Shark Bay on the east side of Wolf Island (Fig. 1.5). We usually saw them swimming along the deep edge of the reef. At Devil's Crown (Floreana Island), on one of our first reconnaissance dives in the islands in 1975, we observed a school of hammerheads swimming directly into our midst. Some large individuals swam directly over our heads, at arm's length—unnerving. The impressive biomass of hammerheads must attest to high abundances of bait fish prey to support these populations. Despite their menacing appearance, Godfrey Merlen, a long-time Galápagos naturalist, noted that there is no record of a hammerhead attack on humans in the Galápagos Islands (Merlen 1988). The website sharkattackdata.com lists six unprovoked and non-fatal shark attacks in the Galápagos Islands over the period 1900–2020. Only a single record of the Galápagos Shark was identified as responsible for one of these attacks.

Fig. 8.1 Scalloped hammerhead sharks (*Sphyrna lewini*) schooling off Wellington Reef, Darwin Island, northern Galápagos Islands. These critically endangered sharks gather in schools of hundreds to thousands at some oceanic islands in the eastern tropical Pacific (11 October 2014). Credit: Viktor Brandtneris

Whale sharks. Massive size, white spotting, and stripes were traits that gave away the presence of the world's largest fish, the Whale Shark (*Rhincodon typus*). It can reach 18 m in length and like baleen whales is a planktivore, feeding on zooplankton and small schooling fishes. We were fortunate to have encountered this shark on a few occasions since it typically occurs in open waters far from coral reefs.

In the early 1970s, Chuck Birkeland observed one individual of *R. typus* in the Gulf of Panama. He followed and photographed it for several minutes. As noted in Chap. 6, I observed a single whale shark at the Uva Island anchorage site, close to the coral reef. My second whale shark encounter occurred during a dive with Joshua Feingold at Banks Bay, Isabela Island. We observed the large shark from below, its white underside in full view. Initially our thoughts were unsettling, until the white spotting came into view. Whale sharks frequented Darwin's Arch at Darwin Island and several divers on our team were fortunate to have seen them on different occasions. Employing satellite tags, Hearn et al. (2013) presented evidence of whale shark long-distance return migrations (1650 km from the west) to the Galápagos Islands and recognized Darwin Island as a major aggregation site.

Satellite tracking studies have revealed the enormous distances traversed in whale shark migrations. Migration routes from the eastern to the western Pacific were found to extend from the Gulf of California, Mexico to the Marshall Islands (Eckert and Stewart 2001), and from the Gulf of Chiriquí, Panama to the Mariana Islands (Guzman et al. 2018). Sharks in both the Mexican and Panamanian routes joined the fast-moving North Equatorial Current (NEC) near Clipperton Atoll. Anne, the affectionately named 7 m long female whale shark tagged in Panama, travelled to Cocos Island, then to the Galápagos Islands before entering the west bound NEC. She travelled a record long, somewhat circuitous path, over approximately 20,100 km in 841 days. In other migratory species, both marine and terrestrial, such long movements are generally related to food availability and environmental conditions suitable for breeding.

Rays. Sting rays were seldom present around our eastern Pacific reef sites, however, in the 1970s manta rays were commonly observed feeding close by the Uva Island reef. On some days schools of up to 10 rays could be seen swimming along the reef front, consuming zooplankton. Uva Bay is a *cul de sac* with winds and currents often moving water masses into the bay, which apparently results in the concentration of plankton along the Uva reef front. It is possible that plankton-feeding whale sharks and manta rays are attracted to similar biomass-rich reef front zones to exploit this food source.

Aníbal Velarde and I once sampled study plots at Uva Island and witnessed a bizarre manta ray behavior. It was a clear and sunny day, and I could see Aníbal sampling about 5 m distant and at 7 m depth. I was up slope at about 4 m depth with my face close to the substrate where I was counting and measuring juvenile corals. The immediate surroundings would darken from time to time, which I attributed to passing clouds. When one dark spell persisted, I glanced down slope to Aníbal to see him pointing excitedly toward me, motioning for me to look up and behind. Good heavens!, the shadow turned out to be an approximately 4 m wide disc of a male manta ray (*Manta birostris*). The ray was attempting to envelope me with its wings and thrusting its underside, with claspers, at my body. (During mating, manta ray claspers are inserted into a female's cloaca to transfer sperm.) I was wearing a black wet suit, covered from head to toe, and perhaps being perceived as a female manta ray? Since this developing situation did not appeal to my better judgment, I gently tapped the ray on a cephalic horn (a fleshy lobe) with my dive knife, with no apparent effect. The ray continued advancing with body contact until I forcefully slapped my dive knife's back dull side across its mouth. It then backed away from me and quickly swam away. This was an unsavory but intriguing encounter indeed.

Sea Turtles

Sea turtles are occasionally found resting in an apparent "deep sleep" in coral reef depressions and recesses. The two turtle species I have identified, the Olive Ridley (*Lepidochelys olivacea*) and the Loggerhead (*Caretta caretta*), were seen digging into *Pocillopora* reef frameworks in Panama (Glynn 2004). Both species were searching for and consuming cryptic sponges attached to the dead basal branches of coral. This foraging behavior destroyed 0.5–1.0 m diameter coral patches per feeding bout. Several excavated patches could sometimes be seen on reef fronts in the Gulf of Chiriquí, with prominent bite marks and sponge removal as telltale evidence of turtle foraging. It would be interesting to determine the long-term impact of this destructive feeding behavior on reef structure and the possibility of relating this to turtle foraging in the past when turtle abundances were higher. Turtle bite marks should be distinguishable from either fish bite marks or unbitten branches resulting from storm damage.

Sea Snakes

A single species of sea snake lives in the eastern tropical Pacific, the Yellow-bellied Sea Snake, *Hydrophis platurus*. It occurs in all tropical seas except for the Atlantic Ocean. Unlike Indo Pacific sea snakes that are reef dwellers, the eastern Pacific species occurs at the sea surface, preying on small planktonic fishes. We have found these snakes commonly associated with drift lines, i.e., sea surface convergence zones where plankton and debris accumulate.

A rare encounter with *H. platurus* in the Pearl Islands, Panama, did not raise immediate concern but later, on reflection of the incident, I was relieved of having avoided a head strike and possible injection of a highly toxic venom. Aníbal and I surfaced from a scuba dive after completing a coral survey and when I turned to face Aníbal he pointed to the back side of my head and remarked excitedly, "Dr. Glynn, there is a sea snake wrapped around your regulator hose." I immediately froze, not wanting to make any sudden movements that might provoke the snake. That was effective, in 2 or 3 min the snake disengaged from the hose and slowly swam away. Aníbal watched all this playing out on the backside of my head and neck while I stared straight ahead remaining motionless. I learned later that envenomation would have been difficult because *H. platurus* has small fangs and needs to chew to release its venom.

Crocodiles

In 2017, we had a somewhat worrisome visit to our study reef on Uva Island. Local fishers warned us that a large female of the American Crocodile (*Crocodylus acutus*) had taken up residence on the Uva Island beach. We were told that some months before our visit divers were obliged to leave the water because of this threatening animal. Fortunately, we were able to complete our work on the reef since there were no signs of the crocodile's presence (e.g., crawl trails on the beach). This crocodilian species is an apex predator in south Florida, the Caribbean, and Central and South America as far south as Peru. Some naturalists regard it as docile and not a threat to larger animals, including humans. Our experiences in Central America are otherwise. There are numerous reports of attacks on humans, some have been documented, and many have not. Some of the attacks have resulted in fatalities. Since arriving in Panama in the late 1960s, local fishers and divers repeatedly warned me to be cautious of this crocodile species. On days of high-water clarity in the Gulf of Chiriquí we could sometimes spot them cruising along the shores of Coiba and Coibita islands. Two recent attacks led to the death of *Caponera*, a dog we befriended on Coibita Island, and a fisher who was illegally poaching conchs (*Strombus*) at Coiba Island. The fisher was returning to his boat, anchored at shallow depth in turbid water near a river mouth, when the crocodile ambushed him. I would advise exercising caution while swimming in estuaries of low water visibility, especially after dark, in tropical America.

Mature American male crocodiles can reach 6 m in length and a weight of 900 kg, approximately equivalent to the size of the Nile (African) and Saltwater (western Pacific) crocodiles. The largest crocodiles observed by us in the eastern Pacific were no longer than about 4 m. *Tito*, a frequent visitor of the Coiba Island's ranger station, is about 3 m in length (Fig. 8.2). American crocodiles inhabit a wide range of coastal environments, including lagoons, mangrove forests, rivers, freshwater lakes and near shore islands. Their diet consists of small mammals, birds, frogs, turtles, and fishes (Fishman and McKinnon 2009). It is possible crocodiles could cause physical damage to coral frameworks in their pursuit of prey, but they are so seldom present in reef areas this threat would be rarely realized.

Fig. 8.2 American crocodile (*Crocodylus acutus*), approximately 3 meters in length, Coiba Island, Gulf of Chiriquí, 19 March 2006. Known as "Tito" by the Coiba Island park rangers, he often appears at the park's headquarters encampment for a fish treat. "Ian, that is close enough!" Photograph courtesy of Ian C. Enochs

Marine Iguanas

Like Charles Darwin, we also marveled at the large numbers of the endemic marine iguana (*Amblyrhynchus cristatus*) present in the Galápagos Islands, especially at Punta Espinosa on Fernandina Island (Figs. 8.3 and 8.4). In walking among the resting iguanas, warming up after feeding in the cool waters, *phst phst* sounds were audible and they emitted jets of salt-charged water from their nostrils. The marine iguana possesses highly efficient salt glands that rid the body of salts ingested while feeding. This allows the iguana to conserve water and prevent dehydration. The iguanas are well adapted for diving and feeding on red and green turf algae. Their diving bouts are usually to depths of five meters or less and for submerged intervals of a few minutes.

Estimates of iguana abundances across the archipelago range from 200,000 to 300,000 individuals. Iguana abundances are highest on the western islands that experience upwelling and elevated algal productivity. During El Niño events, their numbers can be greatly reduced when marine productivity declines. About 60% of the iguanas on Santa Fe Island starved to death during the 1982–83 El Niño disturbance (Laurie 1990). According to Andrew Laurie's study, the magnitude of this impact occurred island wide. These fascinating animals were observed only occasionally in warm water habitats, thus interacting minimally with coral and associated species.

Fig. 8.3 A large aggregation of sunbathing, endemic, Galápagos marine iguanas, *Amblyrhynchus cristatus,* Punta Espinosa, Fernandina Island, Galápagos Islands, 16 January 1975. Photograph courtesy of Charles Birkeland

Fig. 8.4 A Galápagos marine iguana clinging to an algal covered wave-washed rock, Punta Espinosa, Fernandina Island, Galápagos Islands, 17 January 1975. Photograph courtesy of Charles Birkeland

Birds

We encountered numerous iconic bird species during our coral reef studies, e.g., albatrosses, flightless cormorants, and flamingoes; only three are considered here, two piscivores (a booby and a penguin) and a scavenger (a vulture). All three were observed feeding in or near coral habitats, i.e., underwater over coral communities or on the shore. Reviews of El Niño 1982–83 effects on seabirds in and beyond the equatorial eastern Pacific can be found in Glynn (1988) and Duffy (1990).

Nazca Booby. Sea birds were never considered to be a threat to divers until we experienced a bizarre encounter with the Nazca Booby (*Sula granti*)[5] at Wolf Island in the northern Galápagos Islands. The east side of Wolf, at Tiburón Bay (Fig. 1.5), is dominated by steep cliffs that serve as perches for plunge-diving birds. The island shelf drops off rapidly into deep water near the cliffs allowing birds to spot potential prey and dive for fish with minimum effort. Bernhard Riegl and I had just completed deep coral surveys at this site and were biding our time at a 7–8 m safety stop before surfacing. Suddenly, a large school of bait fish appeared that prompted a feeding frenzy with flocks of boobies exploding into the water like missiles. Scores of boobies were forcefully penetrating the water column to a depth of 1 to 2 meters with pointed dagger-like bills, some birds successfully stabbing fish others missing. Our air supply was getting low, and we needed to surface. How could we protect ourselves from this barrage of jabbing sharp bills? We could not conceive of an effective means of protection when surfacing except to arc our arms, hands clasped, over our heads. Better a strike to the arms and hands than to the head. Perhaps our fears were unfounded? The boobies did not strike us—not even a near hit—because they could apparently distinguish between a diver's head and a natural fish food morsel. The Nazca Booby and other booby species are named for their clumsy movements on land. The booby's name is borrowed from the Spanish '*bobo*' (a simpleton or fool), which has no semblance to this bird's aquatic agility and lightning-swift and accurate diving maneuvers.

Boobies and other sea birds that feed primarily in productive upwelling areas of the eastern tropical Pacific experienced high mortalities during the first half of 1983 with the incursion of warm, nutrient-poor El Niño waters (Tovar and Cabrera 1985; Valle Castillo 1985; Duffy 1990). The effects of food deficiencies led to extralimital dispersal, nest desertion, and starvation. Mass migrations of Peruvian and Galápagos boobies arrived in the Gulf of Panama in early 1983. A novel sight, boobies perched on rocks and flying around the STRI marine laboratory. Approximately 7000 Peruvian Boobies and 9000 Blue-footed Boobies were reported to have immigrated into the Gulf of Panama (Smith 1990). Due to differences in their feeding behaviors Peruvian Boobies starved and died whereas most Blue-footed Boobies survived and returned to their normal nesting sites in early 1984.

---

[5] Formerly confused with the Masked Booby (*Sula dactylatra*), the Nazca Booby (*S. granti*) is regarded as a valid species (Pitman and Jehl Jr. 1998). It typically nests on cliffs and steep slopes and occurs in the Galápagos Islands and Malpelo Island, Colombia.

Galápagos Penguin. The endemic Galápagos Penguin (*Spheniscus mendiculus*) is one of the smallest and the world's only penguin species found at and just north of the Equator (Fig. 2.9). Australia's Little Penguin (*Eudyptula minor*) is a few cm shorter and occurs in south Australia and New Zealand. Like the marine iguana, the Galápagos Penguin makes for an exotic sight when seen swimming over corals. On one of our first surveys of the Devil's Crown *Pocillopora* reef, Chuck Birkeland was ecstatic with capturing a photograph of a penguin swimming over live corals. Where else in the world could this possibly occur, perhaps in south Australia?

Feeding on small fishes, we occasionally observed penguins hunting in pocilloporid coral communities at shallow depths. We also once observed several penguins rocket by us at 20 m depth in Tagus Cove, in pursuit of fishes in deeper waters. Galápagos penguins, like marine iguanas, are concentrated in the western islands of Isabela and Fernandina, areas of upwelling and elevated biological productivity. The penguin population size fluctuates dramatically, depending on the severity and frequency of El Niño events. Penguin numbers have been variously estimated at between 3000 to 8000 birds with 600 to 800 breeding pairs (BirdLife International 2006). A 77% decline in penguins and relatively slow recovery were reported following the 1982–83 El Niño event (Valle Castillo 1985; Valle and Coulter 1987).

King Vulture. Vultures, particularly the Black and Turkey vulture species, were often seen feeding on dead fish and other carrion along shorelines in the Gulf of Chiriquí, Panama. During field surveys at the Uva Island coral reef in 2017, Bernhard Riegl began watching a large, high-flying bird over Uva Island (Fig. 8.5). It appeared that this vulture, unlike the common vultures, displayed white feathers on its undersides. Bernhard placed a dead fish on the Uva beach and when the bird descended and came into better view, he identified it as a King Vulture (*Sarcoramphus papa*).

Fig. 8.5 A King Vulture in flight with white underside feathers visible, near Pipeline Road at the observation tower of the Panama Rainforest Discovery Center, 2015. Photograph courtesy of Miguel "Siu" Rios, Fundación Vida Silvestre Panama

Fig. 8.6 Captive King Vultures grooming, Summit Gardens, Gamboa, Panama. Photograph courtesy of Miguel "Siu" Rios, Fundación Vida Silvestre Panama

This bird, portrayed as a god in Mayan codices, is considered one of the most beautiful birds in the world (Fig. 8.6, Panama—Birds & Wildlife Photos; www.PANAMAWILDLIFE.NET). The dead and decaying food source of vultures is repugnant, but these birds perform a vital ecological function that helps to prevent the spread of diseases and enhances the recycling of essential inorganic and organic nutrients. Thanks to Bernhard's keen eye, we now know of another impressive species in the Gulf of Chiriquí (Fig. 8.7).

Marine Mammals

Marine mammal interactions and sightings have involved pinnipeds, dolphins, and whales. We have experienced several up-close encounters with pinnipeds and dolphins in the Galápagos Islands. Two whale species in notably dense pods were observed, one in the Galápagos Islands and the second off Baja California, Mexico.

Pinnipeds. Two endemic pinniped species occur in the Galápagos Islands: the Galápagos Fur Seal, *Arctocephalus galapagoensis*, and the Galápagos Sea Lion, *Zalophus wollebaeki*.[6] Fur seals occur in the productive, upwelling waters of the western islands (western shore of Isabela Island and all around Fernandina Island)

[6] The Galápagos sea lion is here treated as a separate species, closely related to the California and Japanese sea lions. These species have been distinguished based on morphological and behavioral differences (Heath and Perrin 2009).

Fig. 8.7 Bernhard Riegl paying homage to Charles Darwin in Puerto Ayora, Academy Bay, Santa Cruz Island, Galápagos Islands (15 June 2012). The eastward head tilts are signaling the direction to the Charles Darwin Research Sation. Photograph courtesy of Tyler Smith

and we have observed them at Wenman Island in the northern sector of the archipelago. They are nocturnal, diving for fishes and cephalopods that migrate toward surface waters at night. During the day, they are quiescent, sheltering in the shade of rocks and therefore not readily seen. This behavior is an adaptation related to thermoregulation, to avoid over-heating in the equatorial sun. The fur seal was nearly hunted to extinction in the nineteenth Century. Benjamin Morrell, an American sealer, trader, and explorer, harvested many Galápagos fur seals in the early 1800s.[7]

Sea lions were nearly always present at our study sites. They could sometimes be seen hunting singly and capturing fishes. Cooperative hunting has also been reported

[7] It is noted in Morrell's *Narrative of Four Voyages* (1832) that during his first voyage, while remaining in the Galápagos Islands for 2 months, his party harvested about 5000 fur seal skins. Since Morrell was wont to exaggerations, this number could be inflated, but indicative of the large numbers slaughtered. A fur seal hunting party was sent out during his second voyage, again visiting the Galápagos Islands, but they saw few seals. Then it was concluded in Morrell's narrative, "... that we had reaped the harvest in the previous voyage;". It is suspected that fur seal abundances were seriously diminished at that time.

with several individuals steering and stranding ("beaching") fish schools. Numerous instances of cooperative hunting of Amberstripe Scad (*Decapterus muroadsi*), a schooling semipelagic species, was reported by De Roy et al. (2021). These fish were driven from open waters to a natural *cul de sac* embayment and forced onto the beach.

During a survey of a *Pocillopora* community at Santa Fe Island, following the 1982–83 El Niño event, I experienced a frightening encounter with a Galápagos Sea Lion, *Zalophus wollebaeki*. A coral patch inside the north-east anchorage was of special interest because before the El Niño disturbance it exhibited incipient reef development, i.e., beginning to form a coalescent structural framework. Unfortunately, this site was relatively close to a beach with a large colony of sea lions. As a precaution, we anchored our inflatable near the offshore coral site, and I performed the survey measurements quickly in order not to arouse the attention of nearby bull sea lions protecting their harems. I was nearly finished entering coral cover data on my clipboard when a large bull sea lion suddenly approached me and emitted a loud underwater bark. He then began swimming rapid and tight circles around me, continuing to bark. I spun around following his movements the best I could, but he was quick. The lengths of his canine teeth were frightening. He bit into my swim fins several times and I began moving toward our boat. I pulled my arms tightly against my torso and maneuvered my swim fins toward him. I was content to sacrifice the fins instead of some body part. After a while he began targeting my head with wide open mouth and attempting to bite, but I was able to fend him off with a poke to the nose with my dive knife. There were at least two more of these assaults, luckily repelled, before I reached our boat. He finally backed off and returned to his harem on the beach. I was fortunate to have escaped this attack unharmed. After that unsettling encounter, I allowed for greater spacing between sea lion harems and our sampling activities.

While amusing, the playful antics of juvenile sea lions were sometimes annoying during our coral sampling activities. They would bite at and attempt to carry off quadrats, camera strobe lights and other items, sometimes ending in frantic chases by us for equipment retrieval. In a small embayment at Champion Island, a large raft of sea lions was often seen frolicking among colonies of *Porites*. Upon close examination, several coral colonies exhibited typical microatoll morphology with depressed and eroded centers and a living outer ring of coral. Amber Orr and Joshua Feingold observed sea lions poking and laying across the summits of *Porites*, suggesting that this abrasion could contribute to the formation of microatolls (Fig. 1.10). Microatoll colonies are usually attributed to subaerial tidal exposure, sedimentation, and the bioerosive activities of fishes and sea urchins (Glynn 2007). If Amber and Joshua's idea can be substantiated, this would indeed be a novel process in the formation of microatolls, and perhaps unique to the Galápagos Islands.

Darwin Island Dolphins. During one of our early surveys at Darwin Island, we experienced a frightening encounter with a large, fast-moving pod of dolphins, the Long-Beaked Common Dolphin, *Delphinus capensis*. Three of us (Jerry Wellington, Chuck Birkeland, and I) were scuba diving at about 10 m depth in very clear water when we suddenly heard intense squeaking and then caught sight of at least 100 dolphins swimming toward us at high speed. Three or four dolphins in the vanguard were seemingly leading the pod; these appeared to be scouts. The lead dolphins raced towards each of us, ostensibly with the intention of ramming. At the last split second before impact the dolphins suddenly changed course, barely missing us by a hand's length. A hit by one of these torpedo-like dolphins would undoubtedly have resulted in great bodily harm, if not worse. Apparently, we were perceived as threats to the pod and warned accordingly.

Killer Whales. Killer whales or orcas are regularly spotted in the Galápagos Islands, a highly productive region supporting abundant potential prey such as tunas, sharks, pinnipeds, and whales. Our sightings of *Orcinus orca* were rare since these predators occurred more frequently near their prey in the productive waters of the western islands and western island shores (e. g., Isabela and Fernandina islands). Our investigations were concentrated in the warmer waters of the eastern islands where corals were abundant. Once off the northeastern coast of Floreana Island, Joshua spotted two adult orcas swimming nearby. He impulsively leaped into the water and began swimming alongside these imposing animals and the orcas ignored Joshua and continued swimming briskly on their predetermined course. Even though documented predatory attacks on humans are unknown, orcas engage in lively antics—nudging, pushing, tossing—that could result in injuries.

Rake marks on the flukes of humpback whales, observed from feeding grounds in the equatorial eastern Pacific, including the Galápagos Islands, revealed non-lethal orca attacks. Interestingly, Capella et al. (2018) hypothesized that the presence of non-lethal marks on female calves possibly increased their offspring's *a posteriori* survival when birthing, suggesting a maternal memory contributing to avoiding orca attacks. The near lethal orca attacks experienced by young female humpback whales apparently aided them in later life by reducing orca attacks on their offspring.

In the water with orcas and with an uncanny sense of their moods, Tui De Roy, a celebrated Galápagos naturalist, has made some remarkable observations of their feeding habits.[8] Here, to paraphrase from some of Tui's observations, are noted attacks on stingrays and sea turtles. She has watched two male orcas feeding on diamond stingrays (*Dasyatis brevis*) in shallow waters on the south coast of Santa Cruz Island, and one orca killing and eating a marbled stingray (*Taeniura melanospilos*) in James Bay at Santiago Island. In the latter attack, the orca descended over the ray ever so slowly and then smashed in its head with a sudden thrust of its snout. The orca swam away with its prey and presumably dismembered it somehow, possibly to consume only preferred parts. Shortly afterwards, frigate birds (*Fregata magnificens*) swooped down for scraps (as occurred when orcas were preying on turtles). Some orcas revealed spines embedded around their mouths. For stingrays, both ray species are exceptionally large; *D. brevis* can attain a body width of up to 1 m, and *T. melanospilos* up to 1.7 m and a weight of 150 kg. Orcas regularly eat green turtles (*Chelonia mydas*), observed mostly around western Isabela Island. Once, while swimming with orcas after they were feeding on turtles Tui observed

[8] Wildlife photographer, writer, conservationist, Tui has authored several elegant books illustrating the unique Galápagos marine vertebrate fauna (e.g., De Roy 1998, 2020, 2022).

them "playing" with some of the still alive surviving turtles. Tui's observations—"One was dragging a turtle around by its hind flipper while another was swimming along with a flapping turtle whose shell was firmly pressed against its (the orca's) snout which it rotated slightly spinning the turtle around like a frisbee!" Playing behavior is common among several dolphin species and when directed toward potential prey may advance learning skills needed for hunting.

Whales. When aboard M/V *Pirata* in 2000, Captain Lenin Cruz alerted us to whales spouting off our port side on our return to Santa Cruz Island from the northern islands. When approaching one of the whales a huge blunt head, about 1/3 of the total body length, could be seen which is a telltale feature of *Physeter macrocephalus* (Fig. 8.8A). Their off-center blowhole results in a slanted column of warm exhaled air, a condensation cloud also characteristic of sperm whales (Fig. 8.8B). This species is the largest toothed whale, mature males reaching 18 m and females 13 m in length. The pod of whales lay to the east of us with the sun low in the western sky and calm seas, near perfect visibility conditions. Spouting here, spouting there, spouting everywhere, we counted over 50 whales in less than 10 min. They were all dull gray, no white whales of Herman Melville's *Moby Dick* fame. We were all ecstatic over this chance encounter. This large pod likely consisted of several small groups or 'vocal clans' that occasionally come together. The clans are made up of matrilineal lines of females with strong social bonds and unique codas (click sequences). The abundance of sperm whales in the eastern tropical Pacific, especially in the Galápagos Islands, has allowed studies of the social structure and vocal clan dynamics in this iconic species (e.g., Rendell and Whitehead 2003; Cantor et al. 2016).

We were blessed earlier by an unexpected encounter with a large pod of whales in 1994, on this occasion with the world's largest animal. On our return trip to San Diego from Clipperton Atoll, cruising along the remote west coast of Baja California Sur, suddenly our boat's engines went silent. Tim Ekstrom, the young captain of the M/V *Royal Star*, had a strong penchant for protecting marine life. He did not want to disturb or ram any of the basking blue whales (*Balaenoptera musculus*) in our boat's path. We slowly glided among the gobsmackingly beautiful and heart-warming pod of whales. All of our boat's party, on deck and bridge, was awestruck by this unexpected encounter. Tim cut off the ship's radios and requested communication silence from everyone on board so as not to reveal the position of the whale pod. We counted 90 individuals of the whales spouting and within sight. One whale we approached appeared to equal our boat's length, 28 m (92 ft.). They seemed to be resting, as movement was minimal, and no individuals were observed feeding. As we approached the outer bounds of the pod Tim started the boat's engines, and it was slow speed ahead. Animated conversations followed this fortuitous blue whale encounter during our return to San Diego. Blue whales are occasionally encountered in the Galápagos Islands. A single individual was observed, and momentarily frightened Joshua's dive team in their inflatable boat, offshore of Gardner Bay on the north coast of Española Island in 2007 (Fig. 8.9). This whale was alone and observed only briefly before sounding and disappearing.

Fig. 8.8 Sperm whales encountered while traveling between islands on the *Pirata* in the Galápagos Islands, approximately 7 km south of Marchena Island, 5 June 2007, near noontime (12:08 h). (**A**)—dorsal fin visible of whale swimming toward left of photograph. (**B**)—whale spouting at an angle, which is unique to sperm whales. Photographs courtesy of J. S. Feingold

Fig. 8.9 Blue whale sounding near Gardner Island, Española Island (−1.033°S; 89.658°W), 13:42 h, 7 June 2007. The time between breathing and disappearing below the surface was 12 seconds. Photographs and information are courtesy of J. S. Feingold

Avoiding Biohazards

In summary, although we experienced some moments of concern — after completing thousands of hours of research diving — none of our team has ever experienced a flesh bite of concern or a stingray laceration. We followed some basic precautions that helped to prevent offensive or defensive animal attacks, for example: (a)—limit diving time in water of low visibility, (b)—always dive with a facemask to aid in awareness of the surroundings, (c)—employ adequate lighting and backup lights while night diving, (d)—exit the water in the presence of aggressive carnivores and during feeding frenzies, (e)—respect the space (territorial limits) of all wildlife, and (f)—prohibit spearfishing while conducting research dives. To avoid stingray attacks we encouraged our team members to slide or shuffle one's feet while walking in shallow water; also, never grasp a live ray in or out of the water. A ray's tail can be quickly maneuvered to make a forceful stab.

References

Allen GR, Robertson DR (1994) Fishes of the tropical eastern Pacific. Crawford House, Bathurst, Australia, p 332

BirdLife International (2006) Species factsheet: *Spheniculus mendiculus*. http://www.birdlife.org

Cantor M, Whitehead H, Gero S, Rendell L (2016) Cultural turnover among Galápagos sperm whales. R Soc Open Sci 3:160615

Capella JJ, Félix F, Flórez-González L, Gibbons J, Haase B, Guzman HM (2018) Geographic and temporal patterns of non-lethal attacks on humpback whales by killer whales in the eastern South Pacific and the Antarctic peninsula. Endanger Species Res 37:207–218. https://doi.org/10.3354/esr00924

Cortés J, Enochs IC, Sibaja-Cordero J, Hernández L, Alvarado JJ, Breedy O, Cruz-Barraza JA, Esquivel-Garrote O, Fernández-García C, Hermosillo A, Kaiser KL, Medina-Rosas P, Morales-Ramírez A, Pacheco C, Pérez-Matus A, Reyes-Bonilla H, Riosmena-Rodríguez R, Sánchez-Noguera C, Wieters EA, Zapata FA (2017) Marine biodiversity of eastern tropical Pacific coral reefs. In: Glynn PW, Manzello DP, Enochs IC (eds) Coral reefs of the eastern tropical Pacific: persistence and loss in a dynamic environment. Coral reefs of the world 8. Springer, Dordrecht, pp 203–250

De Roy T (1998) Galapagos, islands born of fire. Warwick Publishing, Toronto, p 160

De Roy T (2020) A lifetime in Galapagos. Princeton University Press, Princeton, p 240

De Roy T (2022) A pocket guide to birds of Galapagos. Princeton University Press, Princeton

De Roy T, Espinoza ER, Trillmich F (2021) Cooperation and opportunism in Galápagos sea lion hunting for shoaling fish. Ecol Evol 11:9206–9216

Dominici-Arosemena A, Wolff M (2006) Reef fish community structure in the tropical eastern Pacific (Panamá): living on a relatively stable rocky reef environment. Helgol Mar Res 60:287–305. https://doi.org/10.1007/s10152-006-0045-4

Duffy DC (1990) Seabirds and the 1982-1984 El Niño-Southern Oscillation. In: Glynn PW (ed) Global ecological consequences of the 1982-83 El Niño-Southern Oscillation. Elsevier oceanography series 52, Elsevier, Amsterdam, pp 395–415

Eckert SA, Stewart BS (2001) Telemetry and satellite tracking of whale sharks, *Rhincodon typus*, in the Sea of Cortez, Mexico, and the North Pacific Ocean. Environ Biol Fish 60:299–308. https://doi.org/10.1023/A:1007674716437

Fishman J, McKinnon K (2009) *Crocodylus acutus* (on-line) animal discovery web. http://animaldiversity.org/accounts/*Crocodylus.acutus*/

Glynn PW (1988) El Niño-Southern Oscillation 1982-1983: nearshore population, community, and ecosystem responses. Annu Rev Ecol Syst 19:309–345

Glynn PW (2004) High complexity food webs in low-diversity eastern Pacific reef-coral communities. Ecosystems 7:358–367. https://doi.org/10.1007/s10021-004-0184-x

Glynn PW (2007) Corals. In: Denny MW, Gaines SD (eds) Encyclopedia of tide pools & rocky shores. University of California Press, Berkeley, pp 159–164

Glynn PW, Manzello DP, Enochs IC (eds) (2017) Coral reefs of the eastern tropical Pacific: persistence and loss in a dynamic environment. Coral reefs of the world 8, Springer, Dordrecht, p 657

Glynn PW, Enochs IC, Afflerbach JA, Brandtneris VW, Serafy JE (2014) Eastern Pacific reef fish responses to coral recovery following El Niño disturbances. Mar Ecol Prog Ser 495:233–247. https://doi.org/10.3354/meps10594

Grove JS, Lavenberg RJ (1997) The fishes of the Galápagos Islands. Stanford University Press, Stanford, California, p 863

Guzman HM, Gomez CG, Hearn A, Eckert SA (2018) Longest recorded trans-Pacific migration of a whale shark (*Rhincodon typus*). Mar Biodivers Rec 11:8. https://doi.org/10.1186/s41200-018-0143-4

Hearn AR, Green JR, Espinoza E, Peñaherrera C, Acuña D, Klimley AP (2013) Simple criteria to determine detachment point of towed satellite tags provide first evidence of return migrations of whale sharks (*Rhincodon typus*) at the Galapagos Islands, Ecuador. Anim Biotelemetry 1:1–11

Heath CB, Perrin WF (2009) C–California, Galapagos, and Japanese sea lions: *Zalophus californianus*. In: Wollebaeki Z, Japonicus Z (eds) Encylopedia of marine mammals, 2nd edn. Academic Press, London, pp 170–176. https://doi.org/10.1016/B978-0-12-373553-9.00046-8

Humann P, DeLoach N (1989) Reef fish identification, Galápagos. New World Publications, The University of California Press, Berkeley, California, p 225

Laurie WA (1990) Effects of the 1982-83 El Niño-Southern Oscillation event on marine iguana (*Amblyrhynchus cristatus* Bell, 1825) populations on Galapagos. In: Glynn PW (ed) Global ecological consequences of the 1982–83 El Niño-Southern Oscillation. Elsevier oceanography series 52, Elsevier, Amsterdam, pp 361–380

Merlen G (1988) A field guide to the fishes of Galápagos. Wilmot Books, London, p 60. This book may still be available for purchase at the Libri Mundi bookstore in Quito, Ecuador

Morrell B Jr (1832) A narrative of four voyages to the South Sea, North and South Pacific Ocean, Chinese Sea, Ethiopic and southern Atlantic Ocean, Indian and Antarctic Ocean from the year 1822 to 1831. J. and J. Harper, New York, p 492

Pitman RL, Jehl JR Jr (1998) Geographic variation and reassessment of species limits in the "masked" boobies of the eastern Pacific Ocean. Wilson Bull 110(2):155–170. https://www.jstor.org/stable/4163925

Rendell LE, Whitehead H (2003) Vocal clans in sperm whales (*Physeter macrocephalus*). Proc R Soc Lon B 270:225–231. https://doi.org/10.1098/rspb.2002.2239

Robertson DR, Allen GR (2008) Shorefishes of the tropical eastern Pacific. Smithsonian Tropical Research Institute, Balboa, Panama. Online Information System

Robertson DR, Allen GR (2015) Shorefishes of the tropical eastern Pacific: online information system. Version 2.0. Smithsonian Tropical Research Institute, Balboa, Panamá. https://biogeodb.stri.si.edu/sftep/en/pages

Smith NG (1990) The Gulf of Panama and El Niño events: the fate of two refugee boobies from the 1982-83 event. In: Glynn PW (ed) Global ecological consequences of the 1982-83 El Niño-Southern Oscillation. Elsevier oceanography series 52, Elsevier, Amsterdam, pp 381–393

Thomson DA, Findley LT, Kerstitch AN (1979) Reef fishes of the Sea of Cortez: rocky shore fishes of the Gulf of California. Wiley, New York, p 302

Tovar H, Cabrera D (1985) Las aves guaneras y el fenómeno "El Niño". In: Arntz W, Landa A, Tarazona J (eds) "El Niño", su impacto en la fauna marina. Boletín Instituto del Mar del Perú, volumen extraordinario, Callao, Perú, pp 181–186

Valle Castillo CA (1985) Alteración en las poblaciónes del cormoran no volador, el pingüino y otras aves marinas en Galápagos por efecto de El Niño 1982–83 y su subsecuente recuperación. In: Robinson G, del Pino EM (eds) El Niño in the Galápagos Islands: the 1982–1983 event. Publication of the Charles Darwin Foundation for the Galápagos Islands, Quito, Ecuador, pp 245–258

Valle CA, Coulter MC (1987) Present status of the flightless cormorant, Galapagos penguin and greater flamingo populations in the Galapagos Islands, Ecuador, after the 1982-83 El Niño. Condor 89(2):276–281

Zapata FA, Robertson DR (2007) How many species of shore fishes are there in the tropical eastern Pacific? J Biogeogr 34:38–51

Zhu M, Zhao W, Jia L et al (2009) The oldest articulated osteichthyan reveals mosaic gnathostome characters. Nature 458:469–474. https://doi.org/10.1038/nature07855

Chapter 9
Bioerosion

Abstract Coral reef bioerosion, the fracturing and dissolution of carbonate skeletons by organisms, is inherently high in the biologically productive eastern Pacific region. Disturbances that cause coral mortality can accelerate carbonate erosion by exposing dead surfaces to internal and external bioeroder species. Commonly present internal microscopic bioeroders are cyanobacteria, microalgae, and fungi; among internal macroborers are sponges, polychaetes, and peanut worms, and bivalve mollusks (date mussels). Prominent external bioeroders are hermit crabs, sea urchins (*Diadema* and *Eucidaris*), and fish (pufferfishes and triggerfishes). Destructive but less common coral breakage is caused by sharks and turtles in pursuit of prey. Ocean acidification, a result of anthropogenic-generated CO_2, depresses carbonate skeletal precipitation and internal cementation thus enhancing bioerosive processes. Internal and external bioerosion can cause rapid carbonate loss and reef framework disintegration. Sea urchin bioerosion after the 1982–83 El Niño disturbance amounted to a 0.5 m reduction in vertical framework height of a Panamanian reef over a 16 year period, and complete reef frame destruction of a 1 m high patch reef in the southern Galapagos Islands in <5 years. Paradoxically, the bioerosion of reef frameworks, resulting in increasing cover of dead coral rubble, may not depress but increase cryptic species biodiversity on *Pocillopora* reefs in Panama. Repeated El Niño coral mortality and bioerosion of reef frameworks prevent continued development of coral reefs reaching large size in the Galápagos Islands.

Introduction

To counter the generally accepted early impression that coral reef framework structures were not present in the eastern Pacific, it was first necessary to find structural coral reefs and to demonstrate that they were constructed dominantly by scleractinian corals. This research phase was accomplished as reviewed in "Coral Reef Discoveries" in Chap. 2. The next stage of research required study of the growth history of coral reefs, that is rates of accretion or the accumulation of calcium carbonate structures over time. How rapidly did reef frameworks form and were they

© The Author(s), under exclusive license to Springer Nature Switzerland AG 2025
P. W. Glynn, *Coral Reef Resilience in the Anthropocene*,
https://doi.org/10.1007/978-3-031-81421-1_9

of modern age or constructed in the distant past. Reef growth, however, is also subject to numerous destructive processes of physical, chemical, and biological action. Physical destruction can result, for example, from storm-generated wave action and seismic events, chemical erosion from carbonate dissolution caused by low sea water alkalinity, and biological destruction by organisms involving chemical and/or mechanical erosive processes. Conrad Neumann, an early post-doctoral researcher at RSMAES, is credited with coining the term **bioerosion**, the removal of consolidated mineral or lithic substrate by the direct action of organisms (Neumann 1966). The destructive phase of coral reef development and growth is surveyed in this chapter with primary focus in the eastern Pacific.

Physical Erosion

Physical erosion, owing to mechanical effects impacting carbonate structures, is greatly augmented by preceding erosive activities of organisms and chemical dissolution. Weakened skeletons are much more susceptible to turbulence breakage than pristine colonies untouched by prior erosion. The undercut and weakened bases of framework blocks, and massive colonies that resemble giant mushrooms, are particularly susceptible to toppling (Colgan 1990; Glynn 1990; Lirman et al. 2001). Log damage to reefs is infrequent but has devastating local effects where it occurs. We have observed water-logged tree trunks impact the Uva Island reef. When the trunks come in contact with the reef framework they often become moored and then move up and down with the changing tides crushing corals during each tidal cycle. In one such grounding observed on the Uva reef a log was jostled across the forereef slope for several days leaving a deep trench (1 m wide by 20 m long) of crushed, mostly dead coral fragments. Gary Robinson witnessed the destructive wave action in the Galápagos Islands during the 1982–83 El Niño event. Normally, ocean swells in this area are from the south-southeast quarter; they reversed direction to the northwest during the height of El Niño. This reversal caused the dislodgment of large coral colonies (*Porites, Pavona*) that were deposited by storm turbulence above the high tide line. The reversal of wave assault also resulted in extensive damage to *Pocillopora* frameworks (Robinson 1985). Tropical cyclones usually form north of the equatorial eastern Pacific, at 10°-18° north latitude (Fiedler and Lavin 2017) with greatest effects off the west coast of Mexico (Lirman et al. 2001; Reyes-Bonilla 2003).

Bioeroder Taxa

Bioeroding organisms are taxonomically diverse, ranging from microscopic bacteria, microalgae, and fungi to varied invertebrate and vertebrate metazoans. Depending on the location of their eroding activities bioeroders can be classified as

internal or **external**. Internal bioeroders can be further classified into microscopic (**microborer**) and macroscopic (**macroborer**) groups. Among the commonly encountered microborer organisms in coral skeletons are cyanobacteria, green and red algae, fungi, and foraminifers (Hutchings 2011; Tribollet and Golubic 2011). Several of these groups contain species capable of photosynthesis. Internal macroborers are represented by larger foraminifers, and several kinds of metazoans, e.g., sponges, bryozoans, polychaete worms, sipunculan or peanut worms, bivalve mollusks, and crustaceans (Hutchings 1986, 2011; Glynn and Manzello 2015). It is now necessary to add the records of coral macroborer species discovered by Ana C. Fonseca and Cortés (1998) in massive *Porites* colonies along Pacific Costa Rican shores. These records are a sipunculan and a shrimp-like crustacean, increasing the number of known coral macroborer species to 18. External bioeroders include mollusks (chitons and gastropods), crustaceans (hermit crabs and brachyuran crabs), echinoderms (sea urchins), and fishes (parrotfishes, pufferfishes, trigger fishes, and damselfishes) (Glynn and Manzello 2015).

Internal bioeroders excavate carbonate substrates by a combination of chemical and mechanical processes. Microborers actively erode corals by dissolving the carbonate skeleton and macroborers by employing chemical and mechanical methods (Tribollet and Golubic 2011). Boring sponges are notably active in the eastern Pacific, employing amoebocyte-like etching cells (Schönberg 2008; Carballo et al. 2013). External bioeroders graze on or bite off portions of living and dead coral skeletons, including associated organisms (e.g., filamentous algae and encrusting sponges) that utilize corals as substrates. The radula, a feeding structure employed by gastropods and chitons, is a flexible belt composed of chitinous teeth often enriched with mineral deposits such as magnetite (Lowenstam and Weiner 1989). Aristotle's lantern, the jaw apparatus of sea urchins, is a complex mastigatory organ consisting of five movable calcitic teeth. This structure is remarkable in terms of its maneuverability; when active it can move at different angles to efficiently attack variously angled substrates. Parrotfish teeth consist of fluorapatite, a hard biomineral (Marcus et al. 2017). The teeth of bioeroding fishes that constantly scrape or bite into carbonate skeletons are the products of **biomineralization**, i.e., the accumulation of hardened mineralized tissues lining cutting edges. While the rasping and cutting mouth parts of metazoan bioeroders are very efficient at fracturing substrates, much of the broken carbonate skeleton falls to the bottom. Field and aquarium observations of feeding bioeroders have shown that relatively large amounts of loosened and freed skeletal particles are not ingested. Bioerosion measurements should include not only ingested and defecated carbonates but all eroded substrate material. This has important implications in estimating overall bioerosion rates and potential impacts to reef growth.

Large vertebrates contributing to coral erosion in the eastern Pacific are triggerfishes (Guzmán 1986; Guzmán and Cortés 1989), sea turtles (Glynn 2004, 2008), and sharks (Jiménez 1996–1997). While the feeding activities of these large animals can lead to extensive local damage, their effects appear to be relatively infrequent and therefore difficult to quantify. On a few occasions I have observed Olive Ridley

(*Lepidochelys olivacea*)[1] and Loggerhead (*Caretta caretta*) sea turtles breaking *Pocillopora* frameworks in pursuit of cryptic sponges. The resulting pits were surprisingly large, about 0.5 meters in diameter and depth, and surrounded by excavated live and dead broken coral branches. The turtles were observed grazing on sponges encrusting the dead basal branches of *Pocillopora*. On such occasions, even though I remained at a distance (5–8 meters) to better capture the actual feeding behavior, my presence caused the foraging turtles to depart.

Methods

Our team employed various methods in the field and laboratory to determine the feeding and bioerosion rates of invertebrate and fish corallivores. Because the products of bioerosion are so different in size—ranging from the molecular (by dissolution) to minute grains (by etching) and large grains and framework blocks (from grazing and rooting of coral substrates)—it is necessary to select an appropriate method that will accurately estimate the actual losses. To assess the effects of grazing bioeroders (e.g., sea urchins and pufferfish), actively feeding individuals were removed from carbonate substrates and allowed to defecate over various time intervals. When defecation slowed or ended the resulting calcareous mass was determined and adjusted to 24-h periods and assumed to equal daily bioerosion (Glynn et al. 1972, 1979). Different methods were developed to account for the bioerosion of infaunal organisms. For example, to determine the total carbonate mass lost by a massive coral species, experimental blocks of *Porites* were cut and secured on coral patches undergoing rapid bioerosion by *Eucidaris*. The blocks were collected after ~15 months exposure and sectioned to determine volumetrically the losses due to internal and external bioeroders (Reaka-Kudla et al. 1996). Resistant cathedral limestone blocks, a durable natural stone, were also introduced to the coral patch to account for carbonate gains by settling epilithic, calcifying organisms. Mark Eakin (1992) employed experimental carbonate substrates, and acrylic rods and discs to estimate the external erosion caused by fishes and motile invertebrates. Although not available in our early studies, a micro-computed tomography (μCT) method has been developed to determine accurate erosion and accretion rates by external and internal bioeroding organisms (Silbiger et al. 2014; Enochs et al. 2016a). The microCT methodology has allowed the Enochs team to volumetrically isolate and independently quantify all interacting functional groups instrumental in constructing and eroding carbonate structures (accretion, macroboring, microboring, and grazing). A further advancement involves volumetrically quantifying the eroding taxa using their taphonomic and ichnological signatures in the CT scans (Enochs et al. 2021)[2].

[1] The species name of the Olive Ridley sea turtle is misspelled as "*divacea*" in Glynn (2008), page 197. The correct spelling is *olivacea*.

[2] Taphonomy is the branch of paleontology that studies how organisms decay and are fossilized in the geologic record and ichnology, the study of the marks left in the rock record by organisms. Ian

To obtain estimates of sea urchin bioerosion during brief layovers at remote monitoring sites, we deployed 15-liter bucket enclosures that became known as the "bucket" method. External bioeroders, chiefly sea urchins, were of paramount interest in our studies because of their high abundances and remarkable per capita erosion rates. The bucket method is briefly described here for measurements of bioerosion performed in Panama and the Galápagos Islands.[3] Procedural details can be found in Glynn et al. (1988) and Eakin (1991, 1996). Wide mouth 15-liter buckets with lead weights were secured on the bottom to collect sediments over 24, 48, and 72 h periods. Each deployment utilized (1) empty control buckets, (2) buckets that contained *Pocillopora* colonies or framework blocks, and (3) buckets with five sea urchins and *Pocillopora* colonies or framework blocks. Coarse netting was secured over the buckets' mouths to prevent the escape of the introduced sea urchins. This procedure is best conducted in a calm area absent wave action and turbulence. The carbonate substrates placed in the buckets should be freed of any loose sediment. This is effectively accomplished by vigorously moving submerged framework blocks from side to side until sediment fall ceases. Sediment masses collected in (1), i.e., settled suspended sediments, were subtracted from the masses collected in (2) and (3). The sediment collected in (2), minus (1), was assumed due to internal bioerosion. The sediment collected in (3), minus (1) and (2), was treated as sea urchin bioerosion. Rates were calculated and reported as dry weights (gm per m^2 per day) for all bioeroders. A disadvantage of this method is that it does not account for fish bioerosion because the netting prevents sea urchin escape and the entry of fishes. Methods adopted to assess the effects of other bioeroders can be found in Glynn et al. (1972), Oramas (1979), Achurra Cárdenas and Valdés Araúz (1980), and Glynn (1985, 1988).

B_i = bucket number (i) of sediment

Sea urchin bioerosion = $B_3 - (B_2 + B_1)$

Internal bioerosion = $B_2 - B_1$

In addition to obvious breakage and sediment generation, another manifestation of bioerosion is the loss of vertical relief of reef frameworks. This can result from bioeroders grazing on carbonate surfaces and internal bioeroders weakening subsurface structures and eventually causing subsidence. To measure this erosive effect, iron rebar and PVC (polyvinyl chloride) pipes were hammered into the reef to serve as reference points. Steel wires were strung horizontally between the pipes and the distances were measured from fixed points (knots) along the wire to the reef surface. Such measurements were performed by our team (Glynn 1988; Eakin 1992, 1996)

Enochs' team volumetrically quantified and identified the unique bioeroder markings preserved on their experimental carbonate substrates.

[3] Our team was aware of a researcher who sharply disapproved of the bucket method but refused to explain the reasons why. We found the technique useful for assessing the erosional impacts of sea urchins at remote sites that we were able to visit only briefly. With this method we also obtained estimates of carbonate erosion by internal bioeroders and the dislodgment of carbonate particles by the movements of sea urchin bioeroders. We found that the spine armature of large sea urchins can abrade and loosen carbonate substrates.

in different zones of the Uva reef annually for several years after the 1982–83 El Niño disturbance. Following the initiation of reef recovery, i.e., the beginning of coral recruitment and renewed vertical growth, gaps between the reef surface and the horizontally strung wire (marking the pre-disturbance reef height) began to close, signifying positive carbonate accretion. These were welcome data after so many years of recording framework regression.

Bioerosion Potency

Some bioeroder taxa are particularly destructive of carbonate substrates. The following examples consider the population densities of selected species observed eroding reef carbonates; the rates are adjusted from individual daily erosion and bioeroder densities to meter-square areas per year.[4] Most of the high bioeroder densities noted herein occurred after the 1982–83 El Niño coral mortality event.

Clionaid sponges, largely internal bioeroders, are recognized globally for their damage to coral reefs in areas of high nutrient loading (e.g., Pari et al. 1998). In the Mexican Pacific, two *Cliona* species were found responsible for eroding 4.5 and 5.1 kg $CaCO_3$ m^{-2} y^{-1} following a coral bleaching and mortality event (Nava and Carballo 2008; Carballo et al. 2013).[5] Other internal bioeroders in the eastern Pacific are date mussels (bivalve mollusks), including several species that bore into massive coral skeletons. One species (*Lithophaga laevigata*) preferentially bores into live coral skeletons and is purported to be an obligate symbiont of massive *Porites* colonies (Guzmán 1988; Scott et al. 1988). Like sponges, date mussels are filter feeders and thrive in upwelling, nutrient-rich environments. In the early 1980s, Karl Kleemann introduced date mussels from the Caribbean side of Panama to the Pacific waters of the Gulf of Panama to determine the effects of upwelling on their growth. He found that the growth rates of the Caribbean lithophagine species increased significantly in the Pacific upwelling environment (Kleemann 1990).[6] Some of the

[4]Adjustments for some organisms that are difficult to distinguish as discrete individuals, such as sponges and filamentous algae, are quantified volumetrically or by mass.

[5]The weight of an average size watermelon ranges between 12 to 16 kg. The maximum net accretion, i.e., the mass of coral-generated carbonate adding to *Pocillopora* reef growth in the eastern Pacific, ranges between 10 to 15 kg m^{-2} annually (Alvarado et al. 2017a).

[6]Such a transplantation experiment involving the introduction of non-native species would be greatly discouraged today considering the ill effects demonstrated by numerous introduced non-native marine species. Karl's experiments (Kleemann 1990), conducted in 1981 and 1982, led to the introduction of a non-native Caribbean coral to Pacific Panama, which resulted in considerable taxonomic confusion (Glynn et al. 2016). Karl believed that the *Siderastrea* coral skeleton used in his experiment (as a substrate for the Caribbean date mussels), was dead. Unfortunately, small bits of living coral tissue survived and gave rise to corallith (spherical) colonies that were later discovered and named a new species, *Siderastrea glynni* Budd and Guzman 1994. This unintentional introduction was mentioned to me by Benjamin Grassian, many years later when he was reviewing Kleemann's early publications of work conducted in Panama. I decided that the best approach in

highest densities of *Lithophaga* spp. reported in massive *Porites* hosts ranged from 3060 to 6730 individuals m^{-2} respectively in Panama (Scott et al. 1988) and Costa Rica (Fonseca et al. 2006). Fine-grained sediment captured in traps attached to the undersides of *Porites* at Caño Island, Costa Rica, amounted to 9 kg m^{-2} y^{-1}, likely due primarily to erosion caused by *Lithophaga* spp. (Scott et al. 1988).

The numerous corallivores present on eastern Pacific coral reefs, noted in Chap. 7, contribute importantly to external bioerosion either directly or indirectly. Sea urchins and hermit crabs produce copious amounts of sediment directly while scraping and biting into carbonate skeletons (Glynn et al. 1979; Gilchrist 1985). While feeding on coral tissues, an ovulid gastropod (*Jenneria*) and *Acanthaster* indirectly assist in erosion by leaving dead skeletons more easily penetrated and colonized by diverse bioeroders. Of the grazing invertebrate bioeroders, sea urchins are by far the most destructive at many eastern Pacific coral reef sites. Our studies in the Galápagos Islands recognized the erosion of *Pocillopora* reefs by grazing *Eucidaris* was in close balance with reef growth (Glynn et al. 1979). The likelihood of an indirect effect of overfishing causing an increase in the abundance of *Eucidaris*, especially near centers of human habitation, has received attention and concern (Ruttenberg 2001; Edgar et al. 2010). For example, if fish predators of *Eucidaris* are reduced in numbers by fishing pressure this could lead to a relaxation of predation and increased abundances of sea urchins, overgrazing, elevated bioerosion and loss of coral communities. Several additional invertebrate bioeroders, especially crustaceans and gastropod mollusks, cause high bioerosion rates on eastern Pacific coral reefs (Guzmán 1988; Glynn and Manzello 2015).

In the Galápagos Islands, average sized *Eucidaris* grazing on coral generate about 2.0 gm of sediment per day. This is an order of magnitude higher than the carbonate sediment produced by *Diadema* in Panama, which amounts to about 0.2 gm $CaCO_3$ ind^{-1} day^{-1}. This difference was due primarily to the large size of *Eucidaris* (test diameter, 4.9–5.5 cm) and its grazing efficiency compared to *Diadema* (test diameter, 1.8–2.7 cm). *Eucidaris* densities of 10–50 ind m^{-2} resulted in the erosion of 20–40 kg m^{-2} $yr.^{-1}$. *Diadema* densities of 50–150 ind m^{-2} generated 2.5–10.4 kg m^{-2} $yr.^{-1}$ of carbonate sediment. Coral reef vertical framework erosion by sea urchins after the 1982–83 El Niño event occurred eight times faster in the Galápagos Islands than in Panama (Fig. 9.1).

An informative classification of fish bioeroders on eastern Pacific reefs, recognizing three guilds on eastern Pacific reefs, was reviewed and described by Juan Alvarado and co-authors (2017): (1) herbivores that graze on and abrade carbonate

---

reporting this accidental introduction, in terms of transparency and civility, was to contact Karl and offer him the opportunity to help explain this misstep. Briefly, *S. glynni* was determined to be morphologically identical with *Siderastrea siderea*, a long recognized western Atlantic species. Prior molecular evidence also anticipated the synonymy of the species (Forsman et al. 2005). I was very grateful to Ann Budd and Héctor Guzmán for naming the presumed new species in my honor, but alas, it was necessary to recognize the species as a junior synonym of *S. siderea* (Glynn et al. 2016). Karl assured me that all native Caribbean *Lithophaga* were removed from Pacific waters after his study. None has been reported from the eastern Pacific in the forty years since.

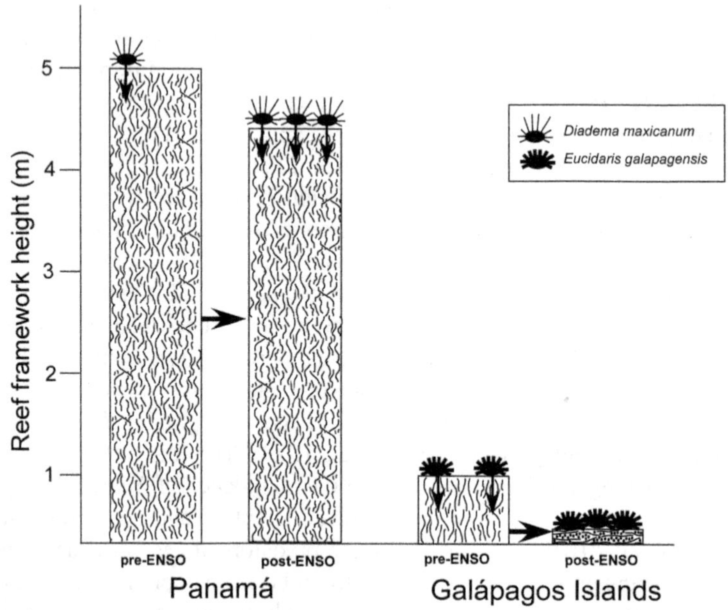

Fig. 9.1 Sea urchin bioerosion of *Pocillopora* patch reefs in Panama and the Galápagos Islands following the 1982–83 El Niño event. Reef-top shaving of the Uva reef in Panama by *Diadema* over 16 years and total destruction of Galápagos reefs by *Eucidaris* in 2–3 years

surfaces, (2) corallivores that graze on and fragment corals, and (3) fish invertivores that fragment corals in their pursuit of associated coral reef fauna. Fish in the third guild, including sharks, triggerfishes and other piscivores, are particularly disruptive in breaking apart coral colonies and reef frameworks in pursuit of prey. For example, three White-tip sharks in pursuit of two triggerfish on a *Pocillopora* reef in Panama generated more than 60 large fragments (average size equaled 11 cm) during an hour-long observation period (Jiménez 1996–97). Another example relates to *Porites evermanni* which is frequently attacked by triggerfish breaking into colonies (by biting) in search of boring bivalves (*Lithophaga* spp.). This poritid species hosts high numbers of boring bivalves and is apparently targeted by triggerfish for this food source. A molecular genetic study has shown that *P. evermanni* is a cryptic species distinct from *Porites lobata* which is infested with fewer boring bivalves and attacked less frequently (Boulay et al. 2014). While parrotfishes and triggerfishes are recognized as important bioeroders in eastern Pacific coral communities, research has been directed primarily to the Guineafowl pufferfish, *Arothron meleagris*, to which we now briefly turn.

Arothron meleagris is ubiquitous on coral reefs in the eastern tropical Pacific, bold in behavior and often seen biting off the branch tips of *Pocillopora*. While the Guineafowl puffer usually preys on *Pocillopora* spp., several other corals, diverse benthic invertebrates, and coralline algae also often form part of its diet (Glynn et al. 1972; Guzmán and Robertson 1989). The gut contents of pufferfish sampled in the Galápagos Islands revealed a mixture of ingested items (Fig. 9.2). This

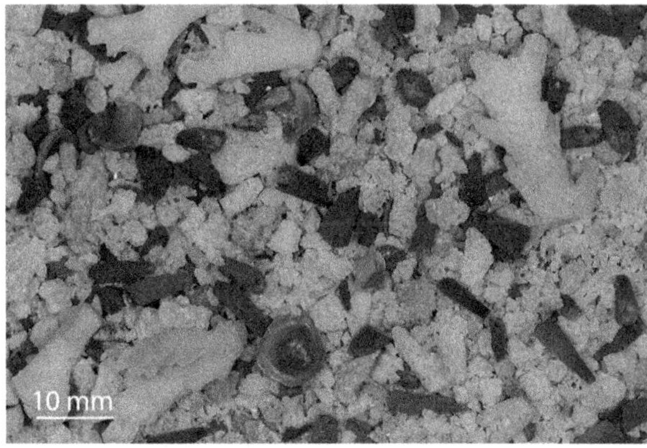

Fig. 9.2 Gut contents of Guineafowl Pufferfish, *Arothron meleagris*. A portion of five combined samples from fish feeding in coral communities in the Baltra Channel, Galápagos Islands (15 August 1986). Mostly *Pocillopora* branch tips are visible (white) and in lesser abundance *Eucidaris* sea urchin spines (dark brown) and gastropod limpet shells (cream-pink)

pufferfish displays two color phases: bright gold-yellow with black splotches, and dark blue purple with numerous white spots. Occasionally some individuals display a combination of the two-color phases. This puffer species is typically present on *Pocillopora* reefs, ranging from a mean abundance of 12 to a maximum reported density of 171.2 ind ha^{-1} (Palacios et al. 2014). The relative fearless behavior of this pufferfish, likely related to its distinctive warning coloration and unpalatability, has allowed an in-depth study of bioerosion's negative and positive facets by a team of Colombian coral reef ecologists (Palacios et al. 2014). By following individual feeding fish, Carlos Muñoz (as part of his doctoral research), could quantify the mass of broken but not ingested coral branches and determine their survivorship. He found a surprisingly high survival rate (>75%) of branches that settled on consolidated reef substrates. A high proportion of these bioeroded branches continued growing and contributed significantly to asexual recruitment. The total rate of *Pocillopora* carbonate destruction by *A. meleagris* on a Gorgona Island coral reef was 1.1 kg m^{-2} y^{-1}, ~ 20% of the annual carbonate production of 5.56 kg m^{-2} y^{-1} (Muñoz Bernal 2017). Thus, although the Guineafowl puffer is a major bioeroder, Carlos has demonstrated quantitatively that this fish is an essential agent of asexual reproduction. This capability was proposed earlier at Gorgona Island and elsewhere in the eastern Pacific for other fishes that fragment corals (Guzmán and López 1991).

Overall assessments show that the proportions of internal or external bioerosion are spatially highly variable. In global comparisons several studies showed internal erosion to be relatively minor, ranging between <0.1% to 20% of total bioerosion (Tribollet and Golubic 2011); these relatively low internal values were mainly by microbioeroders. Studies on the Great Barrier Reef (Tudhope and Risk 1985; Hutchings et al. 2005; Osorno et al. 2005; Tribollet and Golubic 2005), however, showed internal microbioerosion to be predominant, ranging between 20% to 100% of total bioerosion. In the eastern Pacific, internal macrobioeroders have caused significant carbonate losses in *Porites* and other massive coral species. This has been reported for coral reefs in Golfo Dulce and Caño Island, Costa Rica (Cortés 1991; Fonseca and Cortés 1998), and in the Galápagos Islands (Scott et al. 1988). The bioeroders in Costa Rica were active in the 1980s and 1990s and involved primarily bivalve mollusks, *Lithophaga* spp. and *Gastrochaena rugulosa*, a sipunculan peanut worm, and a gebiidean crustacean. External bioerosion in the equatorial eastern Pacific, particularly by sea urchins (*Diadema* and *Eucidaris*) following El Niño associated coral mortality, has caused extensive erosion in Panama (Glynn 1988; Eakin 1992, 1996, 2001) and total reef degradation in the Galápagos Islands (Glynn 1988, 1994; Reaka-Kudla et al. 1996).

Carbonate Chemistry and Calcification

Since publication of Chuck Birkeland's edited book, *Life and Death of Coral Reefs* (1997), the threat of 'ocean acidification' (OA) is better understood. When Chuck approached me to update my 1997 chapter on bioerosion for his new book, *Coral Reefs in the Anthropocene* (2015), I realized that the changing chemistry of the world's oceans—lowering of the carbonate saturation state due to increasing absorption of anthropogenic-generated CO_2—needed to be addressed. Geochemists and climatologists understood early that CO_2 absorption by the world's oceans could modify its chemistry (Broecker et al. 1979). This was brought into focus for coral reefs by Joanie Kleypas et al. (1999) who recognized that depletion of carbonate ions (a reduction of aragonite concentration, Ω_{arag}) could potentially affect coral reef development and their distributions. The reader is referred to the "Ocean Acidification" section in Chap. 4 of this book for a review of the basic chemistry of dissolved inorganic carbon. Several studies quickly followed Kleypas's publication demonstrating the pivotal role of CO_2 absorption and its effects in reducing carbonate concentration and calcification of diverse coral reef organisms (Gattuso et al. 1999; Langdon and Atkinson 2005; Orr et al. 2005; Kuffner et al. 2008; Tribollet et al. 2009; Kroeker et al. 2013).

Fortunately, one of my recent doctoral students, Derek Manzello,[7] was actively researching ocean carbonate saturation state and pH on eastern Pacific coral reefs

[7] Derek Manzello, a stalwart member of our coral reef team, graduated from the University of Miami (UM) with a Ph.D. degree in 2008. He worked at the Atlantic Oceanographic & Meteorological Laboratory (AOML) of the National Oceanic and Atmospheric Administration (NOAA), Miami, Florida campus, from 2008 to 2013, and continued as a Research Oceanographer in the Ocean Chemistry and Ecosystems Division of NOAA from 2014 to 2021. This opportunity was due to the foresight of James C. Hendee, originator of Coral-List list-server and the Coral Health and Monitoring Program (CHAMP). Derek then moved to NOAA's College Park, Maryland bureau in 2021 succeeding Mark Eakin as Coordinator of the Coral Reef Watch (CRW) program. Mark Eakin was one of my first doctoral students, graduating in 1991. He, along with other team members (Fig. 9.3), was instrumental in establishing the Uva Island Beach Club, our field laboratory during the military dictatorship of the "MAN" (Manuel Noriega) in Panama. After a year as a John A. Knauss Sea Grant Fellow, Mark assumed the position of Program Manager in NOAA's Office of Global Programs (now the Climate Program Office). In 2005, after several years serving in assessment initiatives of coral reef bleaching, he became NOAA's Coordinator of the CRW program founded by Al Strong and officially recognized in 1998 (Liu et al. 2006). Mark retired from federal employment in 2020 after a distinguished career with numerous awards for advancing our knowledge of global coral reef bleaching and mortality. Ian Enochs, also a committed researcher in our coral team and Ph.D. graduate of UM in 2010, succeeded Derek as lead investigator of the AOML Coral Program. Ian's doctoral research centered on the biodiversity and trophic

Fig. 9.3 Uva Island "Beach Club", temporary field laboratory erected close to the Uva coral reef study site, Contreras Islands, Gulf of Chiriquí, Panamá (21 February 1989). Juan L. Maté and C. Mark Eakin (standing, left to right), David B. Smith, Peter W. Glynn and José Manuel Gil Lasso (sitting, left to right). Photograph courtesy of C. M. Eakin

(Fig. 4.5). He kindly offered to help by including highlights of his research and relating this to OA studies in other coral reef regions. Eastern tropical Pacific surface waters are among the world's most depleted of carbonate ions due to high biological production and upwelling of relatively low pH waters. The combined action of these conditions can negatively affect carbonate skeleton formation and cementation in reef-associated organisms. Two studies with direct bearings on bioerosion, (a) the harmful effects of high CO_2 and low carbonate saturation on the calcification of crustose coralline algae (Kuffner et al. 2008), and (b) the accelerated rates of bioerosion by endolithic algae (Tribollet and Golubic 2011) and clionaid sponges (Carballo et al. 2013), had piqued the concern of coral reef researchers on the imminent, world-wide negative effects of OA.

Two fundamental processes play vital roles in reef building, calcification and cementation. Calcification of corals and coralline algae produces the building blocks and binding agents of reef growth. Cementation, the biological or chemical precipitation of calcium carbonate, fills voids, binds sediments, and strengthens the rigidity

contributions of the cryptofauna in an eastern Pacific coral reef community. After graduation, Ian continued with coral reef ecosystem research at UM/CIMAS-NOAA/AOML, gaining experience in OA, global warming, and disease studies in the Caribbean, western Atlantic and South Pacific. Under the leadership of Derek and Ian, the AOML Coral Program has grown into a team of 20 with diverse research initiatives in bioerosion, OA, coral diseases, restoration, molecular mechanisms of coral resilience, and coral-related technology development.

of reef frameworks. One of Derek's first studies was to examine submarine cementation in eastern Pacific waters, including Panama and the Galápagos Islands. Only trace amounts of cement were found in the internal pore spaces of eastern Pacific coral skeletons (Manzello et al. 2014). This contrasted markedly with extensive cemented reef crusts and cement-infilled cavities in Caribbean corals, for example in waters of high aragonite saturation on the north coast of Panama (Macintyre and Glynn 1976). Along with crustose coralline algae, well-developed cement crusts and pavements in forereef zones have contributed to the development of wave resistant algal ridges in the San Blas area of Panama (Glynn 1973) and throughout much of the western tropical Atlantic region (Adey and Burke 1976; Adey 1978).[8] Seasonal wind constancy and strength were key conditions favoring algal ridge development. We now also understand the importance of the carbonate saturation state (Ω_{arag}) as it can influence the physical and biological processes of cementation. In consideration of calcification, a study by Ian Enochs and co-workers at Maug Island, northern Mariana Islands, revealed an indirect effect of Ω_{arag} on the erosion of carbonate skeletons. Utilizing a natural pH gradient in a volcanically acidified coral reef area, these researchers reported that OA enhancement promoted the colonization of microborers, especially bioeroding chlorophytes (Enochs et al. 2016b). These two studies call attention to the crucial role of Ω_{arag} and how projected conditions of OA will negatively impact coral reef growth and persistence by the end of the century.

Ravages of Bioerosion

Obvious signs of bioerosion were evident on eastern Pacific coral reefs in the late 1960s, several years before the 1982–83 El Niño disturbance. This was especially noticeable in upwelling environments with undercut and eroded bases of massive coral species and fragile reef frameworks of branching *Pocillopora* colonies (Fig. 9.4). Walking across reef flats at low tide in the Pearl Islands, Gulf of Panama, was tricky due to the poor consolidation of coral branches and the presence of underlying cavities prone to collapse. Unlike coral reef structures in the western Atlantic and Indo-Pacific regions, eastern Pacific coral reefs consist of scant crustose calcareous algal binding and minimal submarine cementation.[9] This condition

[8] Walter Adey, Museum of Natural History, Smithsonian Institution, obtained a surplus Grumman Albatos HU-16 amphibious aircraft from a depot of retired and obsolete military planes to locate and map algal ridge structures in the Caribbean and West Indies. Ground truth studies were performed at several sites for verification, which included coring and C^{14} dating. Those of us involved in large scale coral reef mapping held Walt in high esteem for this innovative approach, before the availability of satellite imagery and remote sensing technology.

[9] Submarine cementation, the chemical precipitation of $CaCO_3$, occurs in coral skeletal and reef framework cavities and between internal sediment grains. This process binds reef frameworks, thus enhancing rigidity and strength.

Fig. 9.4 Sampling a *Pocillopora* coral block for bioerosion aboard RV *Benjamin,* Uva Island, Gulf of Chiriquí, December 1980 (P. W. Glynn, Aníbal Velarde, left to right). Aníbal, my intrepid Panamanian assistant, always exacting and faithful in performing all manner of coral research tasks

was not surprising considering Ray Highsmith's (1980) productivity hypothesis of heightened bioerosion in high nutrient and biologically productive environments. Ray proposed that bioeroder populations would be more abundant and competitive in highly productive environments, such as upwelling centers. *Diadema mexicanum* in the Gulf of Panama are twice the size of those in the non-upwelling Gulf of Chiriquí, and capable of more extensive erosion (Alvarado et al. 2016). Also, since many bioeroder species are suspension or filter feeders, increased trophic resources from the water column, resulting from natural eutrophication and human-induced pollution,[10] would favor their growth and proliferation.[11] Finally, sporadic cold shock conditions accompanying upwelling would also depress coral calcification.

[10] A classic example of pollution causing intense bioerosion and degradation of a fringing reef at Coconut Island, Kāne'ohe Bay, Hawaii, was documented by Paul Jokiel and co-workers. With Paul's permission, a time course cartoon showing dramatic coral recovery following sewage abatement is depicted in Glynn and Manzello (2015).

[11] Another consideration relating to environments of high biological production is the potentially large size achievable by grazing species. For example, in the upwelling Pearl Islands very large and destructive *Diadema* are commonly present on reefs compared to smaller individuals at the non-upwelling Uva Island reef site. The test diameters of *Diadema* at the Saboga Island reef ranged between 3.4 to 3.5 cm, whereas at Uva Island test diameters ranged between 1.8 to 2.3 cm. Median per capita erosion rates of large sea urchins were over twice those of small individuals: 0.47 (± 0.06) versus 0.19 (± 0.04) (Glynn 1988). Median 95% confidence limits are noted in parentheses. Large *Diadema*, with test diameters of 2.9 to 4.8 cm, are also prevalent on coral reefs at the Huatulco upwelling center (Herrera-Escalante et al. 2005; Benítez-Villalobos et al. 2008). Juan Alvarado, from his extensive investigations of *Diadema* in the eastern Pacific, cautions that the positive relationship of increasing size with environmental productivity is not always apparent and

Fig. 9.5 Gerard (Jerry) M. Wellington with an eroded pocilloporid framework block that died during the 1982–83 El Niño event, basalt rock pool, Punta Espinosa, Fernandina Island (21 May 2002). Photograph courtesy of Joshua S. Feingold

Several studies have demonstrated that dead coral skeletons are subject to higher rates of bioerosion than live skeletons (Hutchings 2011; Tribollet and Golubic 2011; Alvarado et al. 2017b). Most internal bioeroder species enter carbonate substrates by way of dead skeletal surfaces, but some boring bivalves (Scott 1988) and clionaid sponges (Schönberg and Wilkinson 2001) can gain entry via live polyp tissues. With over 95% coral mortality observed in the Galápagos Islands at the end of the 1982–83 El Niño disturbance (Glynn et al. 1988), the already intense erosion by *Eucidaris* observed in the 1970s continued (Glynn et al. 1979). After bleaching and death, coral skeletons were soon covered by filamentous algae, which were heavily grazed by high population densities of *Eucidaris*. The iconic *Pocillopora* patch reef inside Devil's Crown experienced rapid bioerosion. This loss was lamented by many Galapageños intimately familiar with the reef. The unique *Pocillopora* reef patches in the basalt rock basins at Punta Espinosa were also experiencing rapid erosion. A few years after the coral mortality event loose *Pocillopora* framework blocks could be found scattered around the basins (Fig. 9.5).

Of the 17 surveyed Galápagos sites in 1975–76 with coral frameworks (Glynn and Wellington 1983), 16 were completely degraded by sea urchins in 10 years or less (Glynn 1994). An experimental study on the Great Barrier Reef also demonstrated rapid erosion, the complete disintegration of coral colonies in 5 years (Morais

that available habitat shelter and the feeding activities of fish and spiny lobster predators likely influence the sizes and abundances of local populations.

et al. 2020). The causative agents of erosion were not reported; parrotfish were suspected, particularly the large, destructive *Bolbometopon muricatum*.

Measuring the shrinking height of the Uva Island *Pocillopora* reef after 1983 was disheartening, with roughly 2 cm losses occurring annually, due primarily to *Diadema* grazing. This degree of erosion, however, did not prepare us for the extraordinarily rapid losses occurring in the Galápagos Islands. Galápagos *Pocillopora* reefs suffered vertical losses of 20–30 cm per year, largely a result of *Eucidaris* grazing (Fig. 9.1). Vertical reef loss at Uva Island, over a period of 16 years, amounted to 8.8% of the pre-El Niño reef height.[12] After more than a decade and a half of monitoring, welcome increases in reef height were finally detected during our reef monitoring, a result of coral recruitment and growth. As of 2017, renewed reef building has not been observed on 11 monitored Galápagos reefs (Glynn et al. 2018). A small incipient patch reef inside the northeast anchorage of Santa Fe Island, however, has recently become re-established at the same site as the pre-existing reef (Ellen Prager, pers. comm.). New coral growth was due entirely to sexual recruitment. It would be interesting to locate the source population(s) of the coral recruits and to monitor this patch reef's course of recovery.

A single Galápagos coral reef, the Wellington reef at Darwin Island, experienced significant erosion but retained much of its primary framework. Jerry Wellington was aware of the Darwin Island coral reef, and we caught a glimpse of it during our 1975 survey. Over the years, due to our limited time at Darwin and generally poor diving conditions (chiefly strong currents), we did not begin to determine the precise location and structure of the reef until 2005 (Glynn et al. 2009).[13] Thanks to Sam (Samuel) Purkis and Jeremy Kerr, who created a bathymetric image of the three-dimensional structure of the reef, we were better able to sample and assess the condition of the reef in 2012 (Glynn et al. 2015). A 1992 survey revealed extensive external erosion of *Porites lobata*. The dome-like tops of several colonies were missing and large *Diadema mexicanum* were nestled in depressions on the summits of colonies with some individuals grazing on eroded surfaces. It was clear that *Diadema* was responsible for this erosion. In addition, numerous sexual recruits of *P. lobata* were observed on the dead areas of massive colonies affected by the 1982–83 El Niño disturbance (Glynn 1994). The sizes of the larger of these young colonies measured 25 cm in diameter, indicating their probable recruitment shortly after 1983. Unlike the southern and central Galápagos Islands, *Eucidaris* was much less abundant at Darwin Island (mean density 1.3 ind m^{-2} versus 10–50 ind m^{-2})

[12] The median thickness of the Uva *Pocillopora* reef, determined by core drilling, was 5.2 m, and that of 11 Galápagos *Pocillopora* reefs 1.0 m, based on rebar (iron rod) probing and visual estimates (see Table 24 in Glynn and Wellington 1983). Reef erosion at Uva Island was based on 7 surveys totaling 139 measurements over 16 years; mean vertical height loss = 2.86 (± 0.65, 95% confidence interval) cm y^{-1}. Galápagos reef decline was observed during annual surveys of 11 patch reefs from 1983 to 1987 with estimated mean height losses of 20–30 cm y^{-1}.

[13] The orientation of Darwin Island is incorrectly illustrated in Fig. 42 in Glynn and Wellington (1983). The island and surrounding submarine shelf should be rotated counterclockwise by about 45°. The reef lies along a NW-SE axis extending from near the island to Darwin's Arch (Glynn et al. 2015).

with most sea urchins sheltered in coral cavities and fissures. The bioerosion we observed during our daylight surveys was caused mainly by *Diadema* and to a lesser extent by *Eucidaris* and fishes (chiefly pufferfishes and trigger fishes). We suspected, but did not verify, that *Eucidaris* mostly emerged at night to graze on exposed dead carbonate surfaces. Since the 2- to 3-meter-high *Porites*/*Pavona* framework towers are still standing we do not believe that internal bioerosion is currently contributing substantially to reef degradation at Darwin Island.[14]

The factors contributing to the rapid recovery of the Wellington Reef are of considerable interest because they may shed light on the enduring persistence of this reef. Possible local conditions supporting reef recovery are: (a) relatively low abundance of echinoids, likely due to predation by abundant fish invertivore populations; (b) rapid recruitment of massive, frame-building coral species; and (c) sea water chemistry. Manzello et al. (2014) described an acidification gradient across the Galápagos Archipelago that correlated with reef development and structural persistence. This gradient extended from south to north with low pH and Ω_{arag} values associated with degraded reefs at southern island sites and elevated pH and Ω_{arag} at Darwin Island. A concerning trend related to present global warming is the projected increase in the frequency of extreme El Niño events (e.g., Cai et al. 2014; Pathirana et al. 2022). If this projection is realized coral mortality would follow and possibly increasing abundances of *Eucidaris*. A modelling study of the Wellington reef gives a high probability of recovery at present *Eucidaris* densities (of about 1 ind m^{-2}), but if sea urchin abundances increase above 4–5 ind m^{-2} reef degradation becomes more probable (Glynn et al. 2017[15]).

Bioerosion can be beneficial to reef growth. As noted earlier, when feeding on *Pocillopora* corals the Guineafowl pufferfish contributes importantly to fragmentation and asexual reproduction. Eastern Pacific triggerfishes also promote the asexual propagation of massive corals such as *Porites* and *Pavona*. At the Saboga Island reef (Pearl Islands, Gulf of Panama) a curious spatial pattern of youthful corals—some spherical, others discoidal or irregularly shaped—attracted our attention. They were reminiscent of a mushroom fairy ring, arranged in a circle. Unlike mushrooms, however, the numerous small corals were unattached and surrounded a large centenarian colony of *Porites lobata*. Inspection of this centrally located colony revealed numerous bite marks by *Pseudobalistes naufragium* (Fig. 7.1). This triggerfish commonly bites off pieces of the coral to expose and eat endolithic boring bivalves, a feeding behavior observed at several reef sites in the eastern Pacific (Glynn et al.

[14] Surveys by Inti Keith (director of the Charles Darwin Research Station) and her co-workers have reported, as of March 2022, that the structural integrity of the Wellington reef is still intact. However, an invasive green alga, *Caulerpa chemnitzia*, has appeared and covered corals, causing mortality that could increase bioerosion (Keith et al. 2022). Inti, in a recent 2022 communication to me, noted that this alga has partially died back.

[15] Peter J. Glynn (my son), a Mathematics Specialist at the Crane Country Day School in Santa Barbara, California, showed an early interest in coral reefs when we resided in the Republic of Panama. He has recently joined our research team, introducing modeling methods to probe the effects of bioerosion on post-bleaching coral reef recovery.

1972, 1982; Guzmán 1986; Guzmán and Cortés 1989; Cantera et al. 2001). Héctor Guzmán (1988) also observed another triggerfish, *Sufflamen verres*, biting into corals on Costa Rican reefs. The numerous bite marks on the enormous colony at Saboga Island matched the dentition of *Pseudobalistes*, and the skeletons of some of the small, loose colonies also revealed bite scars. The numerous *Porites* colonies surrounding the central "donor" colony is an example of how bioerosion can play a positive role affecting reef growth, in this case by spreading asexual propagules and expanding coral cover.

Pesky Damselfishes

An indirect effect of bioerosion that immediately captured our attention involved the aggressive defensive behavior of damselfishes. Any astute observer diving on an eastern Pacific *Pocillopora* reef will soon encounter damselfish defending their territorial lawns. Mark Eakin (1987, 1988), engaged in his doctoral research on the Uva Island reef (Fig. 9.3), demonstrated that damselfish readily eject sea urchins (*Diadema mexicanum*) from their territories, thus significantly reducing the pressure of a potent external bioeroder. It was amusing to watch Mark dangle *Diadema* over damselfish lawns (attached to a line at the tip of a fishing rod) to elicit defensive responses. This procedure was employed to assess the defensive behavior of a range of damselfish sizes as well as sea urchin sizes and their proximity to damselfish lawns. The defensive behavior of damselfishes was also observed on *Pocillopora* reefs in the Galápagos Islands (Glynn and Wellington 1983; Glynn 1988). Interestingly, although this defensive behavior was elicited by different damselfish and sea urchin species at distant sites it was performed similarly. Damselfish ejection involved mouthing and biting sea urchin spines and dragging or swimming with them beyond their defended territories (Fig. 11.4). Damselfish also nudged approaching *Diadema* with their mouths and engaged in "tail beating". In executing tail beating, damselfish would maneuver toward sea urchins tail first and then forcefully whip their tail fin across the leading spines of the approaching sea urchins. Using a similar repertoire of ejection strategies, damselfishes defend corals from *Acanthaster* attacks within and near their territories (Glynn and Colgan 1988). These defensive behaviors were also observed in the South Pacific (Guam, American Samoa), Okinawa, and the northern Indian Ocean (Oman).

The effects of bioeroding fishes, such as parrotfish, are often reduced by damselfish that typically chase the intruding herbivores away from their algal lawns. This defensive behavior by damselfish, however, has been shown to increase bioerosion by other biological agents. In the western Pacific (Great Barrier Reef), higher bioerosion rates, due primarily to clionaid sponges, were reported inside compared with outside damselfish territories (Risk and Sammarco 1982; Sammarco et al. 1986). Bioeroding sponges were less abundant outside of damselfish territories due to the high grazing pressure of fish herbivores. The protected lawn habitat inside damselfish territories offers a refuge for sponges and allows the successful settlement and survival of the early stages of diverse internal bioeroders.

Continuing observations of damselfish–sea urchin interactions on the Uva reef prompted a modeling study to explore an increase in bioerosion due to the ejection behavior of damselfish (Glynn et al. 2020). Ejected *Diadema* tend to form aggregations and these tight-nit groups persist and efficiently erode carbonate surfaces outside of damselfish territories. In simulated post El Niño high *Diadema* abundances, this concentration effect caused additional erosion of 0.204 kg m^{-2} y^{-1}. This amounted to a 20% increase of the annual carbonate erosion attributed to *Diadema* according to Mark Eakin's (1996) budget model.

Bioerosion and Biodiversity

Intuitively, one would think that the degradation of coral reefs resulting from bioerosion would negatively impact biodiversity. This is probably true when bioerosion is excessive, degrading reef frameworks into mostly fine-grained sediments. When live or dead coral skeletons and reef frameworks are separated or broken apart a myriad of cryptic species are invariably exposed. Diverse sessile and motile invertebrates and fishes shelter on internal coral surfaces, in coral rubble, and in voids excavated by internal bioeroders (Jackson and Winston 1982; Choi and Ginsburg 1983; Gischer and Ginsburg 1996; Richter et al. 2001; Scheffers et al. 2010). Bob Ginsburg (1983) and others noted this condition since it contributes significantly to the biodiversity of coral reefs.[16] The biodiversity and biomass of the cryptofauna is now understood to exceed that of the visible surface fauna on virtually all coral reefs. Recent studies in the eastern Pacific also have revealed that the cryptofauna makes up most of a reef's metazoan biodiversity. For example, a meticulous assessment of the motile cryptofauna on a coral reef in the Pearl Islands, Panama, revealed more than 121 species in six phyla, representing predator, herbivore, detritivore, filter feeder and scavenger guilds (Enochs and Manzello 2012). Further, the species richness of the motile cryptofauna is higher among dead vs live coral substrates and increases within successively more eroded reef structures up to a maximum level of intermediate degradation. Extreme levels of erosion resulting in sand and silt sediment without sheltering cavities depress cryptofauna population density and diversity.

[16] Bob Ginsburg defined 'cavity' very broadly to include all open spaces, e.g., (1) intraskeletal cavities; (2) growth, framework, and shelter cavities, such as the spaces beneath shells or coral fragments, and caves and networks of cavities of varying size and shape; (3) interstitial pores of sediments, i.e., the spaces between fine and coarse sand and coral rubble; (4) borings, from minute cellular networks (mm^3) to tubules and irregular chambers (cm^3). In addition, Bob's estimate of the total volume of cavities in a living reef ranged from 50% to 75% and as high as 90% for some coral reefs. From our reef drilling experience, the latter high values would not seem excessive for *Pocillopora* reefs considering their complex interlocking branch frameworks. Often our drill string would suddenly drop a half meter or so when internal cavities were encountered.

In a recent review of the coral rubble microhabitat and its support of a high density and diversity of reef organisms, evidence was presented that this habitat could moderate future impacts resulting from coral framework degradation (Wolfe et al. 2021). For example, many of the organisms, biological processes, and ecological functions of coral reefs are sustained in the rubble habitat. These include primary and secondary production, herbivory, predation, nutrient cycling, bioerosion, and recruitment facilitation. The numerous *Pocillopora* spp. colonies recruited onto the forereef rubble substrates of the Uva reef in 2017 were possibly a manifestation of the suitability of this microhabitat in reef recovery.

Conclusion

Bioerosion is a potent process that has likely played an essential role in shaping the nature of eastern Pacific coral reefs—their small size, youth, patchy occurrence, and low coral diversity. Mitchell Colgan, from his taphonomic (death assemblages) investigations of bioerosion of the Urvina Bay coral community, envisioned a scenario of recurrent intense El Niño events that start coral communities on a cycle of death, bioerosion, and recolonization limiting long-term reef growth. El Niño disturbances, acting in concert with other physical and biological forces have prevented the buildup of substantial coral reef frameworks (Colgan 1990). The destruction of incipient coral reefs in the Galápagos Islands in the 1980s, initiated by a strong El Niño event, which led to catastrophic coral mortality followed by intense bioerosion, would appear to support Mitchell's proposed cycle in this area.

Could there be a causal nexus between El Niño activity (event frequency and intensity), bioerosion and coral reef development across the equatorial eastern Pacific? El Niño 1982–83 ocean warming anomalies were more pronounced in the Galápagos Islands than in core equatorial reef areas (e.g., Costa Rica and Panama),[17] resulting in nearly total coral mortality. Intense bioerosion in the Galápagos Islands degraded and eradicated all but one coral reef in less than 5 years. Most coral reefs elsewhere in the eastern Pacific experienced high coral mortality and bioerosion, but eventually recovered and continued to accrete calcium carbonate. El Niño strength, coral mortality, and bioerosion were less strongly felt at higher latitudes in the equatorial eastern Pacific. Is it possible that the thousand-year-old reefs of the core eastern Pacific region versus the hundred-year-old reefs of Galápagos are a result of El Niño-induced coral mortality followed by rapid bioerosion and delayed reef initiation and growth? If global warming leads to more frequent and intense ENSO activity, as projected in several recent studies, it is likely that mature equatorial reefs will suffer a similar fate as the transient Galápagos coral reefs. Such a scenario is not

[17] Eastern Pacific coral reefs demonstrate mature development from southern Mexico south to Colombia, within the oceanographic zone of tropical surface water (TSW). Coral reefs in the central and southern Galápagos Islands, less well developed, lie in the zone of equatorial surface water (ESW), which is subject to equatorial cold tongue intrusions (Fiedler and Lavin 2017).

unprecedented. Lauren Toth et al. (2012) presented compelling evidence of coral reef loss during the Holocene. They found a 2500-year gap in reef growth in Panama, beginning ~4100 years ago, that was contemporaneous with El Niño activity. Could we again be on course for an extended age of coral reef demise, spearheaded by El Niño disturbances, lethal ocean warming, bioerosion and reef loss?

References

Achurra Cárdenas JB, Valdés Araúz E (1980) Estudios sobre los efectos predatorios de los moluscos *Coralliophila abbreviata* y *Jenneria pustulata* de ambos literales del Istmo de Panama. Tesis, Universidad de Panamá, Escuela de Biología. República de Panamá. 66 pages, 24 plates, 10 figures

Adey WH (1978) Algal ridges of the Caribbean Sea and West Indies. Phycologia 17(4):361–367

Adey WH, Burke R (1976) Holocene bioherms (algal ridges and bank barrier reefs) of the eastern Caribbean. Geol Soc Am Bull 87:95–109

Alvarado JJ, Cortés J, Guzman H, Reyes-Bonilla H (2016) Density, size, and biomass of *Diadema mexicanum* (Echinoidea) in eastern tropical Pacific coral reefs. Aquat Biol 24:151–161. https://doi.org/10.3354/ab00645

Alvarado JJ, Grassian B, Cantera-Kintz JR, Carballo JL, Londoño-Cruz E (2017a) Coral reef bioerosion in the eastern tropical Pacific. In: Glynn PW, Manzello DP, Enochs IC (eds) Coral reefs of the eastern tropical Pacific: persistence and loss in a dynamic environment. Coral reefs of the world 8. Springer, Dordrecht, pp 369–403

Alvarado JJ, Aburto-Oropeza O, Abad R, Barraza E, Brandt M, Cantera J, Estrada P, Gaymer CF, Guzmán-Mora AG, Herlan JJ, Maté JL (2017b) Coral reef conservation in the eastern tropical Pacific. In: Glynn PW, Manzello DP, Enochs IC (eds) Coral reefs of the eastern tropical Pacific: persistence and loss in a dynamic environment. Coral reefs of the world 8. Springer, Dordrecht, pp 565–591

Benítez-Villalobos F, Domínguez-Gómez MT, López-Pérez RA (2008) Temporal variation of the sea urchin *Diadema mexicanum* population density at Bahías de Huatulco, Western Mexico. Rev Biol Trop 56(Suppl 3):255–263

Birkeland C (ed) (1997) Life and death of coral reefs. Chapman & Hall, New York, p 536

Birkeland C (ed) (2015) Coral reefs in the anthropocene. Springer, Dordrecht, p 271

Boulay JN, Hellberg ME, Cortés J, Baums IB (2014) Unrecognized coral species diversity masks differences in functional ecology. Proc R Soc B 281:20131580. https://doi.org/10.1098/rspb.2013.1580

Broecker WS, Takahashi T, Simpson HJ, Peng TH (1979) Fate of fossil fuel carbon dioxide and the global carbon budget. Science 206:409–418

Budd AF, Guzman HM (1994) *Siderastrea glynni*, a new species of scleractinian coral (Cnidaria, Anthozoa) from the eastern Pacific. Proc Biol Soc Wash 107:591–599

Cai W, Borlace S, Lengaigne M, van Rensch P, Collins M, Vecchi G, Timmermann A, Santoso A, McPhaden MJ, Wu L, England MH, Wang G, Guilyardi E, Jin F-F (2014) Increasing frequency of extreme El Niño events due to greenhouse warming. Nat Climate Change 4:111–116. https://doi.org/10.1038/nclimate2100

Cantera JR, Zapata F, Forero P, Francisco V, Jiménez JM, Londoño E, Narvaez K, Neira R, Orozco CA, Toro-Farmer G (2001) Organismos bioerosionadores en arrecifes de Isla Gorgona. In: Barríos LM, López-Victoria M (eds) Gorgona Marina: Contribución al Conocimiento de una Isla Única. Serie Publicaciones Especiales 7. INVEMAR, Magdalena, pp 51–64

Carballo JL, Bautista E, Nava H, Cruz-Barraza JA, Chávez JA (2013) Boring sponges, an increasing threat for coral reefs affected by bleaching events. Ecol Evol 3(4):872–886. https://doi.org/10.1002/ece3.452

Choi DR, Ginsburg RN (1983) Distribution of coelobites (cavity dwellers) in coral rubble across the Florida Reef Tract. Coral Reefs 2:165–172

Colgan MW (1990) El Niño and the history of eastern Pacific reef building. In: Glynn PW (ed) Global ecological consequences of the 1982–83 El Niño-Southern Oscillation. Elsevier oceanography series 52. Elsevier, Amsterdam, pp 183–232

Cortés J (1991) Los arrecifes coralinos de Golfo Dulce, Costa Rica: aspectos geológicos. Revista Geología América Central 13:15–24

Eakin CM (1987) Damselfishes and their algal lawns: a case of plural mutualism. Symbiosis 4:275–288

Eakin CM (1988) Avoidance of damselfish lawns by the sea urchin *Diadema mexicanum* at Uva Island, Panama. In: Proceedings of the 6th international coral reef symposium, Townsville, vol 2, pp 21–26

Eakin CM (1991) The damselfish-algal lawn symbiosis and its influence on the bioerosion of an El Niño impacted coral reef, Uva Island, Pacific Panama. Ph.D. dissertation. University of Miami, p 158

Eakin CM (1992) Post-El Niño Panamanian reefs: less accretion, more erosion and damselfish protection. In: Proceedings of the 7th international coral reef symposium, Guam, vol 1, pp 387–396

Eakin CM (1996) Where have all the carbonates gone? A model comparison of calcium carbonate budgets before and after the 1982–1983 El Niño at Uva Island in the eastern Pacific. Coral Reefs 15:109–119

Eakin CM (2001) A tale of two ENSO events: carbonate budgets and the influence of two warming disturbances and intervening variability, Uva Island, Panama. Bull Mar Sci 69(1):171–186

Edgar GJ, Banks SA, Brandt M, Bustamante RH, Chiriboga Á, Earle SA, Garske LE, Glynn PW, Grove JS, Henderson S, Hickman CP, Miller KA, Rivera F, Wellington GM (2010) El Niño grazers and fisheries interact to greatly elevate extinction risk for Galapagos marine species. Glob Change Biol 16:2876–2890

Enochs IC, Manzello DP (2012) Species richness of motile cryptofauna across a gradient of reef framework erosion. Coral Reefs 31(3):653–661

Enochs IC, Manzello DP, Kolodziej G, Noonan SHC, Valentino L, Fabricius KE (2016a) Enhanced macroboring and depressed calcification drive net dissolution at high-CO_2 coral reefs. Proc R Soc B 283:20161742. https://doi.org/10.1098/rspb.2016.1742

Enochs IC, Manzello DP, Tribollet A, Valentino L, Kolodziej G, Donham EM, Fitchett MD, Carlton R, Price NN (2016b) Elevated colonization of microborers at a volcanically acidified coral reef. PLoS One 11(7):e0159818. https://doi.org/10.1371/journal.pone.0159818

Enochs IC, Toth LT, Kirkland A, Manzello DP, Kolodziej G, Morris JT, Holstein DM, Schlenz A, Randall CJ, Maté JL, Leichter JJ, Aronson RB (2021) Upwelling and the persistence of coral reef frameworks in the eastern tropical Pacific. Ecol Monogr 91(4):e01482. https://doi.org/10.1002/ecm.1482

Fiedler PC, Lavin MF (2017) Oceanographic conditions of the eastern tropical Pacific. In: Glynn PW, Manzello DP, Enochs IC (eds) Coral reefs of the eastern tropical Pacific: persistence and loss in a dynamic environment. Coral reefs of the world 8. Springer, Dordrecht, pp 59–83

Fonseca AC, Cortés J (1998) Coral borers of the eastern Pacific: *Aspidosiphon* (*A.*) *elegans* (Sipuncula: Aspidosiphonidae) and *Pomatogebia rugosa* (Crustacea: Upogebiidae). Pac Sci 52(2):170–175

Fonseca AC, Dean HK, Cortés J (2006) Non-colonial coral macro-borers as indicators of coral reef status in the south Pacific of Costa Rica. Rev Biol Trop 54:101–115

Forsman ZH, Guzmán HM, Chen CA, Fox GE, Wellington GM (2005) An ITS region phylogeny of *Siderastrea* (Cnidaria, Anthozoa): is *S. glynni* endangered or introduced? Coral Reefs 24:343–347

Gattuso JP, Allemand D, Frankignoulle M (1999) Photosynthesis and calcification at cellular, organismal and community levels in coral reefs: a review on interactions and control by carbonate chemistry. Am Zool 39:160–183

Gilchrist SL (1985) Hermit crab corallivore activity. In: Proceedings of the 5th international coral reef congress, Tahiti, vol 5, pp 211–214

Ginsburg RN (1983) Geological and biological roles of cavities in coral reefs. In: Barnes DJ (ed) Perspectives on coral reefs. Australian Institute of Marine Science, Townsville. Brian Clouston, Manuka, pp 148–153

Gischer E, Ginsburg RN (1996) Cavity dwellers (coelobites) under coral rubble in southern Belize barrier and atoll reefs. Bull Mar Sci 58:570–589

Glynn PJ, Glynn PW, Riegl B (2017) El Niño, echinoid bioerosion and recovery potential of an isolated Galápagos coral reef: a modeling perspective. Mar Biol 164:1–17

Glynn PJ, Glynn PW, Maté J, Riegl B (2020) Agent-based model of eastern Pacific damselfish and sea urchin interactions shows increased coral reef erosion under post-ENSO conditions. Ecol Model 423. https://doi.org/10.1016/j.ecolmodel.2020.108999

Glynn PW (1973) Aspects of the ecology of coral reefs in the Western Atlantic region. In: Jones OA, Endean R (eds) Biology and geology of coral reefs, vol 2, Biol 1. Academic Press, New York, pp 271–324

Glynn PW (1985) Corallivore population sizes and feeding effects following El Niño (1982–1983) associated coral mortality in Panama. In: Proceedings of the 5th international coral reef congress, Tahiti, vol 4, pp 183–188

Glynn PW (1988) El Niño warming, coral mortality and reef framework destruction by echinoid bioerosion in the eastern Pacific. Galaxea 7(2):129–160

Glynn PW (1990) Coral mortality and disturbances to coral reefs in the tropical eastern Pacific. In: Glynn PW (ed) Global ecological consequences of the 1982–83 El Niño-Southern Oscillation. Elsevier oceanography series 52. Elsevier, Amsterdam, pp 55–126

Glynn PW (1994) State of coral reefs in the Galápagos Islands: natural vs anthropogenic impacts. Mar Pollut Bull 29(1–3):131–140

Glynn PW (2004) High complexity food webs in low-diversity eastern Pacific reef-coral communities. Ecosystems 7:358–367. https://doi.org/10.1007/s10021-004-0184-x

Glynn PW (2008) Food-web structure and dynamics of eastern tropical Pacific coral reefs: Panamá and Galápagos Islands. In: McClanahan TR, Branch GM (eds) Food webs and the dynamics of marine reefs. Oxford University Press, New York, pp 185–208

Glynn PW, Colgan MW (1988) Defense of corals and enhancement of coral diversity by territorial damselfishes. In: Proceedings of the 6th international coral reef symposium, Australia, vol 2, pp 157–163

Glynn PW, Manzello DP (2015) Bioerosion and coral reef growth: a dynamic balance. In: Birkeland C (ed) Coral reefs in the anthropocene. Springer, Dordrecht, pp 67–97. https://doi.org/10.1007/978-94-017-7249-5_4

Glynn PW, Wellington GM (1983) Corals and coral reefs of the Galápagos Islands. With an annotated list of the scleractinian corals of the Galápagos by John W. Wells. University of California Press, Berkeley, p 330

Glynn PW, von Prahl H, Guhl F (1982) Coral reefs of Gorgona Island, Colombia, with special reference to corallivores and their influence on community structure and reef development. Anales del Instituto de Investigaciones Marinas de Punta de Betín 12:185–214

Glynn PW, Stewart RH, McCosker JE (1972) Pacific coral reefs of Panama: structure, distribution and predators. Geol Rundsch Int J Earth Sci 61(2):483–519

Glynn PW, Wellington GM, Birkeland C (1979) Coral reef growth in the Galápagos: limitation by sea urchins. Science 203:47–49

Glynn PW, Cortés J, Guzmán HM, Richmond RH (1988) El Niño (1982–83) associated coral mortality and relationship to sea surface temperature deviations in the tropical eastern Pacific. In: Proceedings of the 6th international coral reef symposium, Townsville, vol 3, pp 237–243

Glynn PW, Riegl B, Correa AMS, Baums IB (2009) Rapid recovery of a coral reef at Darwin Island, Galápagos Islands. Galapagos Res 66:6–13

Glynn PW, Grassian B, Kleemann KH, Maté JL (2016) The true identity of *Siderastrea glynni* Budd & Guzman, 1994, a highly endangered eastern Pacific scleractinian coral. Coral Reefs 35:1399–1404. https://doi.org/10.1007/s00338-016-1470-8

Glynn PW, Riegl B, Purkis S, Kerr JM, Smith TB (2015) Coral reef recovery in the Galápagos Islands: the northern-most islands (Darwin, Wenman). Coral Reefs 34:421–436. https://doi.org/10.1007/s00338-015-1280-4

Glynn PW, Feingold JS, Baker A, Banks S, Baums IB, Cole J, Colgan MW, Fong P, Glynn PJ, Keith I, Manzello D (2018) State of corals and coral reefs of the Galápagos Islands (Ecuador): past, present and future. Mar Pollut Bull 133:717–733

Guzmán HM (1986) Estructura de la comunidad arrecifal de la Isla del Caño, Costa Rica, y el efecto de perturbaciones naturales severas. M.Sc. thesis. Universidad de Costa Rica, San Pedro, Costa Rica, p 179

Guzmán HM (1988) Distribución y abundancia de organismos coralívoros en los arrecifes coralinos de la Isla del Caño, Costa Rica. Rev Biol Trop 36(2A):191–207

Guzmán HM, Cortés J (1989) Coral reef community structure at Caño Island, Pacific Costa Rica. Pubbl Staz Zool Napoli (I: Mar Ecol) 10:23–41

Guzmán HM, López JD (1991) Diet of the corallivorous pufferfish *Arothron meleagris* (Pisces: Tetraodontidae) at Gorgona Island, Colombia. Rev Biol Trop 39(2):203–206

Guzmán HM, Robertson DR (1989) Population and feeding responses of the corallivorous pufferfish *Arothron meleagris* to coral mortality in the eastern Pacific. Mar Ecol Prog Ser 55:121–131

Herrera-Escalante T, López-Pérez RA, Leyte-Morales GE (2005) Bioerosion caused by the sea urchin *Diadema mexicanum* (Echinodermata: Echinoidea) at Bahías de Huatulco, Western Mexico. Rev Biol Trop 53(Suppl. 3):263–273

Highsmith RC (1980) Passive colonization and asexual colony multiplication in the massive coral *Porites lutea* Milne Edwards & Haime. J Exp Mar Biol Ecol 47:55–67

Hutchings PA (1986) Biological destruction of coral reefs. Coral Reefs 4:239–252

Hutchings PA (2011) Bioerosion. In: Hopley D (ed) Encyclopedia of modern coral reefs: structure, form, and process. Springer, Berlin, pp 139–156

Hutchings PA, Peyrot-Clausade M, Osorno A (2005) Influence of land runoff on rates and agents of bioerosion of coral substrates. Mar Pollut Bull 51:438–447. https://doi.org/10.1016/j.marpolbul.2004.10.044

Jackson JBC, Winston JE (1982) Ecology of cryptic coral reef communities I: Distribution and abundance of major groups of encrusting organisms. J Exp Mar Biol Ecol 64:103–115

Jiménez C (1996) Coral colony fragmentation by whitetip reef sharks at Coiba Island National Park, Panamá. Rev Biol Trop 45(1B):698–700

Keith I, Bensted-Smith W, Banks S, Suarez J, Riegl B (2022) *Caulerpa chemnitzia* in Darwin threatening Galapagos coral reefs. PLoS One 17(8):e0272581. https://doi.org/10.1371/journal.Pone.0272581

Kleemann K (1990) Boring and growth in chemically boring bivalves from the Caribbean, eastern Pacific and Australia's Great Barrier Reef. Senckenbergiana Maritima 21:101–154

Kleypas JA, McManus JW, Meñez LAB (1999) Environmental limits to coral reef development: where do we draw the line? Am Zool 39:146–159

Kroeker KJ, Kordas RL, Crim R, Hendriks IE, Ramajo L, Singh GS, Duarte CM, Gattuso J-P (2013) Impacts of ocean acidification on marine organisms: quantifying sensitivities and interaction with warming. Glob Change Biol 19:1884–1896. https://doi.org/10.1111/gcb.12179

Kuffner IB, Andersson AJ, Jokiel PL, Rodgers KUS, Mackenzie FT (2008) Decreased abundance of crustose coralline algae due to ocean acidification. Nat Geosci 1(2):114–117

Langdon C, Atkinson MJ (2005) Effect of elevated pCO_2 on photosynthesis and calcification of corals and interactions with seasonal change in temperature/irradiance and nutrient enrichment. J Geophys Res 110:1–16

Lirman D, Glynn PW, Baker AC, Leyte Morales GE (2001) Combined effects of three sequential storms on the Huatulco coral reef tract, Mexico. Bull Mar Sci 69(1):267–278

Liu G, Strong AE, Skirving W, Arzayus LF (2006) Overview of NOAA coral reef watch program's near-real-time satellite global coral bleaching monitoring activities. In: Proceedings of the 10th international coral reef symposium, Okinawa, vol 2, pp 1783–1793

Lowenstam HA, Weiner S (1989) On biomineralization. Oxford University Press, New York, p 336

Macintyre IG, Glynn PW (1976) Evolution of modern Caribbean fringing reef, Galeta Point, Panamá. Am Assoc Petrol Geol Bull 60(7):1054–1072

Manzello DP, Enochs IC, Bruckner A, Renaud PG, Kolodziej G, Budd DA, Carlton R, Glynn PW (2014) Galápagos coral reef persistence after ENSO warming across an acidification gradient. Geophys Res Lett 41. https://doi.org/10.1002/2014GL062501

Marcus MA, Amini S, Stifler CA, Sun C-Y, Tamura N, Bechtel HA, Parkinson DY, Barnard HS, Zhang XXX, Isiah Chua JQ, Miserez A, Gilbert PUPA (2017) Parrotfish teeth: stiff biominerals whose microstructure makes them tough and abrasion-resistant to bite stony corals. ACS Nano 11:11856–11865. https://doi.org/10.1021/acsnano.7b05044

Morais J, Morais R, Tebbett SB, Bellwood DR (2020) On the fate of dead coral colonies. Funct Ecol 00:1–13. https://doi.org/10.1111/1365-2435.14182

Muñoz Bernal CG (2017) Modos de reclutamiento coralino y efectos de la depredación por *Arothron meleagris* en corales pocilopóridos de Isla Gorgona. Doctoral thesis. Universidad del Valle, Colombia, p 85

Nava H, Carballo JL (2008) Chemical and mechanical bioerosion of boring sponges from Mexican Pacific coral reefs. J Exp Biol 211(17):2827–2831

Neumann AC (1966) Observations on coastal erosion in Bermuda and measurements of the boring rate of the sponge, *Cliona lampa*. Limnol and Oceanogr 11(1):92–108

Oramas NFA (1979) Estudio ecológico del *Jenneria pustulata* (Mollusca, Gastropoda). Tesis, Universidad de Panamá, Escuela de Biología. República de Panamá. 100 pages, 21 tables, 29 figures

Orr JC, Fabry VJ, Aumont O, Bopp L, Doney SC, Feely RA, Gnanadesikan A, Gruber N, Ishida A, Joos F, Key RM, Lindsay K, Maier-Reimer E, Matear R, Monfray P, Mouchet A, Najjar RG, Plattner GK, Rodgers KB, Sabine CL, Sarmiento JL, Schlitzer R, Slater RD, Totterdell IJ, Weirig MF, Yamanaka Y, Yool A (2005) Anthropogenic ocean acidification over the twenty-first century and its impact on calcifying organisms. Nature 437(7059):681–686. https://doi.org/10.1038/nature04095

Osorno A, Peyrot-Clausade M, Hutchings PA (2005) Patterns and rates of erosion in dead *Porites* across the Great Barrier Reef (Australia) after 2 years and 4 years of exposure. Coral Reefs 24:292–303. https://doi.org/10.1007/s00338-005-0478-2

Palacios MM, Muñoz CG, Zapata FA (2014) Fish corallivory on a pocilloporid reef and experimental coral responses to predation. Coral Reefs 33:625–636. https://doi.org/10.1007/s00338-014-1173-y

Pari N, Peyrot-Clausade M, Le Campion-Alsumard T, Hutchings PA, Chazottes V, Golubic S, Le Campion T, Fontaine MF (1998) Bioerosion of experimental substrates on high islands and atoll lagoons (French Polynesia) after 2 years of exposure. J Exp Mar Ecol Prog Ser 166:119–130

Pathirana G, Oh JH, Cai W, An S-I, Min S-K, Jo S-Y, Shin J, Kug J-S (2022) Explosive increase in convective extreme El Niño events in the CO_2 removal scenario. Res Square. https://doi.org/10.21203/rs.3.rs-1938270/v1

Reaka-Kudla ML, Feingold JS, Glynn PW (1996) Experimental studies of rapid bioerosion of coral reefs in the Galápagos Islands. Coral Reefs 15:101–107

Reyes-Bonilla H (2003) Coral reefs of the Pacific coast of Mexico. In: Cortés J (ed) Latin American coral reefs. Elsevier, Amsterdam, pp 331–349

Richter C, Wunsch M, Rasheed M, Kötter I, Badvan MI (2001) Endoscopic exploration of Red Sea coral reefs reveals dense populations of cavity-dwelling sponges. Nature 413:726–730

Risk MJ, Sammarco PW (1982) Bioerosion of corals and the influence of damselfish territoriality: a preliminary study. Oecologia 52:376–380

Robinson G (1985) Influence of the 1982–83 El Niño on Galápagos marine life. In: Robinson G, del Pino EM (eds) El Niño in the Galápagos Islands: the 1982–1983 event. Charles Darwin Foundation for the Galápagos Islands, Quito, Ecuador, pp 153–190

Ruttenberg BI (2001) Effects of artisanal fishing on marine communities in the Galápagos Islands. Conserv Biol 15:1691–1699

Sammarco PW, Carleton JH, Risk MJ (1986) Effects of grazing and damselfish territoriality on bioerosion of dead corals: direct effects. J Exp Mar Biol Ecol 98(1–2):1–19. https://doi.org/10.1016/0022-0981(86)90072-9

Scheffers SR, Van Soest RWM, Nieuwland G, Bak RPM (2010) Coral reef framework cavities: is functional similarity reflected in composition of the cryptic macrofaunal community? Atoll Res Bull 583:1–24

Schönberg CH (2008) A history of sponge erosion: from past myths and hypotheses to recent approaches. In: Current Developments in Bioerosion, pp 165–202

Schönberg CHL, Wilkinson CR (2001) Induced colonization of corals by a clionid bioeroding sponge. Coral Reefs 20:69–76

Scott PJB (1988) Distribution, habitat and morphology of the Caribbean coral- and rock-boring bivalve, *Lithophaga bisulcata* (d'Orbigny) (Mytilidae: Lithophaginae). J Mollusc Stud 54:83–95

Scott PJB, Risk MJ, Carriquiry JD (1988) El Niño bioerosion and the survival of east Pacific reefs. In: Proceedings of the 6th international coral reef symposium, Townsville, vol 2, pp 517–520

Silbiger NJ, Guadayol Ó, Thomas FIM, Donahue MJ (2014) Reefs shift from net accretion to net erosion along a natural environmental gradient. Mar Ecol Prog Ser 515:33–44. https://doi.org/10.3354/meps10999

Toth LT, Aronson RB, Vollmer SV, Hobbs JW, Urrego DH, Cheng H, Enochs IC, Combosch DJ, Van Woesik R, Macintyre IG (2012) ENSO drove 2500-year collapse of eastern Pacific coral reefs. Science 337(6090):81–84

Tribollet A, Golubic S (2005) Cross-shelf differences in the pattern and pace of bioerosion of experimental carbonate substrates exposed for 3 years on the northern Great Barrier Reef, Australia. Coral Reefs 24:422–434

Tribollet A, Golubic S (2011) Reef bioerosion: agents and processes. In: Dubinsky Z, Stambler N (eds) Coral reefs: an ecosystem in transition. Springer, Dordrecht, pp 435–449. https://doi.org/10.1007/978-94-007-0114-4_25

Tribollet A, Godinot C, Atkinson MJ, Langdon C (2009) Effects of elevated pCO_2 on dissolution of coral carbonates by microbial euendoliths. Global Biogeochem Cycles 23:GB3008. https://doi.org/10.1029/2008GB003286

Tudhope AW, Risk MJ (1985) Rate of dissolution of carbonate sediments by microboring organisms, Davies Reef, Australia. J Sediment Petrol 55:440–447

Wolfe K, Kenyon TM, Mumby PJ (2021) The biology and ecology of coral rubble and implications for the future of coral reefs. Coral Reefs 40(6):1769–1806 https://doi.org/10.1007/s00338-021-02185-9

Chapter 10
Biodiversity: The Seen and Unseen Biota

Abstract Relatively few species compared with other coral reef regions characterized eastern Pacific reef ecosystems. Nonetheless, all ecological functional groups necessary for coral reef formation, growth, and persistence are represented. These include reef building corals and crustose coralline algae, symbiotic zooxanthellae, reef fishes and invertebrates of diverse trophic guilds and ecological functions. Visual surveys, baited traps, and artificial reef structures were employed to help assess coral community diversity. Reported estimates of species richness are: (a) zooxanthellate scleractinian and *Millepora* hydrocorals, ~50 species; (b) symbiotic zooxanthellae, 15 genera, 100 s of species; (c) coral reef fishes, ~500 species; (d) motile cryptic invertebrates, ~290 species. Eastern Pacific reef-building corals and many reef-associated invertebrates and fishes are of Indo-Pacific origin, with virtually no species shared with Caribbean coral reefs. Numerous cryptic metazoan predators likely strongly influence species attempting to recruit and shelter in *Pocillopora* reef communities. These groups vary greatly in abundance during present day coral mortality and reef degradation. How such changes affect biodiversity, ecological functions, and coral reef survival need urgent study.

Introduction

Biodiversity is a term used to sum up all the species present globally or those present in a particular ecosystem such as a rainforest or coral reef.[1] **Species richness** is a commonly used term for the number of species in a specific area or habitat. Such a collection of species is invariably taxonomically broad, including vertebrate and invertebrate animals (metazoans), plants, fungi, and diverse microorganisms such as protists, bacteria, and Archaea. All coral reefs display a great variety of species when viewed superficially, e.g. by swimming over a reef; a reef's true diversity,

[1] There are several levels or types of species or genetic biodiversity, e.g., α-, β-, γ-diversity (local, among communities, and regional, respectively), that are defined and discussed in relation to coral reef ecosystems by Walter Goldberg (2013).

© The Author(s), under exclusive license to Springer Nature Switzerland AG 2025

P. W. Glynn, *Coral Reef Resilience in the Anthropocene*,
https://doi.org/10.1007/978-3-031-81421-1_10

however, concealed within cavities (among colonies or embedded in coral skeletons), sediments and within visible organisms, is unseen by the casual observer. This chapter focuses on the diversity of coral reef metazoans, small and large, visible and hidden (cryptic), one of the primary subjects of our research program in the eastern Pacific. A brief mention is extended to the critically important microbial community associated with reef-building corals. Reef associated plants (e.g., fleshy macroalgae, crustose coralline algae, algal turfs, and rhodoliths) were also briefly introduced in Chap. 1, and considering diversity, more extensively discussed by Cortés et al. (2017) and Fong et al. (2017).

Remarkably, most of Earth's species are unknown or undiscovered. An early quantitative estimate of the number of species (extrapolated from known species of macroscopic metazoans and algae) present in global coral reefs is notably high and divergent, ranging from 618,000 to 9,477,000 (Reaka-Kudla 1997). This estimated range of species richness is markedly high. Other extrapolations, some based on different reefs and methods, have overlapped, and generally supported the extraordinarily high diversity and variance estimates of Marjorie Reaka-Kudla (Small et al. 1998; Bouchet 2006; Mora et al. 2011). A more recent estimate of global coral reef richness by Fisher et al. (2015) proposes 830,000 (550,000—1,330,000, 95% credible limits) eucaryotic species, excluding fungi. The uncertainty associated with these differing estimates relates to the enormous number of undiscovered species, with some estimates >96% of those presently known (Scheffers et al. 2012).[2] Most of these unknown species are small ($\leq$ 1 cm) and cryptic. Of the 34 recognized metazoan phyla, at least 31 are known to contain species inhabiting coral reefs (Paulay 1997; Reaka-Kudla 1997). The ecological functions of reef species are manifold, including structural species (contributing to calcification and reef consolidation), habitat providers, corallivores, detritivores and scavengers, symbionts, parasites, and bioeroders (e.g., Stella et al. 2010; Glynn 2011, 2013; Glynn and Enochs 2011; Cortés et al. 2017). Since coral reefs occupy a very small area of the earth's oceans, only 0.2% by one estimate (Reaka-Kudla 1997), it is possible that coral reefs are the most biologically diverse ecosystem, per unit area, on Earth. A high proportion of the mobile cryptofauna of *Pocillopora* reefs and rubble in Panama could not be identified (Enochs 2012; Enochs and Manzello 2012a) and probably consisted of some undescribed species (Fig. 10.1).

Arriving at a precise number of species in any ecosystem is unattainable. Yet recent advances indicate that it is possible to appreciably narrow the wide range of

[2]The Census of Marine Life, was a 10-year (2000–2010) international effort to catalog Earth's marine biodiversity to help assess our planet's extraordinary species diversity. This program involved over 2700 scientists and countless students from 80 countries, resulting in the naming of >1200 new species (with additional descriptions still in progress), and the creation of the International Ocean Biogeographic Information System (iOBIS) database. We enthusiastically contributed to the marine census in the Soldier Key area in southeastern Florida during the Saltwater Semester at RSMAES. Nancy Knowlton's popular account of the *Citizens of the Sea* (Knowlton 2010) is an excellent introduction to the marine census with exquisite images of marine life, including many coral reef species.

Fig. 10.1 Collage of mobile cryptofauna extracted from coral reef frameworks and rubble (*Pocillopora* spp.) at Uva reef in the Gulf of Chiriquí (2010), and Playa Larga reef in the Gulf of Panama (2008). See Enochs and Hockensmith 2009; Enochs 2012. Courtesy of Ian C. Enochs

estimates presently available. Nancy Knowlton and colleagues of the CReefs[3] project are taking coral reef biodiversity studies to a new level (Knowlton et al. 2010). The latest advances embrace (a) the application of molecular methods identifying organisms by genetic barcoding, and (b) the use of systematic quantitative sampling with standardized units such as autonomous reef monitoring structures (ARMS) for monitoring changes in species diversity. The CReefs selected study sites occur widely across the Pacific (including Panama), Indian Ocean, and Western Atlantic coral reef regions.

A general concern regarding the biodiversity of coral reefs, summarized by Marjorie Reaka-Kudla (2005), is the high extinction risk of the primarily unknown metazoans of coral reefs, namely the small cryptobiome. The life history traits of these species—low population densities, limited geographic ranges, and habitat specialists—all contribute to their rarity and vulnerability to disturbances. It is possible that tens of thousands of these species have already perished since the recent

[3] CReefs is the coral reef component of the Census of Marine Life, an international cooperative network launched by the Australian Institute of Marine Science (AIMS) in 2005. Three institutional nodes were hosted over a 10-year period, namely AIMS, the Smithsonian Institution, and the Pacific Island Fisheries Center of NOAA (see aims.gov.au/docs/about/partnerships/creefs.html). Taxonomists from more than eighty countries were brought together to assess and review the diversity, distribution and abundance of present-day marine life.

world-wide degradation and loss of coral reefs provoked attention in the early 1980s. Jeremy Jackson proffered an estimate of coral reef decline during the Anthropocene, involving degradation and loss of 60% of known global coral reefs, not long ago (Jackson 2008). In addition, Carpenter et al. (2008) warned that one-third of the world's reef-building corals face extinction due to present day disturbances. Such ongoing and potential losses, with accompanying Centinelan extinctions,[4] add to our uncertainty of biodiversity in natural ecosystems and ignorance of the ecological functions of forever-missing species.

Why should we care about the biodiversity of coral reefs? Something is intriguing about species-rich communities that arouses our curiosity: the immense variety of species' body plans, their behaviors, evolution and diverse phylogenies, and ecological interactions and functions. Species diversity, environmental disturbances, and the complexity of species interactions relating to community stability have been at the forefront of ecological investigations for several decades (e.g., Paine 1966; Dayton 1971; May 1973). Joe (Joseph H.) Connell advanced a compelling hypothesis in 1978, the "intermediate disturbance hypothesis" posited that moderate (intermediate) disturbance regimes could elevate community diversity (Connell 1978). Factors influencing community structure and stability, particularly in tropical forests and coral reefs, were a topic of intense ecological research and a focus of study at the Smithsonian Tropical Research Institute (STRI). As noted in Chap. 7, "Corallivores and General Predation", Jim Porter and I were engaged in investigating the small scale or local effects of *Acanthaster* predation on coral community diversity. In the context of (a) intermediate disturbance elevating diversity, (b) diversity underpinning high complexity, and (c) high complexity promoting community stability, species diversity is a critical ecological cornerstone demanding attention.

How will the Anthropocene declines in coral reef diversity affect their resilience and persistence? Much has been written about the value of coral reefs, their worth in large measure attributable to their extraordinarily high biodiversity. Some 500 million coastal people depend on coral reefs for food, protection from natural disasters, and livelihoods.[5] Surely the persistence of coral reefs should be of great concern to humankind and demand top priority in safeguarding this rich and bountiful ecosystem.

In this chapter, I recount research that my students, colleagues and I have pursued to understand better the diversity and functional roles of eastern Pacific coral

[4] E. O. Wilson called attention to "Centinelan extinctions", referring to the extinction of species following the destruction of their habitats before they become known to science. Such an event occurred in western Ecuador in the 1980s, after the loss of 90 species of endemic plants to clear-cutting agriculture at Centinela Ridge (Wilson 1992). "The one process ongoing…that will take millions of years to correct is the loss of genetic and species diversity by the destruction of natural habitats. This is the folly our descendants are least likely to forgive us."—E. O. Wilson, Harvard Magazine, January–February 1980.

[5] Chuck Birkeland gives an insightful perspective on the value of coral reefs from his extensive experiences across the Pacific Ocean region (Birkeland 2015). A recent estimate of the commercial value of global coral reefs, in terms of goods and services, is US$375 billion per year (The Nature Conservancy 2022).

reef species. Our studies focused on the ecological engineers of coral reefs, i.e., zooxanthellate reef-building corals (scleractinian corals and hydrocorals, both calcifying cnidarian taxa) and associated visible and cryptic motile fauna.[6] A potentially rich cryptic sessile fauna, prominent in reef regions globally (Vasseur 1974; Jackson 1977), was not investigated by our team but requires attention. Common occurring sessile biota firmly affixed to coral substrates, include microbial biofilms (Remple et al. 2021), coralline algae, sponges, hydroids, azooxanthellate corals, polychaete worms, bivalve mollusks, bryozoans, barnacles, and tunicates. Jorge Cortés et al. (2017), in a comprehensive review of eastern Pacific coral reef diversity, should be consulted for progress on this topic in general as well as information on reef-associated meiofauna, rhodolith communities, and reef zooplankton communities.[7] Additional information on eastern Pacific biodiversity and the function of coral reef macrobiota and cryptic fauna is available in Zapata and Robertson (2007), Glynn (2011, 2013), Glynn and Enochs (2011), and Fong et al. (2017).

Assessing Biodiversity

Reef-building (zooxanthellate) corals. Corals and fishes are the two animal groups first observed on coral reefs, also the best known scientifically. These two metazoan taxa were the first we attempted to identify. Fortunately, several competent researchers were available to assist us in this task. The systematic publications of invertebrate paleontologist Professor J. Wyatt Durham were beneficial in the early stages of our coral identifications. I began a lengthy correspondence with Prof. Durham, curator of the Museum of Paleontology, Department of Paleontology, University of California (UC) at Berkeley. I had the great pleasure of meeting with Prof. Durham in his laboratory and the paleontology museum at UC Berkeley just before his retirement. He thanked me for the several coral specimens in the museum's collection I sent over the years for species identification and verification. He encouraged our studies and held firmly to the belief that structural coral reefs would eventually be found in the eastern tropical Pacific. During my visit, we engaged in lengthy discussions of coral systematics and the types of coral reef formations we

[6] Our only experience with non-protistan (mainly bacteria) microorganisms involved searching for possible infectious agents in corals that bleached and died during the 1982–83 El Niño event. I teamed up with Esther C. Peters and Len (Leonard) Muscatine employing field transplantation experiments and histopathological examination of necrotic and normally-pigmented coral tissues in search of pathogens (Glynn et al. 1985). From this study we concluded that high thermal stress was responsible for the loss of zooxanthellae and coral tissue deterioration and not from the effects of microbial infectious agents.

[7] Héctor Guzmán and co-workers initiated studies on the meiofauna and zooplankton communities associated with Costa Rican coral reefs (Guzmán et al. 1987; Guzmán and Obando 1988). Rhodolith (calcareous red algae often of spherical form) communities, commonly adjacent to eastern Pacific reefs, support a rich assemblage of cryptic organisms (Reyes-Bonilla and López-Pérez 2009; Solano-Barquero et al. 2022).

discovered in Panama and the Galápagos Islands. Prof. Durham was delighted to hear of these discoveries, the nature of the reef structures, their coral faunas, and vindication of his expectation of their existence in the eastern Pacific. Prof. Durham kindly offered many useful tips on identifying problematic species while examining the coral collections. John W. Wells, Prof. of Geology at Cornell University, also kindly shared his extensive knowledge of Pacific corals and contributed a systematic account of the corals of the Galápagos Islands to the Galápagos coral book authored by Jerry Wellington and me (Glynn and Wellington 1983). We became his devoted students during our time together in the Galápagos Islands. Later, help with coral taxonomy was offered by J. E. N. (Charlie) Veron, former Chief Scientist, Australian Institute of Marine Science, and Stephen Cairns, former chair of the Department of Invertebrate Zoology, curator, and Research Zoologist Emeritus, U.S. National Museum of Natural History, Smithsonian Institution.

In naming the corals and reef-associated species (invertebrates and fishes), it soon became apparent that many, for some taxa most, were well known Indo-Pacific species. Virtually no eastern Pacific reef species occur in the Caribbean region today. The Central American land bridge has been an effective biotic barrier since its formation about 3 million years ago (mya). Long distance west to east dispersal across the Pacific Ocean, enhanced during El Niño events and oceanic island steppingstones, has served to populate western American tropical shores with an essential component of Indo-Pacific coral reef species.

When we began to take stock of the species richness of eastern Pacific corals in the early 1970s, relatively few reef-building (zooxanthellate) species were recognized (< 30 species region wide), and the available taxonomic literature was limited.[8] About a dozen species were known from Panama and listed from various collection sites by Jim Porter in 1972. A notable increase in new records and the description of new coral species occurred in the early 2000s, with 42 species of zooxanthellate scleractinian corals listed by Héctor Reyes-Bonilla (2002). This number increased to 47 by 2010–2015; 26 species each were known from the Panama and Galápagos coral faunas (Glynn et al. 2017). At least 11 morphospecies of *Pocillopora* were recognized and described, two at the turn of the present century: *Pocillopora inflata* (Glynn 1999) and *Pocillopora effusa* (Veron 2000a). [According to the World Register of Marine Species (WoRMS), the correct grammatical agreement of the specific epithet is *effusa*, not *effusus*.] The task of

[8]The following references, a few with only scant information, were the principal taxonomic sources available to us: Verrill (1868-70), Crossland (1927), Durham and Barnard (1952), Squires (1959), and Durham (1966). Prof. Verrill, a renown invertebrate zoologist at Yale University, did not engage in field work in Panama but studied and named several species from material sent to him by collectors. Only 3 to 5 zooxanthellate coral species were noted in Crossland's *S.Y. St. George Expedition's* results; surveys were planned to include Panama, Colombia (Gorgona Island), the Galápagos Islands, and islands in the south-central Pacific. Despite Prof. Cyril Crossland's notoriety as one of his day's most accomplished coral reef scientists, the expedition's results were insubstantial and disappointing. His observations were confined to corals exposed at low tide and to material retrieved from dredging. A diver-collector was available on this expedition, but evidently of limited ability.

distinguishing between the dozen or so putative species of *Pocillopora* has been a long-time challenge due to the highly variable morphology of this taxon (Veron 1995).

With the emergence of genetic markers, morphological traits, and unique zooxanthella host associations, it is now possible to establish objective delineations among *Pocillopora* spp. and other closely related species. A genetic analysis of *Pocillopora* morphospecies combined with symbiont specificity indicated the existence of just three genetic lineages or species of *Pocillopora*[9] in the eastern Pacific region (Pinzón and LaJeunesse 2011). The most ecologically common of these was designated ***Pocillopora* type 1** corresponding to the valid species *Pocillopora grandis* (= *Pocillopora eydouxi*) with morphologies encompassing that of *Pocillopora damicornis, Pocillopora verrucosa, Pocillopora meandrina, Pocillopora capitata, Pocillopora elegans,* and *P. eydouxi* (Fig. 10.2). Synonymous with the invalid name *P. eydouxi*, this species bears a noble name, *grandis* (Lt. great, grand, lofty). It is widespread in the eastern Pacific, from the Gulf of California to the Galápagos Islands. The second most common species restricted to more equatorial latitudes is ***Pocillopora* type 3** corresponding to the valid species *P. verrucosa*, but whose colonies also have the appearance of *P. verrucosa*, as well as *P. damicornis*, and *P. meandrina.*[10] *Pocillopora verrucosa* occurs in the southern or equatorial eastern Pacific, in Panama, Colombia, and the Galápagos Islands. A third and rare genetic lineage closely related to *P. ligulata* is ***Pocillopora* type 2**, which may be synonymous with *P. inflata* (Glynn 1999). It was found only at Clipperton Atoll but may also occur at Malpelo and Cocos Islands. Lamentably, our long-time faithful acquaintance, *P. damicornis* (*sensu stricto*), does not appear to occur in the eastern Pacific. The distributional range of *P. damicornis* extends from Hawaii across the tropical Pacific to the mid/eastern region of the Indian Ocean.

Molecular genetic data allow insights into coral phylogeny or the evolutionary pathways of species' descent over time. A relevant study by Johnston and coworkers (2017) reported that living *Pocillopora* species evolved from a common ancestral lineage within the last ~3 million years. Fossil evidence indicates that *Pocillopora* originated during the Eocene (56–33.9 Ma) and was one of the dominant genera in the Caribbean during the Miocene-Pliocene and most of the Pleistocene epochs (McNeill et al. 2012).[11] *Pocillopora palmata*, a large, robust

---

[9] Following the Biological Species Concept, namely that species are members of populations that actually or potentially interbreed in nature; their identities are not based on similarities in appearance.

[10] Data provided by Pinzón et al. (2013) and Johnston et al. (2017) have allowed relating existing scientific names with actual genetically defined lineages. As a cautionary note, the validity of the species *P. inflata* has been questioned by changes in the skeletal morphology of coral colonies in the Gulf of California (Paz-García et al. 2015). Several tagged colonies of *P. grandis* with the '*damicornis*' morphology were observed to switch to the colony morphology typical of *P. inflata* over a 44-month period. Then, 7 months later, the altered colonies were observed recovering to the initial '*damicornis*' morphology. These transformations occurred during changes in storm frequency and chlorophyll-a concentration.

[11] For further discussion, including paleontological evidence, the reader is referred to Budd (2000), Toth et al. (2015), and López-Pérez (2017).

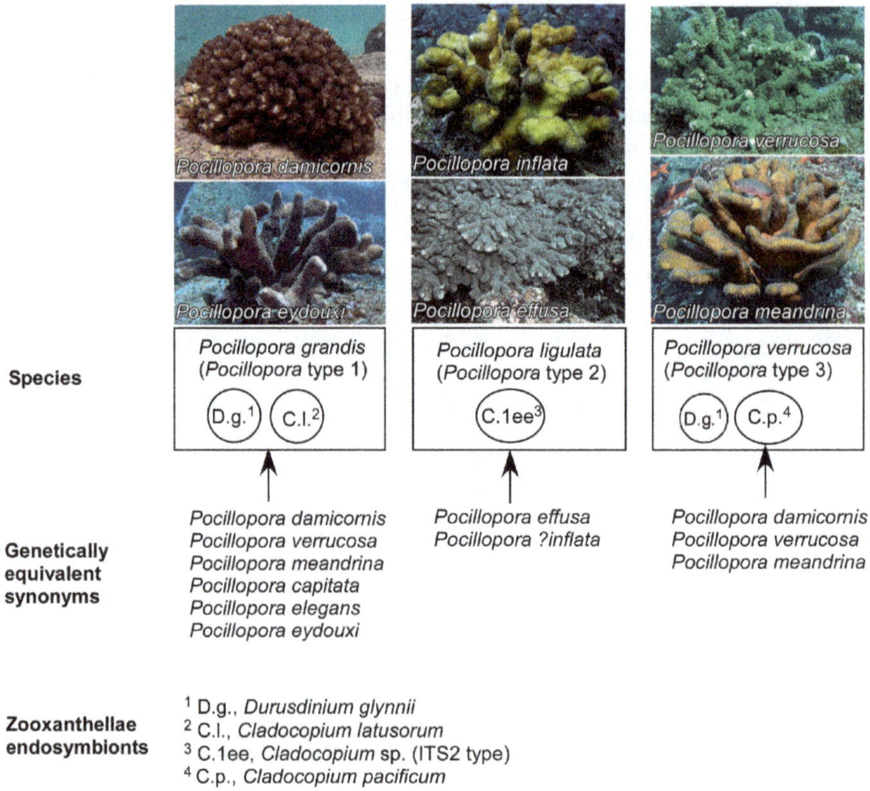

Fig. 10.2 Eastern Pacific *Pocillopora* species, dinoflagellate endosymbiont associates and genetically equivalent coral species synonyms. Photographs from Galápagos Islands of live, morphologically determined *Pocillopora* colonies aligned with genetic species identifications to demonstrate the marked differences in appearance. *P. damicornis*, North Seymour Island, 5 meters depth, 10 May 2023, J. S. Feingold. *P. eydouxi*, Wolf Is., 8 m, 19 May 2006, J. C. Moncayo (aka Macaron). P. inflata, Wolf Is., 10 m, 6 February 2012, J. S. Feingold. *P. effusa,* Wolf Is., depth?, 19 May 2006, J. C. Moncayo. *P. verrucosa*, Wolf Is., 10 m, 19 May 2006, J. C. Moncayo. *P. meandrina*, Wolf Is., 9 m, 6 February 2012, J. S. Feingold. Sources: Pinzón and LaJeunesse (2011), Cunning et al. (2013), LaJeunesse et al. (2018), and Turnham et al. (2021)

species, was the sole surviving western Atlantic *Pocillopora* by the end of the Pleistocene (Geister 1977; Pandolfi and Jackson 2001). Some of the colonies attained gigantic proportions compared with eastern Pacific species. Don (Donald F.) McNeill, Department of Geology, University of Miami, has shown me specimens and photographs of *Pocillopora crassaramosa* colonies from the Arroyo Bellaco formation (Dominican Republic) with 8 cm diameter stems and heights of 1 to 2 m (Fig. 10.3). The most recent living *P. palmata* (~80 kya) are known from coral deposits in Barbados (Pandolfi et al. 2001) and the Dry Tortugas, Florida Keys (Toth et al. 2015). The Caribbean extinction of *Pocillopora* is puzzling, considering its widespread occurrence and dominance in contemporary marginal reef

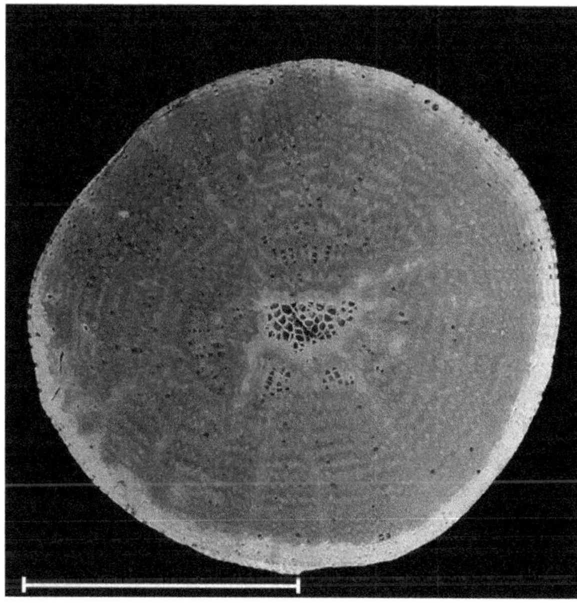

Fig. 10.3 *Pocillopora crassaramosa*, cross section of branch 8.1 cm maximum diameter, scale bar = 4 cm. Late Miocene, Arroyo Bellaco, Cibao Basin, northern Dominican Republic. Courtesy of Donald F. McNeill

environments. Genetic evidence has also revealed that broadcast spawning *Pocillopora* species have recently evolved from a brooding species lineage (Pinzón et al. 2013). Contemporary broadcast spawning *Pocillopora* demonstrate high dispersal potential and genetic connectivity within the eastern Pacific and across the east Pacific barrier (Pettay and LaJeunesse 2013). This raises the question: Did *P. palmata* belong to the reproductive brooding lineage that disadvantaged the species to cope with changing environmental conditions?

Other eastern Pacific species-rich coral genera are *Porites* (10 species) and *Pavona* (8 species). *Porites arnaudi*, with large colonies bearing distinctive tiered plates and descending frilled skirts, was described from Clipperton Atoll by Héctor Reyes-Bonilla and Carricart-Ganivet (2000). Recognition of the unique colony morphology, followed by detailed study of the skeletal microstructure, allowed these workers to name the species; in honor of Captain Ramón de Arnaud Sr., the last Commander in Chief of the Mexican Navy garrison on Clipperton Atoll.[12] New records of *Porites*, including several central/western Pacific species, were reported from the Revillagigedo Archipelago off the western Mexican coast (Ketchum and Reyes Bonilla 2001). These species occurrences suggest that the Mexican oceanic islands (and Clipperton Atoll) could serve as stepping-stones allowing the migration of Indo-Pacific species to the eastern Pacific region. Since three separate series of *Porites* colonies could not be identified (by morphological analysis, in the

[12] Marooned on Clipperton with his men and family, and facing certain death from starvation, he drowned in attempting to cross the reef to reach a real or imaginary ship offshore. A fascinating account of the travails of the early Mexican settlers on Clipperton was written by J. M. Skaggs (1989).

aforenoted publication) further study is necessary. The long-recognized small and brooding *Porites panamensis* is an Atlantic relic species more closely related to *Porites* in the Caribbean than those that evolved in the eastern Pacific. Even its algal symbiont species are more similar to the Caribbean taxa than those in the Pacific corals (see Chap. 3, "Origin of the Eastern Pacific Coral Reef Province").

Porites rus, a well-known Indo-Pacific species, was found as recently settled juveniles at Sámara, Costa Rica after the 1982–83 El Niño disturbance (Cortés and Murillo 1985). We speculated that its larvae could have dispersed to the eastern Pacific during the influx of El Niño Central Pacific waters. This discovery was fascinating because *P. rus* is an important ecological engineer, a prominent frame-building coral in the western Pacific. Reef building would certainly benefit from the successful establishment of this species, so we reasoned. A few years later Jorge Cortés, Carlos Jiménez and I visited the original discovery site to assess the condition of this recent arrival. The three of us searched thoroughly for living or dead colonies of *P. rus,*[13] but without success. The usual live colonies of eastern Pacific *Pocillopora* and *Porites* species were present, signifying healthy environmental conditions in this habitat. Alas, the occurrence of *P. rus* in the eastern Pacific, shown in Charlie Veron's *Corals of the World* treatise (Veron 2000b, volume 3), should be deleted. Its presence in Costa Rica was apparently short-lived; it did not become established.

Boulay et al. (2014) discovered a classic case of hidden diversity from genetic data in a study of *Porites*. What was long recognized as *Porites lobata*, the chief, massive, reef-building coral of the eastern Pacific region, is two species. Both species are nearly identical morphologically but genetically distinct, with notable ecological differences. Sampling of *Porites* populations at several eastern Pacific sites showed that *P. evermanni* is predominant along continental shores and *P. lobata* at offshore island sites. Also, *P. evermanni* was abundant where triggerfishes broke into colonies to prey on boring bivalves, principally *Lithophaga* spp. This destructive feeding behavior seems to have enhanced asexual propagation, favoring the proliferation of *P. evermanni* in suboptimal environments. In addition, evidence has indicated that *P. evermanni* is more resistant to high temperature than *P. lobata* even though both species likely host two closely related zooxanthella species of *Cladocopium*. These attributes could be pivotal in allowing poritid corals to expand the range of reef development and associated diversity into marginal coastal environments.

Pavona chiriquiensis was described by Juan Maté, Tom (Thomas) A. Stemann and me in 2001 (Glynn et al. 2001). The recognition of this species has an interesting history. From the very beginning of viewing live colonies of *Pavona varians*, some colonies appeared to be slightly different in color and skeletal structure

[13] John Wells verified the identity of this species; the skeleton of a newly settled specimen is cataloged in the Museum of Zoology, University of Costa Rica, San Pedro, San José. Costa Rica.

than others. These slightly different colonies mostly grew horizontally, spreading over flat basalt substrates. In contrast, the putative *P. varians* colonies tended to produce leaf or plate-like extensions lifting away from carbonate substrates. When we began to investigate the reproductive biology of corals in the mid-1980s, and were aware of the aforenoted differences, we referred to the deviant *Pavona* colony as *Pavona* sp. a. As referred to in Chap. 6 (Coral Reproduction), *Pavona* sp. a spawns during the same lunar days as *P. varians*, but in the early evening hours instead of at daybreak. When describing this species, we uncovered numerous differences from *P. varians*, including elegantly colored oral discs and polyp mouths, ranging from white to silvery white, green, yellow and turquoise. *Pavona chiriquiensis* is widely distributed throughout the equatorial eastern Pacific and has also been reported far west from Palmyra Island in the Northern Line Islands, south of Hawaii. *Gardineroseris planulata*, the only species in the genus (monospecific), is uncommon but widespread across the eastern Pacific and Indo-Pacific regions.

Zooxanthellate fire corals. On our first reconnaissance dive of the Uva reef we were pleasantly surprised to find three species of the fire coral *Millepora,* never before reported from fossil or living eastern Pacific coral assemblages (Glynn 1972; Glynn and de Weerdt 1991). The most abundant of these, the thinly branching species, *Millepora intricata,* formed extensive cover on the Uva forereef and other reef sites in the Gulf of Chiriquí, providing shelter for numerous invertebrates and fishes. When large piscivores threatened juvenile triggerfish (*Sufflamen verres*) they commonly dove into fire coral colonies. Chap. 5 ("El Niño Southern Oscillation") notes that two fire coral species experienced local extinctions during the 1982–83 El Niño disturbance. *Millepora intricata* has fully recovered since 1982–83 and again shelters diverse fish and invertebrates among its dense network of branches. *Millepora platyphylla* and *Millepora boschmai* have not reappeared in Panama (Glynn 2012). Yet, five colonies of *M. platyphylla* have been discovered at Clipperton Atoll (Boissin et al. 2019), and *M. boschmai* is now known to occur in Indonesia (Razak and Hoeksema 2003). Whether fire corals will become re-established on eastern Pacific coral reefs is problematic considering their high susceptibility to bleaching mortality and projected increasing El Niño activity.

Symbiodiniaceae (zooxanthellae). Before leaving the reef-building corals, it is necessary to introduce an entirely new level of hidden diversity vital to coral reefs' existence. Great strides have been made in the past two decades in revealing the diverse communities of bacteria and viruses, Archaea, endolithic algae, fungi, and protists that constitute the coral holobiont (Ainsworth et al. 2010; Rohwer and Youle 2010). Many of these associated taxa have been shown to increase the adaptive capacity of their coral hosts. There is even evidence that viruses play a role in coral therapy, disease, thermal tolerance, and host evolution (Voolstra et al. 2021).

Molecular studies have revealed that the endosymbiotic dinoflagellate protists (zooxanthellae), Family Symbiodiniaceae, that associate with reef-building corals consist of at least seven genera of phototrophic symbionts (LaJeunesse et al. 2018).[14] As noted in Chap. 5, Todd LaJeunesse suggested that the eventual number of symbiotic dinoflagellate genera may reach 15 or more. As for species richness per genus, it is presently impossible to be precise, but Todd suggests that hundreds of species occur in one genus (*Cladocopium*), and many (5–60 species per genus) in other genera.[15] Generally, a single species of zooxanthella per coral host is the rule, but some corals harbor two species, e.g., in the genus *Pocillopora*. Zooxanthellae densities are incredibly high in many coral species, with 1 to 3 million cells per cm^2 of surface tissue. This staggering degree of cryptic species diversity is critical for the survival of reef-building corals in terms of nutritional needs, calcification, and protection from environmental extremes. Some species of the zooxanthella genus *Durusdinium*, for example, are resistant to heat stress and cold-water events and assist in coral recovery after severe thermal disturbances (LaJeunesse et al. 2010; Baker et al. 2017).

Reef Fish Diversity

Regionally, about 1300 species of shore fishes occur in the tropical eastern Pacific (Zapata and Robertson 2007; Robertson and Cramer 2009), and ~ 500 of these occupy coral and shallow rock habitats (Robertson and Allen 2015; Cortés et al. 2017). Compared with the coral reef fish faunas of most Indo-Pacific reefs, the fish communities of eastern Pacific reefs are much less diverse, but it could be argued, not depauperate (i.e., extremely impoverished). Any one reef may support 50–100+ fish species, including those visible and of cryptic behavior (Zapata and Morales 1997; Dominici-Arosemena et al. 2005; Dominici-Arosemena and Wolff 2006; Benfield et al. 2008; Alzate et al. 2014; Glynn et al. 2014; Robertson and Allen 2015). Those visible fishes are variously involved in grazing, foraging for prey,

[14] Since recognized as mutualists in animal and protist hosts in the late nineteenth century by Sir Patrick Geddes, zooxanthellae were generally regarded as a single species, namely *Symbiodinium microadriaticum*. This view was held until the early 1990s when molecular data revealed genetically diverse zooxanthella populations (Rowan and Powers 1991). Initially genetically distinct zooxanthellae were referred to as "types", or were alphanumerically designated as clades, e.g., C42a, D1a. Clade designations refer to groups of organisms believed to have evolved from a common ancestor; clades are also regarded as monophyletic groups. Estimates from molecular dating indicate that these micro-algal symbionts originated and diversified with corals ~140–200 Ma. Todd LaJeunesse and co-workers formerly described new genera and species, thus validating the current use of binomial scientific names (LaJeunesse et al. 2018). For example, C42a is now *Cladocopium latusorum* and D1a is *Durusdinium trenchii*. Parkinson and LaJeunesse (2021) authored an engaging popular article on the history of research and the changing nomenclature of coral zooxanthella symbionts, which is well worth reading.

[15] In addition to the original genus *Symbiodinium*, newly proposed genera of the Symbiodiniaceae that are most commonly symbiotic with reef-associated cnidarians, sponges, foraminiferans, flatworms, and mollusks are *Breviolum*, *Cladocopium*, *Durusdinium*, *Effrenium*, *Fugacium*, and *Gerakladium*.

defending territories, engaged in reproductive behavior (spawning, egg laying), among other things.

In the late 1960s, rumors were circulating among ichthyologists of the occurrence of several unreported Indo-West Pacific coral reef fish species in the Gulf of Chiriquí, Panama. This encouraged Ira Rubinoff from STRI to organize and help finance fish collections in Chiriquí coral reef habitats in 1970 and 1971. The collections, at 16 key localities, amounted to the documentation of 165 species, with 24% of the species belonging to the Indo-West Pacific fauna (Rosenblatt et al. 1972)[16]. Like the eastern Pacific coral fauna, the coral reef fish fauna is a mixture of endemic and immigrant Indo-Pacific species, resulting from isolation and sporadic immigration across the eastern Pacific barrier. Prominent among the Indo-Pacific species were predators (muraenid or moray eels, jacks, and wrasses), herbivores (parrotfishes and surgeonfishes), and corallivores (puffers). Of the ten most abundant Uva reef fishes surveyed from 1980 to 2010, four (one corallivore and three herbivores) belonged to the Indo-Pacific fauna (Glynn et al. 2014). Two key bioeroding triggerfishes, *Pseudobalistes naufragium* and *Sufflamen verres*, are eastern Pacific endemic species. Some Indo-Pacific species have important ecological interactions on coral reefs, with direct and indirect effects on biodiversity (Glynn et al. 1972; Glynn et al. 2014; Palacios et al. 2014). For example, at sites in Panama where two Indo-Pacific parrotfish species (*Scarus ghobban* and *Scarus rubroviolaceus*) comprised 70–90% of the roving fish herbivore biomass (Smith 2005[17]), exclosure studies showed strong herbivore effects on algal communities, even when invertebrate herbivore abundance was low (Fong et al. 2006; Smith 2006; Smith et al. 2010). In particular, *S. ghobban* can take over 30 bites per minute on reef substrates (Smith 2008). This suggests that these Indo-Pacific herbivores exert important top-down control of algae and limit the profusion of macroalgal species onto coral reefs.

Following the 1982–83 El Niño event, several new records of Indo-Pacific species were reported in the eastern Pacific. Considering the west-to-east advection of tropical waters during such events (Wyrtki 1975) it seemed plausible that central

---

[16] Rotenone, a potent crystalline insecticide, was used in making these collections. Rotenone interrupts aerobic cellular respiration, depriving fish of oxygen. It is effective in exposing and killing cryptic coral reef fishes. It also kills other organisms, including various reef invertebrate taxa and corals. The use of lethal chemicals has been defended by Robertson and Smith-Vaniz (2008, 2010). Considering our understanding of their deleterious effects, we believe they should be used only in modest amounts in small areas by competent and responsible fish biologists. Their defense of rotenone use also applies to clove oil, a natural anesthetic, and various poisons. The employment of non-lethal ARMS (simulated coral frameworks) and baited traps have also proved helpful as alternative methods in revealing the presence of cryptic reef species.

[17] Tyler Smith, a graduate of Western Washington University, joined our eastern Pacific coral team at RSMAES in 1999. He began with field research at the Uva reef and then focused his doctoral studies on the ecology of parrotfishes in the upwelling environment of the Pearl Islands. Soon after receiving the Ph.D. degree in 2005, he joined the Center for Marine and Environmental Studies at the University of the Virgin Islands where he is now a Research Professor. Tyler is pursuing Caribbean studies concerning coral reef resilience, coral thermal stress, mesophotic coral ecosystems, and ciguatera poisoning. He plans to continue with coral studies in Panama with Juan Maté and the Galápagos Islands with Viktor Brandtneris.

Pacific planktonic larvae and drifting organisms could be dispersed into the eastern Pacific region.[18] Jerry Ault and I hypothesized that transpacific species introductions would be favored during such events (Glynn and Ault 2000). Several instances of Indo-Pacific arrivals were reported by investigators following this event, including corals, mollusks, echinoderms, and fishes, all with long-lived larval stages. Previous studies supported this view, including molecular analyses demonstrating high genetic connectivity between transpacific fish and echinoderm populations (Rosenblatt and Waples 1986; Lessios et al. 1998).

Despite these findings and more recent evidence supporting long distance dispersal (Lessios and Robertson 2006; Lessios and Baums 2017), a modeling study failed to show ocean surface currents reaching into the EP over 14.5-years, including the extreme 1997–98 El Niño event (Wood et al. 2016). This study employed a biophysical coral larval dispersal model, which failed to demonstrate eastward, cross transpacific connections. This contrary result was unexpected and perhaps best unveils the favorable conditions and events that must be met for the successful transport and establishment of immigrant species across the eastern Pacific barrier, for example: very strong El Niño events, favorable current trajectories and velocities, timing of spawning, available drift materials for rafting, suitable habitat at arrival site, and ability of immigrant species to interact successfully with novel resident species. Indeed, the stars need to be properly aligned to allow for successful transpacific immigration events.

From recent comprehensive documentation of 1100+ eastern tropical Pacific shore fish distributions by Ross Robertson,[19] Jerry Allen, and colleagues, the eastern

[18] Considering the variable nature of El Niño events, in terms of strength and spatial extent, the southeasterly trade winds invariably weaken as does the North Equatorial Current (NEC) at the surface (Wang et al. 2017). At such times, the surface current can reverse and flow from west to east. Eastward flow is enhanced by the zonal sea level slope—higher in the west and lower in the east—which propels surface currents down slope. The eastward moving surface water continues to warm, assisted by Kelvin wave propagation that forces the thermocline deeper. The Equatorial Under Current (EUC), flowing from west to east below the NEC, has also been considered a possible corridor for transporting Central Pacific propagules to the eastern Pacific (Richmond 1990). A recent study conducted just west of the Galápagos Islands, spanning the 2014–16 strong El Niño event, revealed eastward velocities of the EUC extending from the core to about 30 m depth and temperatures of ~22 °C (Rudnick et al. 2021). Tropical pelagic larvae, including some corals, would likely survive these conditions.

[19] Impressed with his study of sex reversal in a wrasse (Robertson 1972), Ira Rubinoff and Martin Moynihan asked me to meet with Ross Robertson to entertain the idea of joining STRI as a research scientist. We met aboard the M. V. *Marco Polo*, cruising the Great Barrier Reef, during the Second International Symposium on Coral Reefs to discuss an offer and research possibilities in the eastern tropical Pacific. Ross joined the STRI bureau in 1975 and has continued as a fish field biologist extraordinaire for nearly 50 years. Ross presently continues with research as a Senior Scientist Emeritus at STRI. His daily bicycle rides that begin at 3 AM, the coolest time of day in Panama, contribute to his vigor and intellectual acumen. Among his extensive range of contributions, totaling 330+ publications and still counting, he published with Jerry (Gerald R.) Allen, "Fishes of the Tropical Eastern Pacific" (Allen and Robertson 1994) and an invaluable online resource (Robertson and Allen 2015) of the eastern tropical Pacific fish fauna. In addition to Ross's eastern Pacific studies his research has extended into the Caribbean and other tropical biogeographic regions. Notable

Pacific region can be divided into three biogeographic provinces: **Cortez** (Gulf of California and lower Pacific Baja California), **Panamic** (Magdalena Bay, ~25°N to the Gulf of Guayaquil, ~4°S), and **Ocean Island** (Revillagigedo Islands, Clipperton Atoll, Cocos Island, Malpelo Island, and Galápagos Islands) (Robertson and Cramer 2009). Unlike the close affinity of the Rapa Nui coral fauna with the eastern Pacific (Glynn et al. 2007), dispersal routes for fishes seem to be mainly from the central and south Pacific (Delrieu-Trotten et al. 2019). Regarding fish species richness, how do the 165 species collected by Rosenblatt's team in the Gulf of Chiriquí compare with other sites in these three provinces. In an analysis of latitudinal diversity gradients along continental shores, Mora and Robertson (2005)[20] found that fish species associated with coral or rocky reefs peaked in species richness at the equatorial, Panama/Costa Rica border and secondarily in the subtropical lower Gulf of California. Coral reef development is regionally most remarkable in the equatorial eastern Pacific, in western Panama (Chiriquí) and Costa Rica (Cortés et al. 2017; Glynn et al. 2017). Habitat diversity, including coral abundance, is also high on the lower Baja California Peninsula and adjacent mainland sites (Fernández-Rivera Melo et al. 2018). As for the causation of this diversity pattern, Mora and Robertson (2005) proposed mechanisms based on variations of two long-held tenets of biogeographers: "center of origin" and "center of accumulation". They hypothesized that the center of origin hypothesis involves frequent habitat fragmentation and isolation resulting in high levels of speciation. High habitat diversity and environmental variability would also promote species accumulation, including endemic and immigrant Indo-Pacific species. Compared with continental provinces, the fish fauna of the Oceanic Island Province is small, composed of different functional groups, exhibits higher endemism, and has more transpacific species. Other possible factors affecting species immigration and diversity are the North Equatorial Counter Current which flows into Panama and Costa Rica (Glynn and Ault 2000; Guzmán and Cortés 2007) and the Revillagigedo Islands, suggested biogeographic stepping stones (way stations) from the central Pacific to the Cortez Province (Ketchum and Reyes Bonilla 1997, 2001).

The diversity and trophic structures of fishes were observed to change at the Uva reef over 26-years (1984–2010) of coral recovery (Glynn et al. 2014). The abundance of species that preyed upon invertebrates and fishes living within the interstices of pocilloporid corals and associated reef frameworks was particularly affected. The overall trend was asymptotic, with low fish consumer diversity at low

among his Caribbean work was the founding and operation of the San Blas field station. This laboratory supported scores of coral reef researchers until its closure in 1998. Thanks to Ross's generosity, fellowships are now offered to support field-oriented research on neotropical shore fishes at the predoctoral level and deep reef fish research for early career fish biologists.

[20]This study of the regional diversity patterns of shore fishes in the eastern tropical Pacific employed advanced analytical methods on species with different range sizes. The authors assessed how the mid-domain effect and environmental variables (energy supply, ecological variability, and habitat heterogeneity) predicted the latitudinal diversity gradients. The interested reader is referred to Zapata et al. (2003) for an in-depth review of the mid-domain effect and models.

coral cover, high diversity at intermediate coral cover, and again low fish diversity at maximum recovery (highest coral cover dominated by *Pocillopora*).

Determining what drove these trends is challenging since one or even several different ecological functions may be at play. Prey availability may be a factor as higher biomass of cryptofauna is associated with live versus dead coral substrates on the Uva Island reef (Enochs and Hockensmith 2009[21]; Enochs 2012). Prey diversity could also influence the diversity of predators, as dead coral substrates of intermediate degradation possess diverse associated cryptic communities. This is presumably due to greater niche diversity, e.g., food and shelter type, of degraded rubble habitats (Enochs and Manzello 2012a, b). "Diversity begets diversity", a common refrain among ecologists. A reduction in prey diversity may therefore occur in homogeneous stands of contiguous *Pocillopora*; the abundance of aggressive territorial crustaceans may also influence some species from taking up residence in such microhabitats (Enochs and Hockensmith 2009; Enochs and Manzello 2012a).

The structure may also play a role in sheltering capacity. When *Pocillopora* cover reached 30% and higher it may have become more difficult for fish consumers to locate and extract prey, especially cryptic prey species, from dense and rigid coral thickets (incipient frameworks). Enochs et al. (2011) found higher cryptic biomass in low porosity, tightly packed shelters. Other fish trophic guilds whose species can obtain food from outside reef structures, such as herbivores, planktivores, and corallivores, did not exhibit significant changes in diversity during coral recovery. This result supports the long-held view that coral reef structures provide shelter to numerous species seeking protection from epibenthic predators, thus contributing significantly to the generation of diversity "hot spots".

Since zooxanthellate (hermatypic) corals require light, most species in this group are exposed on reef surfaces and thus readily observed.[22] Azooxanthellate

---

[21] Still an undergraduate in the Marine Science Program (UM), Gena Hockensmith approached me enthusiastically about wanting to engage in coral reef research. She joined our research team and soon was helping with fieldwork in Panama. Gena became interested in the abundance, diversity and behavior of invertebrates associated with living and dead *Pocillopora* colonies. At the time, we wondered how El Niño induced coral mortality might affect these invertebrate guests. She researched this subject, culminating in an M.A. degree thesis in 2008, entitled "A Study of the Vagile Invertebrate Communities Inhabiting Living and Dead *Pocillopora damicornis* colonies in the Eastern Tropical Pacific". During Gena's field work, she found it convenient to process coral samples on the Uva Island beach. On one occasion, I was concerned about her safety. From our vantage point on the reef, I spied three local fishers, unknown to us, gathered closely around her on the beach. This situation prompted a quick visit from us. Upon approaching the gathering, we were relieved that the fishers were sincerely fascinated with the sampling work in progress. She had instructed the men to help her with the sorting and measurements; they were all temporarily laboring as Gena's assistants, with genuine interest and efficiency, Citizen Science.

[22] A notable exception is *Leptoseris incrustans* in the northern Galápagos Islands. This species typically occurs in caves and on the shaded sides of boulders between 10–20 m depth. Curiously, colonies commonly shared their habitat with moray eels, which made their study uncomfortable. I thank Michel Pichon for identifying this coral species. Jorge Cortés informed me that *Leptoseris scabra* occurs similarly, present under large colonies of *Porites lobata*, at Isla del Coco.

(ahermatypic) corals, however, do not require a photosynthetic energy source and do not host symbiotic algae. These corals are usually cryptic, inhabiting the dark recesses of reef frameworks and mesophotic habitats. They are as species rich as zooxanthellate corals with 47 species reported in the eastern tropical Pacific (Cortés et al. 2017). As noted above, many coral reef fishes are visible except for small eels and other cryptic species that inhabit coral rubble and internal reef cavities. Visual fish species counts have been shown to underestimate total local species richness by 28 to 36% (Glynn 2006; Alzate et al. 2014).

In Search of the Cryptofauna

A commonly employed collection method for cryptic macrofauna inhabiting *Pocillopora* involves enclosing a coral colony in a plastic bag to avoid escape and extracting the animals at the surface, in the field or laboratory. Larry (Lawrence G.) Abele successfully used this method in Panama (Abele 1974, 1976)[23] and it is commonly employed in recent studies (e.g., Alvarado and Vargas-Castillo 2012; Salas-Moya et al. 2021; Alonso-Domínguez et al. 2022). A rich crustacean fauna is associated with *Pocillopora* spp. across the Indo-Pacific region from the eastern Pacific to the western Indian Ocean and the Red Sea. Larry recorded 55 species in the Pearl Islands, Gulf of Panama, and this richness is observed elsewhere in the eastern Pacific. For example, Alonso-Domínguez et al. (2022) reported 88 species of decapod and stomatopod *Pocillopora* associates from four localities over 2 years from the central Mexican Pacific coast.

We employed baited traps and coral rubble substrates to assess the diversity and abundance of cryptic motile fauna in reef frameworks. The baited traps, with fresh strips of fish muscle, were small and could be positioned within reef frameworks. Most of the animals captured in the traps were scavengers, with occasional predators presumably attracted to the trapped scavengers (Glynn 2011, 2013). *Pocillopora* rubble structures were designed to be placed in different reef habitats and retrieved to sample the attracted species. These consisted of open-topped cylinders of plastic mesh netting referred to as ARFs (artificial reef framework units) in Ian Enochs studies in the Pearl Islands (Enochs et al. 2011). We also constructed large (25–30 liter volume) netted bags of *Pocillopora* rubble for deployment on the Uva reef. These were referred to as SRFs (simulated reef frame units) and were collected and sampled every 6 months over 2 years (Glynn 2006). Like the ARMS (autonomous

[23] Larry Abele performed his doctoral research in Panama investigating the effects of fluctuating environmental conditions on the species richness of the crustacean fauna of *Pocillopora*. Larry was awarded a STRI predoctoral fellowship to support this research. His penetrating intellectual curiosity and enthusiasm enlivened the research atmosphere of our nascent marine laboratory. He received his Ph.D. degree from the University of Miami in 1972 and joined the Department of Biological Sciences at Florida State University (FSU) in 1973. He quickly advanced through the academic ranks and became Provost at FSU, serving from 1994 to 2010.

reef monitoring structures) methodology (Brainard et al. 2009), this allowed us to simulate and sample reef cavities without damaging coral colonies or reef frameworks. Also, the coral branches offered natural surfaces (limestone and biofilm communities) for recruiting organisms. The traps and coral rubble structures revealed an astonishing array of cryptic metazoans, both fishes and invertebrates. A dedicated team effort was required for the extraction, preservation, and identification of the rich assortment of organisms sheltering in the SRFs (Fig. 10.4)

Other reef species that are cryptic during daylight hours, some quite large, emerge after sunset. We performed night diving surveys to determine the presence of such species and observed several. Two examples of particular interest were a sea cucumber and sea hare, species seldom seen on the reef during the day. Numerous individuals of the sea cucumber, *Euapta godeffroyi*, up to one meter in length, emerged from the reef after sunset and crawled along the sand/rubble bottom (Fig. 10.5). The circumoral tentacles deftly capture sediment and detritus, which is ingested for its nutritional value. Some of these sea cucumbers cautiously extended their bodies about halfway out of the reef. A gentle finger tap would trigger a rapid retraction and disappearance into the sheltering reef. The other nocturnally active species was *Dolabella auricularia*, the largest known tropical sea hare. Most individuals on the Uva reef were 10–15 cm in length and were observed grazing after

Fig. 10.4 Eastern Pacific coral reef research team at Coibita Island Beach Club, Gulf of Chiriquí (August 2007). Front row, left to right, Franco Camarena (boat captain), Ranjan Muthukrishnan, Adrienne Simoes Correa, Peter Glynn, Peggy Fong, Tyler Smith, Juan Maté; back row, left to right, 'Potin' (Olmedo Atencio), Ian Enochs, Derek Manzello

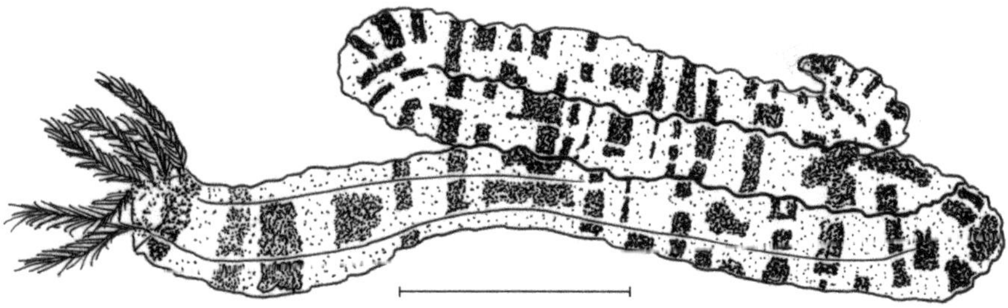

Fig. 10.5 *Euapta godeffroyi*, a sea cucumber detritivore, scale bar = 5 cm

sunset. Associated with *Euapta* were minute gastropods, *Melanella* cf. *exilis*, also reported from Clipperton Atoll (Kaiser 2007), and with *Dolabella* the pearlfish, *Carapus mourlani* (Glynn et al. 2008). In the next chapter, we revisit these closely associated species relationships (symbioses).

Although we often referred to the baited traps as "scavenger traps," some entrapped animals were likely attracted to the scavengers feeding on the bait. During each survey, we embedded about twenty baited traps in natural reef frame cavities before sunset and collected them soon after sunrise the following morning. Each trap had a 50 cm long line and float to ease its location during collection. During retrieval, the trap opening was closed, and the entrapped animals were tentatively identified, counted and cataloged. Captured animals included flatworms (Polycladida), polychaete worms, crustaceans (mysids, ostracods, hermit, and brachyuran crabs), mollusks (gastropods and bivalves), and muraenid eels (Glynn 2013). The traps were phenomenally successful in terms of the variety of taxa captured and the number of individuals. We are grateful to Irving Bethancourt, a team volunteer from the University of Panama, who designed the traps. Irving is presently (as of 2024) the Scientific Coordinator for marine research at STRI's Naos Research Complex.

A cirolanid isopod (*Cirolana parva*) often dominated the collections with 10 to 20 individuals captured in each trap (Fig. 10.6). Remarkably, isopods in the genus *Cirolana* also were the most trapped animals on a reef flat in Australia (Keable 1995). On the Great Barrier Reef in eastern Australia, gnathiid isopods are one of the most prevalent and abundant taxa of ectoparasites, with cirolanid (flabelliferan) micropredators in second place (Jones and Grutter 2007). These isopods negatively impact the health and behavior of fish larvae recruited to coral reefs. Gnathiid isopods were not observed on the Uva reef, but cirolanid micropredators may have a similar negative effect on cryptic species recruits. It was my good fortune to have Carmen Glynn's help in collecting, sorting, and performing prey-choice experiments with isopods (Glynn and Glynn 1974, Fig. 10.7).[24]

[24] Considering the ubiquity and abundance of other isopod taxa that might engage in critically important early life history interactions of recruiting organisms, this is a research area worthy of continuing study. Isopod species inhabiting coral reef habitats (e.g., *Apanthuroides millae*) and

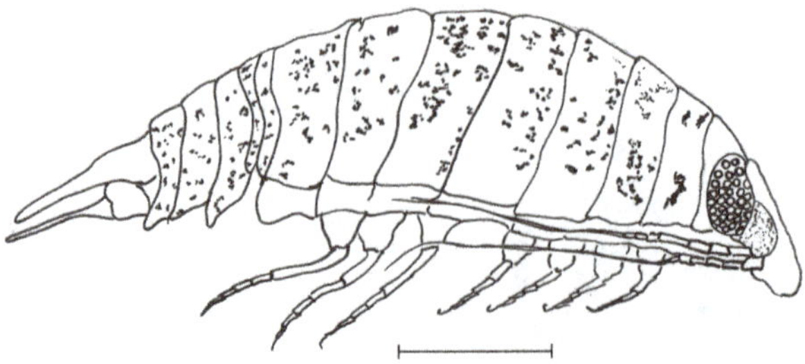

Fig. 10.6 *Cirolana parva*, scale bar = 1 mm

Fig. 10.7 Carmen S. Quiñones Acosta de Glynn, Reading Specialist, Learning Center, Gulliver Academy (1987–2005), Miami, Florida. Wife, mother, soulmate, confidant, research assistant, "engineer", love of my life

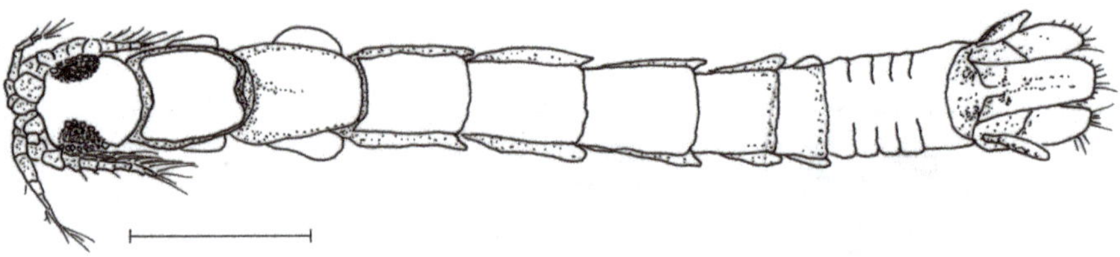

Fig. 10.8 *Apanthuroides millae*, scale bar = 5 mm

In retrieving coral rubble bags (SRFs) at the Uva reef, we encountered green-colored, smooth skinned, flaccid worms (Fig. 10.11). These 2–5 cm-long worms were found beneath the SRFs in a silt/mud sediment microhabitat. Also present in

sand beach environments (e.g., *Ancinus panamensis* and *Heteroserolis tropica*), could well influence the species composition of their respective communities (Figs. 10.8, 10.9, and 10.10).

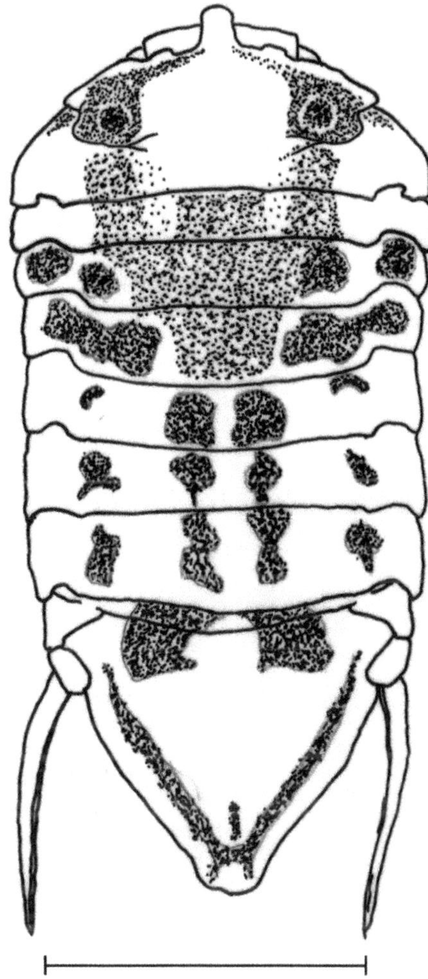

Fig. 10.9 *Ancinus panamensis*, scale bar = 1 mm

the sediment samples were fleshy, greenish colored gutter-like straps, which we later determined were proboscises that had separated from the worms, probably a spoon or echiuran worm, which rarely occurs in the eastern Pacific region. We contacted Prof. Ramlall, Biseswar, University of Kwa-Zulu-Natal, South Africa, a renowned authority on echiuran systematics, to help identify this species. He confirmed it was a new species and now carries the awkward epithet, *Anelassorhynchus panamensis* (Biseswar and Glynn 2016). Considering its high abundance (5–30 ind. m^{-2}) and sediment-packed intestines, this species likely plays a key role in recycling reef base sediments and nutrient regeneration on the Uva reef.

The occurrence of pygmy octopuses in the SRF collections was also unexpected. None of our team had ever previously encountered these small octopuses. Large

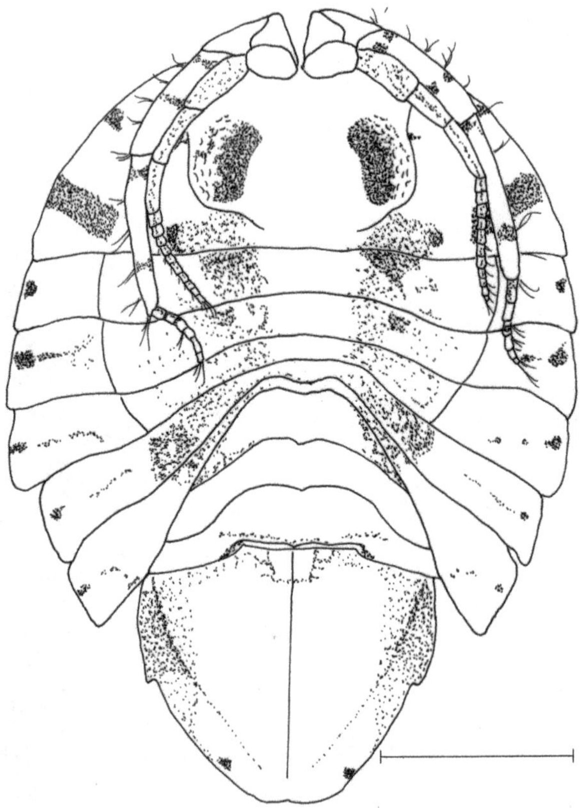

Fig. 10.10 *Heteroserolis tropica*, scale bar = 1 mm

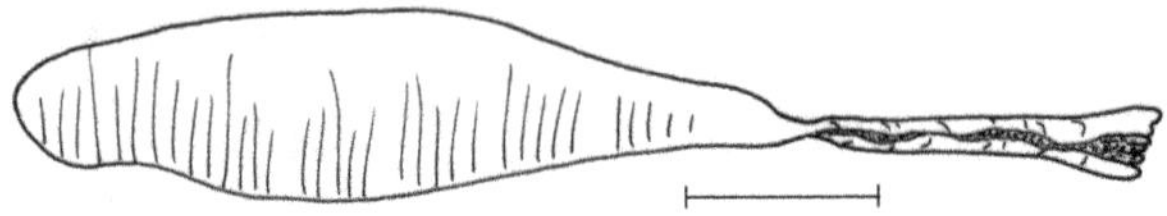

Fig. 10.11 *Anelassorhynchus panamensis*, scale bar = 1 cm

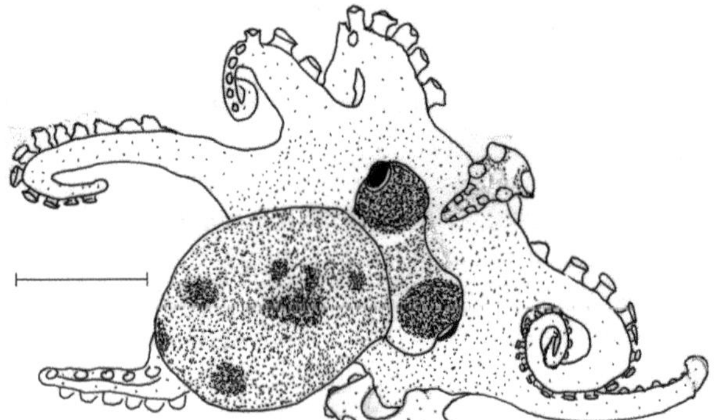

Fig. 10.12 *Octopus* sp., an undescribed pygmy octopus, scale bar = 1 cm

octopuses were observed on the reef, but never this minuscule species. Their bodies (visceral hump) are not much larger than a thumbnail (Fig. 10.12). Since they were adept at avoiding detection and escaping, I cupped one individual in my hand to show to our sorting team to alert them of their presence. Suddenly, I felt a sharp pain in the palm of my hand. The tiny, agitated octopus deservedly bit me. I then began to feel a numbing sensation in my hand. I had been envenomated, and this persisted for about one hour. Considering the common occurrence of this octopus, we believe it is an important predator in cryptic *Pocillopora* communities. Eric Hochberg of the Santa Barbara Natural History Museum (SBNHM) concluded that this octopus was probably an undescribed species in the genus *Octopus*.[25]

A comparison of cryptic reef fish densities and species richness demonstrated a highly significant difference in the DCR and SRF coral treatments (Glynn 2006). The DCR treatment, an essentially two-dimensional dead coral rubble substrate, was designed to approximate patches of eroded *Pocillopora* corals, resulting from an El Niño disturbance. The SRF treatment was designed to mimic a dead three-dimensional reef framework. A statistical summary of four surveys over 18 months revealed mean cryptic fish densities ranging from 0 to 4.4 fishes m^{-2} in the DCR treatment and from 41.2 to 111.2 fishes m^{-2} in the intact coral framework treatment (SRF). Species richness ranged from 0 to 6 species and 9 to 14 species in DCR and SRF substrates, respectively. These differences in cryptic reef fish abundances and diversity suggest the degree of change possible following the loss and erosion of reef framework structures. Prominent among the cryptic fishes were muraenid eels, carapids, antennariids, and bythitids, the majority small (8–15 cm total length) and all carnivorous.[26]

Ian Enochs' studies of the motile cryptic fauna of coral reefs in the Pearl Islands have revealed insights on variations of biodiversity in relation to live coral cover, framework types, reef zonation, and degree of coral reef degradation.[27] The

[25] The specimens I sent to Eric are presently in the collections of the Santa Barbara Natural History Museum: SBNHM 464721–464,728; 8 lots, 9 specimens; 1 lot from September 2003; 6 lots from March 2004; and 1 lot from March 2005. We are hopeful that someday a cephalopod specialist will re-study this octopus and document its taxonomic status.

[26] The cryptic fishes collected from the coral sediment bags (SRF), not observed in visual surveys, were muraenid eels (*Gymnothorax panamensis*, *Gymnothorax undulatus*, *Muraena lentiginosa*), carapids (*Encheliophis* n. sp., *Encheliophis vermicularis*), antennariids (*Antennarius sanguineus*, *Antennarius coccinea*, *Antennarius strigatus*), bythitids (*Ogilbia* sp., *Oligopsus* sp.), scorpaenids (*Scorpaenodes xyris*, scorpaenid sp.), and a labrisomid (*Starksia fulva*).

[27] Ian C. Enochs was recruited to our coral reef research team after graduating *cum laude* from UM in 2006 with a B. S. degree in Marine Science and Biology. He labored many days in the field (Uva reef) and laboratory before concentrating his graduate research on the cryptic fauna of coral reefs in the Gulf of Panama. Ian was fascinated with the profusion of cryptic, motile animals inhabiting reefs, including their ecological functions and species interactions. He contributed significantly to our team's studies of coral recovery, reproduction, and symbiotic relationships before engaging in his doctoral research. He is presently a Research Ecologist with NOAA's Atlantic Oceanographic and Meteorological Laboratory (AOML) and Adjunct Assistant Professor in the Department of Marine Biology and Ecology at RSMAES (UM). Enochs assumed leadership of the AOML Coral Program in 2021 focusing on investigating the responses of corals and reef biota to global change.

extraordinarily high species richness of cryptic species associated with pocilloporid corals, abundantly documented by early workers, and reviewed in Cortés et al. (2017), was also reported in Ian's studies (Fig. 10.1). For example, a total of 289 Operational Taxonomic Units (OTUs)[28] from six phyla were identified from both live *Pocillopora damicornis* (sensu lato) and dead pocilloporid reef frameworks in the Pearl Islands, Gulf of Panama (Enochs and Manzello 2012b). These studies are particularly discerning because they were designed to address experimentally how species richness is affected by live and dead coral cover and the degree of framework deterioration. Baskets of live and dead coral rubble (ARFs) were the experimental units of choice. The ARF designs allowed testing for porosity or internal habitat space and current flow, which are critical to the cryptofauna. ARF replicates were also deployed across reef zones of varying degradation. Collecting and extracting the motile cryptic fauna was labor intensive, requiring many hours of our team's labors (Fig. 10.4).

An unexpected trend of species richness in dead versus live coral frameworks demonstrated nearly double the number of cryptic taxa in dead compared with live coral substrates. Estimated species richness in dead pocilloporid frameworks ranged from 261 to 370 OTUs, and in live coral substrates, from 112 to 219 OTUs (Enochs and Manzello 2012a). Live *Pocillopora* corals supported a rich obligate crustacean fauna of high biomass. Still, these species were absent and replaced by diverse, nonsymbiotic crustaceans of lower biomass on dead coral substrates (Enochs et al. 2011). Ian's investigations of coral mortality and framework erosion demonstrated high cryptofauna densities in reef frameworks of intermediate erosion and degradation (Enochs 2012; Enochs and Manzello 2012a, b). Dead coral rubble contained the greatest number of species (227–320 OTUs). An abundance of shelters in eroded reef structures was critical for supporting diverse, motile cryptofauna populations. The lowest estimated richness (47–115 OTUs) occurred in reef zones of high live coral cover and greatest vertical relief.

He succeeded Derek Manzello (former team leader of the Coral Program) and Jim (James) Hendee, "retired" Director of AOML's Ocean Chemistry and Ecosystems Division, and originator of CHAMP (Coral Health and Monitoring Program), an umbrella program funding all aspects of coral conservation at AOML, and which became the Coral Program. Jim continues to assist in his pioneering project to monitor and model in near real-time atmospheric and oceanographic parameters affecting marine environmental events of coral ecosystems (Gramer et al. 2023). Other activities under CHAMP that Jim initiated were the Coral-List listserve, a forum dedicated to coral reef research, and the very first coral-related website in the world. The International Coral Reef Society now manages NOAA's Coral listserve and is moderated by C. Mark Eakin. Ian and his team are presently researching the impacts of ocean warming and ocean acidification on the persistence of coral reef framework structures that provide essential habitat supporting ecosystem biodiversity and services.

[28] An Operational Taxonomic Unit (OTU) is a working or pragmatic term used to group individual organisms by similarity (morphological or genetic) and serves as a working proxy for "species". For example, suppose a shrimp can be classified as belonging to the genus *Alpheus*, but does not match the available descriptions of any known species of *Alpheus*. In this case, OTU status can be provisionally assigned. Such problematic specimens can be preserved and cataloged for later study and then identified or described by traditional morphological or DNA methodologies.

These results do not aline with some of my studies that have shown or predicted declines in species richness with coral reef framework degradation. For example, as noted above the diversity of cryptic reef fishes declined by about one-half in experimental two-dimensional rubble (DCR) treatments compared with intact three-dimensional reef frame (SRF) treatments. In addition, of the thirty invertebrate phyla associated with coral reefs, at least 27 contain cryptic species that require shelter in the spaces of carbonate substrates (Glynn 2011). My prediction of tandem reef coral and biodiversity decline followed these observations. Indeed these differences will stimulate additional research on this subject. The results of Enoch's team challenge the paradigm that abundant live coral is necessary to support maximum reef biodiversity. Their research has unequivocally demonstrated that motile cryptic biodiversity should persist independently of live coral cover and intact reef coral frameworks. Live *Pocillopora* corals are necessary to support obligate endo- and ectosymbiotic, Symbiodiniaceae and crustaceans, respectively, that depend on live scleractinian coral hosts. Further, abundant populations of actively calcifying reef-building corals are essential to generate carbonate frameworks and eroded sediments to replace present day and projected future losses.

Moving Forward

Following the UN Decade on Ecosystem Restoration (2021–2030), it is encouraging to observe increasing focus and implementation of measures to mitigate biodiversity loss. Namely, members of the 15th Conference of the Parties (COP 15) to the United Nations Convention on Biological Diversity (CBD) held in Montreal on December 2022 agreed to "halt and reverse" biodiversity loss by the end of the decade.[29] To implement this objective, nearly 200 countries supported four goals, including the "30–30" plan to conserve 30% of the world's land and 30% of the ocean by 2030 and fully support countries' actions to manage the oceans and sponsor coral restoration initiatives sustainably. To finance this plan,"developed" countries approved the mobilization of $30bn for "developing" countries by 2030. This is unquestionably daunting, but progress toward tracking and understanding the reasons for change in species richness will benefit efforts to halt and conserve coral

[29] I am pleased to acknowledge David Obura's conservation work as it relates to the accomplishments of COP 15. David joined our coral reef team in 1989 after graduating with a Bachelor of Arts degree with Honors in Biology from Harvard University. He was granted the Ph.D. degree by UM in 1995. Obura assisted with coral studies in the Galápagos Islands while his doctoral research focused on the effects of river-borne sediments on Kenyan coral reefs. An Africa Dissertation Internship Award from the Rockefeller Foundation supported this research. In recent years David's studies have focused on coral reef conservation and management in the western Indian Ocean. He was the Founding Director of CORDIO East Africa. While at RSMAES we soon discovered that David was an accomplished long-distance runner. In the late afternoon, he would often be seen running along the Rickenbacker Causeway. His race running successes in Miami and the Galápagos Islands became a part of our coral team's lore.

reef biodiversity. Of significant help in this effort are comprehensive inventories of known ecosystem biodiversity at specific locations, listings of active taxonomic authorities, and the collections of taxa available for study. *Marine Biodiversity of Costa Rica, Central America*, edited by Wehrtmann and Cortés (2009) is a *tour de force* compendium offering such valuable information. Additional treatments of this genre, concentrated in other coral reef regions of the world, could significantly advance the goals of COP15.

The enormous losses of global coral reefs over the past three decades, due primarily to bleaching and mortality from heat stress, are unprecedented (e.g., Heron et al. 2017). Although these losses usually focus on corals and fishes, we must be cognizant of changes in abundance and disappearance of associated cryptic coral reef organisms, which contribute overwhelmingly to biodiversity and the performance of essential roles in ecosystem function and persistence (Przeslawski et al. 2008; Hoegh-Guldberg 2012). Very little is known of the extinction risks and changes affecting this mostly unseen biodiversity of coral reefs, an urgent concern in need of attention.

References

Abele LG (1974) Species diversity of decapod crustaceans in marine habitats. Ecology 55:156–161

Abele LG (1976) Comparative species composition and relative abundance of decapod crustaceans in marine habitats of Panama. Mar Biol 38:263–278

Ainsworth TD, Thurber RV, Gates RD (2010) The future of coral reefs: a microbial perspective. Trends Ecol Evol 25(4):233–240

Allen GR, Robertson DR (1994) Fishes of the tropical eastern Pacific. Crawford House, Bathurst, Australia, p 332

Alonso-Domínguez A, Ayón-Parente M, Hendrickx ME, Riós-Jara E, Vargas-Ponce O, del Esqueda-González MC, Rodríguez-Zaragoza FA (2022) Taxonomic diversity of decapod and stomatopod crustaceans associated with pocilloporid corals in the central Mexican Pacific. Diversity 14(2):72. https://doi.org/10.3390/d14020072

Alvarado JJ, Vargas-Castillo R (2012) Invertebrados asociados al coral constructor de arrecifes *Pocillopora damicornis* en Playa Blanca, Bahía Culebra, Costa Rica. Rev Biol Trop 60(Suppl 2):77–92

Alzate A, Zapata FA, Giraldo A (2014) A comparison of visual and collection-based methods for assessing community structure of coral reef fishes in the tropical eastern Pacific. Rev Biol Trop 62(Suppl 1):359–371

Baker AC, Correa AMS, Cunning R (2017) Diversity, distribution and stability of *Symbiodinium* in reef corals of the eastern tropical Pacific. In: Glynn PW, Manzello DP, Enochs IC (eds) Coral reefs of the eastern tropical Pacific: persistence and loss in a dynamic environment. Coral reefs of the world 8. Springer, Dordrecht, pp 405–420

Benfield S, Baxter L, Guzmán HM, Mair JM (2008) A comparison of coral reef and coral community fish assemblages in Pacific Panama and environmental factors governing their structure. J Mar Biol Assoc U K 88:1331–1341

Birkeland C (2015) Coral reefs in the anthropocene. Springer, Dordrecht, pp 1–15

Biseswar R, Glynn PW (2016) A new coral reef associated species of Echiura, *Anelassorhynchus panamensis* from the eastern Pacific Ocean. Proc Biol Soc Wash 129:56–65

Boissin E, Pogoreutz C, Pey A, Gravier-Bonnet N, Planes S (2019) *Millepora platyphylla* (Cnidaria, Hydrozoa) range extended back to the eastern Pacific, thanks to a new record from Clipperton Atoll. Zootaxa 4668(4):599–600

Bouchet P (2006) The magnitude of marine biodiversity. In: Duarte CM (ed) The exploration of marine biodiversity. Scientific and technological challenges. Fundación BBVA, Bilbao, pp 31–64

Boulay JN, Hellberg ME, Cortés J, Baums IB (2014) Unrecognized coral species diversity masks differences in functional ecology. Proc R Soc B 281:20131580. https://doi.org/10.1098/rspb.2013.1580

Brainard R, Moffitt R, Timmers M, Paulay G, Plaisance L, Knowlton N, Caley J et al (2009) Autonomous Reef Monitoring Structures (ARMS): a tool for monitoring indices of biodiversity in the Pacific Islands. In: 11th Pacific science inter-congress: And 2nd symposium on French research in the Pacific: 02–06 March 2009, Tahiti, French Polynesia: Pacific countries and their ocean facing local and global changes. Pacific Science Association, Honolulu, HI

Budd AF (2000) Diversity and extinction in the Cenozoic: history of Caribbean reefs. Coral Reefs 19:25–35

Carpenter KE, Abrar M, Aeby G, Aronson RB, Banks S, Bruckner A, Chiriboga A, Cortes J, Delbeek JC, DeVantier L, Edgar GJ (2008) One-third of reef-building corals face elevated extinction risk from climate change and local impacts. Science 321(5888):560–563

Connell JH (1978) Diversity in tropical rainforests and coral reefs. Science 199:1302–1310

Cortés J, Murillo MM (1985) Comunidades coralinas y arrecifes del Pacífico de Costa Rica. Rev Biol Trop 33:197–202

Cortés J, Enochs IC, Sibaja-Cordero J, Hernández L, Alvarado JJ, Breedy O, Cruz-Barraza JA, Esquivel-Garrote O, Fernández-García C, Hermosillo A, Kaiser KL, Medina-Rosas P, Morales-Ramírez A, Pacheco C, Pérez-Matus A, Reyes-Bonilla H, Riosmena-Rodríguez R, Sánchez-Noguera C, Wieters EA, Zapata FA (2017) Marine biodiversity of eastern tropical Pacific coral reefs. In: Glynn PW, Manzello DP, Enochs IC (eds) Coral reefs of the eastern tropical Pacific: persistence and loss in a dynamic environment. Coral reefs of the world 8. Springer, Dordrecht, pp 203–250

Crossland C (1927) The expedition to the South Pacific of the SY 'St. George'. Marine ecology and coral formations in the Panama region, the Galapagos and Marquesas Islands, and the atoll of Napuka. Trans R Soc Edinb 55:531–554

Cunning R, Glynn PW, Baker AC (2013) Flexible associations between *Pocillopora* corals and *Symbiodinium* limit utility of symbiosis ecology in defining species. Coral Reefs 32(3):795–801. https://doi.org/10.1007/s00338-013-1036-4

Dayton PK (1971) Competition, disturbance, and community organizations: the provision and subsequent utilization of space in a rocky intertidal community. Ecol Monogr 41:351–389

Delrieu-Trotten E, Brosseau-Acquaviva L, Mona S, Neglia V, Giles EC, Rapu-Edmunds C, Saenz-Agudelo P (2019) Understanding the origen of the most isolated endemic reef fishes of Rapa Nui. J Biogeogr 46:723–733. https://doi.org/10.1111/jbi.13531

Dominici-Arosemena A, Wolff M (2006) Reef fish community structure in the tropical eastern Pacific (Panamá): living on a relatively stable rocky reef environment. Helgol Mar Res 60:287–305. https://doi.org/10.1007/s10152-006-0045-4

Dominici-Arosemena A, Brugnoli-Olivera E, Cortés-Nuñez J, Molina-Ureña H, Quesada-Alpizar M (2005) Community structure of eastern Pacific reef fishes (Gulf of Papagayo, Costa Rica). Tecnociencia 7:19–41

Durham JW (1966) Coelenterates, especially stony corals, from the Galápagos and Cocos Islands. In: Bowman RI (ed) The Galápagos, Proceedings of the symposium of the Galápagos International Scientific Project. University of California Press, Berkeley and Los Angeles, pp.123–135

Durham JW, Barnard JL (1952) Stony corals of the eastern Pacific collected by the *Velero III* and *Velero IV*. Allan Hancock Pac Exp 16(1).1–110

Enochs IC (2012) Motile cryptofauna associated with live and dead coral substrates: implications for coral mortality and framework erosion. Mar Biol 159(4):709–722

Enochs IC, Hockensmith G (2009) Effects of coral mortality on the community composition of cryptic metazoans associated with *Pocillopora damicornis*. In: Proceedings of the 11th international coral reef symposium, Ft. Lauderdale, Florida, vol 2, pp 1375–1379

Enochs IC, Manzello DP (2012a) Responses of cryptofaunal species richness and trophic potential to coral reef habitat degradation. Diversity 4:94–104. https://doi.org/10.3390/d4010094

Enochs IC, Manzello DP (2012b) Species richness of motile cryptofauna across a gradient of reef framework erosion. Coral Reefs 31(3):653–661

Enochs IC, Toth LT, Brandtneris VW, Afflerbach JC, Manzello DP (2011) Environmental determinants of motile cryptofauna on an eastern Pacific coral reef. Mar Ecol Prog Ser 438:105–118. https://doi.org/10.3354/meps09259

Fernández-Rivera Melo FJ, Reyes-Bonilla H, Ramírez-Ortiz G, Alvarez-Filip L (2018) Latitudinal variation in structure and function of conspicuous reef fish assemblages along the western Gulf of California. Revista Mexicana de Biodiversidad 89:1154–1166

Fisher R, O'Leary RA, Low-Choy S, Mengersen K, Knowlton N, Brainard RE, Caley MJ (2015) Species richness on coral reefs and the pursuit of convergent global estimates. Curr Biol 25(4):500–505

Fong P, Smith TB, Muthukrishnan R (2017) Algal dynamics: alternate stable states of reefs in the eastern tropical Pacific. In: Glynn PW, Manzello DP, Enochs IC (eds) Coral reefs of the eastern tropical Pacific: persistence and loss in a dynamic environment. Coral reefs of the world 8. Springer, Dordrecht, pp 339–367

Fong P, Smith TB, Wartian MJ (2006) Epiphytic cyanobacteria maintain shifts to macroalgal dominance on coral reefs following ENSO disturbance. Ecology 87:1162–1168

Geister J (1977) Occurrence of *Pocillopora* in Late Pleistocene Caribbean coral reefs. In: Second symposium international sur les coraux et récifs coralliens fossils. Paris, Sept 1975 B. R. G. M. (Paris) Mem. 89, pp 378–388

Glynn PW (1972) Observations on the ecology of the Caribbean and Pacific coasts of Panamá. In: Jones ML (ed) The Panamic biota: some observations prior to a sea-level canal. Bull Biol Soc Wash 2:13–30

Glynn PW (1999) *Pocillopora inflata*, a new species of scleractinian coral (Cnidaria: Anthozoa) from the tropical eastern Pacific. Pac Sci 53(2):168–180

Glynn PW (2006) Fish utilization of simulated coral reef frameworks versus eroded rubble substrates off Panamá, eastern Pacific. In: Proceedings of the 10th international coral reef symposium, Okinawa, vol 1, pp 250–256

Glynn PW (2011) In tandem reef coral and cryptic metazoan declines and extinctions. Bull Mar Sci 87(4):767–794. https://doi.org/10.5343/bms.2010.1025

Glynn PW (2012) Global warming and widespread coral mortality: evidence of first coral extinctions. In: Hannah L (ed) Saving a million species: extinction risk from climate change. Island Press, Washington, DC, pp 103–119

Glynn PW (2013) Fine-scale interspecific interactions on coral reefs: functional roles. In: Lang MA, Marinelli RL, Roberts SJ, Taylor PR (eds) Research and discoveries: the revolution of science through scuba. Smithsonian contributions to the marine sciences, no. 39. Smithsonian Institution Scholarly Press, Washington, DC, pp 229–248

Glynn PW, Ault JS (2000) A biogeographic analysis and review of the far eastern Pacific coral reef region. Coral Reefs 19:1–23

Glynn PW, Enochs IC (2011) Invertebrates and their roles in coral reef ecosystems. In: Dubinsky Z, Stambler N (eds) Coral reefs: an ecosystem in transition. Springer, Dordrecht, pp 273–325. https://doi.org/10.1007/978-94-007-0114-_18

Glynn PW, Glynn CS (1974) On the systematics of *Ancinus* (Isopoda, Sphaeromatidae), with the description of a new species from the tropical eastern Pacific. Pac Sci 28(4):401–422

Glynn PW, de Weerdt WH (1991) Elimination of two reef-building hydrocorals following the 1982–1983 El Niño warming event. Science 253:69–71. https://doi.org/10.1126/science.253.5015.69

Glynn PW, Wellington GM (1983) Corals and coral reefs of the Galápagos Islands. With an annotated list of the scleractinian corals of the Galápagos by John W. Wells. University of California Press, Berkeley, p 330

Glynn PW, Maté JL, Stemann TA (2001) *Pavona chiriquiensis*, a new species of zooxanthellate scleractinian coral (Cnidaria: Anthozoa: Agariciidae) from the eastern tropical Pacific. Proc Biol Soc Wash 10:210–225

Glynn PW, Peters EC, Muscatine L (1985) Coral tissue microstructure and necrosis: relation to catastrophic coral mortality in Panama. Dis Aquat Org 1:29–37

Glynn PW, Stewart RH, McCosker JE (1972) Pacific coral reefs of Panama: structure, distribution, and predators. Geol Rundsch Int J Earth Sci 61(2):483–519

Glynn PW, Enochs IC, McCosker JC, Graefe AN (2008) First record of a pearlfish, *Carapus mourlani*, inhabiting the aplysiid opisthobranch mollusc *Dolabella auricularia*. Pac Sci 62(4):593–601

Glynn PW, Enochs IC, Afflerbach JA, Brandtneris VW, Serafy JE (2014) Eastern Pacific reef fish responses to coral recovery following El Niño disturbances. Mar Ecol Prog Ser 495:233–247. https://doi.org/10.3354/meps10594

Glynn PW, Wellington GM, Riegl B, Olson DB, Borneman E, Wieters EA (2007) Diversity and biogeography of the scleractinian coral fauna of Easter Island (Rapa Nui). Pac Sci 61:67–90

Glynn PW, Alvarado JJ, Banks S, Cortés J, Feingold JS, Jiménez C, Maragos JE, Martínez P, Maté JL, Moanga DA, Navarrete S, Reyes-Bonilla H, Riegl B, Rivera F, Vargas-Ángel B, Wieters EA, Zapata FA (2017) In: Glynn PW, Manzello DP, Enochs IC (eds) Coral reefs of the eastern tropical Pacific: persistence and loss in a dynamic environment. Coral reefs of the world 8. Springer, Dordrecht, pp 107–176

Goldberg WM (2013) The biology of reefs and reef organisms. The University of Chicago Press, Chicago, p 401

Gramer LJ, Soden M, Hendee JC (2023) Operational ecoforecasting for coral reefs using artificial intelligence and integrated near realtime environmental data. Bull Mar Sci 99(3):379–394

Guzmán HM, Cortés J (2007) Reef recovery 20-yr after the 1982-83 El Niño massive mortality. Mar Biol 151:401–411

Guzmán HM, Obando VL (1988) Diversidad y abundancia diaria y estacional del zooplancton marino de la Isla del Caño, Costa Rica. Rev Biol Trop 36:139–150

Guzmán HM, Obando VL, Cortés J (1987) Meiofauna associated with a Pacific coral reef in Costa Rica. Coral Reefs 6:107–112

Heron SF, Eakin CM, Douvere F, Anderson K, Day JC, Geiger E, Hoegh-Guldberg O, Hooidonk R, Hughes T, Marshall P, Obura D (2017) Impact of climate change on world heritage coral reefs. In: A first global scientific assessment. UNESCO World Heritage Centre, Paris

Hoegh-Guldberg O (2012) Coral reefs, climate change, and mass extinction. In: Hannah L (ed) Saving a million species. Extinction risk from climate change. Island Press, Washington, DC, pp 261–283

Jackson JBC (1977) Competition on marine hard substrata: the adaptive significance of solitary and colonial strategies. Am Nat 111:743–767

Jackson JBC (2008) Evolution and extinction in the brave new ocean. Proc Nat Acad Sci USA 105(Suppl 1):11458–11465

Johnston EC, Forsman ZH, Flot J-F, Schmidt-Roach S, Pinzón JH, Knapp ISS, Toonen RJ (2017) A genomic glance through the fog of plasticity and diversification in *Pocillopora*. Nat Sci Rep 7:5991. https://doi.org/10.1038/s41598-017-06085-3

Jones CM, Grutter AS (2007) Variation in emergence of parasitic and predatory isopods among habitats at Lizard Island, Great Barrier Reef. Mar Biol 150:919–927. https://doi.org/10.1007/s00227-006-0416-2

Kaiser KL (2007) The recent molluscan fauna of Île Clipperton (tropical eastern Pacific). Festivus 39:1–162

Keable SJ (1995) Structure of the marine invertebrate scavenging guild of a tropical reef ecosystem: field studies at Lizard Island, Queensland, Australia. J Nat Hist 29:27–46. https://doi.org/10.1080/00222939500770021

Ketchum JT, Reyes Bonilla H (1997) Biogeography of hermatypic corals of the Archipelago Revillagigedo, Mexico. In: Proceedings of the 8th international coral reef symposium, Panama, vol 1, pp 471–476

Ketchum JT, Reyes Bonilla H (2001) Taxonomía y distribución de los corales hermatípicos (Scleractinia) del Archipiélago de Revillagigedo, México. Rev Biol Trop 49(3–4):803–848

Knowlton N (2010) Citizens of the sea, wondrous creatures from the census of marine life. National Geographic Society, Washington, DC, p 209

Knowlton N, Brainard RE, Fisher R, Moews M, Plaisance L, Caley MJ (2010) Coral reef biodiversity. In: McIntyre AD (ed) Life in the World's oceans. Blackwell, London, pp 65–77

LaJeunesse TC, Parkinson JE, Gabrielson PW, Jeong HJ, Reimer JD, Voolstra CR, Santos SR (2018) Systematic revision of Symbiodiniaceae highlights the antiquity and diversity of coral endosymbionts. Curr Biol 28:1–11. https://doi.org/10.1016/j.cub.2018.07.008

LaJeunesse TC, Smith R, Walther M, Pinzón J, Pettay DT, McGinley M, Aschaffenburg M, Medina-Rosas P, Cupul-Magaña AL, López Pérez A, Reyes-Bonilla H, Warner ME (2010) Host-symbiont recombination versus natural selection in the response of coral-dinoflagellate symbioses to environmental disturbance. Proc R Soc B 277:2925–2934. https://doi.org/10.1098/rspb.2010.0385

Lessios HA, Baums IB (2017) Gene flow in coral reef organisms of the tropical eastern Pacific. In: Glynn PW, Manzello DP, Enochs IC (eds) Coral reefs of the eastern tropical Pacific: persistence and loss in a dynamic environment. Coral reefs of the world 8. Springer, Dordrecht, pp 477–499

Lessios HA, Robertson DR (2006) Crossing the impassable genetic connections in 20 reef fishes across the eastern Pacific barrier. Proc R Soc B Biol Sci 273:2201–2208

Lessios HA, Kessing BD, Robertson DR (1998) Massive gene flow across the world's most potent marine biogeographic barrier. Proc R Soc Lond B 265:583–588

López-Pérez A (2017) Revisiting the Cenozoic history and the origin of the eastern Pacific coral fauna. In: Glynn PW, Manzello DP, Enochs IC (eds) Coral reefs of the eastern tropical Pacific: persistence and loss in a dynamic environment. Coral reefs of the world 8. Springer, Dordrecht, pp 39–57

May RM (1973) Stability and complexity in model ecosystems. Princeton University Press, Princeton, New Jersey

McNeill DF, Klaus JS, Budd AF, Lutz BP, Ishman SE (2012) Late Neogene chronology and sequence stratigraphy of mixed carbonate-siliciclastic deposits of the Cibao Basin, Dominican Republic. Geol Soc Am Bull 124(1/2):35–58. https://doi.org/10.1130/B30391.1

Mora C, Robertson DR (2005) Causes of latitudinal gradients in species richness: a test with fishes of the tropical eastern Pacific. Ecology 86(7):1771–1782. https://doi.org/10.1890/04-0883

Mora C, Tittensor DP, Adl S, Simpson AGB, Worm B (2011) How many species are there on earth and in the ocean? PLoS Biol 9(8):e1001127. https://doi.org/10.1371/journal.pbio1001127

Paine RT (1966) Food web complexity and species diversity. Am Nat 100(910):65–75

Palacios MM, Muñoz CG, Zapata FA (2014) Fish corallivory on a pocilloporid reef and experimental coral responses to predation. Coral Reefs 33:625–636. https://doi.org/10.1007/s00338-014-1173-y

Pandolfi JM, Jackson JBC (2001) Community structure of Pleistocene coral reefs of Curaçao, Netherlands Antilles. Ecol Monogr 71(1):49–67

Pandolfi J, Jackson JBC, Geister J (2001) Geologically sudden extinction of two widespread late Pleistocene Caribbean reef corals. In: Jackson JBC, Lidgard S, McKinney K (eds) Evolutionary patterns: growth, form, and tempo in the fossil record. University of Chicago Press, Chicago, pp 120–158

Parkinson JE, LaJeunesse TC (2021) Solar symbionts. Coral Mag 18(5):28–49

Paulay G (1997) Diversity and distribution of reef organisms. In: Birkeland C (ed) Life and death of coral reefs. Chapman & Hall, New York, pp 298–353

Paz-García DA, Hellberg ME, García-de-León FJ, Balart EF (2015) Switch between morphologies of *Pocillopora* corals. Am Nat 186(3):434–440. https://doi.org/10.1086/682363

Pettay DT, LaJeunesse TC (2013) Long-range dispersal and high-latitude environments influence the population structure of a "stress-tolerant" dinoflagellate endosymbiont. PLoS One 8:e79208

Pinzón JH, LaJeunesse TC (2011) Species delimitation of common reef corals in the genus *Pocillopora* using nucleotide sequence phylogenies, population genetics and symbiosis ecology. Mol Ecol 20:311–325. https://doi.org/10.1111/j.1365-294X.2010.04939.x

Pinzón JH, Sampayo E, Cox E, Chauka LJ, Chen CA, Voolstra CR, LaJeunesse TC (2013) Blind to morphology: genetics identifies several widespread ecologically common species and few endemics among Indo-Pacific cauliflower corals (*Pocillopora*, Scleractinia). J Biogeogr 40:1595–1608

Porter JW (1972) Ecology and species diversity of coral reefs on opposite sides of the Isthmus of Panama. In: Jones ML (ed) The Panamic biota: some observations prior to a sea-level canal. Bull Biol Soc Wash 2:89–116

Przeslawski R, Ahyong S, Byrne M, Worheide G, Hutchings P (2008) Beyond corals and fish: the effects of climate change on noncoral benthic invertebrates of tropical reefs. Glob Chang Biol 14:2773–2795. https://doi.org/10.1111/j.1365-2486.2008.01693.x

Razak TB, Hoeksema BW (2003) The hydrocoral genus *Millepora* (Hydrozoa: Capitata: Milleporidae) in Indonesia. Zoologische Verhandelingen 345:313–336

Reaka-Kudla ML (1997) The global biodiversity of coral reefs: a comparison with rainforests. In: Reaka-Kudla ML, Wilson DE, Wilson EO (eds) Biodiversity II: understanding and protecting our biological resources. Joseph Henry Press, Washington, DC, pp 83–108

Reaka-Kudla ML (2005) Biodiversity of Caribbean coral reefs. In: Miloslavich P, Klein E (eds) Caribbean marine biodiversity. Des Tech Publishers, Lancaster, Pennsylvania, pp 259–276

Remple KL, Silbiger NJ, Quinlan ZA, Fox MD, Kelly LW, Donahue MJ, Nelson CE (2021) Coral reef biofilm bacterial diversity and successional trajectories are structured by reef benthic organisms and shift under chronic nutrient enrichment. NPJ Biofilms Microbiomes 7:84. https://doi.org/10.1038/s41522-021-00252-1

Reyes-Bonilla H (2002) Checklist of valid names and synonyms of stony corals (Anthozoa: Scleractinia) from the eastern Pacific. J Nat Hist 36:1–13

Reyes-Bonilla H, Carricart-Ganivet JP (2000) *Porites arnaudi*, a new species of stony coral (Anthozoa: Scleractinia: Poritidae) from oceanic islands of the eastern Pacific. Proc Biol Soc Wash 113:561–571

Reyes-Bonilla H, López-Pérez RA (2009) Corals and coral reef communities in the Gulf of California. In: Johnson ME, Ledesma-Vázquez J (eds) Atlas of coastal ecosystems in the Western Gulf of California: tracking limestone deposits on the margin of a Young Sea. University of Arizona Press, Tucson, pp 43–55

Richmond RH (1990) The effects of the El Niño-Southern Oscillation on the dispersal of corals and other marine organisms. In: Glynn PW (ed) Global ecological consequences of the 1982–83 El Niño-Southern Oscillation. Elsevier oceanography series 52. Elsevier, Amsterdam, pp 127–140

Robertson DR (1972) Social control of sex reversal in a coral reef fish. Science 177:1007–1009

Robertson DR, Allen GR (2015) Shorefishes of the tropical eastern Pacific: online information system. Version 2.0. Smithsonian Tropical Research Institute, Balboa, Panamá. https://biogeobd.stri.si.edu/sftep/en/pages

Robertson DR, Cramer KL (2009) Shore fishes and biogeographic subdivisions of the tropical eastern Pacific. Mar Ecol Prog Ser 380:1–17. https://doi.org/10.3354/meps07925

Robertson DR, Smith-Vaniz WF (2008) Rotenone: an essential but demonized tool for assessing marine fish diversity. Bioscience 58(2):165–170

Robertson DR, Smith-Vaniz WF (2010) Use of clove oil in collecting coral reef fishes for research. Mar Ecol Prog Ser 401:295–302. https://doi.org/10.3354/meps08374

Rohwer F, Youle M (2010) Coral reefs in the microbial seas. Plaid Press, Granada Hills, California, p 201

Rosenblatt RH, Waples RS (1986) A genetic comparison of allopatric populations of shore fish species from the eastern and Central Pacific Ocean: dispersal or vicariance? Copeia 1986:275–284. https://doi.org/10.2307/1444988

Rosenblatt RH, McCosker JE, Rubinoff I (1972) Indo-West Pacific fishes from the Gulf of Chiriquí, Panamá. Contrib Sci Nat Hist Mus 234:1–18

Rowan R, Powers DA (1991) Molecular genetic identification of symbiotic dinoflagellates (zooxanthellae). Mar Ecol Prog Ser 71:65–73

Rudnick DL, Brechner Owens W, Shaun Johnston TM, Karnauskas KB, Jakoboski J, Todd RE (2021) The equatorial current system west of the Galápagos Islands during the 2014-16 El Niño as observed by underwater gliders. J Phys Oceanogr 51:3–17. https://doi.org/10.1175/JPO-D-20-0064.1

Salas-Moya C, Vargas-Castillo R, Alvarado JJ, Azofeifa-Solano JC, Cortés J (2021) Decapod crustaceans associated with macroinvertebrates in Pacific Costa Rica. Mar Biodiver Rec 14:6. https://doi.org/10.1186/s41200-020-00199-w

Scheffers BR, Joppa LN, Pimm SL, Laurance WF (2012) What we know and don't know about Earth's missing biodiversity. Trends Ecol Evol 27(9):501–510

Skaggs JM (1989) Clipperton: A history of the Island the world forgot. Walker, New York, p 318

Small AM, Adey WH, Spoon D (1998) Are current estimates of coral reef biodiversity too low? The view through the window of a microcosm. Atoll Res Bull 458:1–20

Smith TB (2005) The dynamics of coral reef algae in an upwelling system. PhD dissertation. University of Miami, Coral Gables, p 156

Smith TB (2006) Benthic algal production in an upwelling coral reef environment (Gulf of Panamá): evidence for control by fish herbivores. In: Proceedings of the 10th international coral reef symposium, Okinawa, vol 1, pp 854–861

Smith TB (2008) Temperature effects on herbivory for an Indo-Pacific parrotfish in Panamá: implications for coral-algal competition. Coral Reefs 27:397–405

Smith TB, Fong P, Kennison R, Smith J (2010) Spatial refuges and associational defenses promote harmful blooms of the alga *Caulerpa sertularioides* onto coral reefs. Oecologia 164:1039–1048

Solano-Barquero A, Sibaja-Cordero JA, Cortés J (2022) Macrofauna associated with a rhodolith bed at an oceanic island in the eastern tropical Pacific (Isla del Coco National Park, Costa Rica). Front Mar Sci 9:858416. https://doi.org/10.3389/fmars.2022.858416

Squires DF (1959) Reports of the Puritan-American Museum of Natural History expedition to Western Mexico, 7. Corals and coral reefs in the Gulf of California. Bull Am Mus Nat Hist 118(art. 7):367–432

Stella JS, Jones GP, Pratchett MS (2010) Variation in the structure of epifaunal invertebrate assemblages among coral hosts. Coral Reefs 29:957–973. https://doi.org/10.1007/s00338-010-0648-8

The Nature Conservancy (2022) Reef resilience network. Retrieved 2022, from https://reefresilience.org/

Toth LT, Kuffner IB, Cheng H, Edwards RL (2015) A new record of the late Pleistocene coral *Pocillopora palmata* from the Dry Tortugas, Florida reef tract, USA. PALAIOS 30:827–835. https://doi.org/10.2110/palo.2015.030

Turnham KE, Wham DC, Sampayo E, LaJeunesse TC (2021) Mutualistic microalgae co-diversify with reef corals that acquire symbionts during egg development. ISME J 15(11):3271–3285

Vasseur P (1974) The overhangs, tunnels, and dark reef galleries of Tulear (Madagascar) and their sessile invertebrate communities. In: Proceedings of the 2nd international coral reef symposium, Australia, vol 2, pp 143–159

Veron JEN (1995) Corals in space and time, the biogeography & evolution of the scleractinia. Comstock/Cornell, Ithaca, p 321

Veron JEN (2000a) Corals of the world, vol 2. Australian Institute of Marine Science, Townsville, p 429

Veron JEN (2000b) Corals of the world, vol 3. Australian Institute of Marine Science, Townsville, p 490

Verrill AE (1868) Review of the corals and polyps of the west coast of America. Trans Conn Acad Arts Sci 1(6):377–567

Voolstra CR, Suggett DJ, Peixoto RS, Parkinson JE, Quigley KM, Silveira CB, Sweet M, Muller EM, Barshis DJ, Bourne DJ, Aranda M (2021) Extending the natural adaptive capacity of coral holobionts. Nat Rev Earth Environ 2:747–762

Wang C, Deser C, Yu J-Y, DiNezio P, Clement A (2017) El Niño and Southern Oscillation (ENSO): a review. In: Glynn PW, Manzello DP, Enochs IC (eds) Coral reefs of the eastern tropical Pacific: persistence and loss in a dynamic environment. Coral reefs of the world 8. Springer, Dordrecht, pp 85–106

Wehrtmann IS, Cortés J (eds) (2009) Marine biodiversity of Costa Rica, Central America. Monographiae Biologicae 86. Springer, Dordrecht, p 538

Wilson EO (1992) The diversity of life. W. W. Norton, New York, p 424

Wood S, Baums IB, Paris CB, Ridgwell A, Kessler WS, Hendy EJ (2016) El Niño and coral larval dispersal across the eastern Pacific marine barrier. Nat Commun 7:12571. https://doi.org/10.1038/ncomms12571

Wyrtki K (1975) El Niño—the dynamic response of the equatorial Pacific Ocean to atmospheric forcing. J Phys Oceanogr 5(4):572–584

Zapata FA, Morales YA (1997) Spatial and temporal patterns of fish diversity in a coral reef at Gorgona Island, Colombia. In: Proceedings of the 8th international coral reef symposium, Panama, vol 1, pp 1029–1034

Zapata FA, Robertson DR (2007) How many species of shore fishes are there in the tropical eastern Pacific? J Biogeogr 34:38–51

Zapata FA, Gaston KJ, Chown SL (2003) Mid-domain models of species richness gradients, assumptions, methods and evidence. J Anim Ecol 72:677–690

Chapter 11
Symbiosis: Microorganisms to the Megafauna

Abstract The three classical types of symbiotic relationships, i.e., those species living in close, heterospecific associations—commensalism, mutualism, and parasitism—are examined from coral reef environments in the Caribbean (Puerto Rico), eastern Pacific, and southeastern Florida. Recent studies show that many formerly regarded commensal relationships, where one partner benefits and the other is unaffected, are more accurately classified as examples of mutualism or parasitism. Both symbiotic partners benefit in mutualistic relationships, often revealing exceptional coevolutionary adaptations. Exemplary cases are (a) coral crustacean guards, (b) shrimp and fish cleaner-client relationships, and (c) damselfish-algal lawn interactions. Parasitism, wherein one partner benefits at the expense of another, has not received sufficient attention in eastern Pacific coral reef studies. Parasites, pathogens, and diseases are interdependent, and their effects or impacts largely depend on abiotic and biotic environmental stressors. The cause(s) of coral bleaching and mortality were vigorously investigated in the early 1980s, following the first reported worldwide reef disturbance. Diseases affecting corals (white syndrome, "trematodiasis") and sea urchins (a ciliate protozoan) are of concern relative to eastern Pacific coral reefs. Lithophagine date mussels, long regarded as parasites of corals, benefit their hosts by providing a nitrogen source that increases growth and may best be regarded as mutualists.

Introduction

One of the great fascinations of coral reefs is the profusion of species that live in close association with each other, in large part a result of the extraordinarily high biodiversity of coral reefs. Species crowd reefs, and if species living in close and intimate relationships gain a selective advantage, i.e., an increase in fitness, such associations evolve. **Symbiosis** is the term used to designate close heterospecific associations of two or more species living together, often for protracted periods. These relationships involve a **host** species which provides a habitat and sustenance for a **guest** species. The host is usually larger than the guest(s), which may number

© The Author(s), under exclusive license to Springer Nature Switzerland AG 2025
P. W. Glynn, *Coral Reef Resilience in the Anthropocene*,
https://doi.org/10.1007/978-3-031-81421-1_11

hundreds to thousands of individuals in a given relationship. In the original definition of symbiosis, as "phenomena of dissimilar organisms living together" (de Bary 1879), no distinction was made regarding harm or benefit to the partners. This early interpretation, from a botanical perspective, involved lichens, symbiotic associations of algae and fungi.[1]

Three general classes have been traditionally recognized in the study of symbioses, namely, commensalism, mutualism, and parasitism. **Commensalism** is a relationship where one partner benefits and the other is unaffected. In a **mutualism**, both partners derive a benefit from the association. **Parasitism** is an association in which one partner benefits at the expense of the other. Continuing research has demonstrated the difficulty of assigning particular symbiotic relationships to these general classes. Peter Castro (1988, 2015),[2] in the context of coral reef symbioses, has called attention to the nuances and often weak and shifting boundaries separating these relationships. Leung and Poulin (2008) argued that many examples of symbiosis are highly plastic, never fixed, and easily pushed from one traditional category to another, depending on environmental conditions, life history traits, ontogenetic (developmental) stage, *inter alia*. Likely, most of the interspecific associations encountered in our coral reef studies and discussed here are symbioses, but which type will require further study. The use of metabolites, a currency employed in classifying symbioses, should continue to be advantageous in revealing nutritional relationships and interdependencies. For example, commensal symbionts are associated with their hosts without exhibiting metabolic dependency, mutualists exhibit reciprocal metabolic dependency, and parasites are metabolically dependent on their hosts. As more information on the biology of symbionts is attained, many of the early partnership assignments will require revision.

Cognizant of the difficulties in making correct assignments to the three types of symbioses, particularly in the absence of critical biological information, the associations presented here are grouped under the three traditional categories. This arrangement is convenient for accessing the older literature and helpful in searching

---

[1] A brief, informative account of de Bary's 1878 address on symbiosis, academic life, and research achievements is available in Paracer and Ahmadjian 2000.

[2] Peter Castro was the first Puerto Rican student with whom I seriously discussed coral reef biology. We enjoyed lively discussions and informative field experiences on the coral reefs offshore of La Parguera (Magueyes Island), the location of the University of Puerto Rico's (UPR Mayagüez) marine laboratory. The western UPR campus is now referred to as the *Recinto Universitario de Mayagüez* or RUM and the field station as Magueyes Marine Laboratory. Upon graduation from UPR Mayagüez, Peter entered the graduate program at the University of Hawaii where he earned M.S. and Ph.D. degrees in Marine Zoology, specializing in parasitology and the biology of symbiosis. He joined the Biological Sciences Department faculty at the California State Polytechnic University, Pomona in 1972 where he remained a dedicated teacher and researcher for four decades. He lightened his teaching load in 2003, was appointed Professor Emeritus and devoted his time to revising the enormously popular text, *Marine Biology* (2024, 12th edition), with coauthor Michael E. Huber (Castro and Huber 2024). In addition, he is serving as Editor-in-Chief of the Journal of Crustacean Biology. Peter established and funded several scholarships in the environmental sciences, including the Rachel Carson Environmental Science Scholarship and Peter Castro scholarships supporting studies in evolution, biodiversity, and faculty research.

for recently reported studies of symbiotic relationships.[3] The following terminology and definitions are used here to describe the various symbiotic associations. **Ectosymbionts** refers to guest organisms that reside on external surfaces and **endosymbionts** to those located internally, e.g., in body cavities, skeletons, tissues, and cells. Symbiont **persistence** relates to the time that species associations occur and can vary from permanent[4] to intermittent, e.g. during each species'entire life cycle or only during short periods, such as cleaner shrimp briefly visiting a parasitized fish. Most endoparasites leave their host as larvae or spores at some stage to find and infect new hosts. Organisms that can exist either in a symbiotic relationship or outside of it, i.e. not dependent on a living host, are termed **facultative symbionts**. This type of relationship exists to different degrees, e.g., some facultative symbionts associate with a specific host species whereas others are undemanding (promiscuous) and may coexist with different species. **Obligate symbionts** require a living host and will not survive without one. These symbionts can be exacting, requiring a unique host species or a range of related species with similar ecological functions, but all living. The coral holobiont is a quintessential example, hosting various endosymbiotic algal species (zooxanthellae). Both coral host and zooxanthellae are mutually co-dependent.

Metazoan symbioses, i.e. invertebrates and fishes, are a principal focus of this chapter. It is nevertheless essential to remember that reef-building corals, hosting diverse zooxanthellae and other microbial entities, represent an exemplary example of symbiosis (see Chaps. 5, 10). There is strong evidence that ancestral corals hosted photosynthetic endosymbionts as early as the Late Triassic, about 212 Ma (Frankowiak et al. 2016). This relationship promoted the formation of structural coral reefs and concomitant living conditions for the great variety of life forms on coral reefs today. The critical functions of zooxanthellae and other microbial entities in coral symbioses, in terms of calcification, nutrition, immunity, resilience and persistence of interspecific relationships, are addressed in a rich contemporary literature (e.g., Baker 2003; Rohwer and Youle 2010; Stambler 2011; Krediet et al. 2013; Webster and Reusch 2017). Going back much further in the evolutionary history of life, the theory of symbiogenesis posits the origin of eukaryotic cells by the acquisition of organelles (e.g., flagella, mitochondria, and chloroplasts) via a

[3]Widely recognized early sources of information on tropical, marine symbioses, including coral reef examples, can be found in Dales (1957, 1966), Muscatine (1961), Feder (1966), Robertson (1970), Vernberg (1974), Patton (1976), and Castro (1988)

[4]I thank Peter Castro for bringing a marine nematode species to my attention, a rare example of an endoparasite that remains with its host permanently (Oksanen et al. 2022). This association relates to several species of *Trichinella* that inhabit the intestines of terrestrial and marine mammals. Polar bears, arctic foxes, walruses and several other Arctic and subarctic mammals have these adult parasitic worms. The larvae hatch from eggs that develop inside the uterus of nematode females and then migrate from the intestine to skeletal muscles in the host. There they encyst and the next host becomes infected when eating the muscle tissues of the initial host's prey. Seals and walruses are among the marine mammals carrying the parasite, and humans (predominantly Indigenous Peoples of the North), become infected by eating raw or improperly cooked game.

symbiotic incorporation and coevolution of prokaryotic free-living microbes (e.g., Margulis 1996, 1998; Paracer and Ahmadjian 2000).[5]

In this chapter I share our observations and studies of some examples of the great variety of symbioses encountered on coral reefs. Emphasis is placed on eastern Pacific coral reefs with examples from Puerto Rican reefs and octocoral communities in southeast Florida. Some of these associations will hopefully reveal new and valuable information on the biology of the symbionts and their functional interrelationships.

Commensalism

The numbers and variety of reported marine commensal relationships are remarkable (e.g., Dales 1957; Bruce 1976; Castro 1988). In the early reviews of commensalism, 'hydroid-invertebrate/fish' and 'sea anemone/crab' partnerships were abundantly cited, each consisting of nearly 30 distinct symbiotic relationships. Overall, host taxa included all major animal groups, ranging from sponges and cnidarians to mollusks, crustaceans, echinoderms, and fishes. Most of these examples reported on the effects of one partner, usually the guest symbiont, benefiting from feeding with or on the host. Often it was assumed that no harm or benefit befell the host. Commensalism originates from the Latin word *commensalis*, derived from *com* (together) and *mensa* (table), referring to a relationship in which both partners feed together and share their food. Some recent studies show that formerly recognized classes of symbiosis, e.g., "commensal" relationships, should more accurately be considered examples of mutualism or parasitism.

Multispecies group foraging by fishes is often observed on coral reefs where mixed species schools enhance prey capture. Hunting behavior of mixed species groups of piscivores are common in Costa Rica, on coastal coral reefs (Auster et al. 2019) and at depths to ~300 m at oceanic Cocos Island (Auster et al. 2016). From personal observations at Uva reef, schools of jacks (*Gnathanodon speciosus*) were often seen lifting with their mouths and jostling falling *Pocillopora* colonies in the water column to facilitate the capture of crustacean symbionts. The **focal** jack feeding behavior attracts wrasses, hawk fish, and other fish **followers** that also capture crustacean prey. Triggerfishes (*Pseudobalistes naufragium* and *Sufflamen verres*) excavating coral substrates in search of prey also attract diverse fish predators.

[5] I was privileged to meet Lynn Margulis and her student, David Bermudes, in 1987 at the First Eilat Symposium on Marine Symbioses. They presented a provocative hypothesis positing the sudden appearance of new species and possibly higher taxa via symbiont acquisition and integration (Bermudes and Margulis 1987). Their explanation of evolution was iconoclastic because it challenged Darwin's gradualist tenet of the accumulation of mutations as the fundamental driver of evolution. This view of eukaryotic cellular evolution was first proposed by Dr. Margulis in 1967 (Sagan 1967) when married to Carl Sagan, the celebrated American astronomer and planetary scientist. The endosymbiotic origin of cells has received significant support from molecular and genetic studies (Mukherjee 2022).

These examples qualify as commensal interactions wherein the opportunistic fish followers would benefit from increased prey exposure and capture, and the focal foragers would not be affected. In the feeding melees the diverse strategies of prey capture by fish followers may benefit the focal excavators by enhancing access to shelter sites and exposing potential prey. Facilitative, symbiotic interactions play important roles in community ecology (e.g., Sousa 2001; Stachowicz 2001).

Isopods living in close association with intertidal chitons first caught my attention in Puerto Rico in 1965 while sampling the molluscan hosts to determine their reproductive condition. A high rate of co-occurrence was generally observed with numerous isopods concentrated along the pallial grooves, and often wedged between the chiton's gill filaments (Fig. 11.1). The guts of the isopod species, *Dynamenella perforata*, were usually packed with algae, the same plant species being scraped from the substrate and consumed by the chiton, *Acanthopleura granulata*. Since no damage to the chiton's gills was apparent and the isopods benefited from the shelter and food source of the mollusk, this heterospecific relationship was regarded as a commensalism (Glynn 1968).[6] More specifically, this appeared to be a facultative symbiotic relationship because *D. perforata* was also observed sheltering under another chiton species, *Chiton tuberculatus*. Not investigated in this study was the possibility that the isopods were benefiting the chiton host by removing excess algal

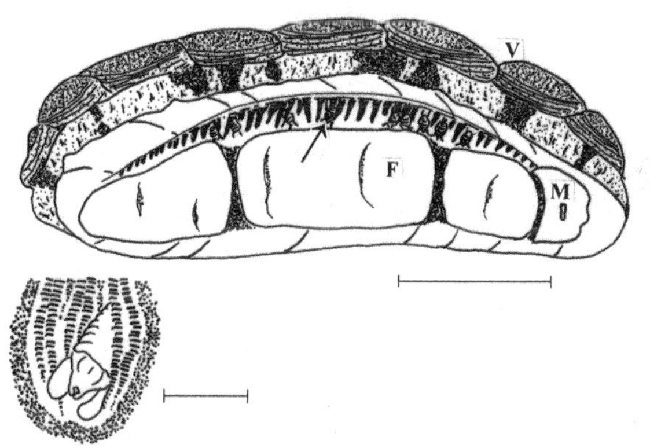

Fig. 11.1 Tilted lateroventral view of *Acanthopleura granulata* hosting isopods (*Dynamenella perforata*) snuggled among its gill filaments. Arrow points to one of several isopod symbionts. The enlarged inset drawing illustrates an isopod's orientation within the gill chamber. Lettered are chiton's foot (F), mouth (M), and calcareous skeletal valves (V). Scale bars = 2 cm (chiton) and 2 mm (isopod)

[6] Publication of this study, one of my first, was not without trepidation and added concern. Several weeks after submission of the study to the Bulletin of Marine Science (BMS), the BMS editorial office informed me that there had been a fire with the loss of numerous manuscripts. The letter contained both good and bad news. The good news was that the study was accepted for publication; the bad news, several of the original illustrations were destroyed by the fire. This was before the availability of copy machines and digital imaging. Fortunately, photographic copies of all materials were made before posting the original manuscript submission. Happily, I was able to reproduce and supply the requested figures posthaste.

fragments and other potential contaminants (organisms and detritus) from the gill filaments. This is still an unresolved possibility.

Coincidentally, this same isopod-chiton symbiosis was discovered in the Bahamas in a parallel study in 1967 (Brattegard 1968). This study concluded that the isopods chiefly sheltered with chitons to avoid desiccation and possibly predation. Our investigations demonstrated that the isopods could also benefit from a food source supplied by the chiton host. We performed gut analyses on isopods associated with feeding chiton hosts and found that the guts of several individuals were packed with the same algal species that the chitons were ingesting. Further, the guts of isopods examined from non-grazing chiton hosts were devoid of algae. In keeping with the term "commensal", both partners were indeed feeding from the same table and benefiting from seemingly identical diets. The isopods, however, were feeding on uneaten food scrapings that the chiton host was voiding.

While this study focused chiefly on the *Dynamenella/Acanthopleura* association, a surprising number of similar isopod/chiton and other invertebrate/chiton relationships were observed in Puerto Rico. In addition to the *Dynamenella/Acanthopleura* and *Dynamenella/Chiton* symbioses discussed above, three additional isopod/chiton partnerships were discovered in the coral reef intertidal zone (Glynn 1968). These involved three different isopod species in specific associations with *C. tuberculatus* and *Chiton marmoratus*. The inhalant chamber of the pallial groove, along which uneaten food particles move, was the most frequented site of the isopods. Other metazoans found sheltering in the pallial groove and spaces under the chiton's foot included a foraminiferan, two species of harpacticoid copepods, and a springtail (Subclass Collembola, closely related to insects). Caribbean chitons are important hosts in intertidal reef zone habitats, providing shelter (from desiccation and predators) and a food source for diverse symbionts.

Compared with the Indo-West Pacific and Eastern Tropical Pacific regions, coral-crab symbioses are uncommon in the tropical western Atlantic. When extensive and healthy stands of *Acropora palmata* were omnipresent in Puerto Rico in the 1960s, however, a coral-crab association was commonly observed on the reefs offshore of La Parguera. This association involved the small trapezioid crab, *Domecia acanthophora*, and one of its most frequent coral hosts, *Acropora palmata*. Close inspection of the spreading fronds of *Acropora* would often reveal *Domecia*, at rest in crevices or pits or scurrying over the coral's surface. Like the well-known Indo-Pacific gall crabs (Cryptochiridae), *Domecia* also forms pit-like dwellings in the coral skeleton (Hoeksema and García-Hernández 2020). From a study of the crab's mouthparts and gut contents, Wendell Patton (1967) concluded that *Domecia* feeds on organic detritus captured from the water and possibly on its host's copious mucus secretions. Since *Domecia* received shelter and an enhanced opportunity to feed, with no apparent harm to its coral host, the partnership was considered an example of commensalism. Because *Domecia* is not solely dependent on live *Acropora palmata* but also occurs in other associations with or without live coral, e.g., other *Acropora* species and *Millepora complanata* (Hoeksema and García-Hernández 2020), it is more precisely a facultative commensal. This kind of nonrestrictive (facultative) relationship may shed light on the origin of coral-crab symbioses. Other coral-crab relationships that involve more specific and directed interactions, likely via coevolutionary processes that result in joint beneficial effects, are considered below under "Mutualisms".

The pearl fish/sea hare association we investigated in Panama is perhaps closer to a strict commensal symbiosis, with the fish symbiont receiving shelter in the mollusk's mantle cavity and the mollusk host experiencing no apparent harm. The pearl fish, *Carapus mourlani,* widely distributed in the tropical Indo-West Pacific region, was observed at the Uva Island coral reef sheltering in the mantle cavity of *Dolabella auricularia*, an opisthobranch mollusk (Glynn et al. 2008). This was the first reported association of the fish symbiont with a sea hare (Fig. 11.2). Pearlfish usually form partnerships with echinoderms. The reported hosts of pearl fish in Costa Rica are sea stars and sea cucumbers (Chacón-Monge et al. 2021).

Our most informative discoveries of the pearl fish/sea hare symbiosis were made from aquarium observations of the nocturnal feeding behavior of *C. mourlani*. We maintained this symbiosis over three months by supplying freshly collected algae with naturally associated crustaceans. A remote, night-vision, digital video recorder and an infrared (IR) light emitting diode (LED) illuminator were positioned over the aquarium and programmed to turn on automatically at low light levels, beginning before sundown and ending after sunrise. A low-light IR sensitive camera was connected to a DVR and positioned to record the symbiont's activities. Video files were uploaded to a computer for review of nocturnal activities. On several nights, the IR video recordings showed *C. mourlani* feeding on swimming crustaceans (amphipods, ostracods, and mysids) that had emerged from the algae. Details of prey detection, capture and ingestion were visible in the video recordings. For example, pearlfish detected their crustacean prey in nearly total darkness, seemingly from the swimming motion of the prey. Crustacean prey was grasped, crushed, and the body parts ingested. Some amphipods were captured, chewed, and then rejected. These aquarium observations suggested that the grazing activities of *Dolabella* in the field might cause an escape behavior of crustaceans that *Carapus* would more easily detect. This hypothesized foraging behavior would be akin to hamlet fish that follow grazing parrotfish and surgeon fish schools. The carnivorous hamlets, feeding on invertebrates and fishes flushed from algal turfs, benefit from the activities of the grazing herbivore schools (Ogden and Buckman 1973). This symbiosis, an example

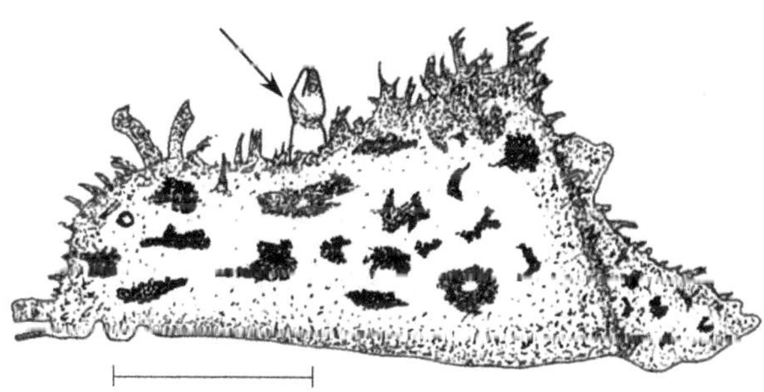

Fig. 11.2 *Dolabella auricularia* hosting *Carapus mourlani* (arrow), scale bar = 4 cm

of commensalism, exemplifies the benefit from eating at the same table, in this case from the same mat of benthic algae.

Before moving on to other types of symbioses, as a precautionary measure it is instructive to be fully aware of the interactions and effects used to determine "commensal" symbiotic relationships. An example of a perceived commensal relationship between a benthic ctenophore and its octocoral hosts illustrates this difficulty. This case involves the ctenophore, *Coeloplana waltoni*, an obligate symbiont that lives and captures zooplankton with its tentacles from the branches of several octocoral species off southeastern Florida (Glynn et al. 2014) (Fig. 11.3). In 2015, when I was approaching retirement, several undergraduate students in the marine invertebrate course I was teaching became interested in assisting with investigations of the *Coeloplana*/octocoral symbiosis.[7] During our study of the feeding behavior of *Coeloplana* we observed ctenophore tentacles entering and pilfering food captured by octocoral polyps (Glynn et al. 2017b). On other occasions, we observed octocoral host polyps pilfering the ctenophore's food (Glynn et al. 2018). Perhaps the *Coeloplana* symbiont benefited nutritionally under certain conditions and the octocoral host at other times. In the first instance, food capture was like that described in a hydroid-crustacean (caprellid amphipod) relationship, a case of kleptocommensalism (Bavestrello et al. 1996).

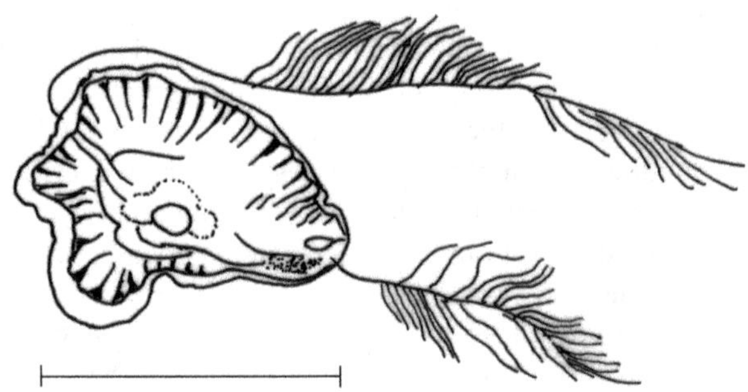

Fig. 11.3 *Coeloplana waltoni* with extended feeding tentacles, scale bar = 2 mm

[7] Rumors began circulating in the marine biological community that benthic ctenophore research was underway at RSMAES. I was invited to attend a two-day conference (Ctenopalooza) on ctenophore research at the Whitney Laboratory, University of Florida, in St. Augustine. This two-day meeting in 2016, attended by 75 researchers, was the first international meeting dedicated to ctenophore biology. Thankfully, this invitation included my student assistants, Brian Coffman, Tayla N. Forston, Michael P. C. Fuller, and Shannon G. Moorhead, who all received registration discounts and travel stipends. At the closing banquet ceremony, to my delight the first-place poster prize was awarded to our team's student poster presentation of benthic ctenophore ecology. The notoriety and interest that followed from our laboratory's studies attracted new student recruits and continuing ctenophore research to 2021.

Mutualism

The reciprocal adaptive responses of closely interacting species, leading to symbiotic mutualisms, often exhibit fascinating results on coral reefs. Two such mutualisms, coral host defense by crustacean guards and cleaning symbioses, are examined here. Both of these symbiotic relationships occur commonly in the eastern tropical Pacific. The mutualism of crustacean guards, first investigated in the eastern tropical Pacific (Panama) in the early 1970s[8] is reviewed, with additional personal observations from various areas in the Indo-West Pacific region. An overview of shrimp and fish cleaning symbioses is from interactions observed in Puerto Rico and the eastern tropical Pacific. Information and insights by others in this popular field of study are accredited where relevant. Also considered is the symbiotic relationship between date mussels (*Lithophaga* spp.) and the corals they bore into because of their prevalence and importance in bioerosion and asexual propagation in the eastern tropical Pacific. Finally, mention is made of Mark Eakin's convincing argument that damselfishes and their algal lawns should be viewed as an example of mutualism (Eakin 1987).

Coral crustacean guards. One of my most enjoyable and satisfying research experiences was observing and documenting the behavior and interactions of crabs and shrimp that protect their coral hosts. The crustacean defensive behaviors are fascinating in and of themselves, but of added consequence are their ecological effects on coral survival and community structure. The defensive behaviors of *Trapezia* spp. crab and *Alpheus lottini* shrimp guards (now known to involve several cryptic species, e.g., Rouzé et al. 2017) were introduced in the section, "*Acanthaster* and Coral Community Diversity", in Chap. 7.[9] Several branching pocilloporid coral species are engaged in symbiotic relationships mutually beneficial to coral hosts and crustacean guards alike. Initially, this symbiotic relationship was viewed as parasitism or commensalism (Knudsen 1967; Robertson 1970), and more recently mutualism (Glynn 1983a, b; Castro 1988; Pratchett 2001; Stewart et al. 2006; McKeon and Moore 2014). In-depth studies have revealed that corals, species in the families Pocilloporidae and Acroporidae, provide crustacean symbionts with a

[8] The first published accounts of crab guards defending their coral hosts from *Acanthaster* attack were observed in the southwestern Pacific region (Pearson and Endean 1969; Weber and Woodhead 1970). Crustacean-host defensive interactions were first quantified in Panama in 1974 (Glynn 1976). These included **up-down jerking** (Fig. 7.8), **nipping**, i.e., pinching *Acanthaster's* spines and tube feet (Fig. 7.9), causing but **resisting** sea star retreat (Fig. 7.10), and **snapping** by pistol shrimp guards (Fig. 7.11). These various defensive responses were observed and recorded during standardized 3-minute intervals (see Chap. 7).

[9] *Trapezia* spp., family Trapeziidae, are obligate symbionts of corals in the family Pocilloporidae (*Pocillopora*, *Stylophora*, and *Seriatopora*), and *Tetralia* spp., family Tetraliidae, are obligate symbionts of the family Acroporidae (*Acropora, Montipora*). *Tetralia* spp., formerly members of the Trapeziidae, are presently assigned to the Tetraliidae, an independent family erected by Castro et al. (2004).

habitat that offers protection from predators and a food source[10]; also, crustacean symbionts "sanitize" their hosts, i.e., assist in shedding contaminants, microorganisms, and larval settling stages (Glynn 1983a), and protect their hosts from corallivores (Glynn 1976, 1987; Pratchett 2001; McKeon and Moore 2014).

Two topics of particular interest were the global similarities of the crustacean-coral defensive repertoires, and the parallels with the iconic ant-*Acacia* protection mutualism. As noted in Chap. 7, the defensive behaviors of coral guards, *Trapezia* spp. crabs and *Alpheus lottini* shrimp, were remarkably similar across vast expanses of the Indo-West Pacific and Eastern Tropical Pacific regions. The crustacean defensive responses were virtually identical in type, strength, frequency, and escalation during corallivore attacks. No novel differences in defensive behavior across geographic sites were ever observed. Perhaps the capacity of the guards to undergo long distance dispersal, promoting genetic continuity, and their limited morphological structures (pincers in crabs and shrimp, and a snapping claw in shrimp) constrain development of different offensive behaviors?

The ant-*Acacia* mutualism, first studied in detail by Daniel Janzen in the lowlands of Mexico and Central America (Janzen 1966, 1974), involves pseudomyrmecine ants which live on and defend bull's-horn A*cacia* plants. The host plants provide the ant guests with a habitat niche in their swollen thorns, sugars from nectaries, and oils, sugars, and proteins from structures (Beltian bodies) in leaflet tips. Worker ants are omnipresent on the surface of plants and are quickly alerted, becoming extremely aggressive toward approaching herbivores, small or large. Whereas defensive behaviors in crustacean guards involve visual displays and physical contact, ant guards have a repulsive acid taste, perform aggressive biting, stinging, and group defense. Usually, all larger crustacean guards assist in colony defense. Ant guard mimicry[11] performed by diverse arthropods, is commonly observed in acacias. Diverse arthropod taxa, including jumping spiders, katydids, true bugs, and beetles, often imitate the constant waving of ant antennae and the jerky and zigzag movements of ants (Ants Mimicry—Myrmecomorphy; Cushing 2012; Styrsky 2014). Mimicry in crustacean guards has not been reported, which is

[10] A recent study at Eilat, Gulf of Aqaba, demonstrated that *Trapezia* inhabiting *Stylophora* move to peripheral colony branches at night where they feed on demersal plankton (Shmuel et al. 2022). An earlier study by Israeli investigators also reported a novel *Trapezia* diet which included coral host tissues (Rinkevich et al. 1991). Moreover, in the Hawaiian Islands, Stimson (1990) demonstrated an astonishing variation of diet in *Trapezia* that included "fat bodies", lipid-rich morsels produced by *Pocillopora* harboring crab symbionts. These studies illustrate notable dietary differences and underline the folly of assuming that closely related species necessarily share similar feeding strategies and diets.

[11] Mimicry is a type of symbiosis in which an organism, the **mimic**, seeking protection or some other advantage, evolves to resemble in morphology, color, or behavior a more aggressive or protected organism, the **model**. A third organism, for example a potential predator, the **dupe**, is deceived due to the vulnerable mimic posing as the threatening model. See Norman et al. (2001) for a remarkable case of mimicry in an Indo-Malayan octopus, reported to mimic multiple venomous models (a flatfish, lionfish, and sea snake), thus duping predatory fishes.

not surprising considering their high palatability (Castro 1978) and apparent absence of molecular deterrents; they are inherently poor models.

This brings us to the question of which characteristics are shared by "super symbiont hosts". Two cnidarian taxa hosting abundant and diverse symbionts offer extensive habitat space via branching colony morphologies, namely scleractinian corals, acroporid and pocilloporid species (e.g., Patton 1976; Alvarado and Vargas-Castillo 2012), and octocorals, alcyonarians (e.g., Bayer 1961; Goh et al. 1999; Sánchez 2016). In a recent survey conducted in Costa Rica, 74% of 100 crustacean symbionts were associated with cnidarians and other invertebrate hosts (Salas-Moya et al. 2021). The majority of the symbionts were collected from scleractinian (*Pocillopora*) corals and several from octocorals and antipatharians (black corals). These cnidarian hosts are complex, multi-branching colonies that are exposed to currents carrying plankton and particulate organic matter (potential food sources). In addition, many scleractinian and octocoral cnidarians host zooxanthellae that produce energy-rich photosynthates and copious mucus, potentially supplementing the symbiont's food supply.[12] Black corals do not host photosynthetic symbionts. Finally, symbionts receive some associational protection from their host's cnidocytes (stinging cells).

Origin of crustacean guard mutualisms. A chance encounter of a non-trapezioid crab defending its coral host may offer a hint of an evolutionary pathway leading to the origin of the *Pocillopora Trapezia* mutualism. This initial observation was made in 1987 at the Uva Island coral reef in Panama. While documenting the feeding behavior of *Acanthaster* preying on *Gardineroseris* colonies, a large individual sea star began to settle onto the coral's tissues to feed. These were remnant patches of live coral that remained on several colonies after the 1982–83 El Niño bleaching event. Suddenly, the sea star began to experience forceful up and down movements. The sea star quickly moved away from the live coral patch. As *Acanthaster* was retreating, a small crab (thumbnail sized) nipped and cut the sea star's tube feet. Examination of this crab species' shelter sites on other *Gardineroseris* colonies revealed live coral tissues surrounding the crab's lair—a depression or pit in the coral skeleton into which the crab retreated when disturbed. Is it possible that this crab, like *Trapezia* spp., is commonly effective in defending its coral host? It appears that this relationship could benefit both partners, the crab protecting its coral host and the coral offering the crab shelter (and possibly mucus food), seemingly a case of mutualism. The crab species, *Daira americana*, Family Dairidae, was identified by decapod crustacean systematist Ray (Raymond B.) Manning, U. S. National Museum of Natural History. *Daira* is known to inhabit non-coral and dead coral cavities and is therefore a facultative symbiont of live coral. It is unknown if *Daira* modifies the coral skeleton in the formation of its dwelling. The Uva Island

[12] In contrast with most tropical regions, the shallow water octocorals of the eastern Pacific do not host zooxanthellae (Van Oppen et al. 2005). Many years ago, Ted (Frederick M.) Bayer related this unusual absence of zooxanthellae to Oda (Odalisca) Breedy who has also not observed photosynthetic symbionts in eastern Pacific octocorals. In Oda's words, "This is a most interesting and difficult question."

Gardineroseris colonies exhibited robust recovery, chiefly by asexual re-sheeting of live patches (Fong and Glynn 1998, 2001), following the 1982–83 and 1997–98 El Niño bleaching events. It is likely that *D. americana* contributed to this recovery, but to what extent is unknown. Further study is necessary to determine if *D. americana* conferred protection is of ecological relevance as it is for *Trapezia* and *Alpheus*. It would also be interesting to study the diet of *Daira* to determine if coral mucus, entrapped detritus, or coral tissues are consumed, thus contributing to the benefit of the crustacean crab symbionts. Could it be that a food source and a protective habitat niche may have been critical resources in play on the evolutionary pathway to mutualism for other coral-crustacean associates? As noted earlier, the Caribbean commensal crab, *Domecia acanthophora*, receives protection and a mucus food source but is not known to defend or provide any other benefit to its coral host. This association needs further study.

Cleaner-client behavior. Cleaning symbiosis, performed by shrimp and fish cleaners, involves the removal of ectoparasites and diseased and necrotic tissues from invertebrate and vertebrate clients. Early researchers regarded this symbiosis as a quintessential case of mutualism, in which the cleaner benefits nutritionally and the client from healing and improved health. This view, however, has changed dramatically during the last few decades from continuing studies of marine parasites, long-term experimental data on cleaner community effects, and observations on numerous new symbiotic interactions.

With the advent of scuba technology in the 1950s, discoveries of cleaning symbioses in the marine environment, especially on coral reefs, gained considerable attention (e.g., Eibl-Eibesfeldt 1955; Limbaugh 1961; Limbaugh et al. 1961). Still in high school in 1950, I was aware of Conrad Limbaugh's exploits and remarkable discoveries. I made a pilgrimage from Coronado, California, my hometown, to La Jolla (< 30 km distant) to meet Conrad. I did meet him, but only briefly, not long enough to discuss his research. My interest in cleaning behavior continued when I joined Don Abbott's laboratory at the Hopkins Marine Station. There I met Howie (Howard)) Feder who championed Conrad's work in a comprehensive review of the subject (Feder 1966).[13] Conrad was the first to discover symbiotic cleaning behavior in the eastern Pacific, with documentation of cleaning interactions off southern California, from the Gulf of California, the west Mexican coast, and Clipperton Atoll.

The clients of reef cleaners, fishes (e.g., snappers, grunts, groupers, and eels) and pelagic fish megafauna (e.g., sharks, manta rays, whale sharks, and sunfish) generally arrive at a "cleaning station" and pose to the cleaners to present body areas needing attention. The brightly colored shrimp and fish cleaners perform ritualistic displays to attract host clients. Under the right conditions, and with "patience", shrimp cleaners can be observed performing exquisite "dances" directed toward

---

[13] Conrad's untimely death in 1960, presumably a result of losing direction while scuba diving in an underground river in southern France, was a shock to many familiar with his unprecedented and exciting discoveries. Howie Feder, Conrad's brother-in-law, took it upon himself, with Denton Limbaugh (Conrad's father), to include posthumously Conrad's extensive and detailed field observations in a review article of cleaning symbiosis in the marine environment (Feder 1966).

potential clients, with delicate body movements and gentle waving of chelipeds and antennae. Sometimes a crush of clients collects at a cleaning station, forming an orderly queue awaiting their turn.

My first encounter with a tropical cleaning shrimp, *Stenopus hispidus*, the barber shrimp cleaner, was on prominent graphic display at the Institute of Marine Biology (IMB), University of Puerto Rico, Mayagüez. This red banded shrimp, a commonly encountered cleaner in the Indo-West Pacific and wider Caribbean regions was pictorially omnipresent, the emblem of IMB. As advertised, it was abundantly present on coral reefs off La Parguera and along the nearshore banks of mangrove canals. I frequently observed the barber shrimp, often in pairs, cleaning moray eels and snappers. The most common reef fish cleaning behavior observed by our team in the eastern Pacific involved shrimp cleaning moray eels, the cleaners moving gingerly over eel's heads and into their mouths and gill chambers. Pelagic megafauna, including manta-ray clients and angelfish cleaners, were often seen at oceanic island sites. For example, angel fish (*Holacanthus clarionensis*) cleaning manta rays (*Manta birostris*) was a common symbiosis in the oceanic Revillagigedo Islands. A high frequency of cleaning interactions targeting top predators, such as groupers, snappers, sharks, and rays, was reported at the oceanic island of Malpelo, a World Heritage site (Quimbayo et al. 2016). It was suggested that this marine protected area, with abundant active cleaners and large predator clients, is likely a reflection of effective conservation.

Despite the intuitively sound and straight forward reasoning invoked to explain the origin, evolution and stability of cleaning symbiosis, the dynamics of these interactions can be highly complex, dependent on ecological, behavioral, and social processes. Recent evidence suggests that cleaning symbiosis can range from mutualistic to parasitic outcomes mediated by cleaner and client signals. For key questions and advances in our understanding of cleaning symbiosis, the interested reader is referred to Losey Jr. (1987), Côté (2000), Grutter (2005), and Caves (2021).

Damselfish symbionts. Besides the significant effects of the damselfish-algal lawn mutualism on coral community functional processes—e.g., nutrient concentration, primary production, larval settlement, and benthic species composition—damselfish defenses against sea urchin bioeroders and corallivores also exercise a strong influence. Here, I briefly point out two damselfish interactions that our team investigated in the eastern Pacific and at sites in the Indo-West Pacific region.

Damselfishes play a vital role in ejecting sea urchin bioeroder species from their algal lawn territories, thus limiting the amount of carbonate erosion (Glynn et al. 1979; Glynn and Wellington 1983; Eakin 1988, 1992; Glynn 1988). The variety of damselfish defensive movements can begin with inspection, spine nipping, nudging, and then lifting, swimming with a sea urchin (by mouthing a spine) and ejecting the intruder beyond its territory, sometimes up to 1–1.5 m away (Fig. 11.4). During normal, non-ENSO conditions sea urchins are moderately abundant on eastern Pacific coral reefs and net carbonate accretion is positive, allowing for reef growth. ENSO events, however, can cause high coral mortality and are sometimes associated with increased abundances of sea urchins, resulting in rapid erosion and loss of coral frameworks. This disruptive scenario occurred in the Galápagos Islands and

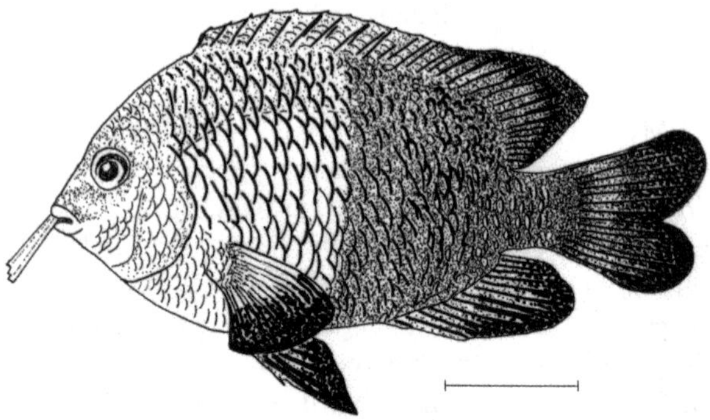

Fig. 11.4 *Stegastes acapulcoensis*, ejecting a sea urchin spine from its territory, scale bar = 3 cm

Panama during and following the 1982–83 El Niño event (Fig. 9.1). Galápagos patch reefs in the central and southern sectors of the archipelago were eroded entirely, reduced to rubble and sand sediments (Glynn 1994). In Panama, reef growth experienced a temporary shutdown with substantial framework losses (Eakin 1988, 1992; Glynn 1988; Fong et al. 2017).

Interestingly, a post-ENSO investigation in Panama found that territorial damselfish may sometimes enhance bioerosion. Damselfish ejection behavior can cause sea urchin aggregations and increased bioerosion under some circumstances. For example, when sea urchins are ejected into reef depressions or furrows, they often clump together and behave as a social unit (Pearse and Arch 1969).[14] Echinoid aggregations composed of large numbers of individuals that remain in the same area for several days can result in significant carbonate erosion. A post-ENSO nuanced approach by our team revealed that the damselfish ejection response can cause an increase in coral erosion. For example, on the Uva reef, this amounted to 0.20 kg $CaCO_3$ m^{-2} y^{-1} (Glynn et al. 2020) or a 20% increase in Mark Eakin's annual erosion budget analysis (Eakin 1996).

Damselfish symbionts also forcefully defend their territories from invertebrate and fish corallivores. The repertoire of defensive behaviors toward *Acanthaster* includes (a) displacement activities: inspection and rolling against the substrate; and escalation to (b) physical contact: body brushing, tail beating, nipping spines and

[14] John Pearse, a close friend and graduate student peer at the Hopkins Marine Station (Stanford University) in the late 1950s, was a leading authority on marine invertebrates, including tropical echinoderms. Vicki Pearse, daughter of Mildred and Ralph Buchsbaum, was likewise active in marine invertebrate research and was the founding editor of the quarterly journal, *Invertebrate Biology*. The seed of their half-century marriage and collaboration germinated at Hopkins Marine Station (HMS). John and Vicki were jointly awarded the Lifetime Achievement Award of the Western Society of Naturalists in 2008, and both were Fellows of the California Academy of Sciences. Upon retirement, after distinguished careers at the University of California Santa Cruz, they moved into the Buchsbaum home on Ocean View Boulevard, Pacific Grove, overlooking HMS and Monterey Bay. The Buchsbaum-Pearse home, a lyceum for marine biologists, always offered warm hospitality and lively discussions of discoveries and advances in the field. Their models for generous hosting, in both their Santa Cruz and Pacific Grove homes, were Don and Izzie (Isabella) Abbott, esteemed professors at HMS.

tube feet (Glynn 1988; Glynn and Colgan 1988). Fishes intruding on damselfish territories are usually chased, and sometimes bitten. These activities promote coral protection and survival and result in higher coral cover, species diversity, and longevity of pocilloporid species within the territorial boundaries of damselfish lawns (e.g., Glynn and Colgan 1988; Gochfeld 2010; Pruitt et al. 2018; Honeycutt et al. 2023).

Parasitism

Parasites, pathogens, and diseases are not subjects often reported respecting eastern Pacific coral reef research (Cortés 2003; Glynn and Enochs 2011; Rodríguez-Villalobos and Reyes-Bonilla 2019). In a recent review of the metazoan marine parasites of Costa Rica, 101 species were listed as occurring on eastern Pacific vertebrate hosts, primarily trematode and cestode gut endoparasites in sea turtles, fishes, elasmobranchs, and dolphins (Solano-Barquero et al. 2023). Invertebrate host species were crustaceans, such as crabs, barnacles, lobster, and shrimp. A few of the listed host species are coral reef associates; however, beyond taxonomic assignments little information of ecological interest was noted. Regarding coral disease outbreaks resulting in widespread mass mortality, the eastern tropical Pacific region stands alone globally in not having experienced such devastation. Parasites live on or in other organisms, the hosts, and when they cause a disease (a structural or functional disorder) they become known as pathogens. Abiotic toxicants or toxic pollutants, such as heavy metals, petroleum, and synthetic organic chemicals, are also classified as pathogenic or etiologic agents (Peters 2015). The distinction between abiotic and biotic effects can be difficult to determine in some cases, such as mortality events associated with water blooms (red tides). For example, dense populations of dinoflagellates can produce copious amounts of mucus that smother organisms, thus causing anoxic stress and death. Such an event caused localized high coral mortality in Costa Rica and Panama in 1985 (Guzmán et al. 1990).

Regarding parasitic symbioses, I digress to discuss our initial attempts to understand the cause(s) of coral bleaching in the early 1980s. Of necessity, this research led to consideration of possible parasitic, pathogenic infections, and non-organismic (abiotic) agents such as chemical pollutants.

Research on parasites and diseases were areas beyond our expertise. In our initial concerns with coral bleaching and mortality during the 1982–83 El Niño event, we teamed up with Esther Peters and Len (Leonard) Muscatine for help in understanding if this widespread coral mortality event was due to a biotic or abiotic pathogen. We sampled affected and normal-appearing corals in the Gulf of Chiriquí, Panama, a region subject to intense agriculture, during and immediately after coral bleaching and mortality. Histopathological analysis did not reveal any suspect biotic pathogens that could be responsible for coral tissue necrosis and mortality (Glynn et al. 1985). Since I had just joined the faculty of the UM marine school (RSMAES), this was an opportune time to team up with Gene (Eugene) Corcoran, a marine school chemist who was routinely analyzing tissues for trace amounts of diverse pollutants. At that time (summer 1983), I was invited to join the instructional staff of a graduate research training course on Coconut Island, University of Hawaii. There I could relate the details of the dramatic coral mortality event in the eastern Pacific. I was asked to brief the course participants, many with extensive experience in coral reef disturbance events, on the widespread coral mortality underway at that time (1983) in the eastern Pacific. We hypothesized, following detection of trace amounts of pesticide (herbicide) residues in bleached and moribund Panamanian corals, that these chemicals could be responsible for the mortality event in progress (Glynn et al. 1984). The stage was set for performing pesticide tolerance experiments, which were soon underway in the Coral Reef Population Biology course on Coconut Island.

We began testing the tolerance of Hawaiian colonies of *Pocillopora damicornis*, the same morphospecies affected in Panama. Controlled tolerance experiments of herbicides, at the same concentrations found in affected corals, caused bleaching and tissue mortality as observed naturally on coral reefs in 1983.[15] The intensity of bleaching and mortality, however, also closely corresponded spatially (geographically) with anomalously high sea water temperatures (Glynn 1984). This thermal relationship, i.e., the abnormally elevated sea temperatures vis-à-vis coral mortality, was observed across the equatorial eastern Pacific, from Costa Rica, Panama, Colombia, and Ecuador, including the Galápagos Islands (Glynn et al. 1988). At that time, agricultural activities and polluted effluents were negligible in the Galápagos. Additionally, widespread coral bleaching and mortality events were reported globally in 1982–83, several occurring during anomalous sea-warming conditions (Brown 1987). Paul Jokiel, a resident scientist at the Coconut Island laboratory (University of Hawaii), argued that prolonged high sea temperature associated with the 1982–83 El Niño event could be responsible for the eastern Pacific coral bleaching. Paul's previous research on high temperature stress and coral mortality in Hawaii, with Steve Coles (Jokiel and Coles 1974, 1977), offered convincing parallels with the bleaching syndrome in Panama. Moreover, an El Niño simulation experiment performed by Luis D'Croz and me in 1985 duplicated the coral bleaching responses observed naturally in Panama during 1983 (Glynn and D'Croz 1990).

In Miami, Sam (Samuel) Snedaker, Alina Szmant, Darren Rumbold, and Julie Firman[16] joined our study of pollutants in corals and reef-associated organisms in

[15] A phenoxy acid, 2, 4-D, was the principal herbicide employed in the laboratory tolerance tests. It was selected because it is a commonly used herbicide in Panama and was detected in bleached coral tissues (10–20 ppb) during the bleaching/mortality event in 1983. This compound mimics the plant growth hormone auxin, causing uncontrolled growth, coral bleaching and death. It is readily available in gardening supply stores, sold under the trade name "Weed-B-Gon".

[16] Julie C. Firman, an early member of our coral reef research team, concentrated her doctoral studies on the ecotoxicology of pesticides in reef-building corals in the Florida Keys (Firman 1995; Peters et al. 1997). Ecotoxicology is the scientific discipline that focuses on the effects of toxic chemicals harmful to the environment and organisms. Due to a hiring hiatus in Julie's area of expertise (coral ecotoxicology), it was necessary for her to consider other employment options, which took her to the salmon streams of Oregon. Julie is a Natural Resource Specialist in the

the Florida Keys (e.g., Glynn et al. 1995; Snedaker et al. 1995). We found organic residues and heavy metal burdens in several organisms. It was difficult to tie these pollutants directly to declining coral health, however, because of confounding conditions, e.g., sea warming and diseases, implicated in bleaching and mortality in south Florida. With continuing ENSO thermal disturbances in the eastern Pacific (1986–87, 1990–92) and ongoing coral reef recovery studies related to the 1982–83 event, it was necessary to suspend our monitoring of coral reef contaminants in Florida. Also, some UM administrators expressed concern over the high levels of potentially harmful contaminants we reported in corals and reef-associated seafood species. We were encouraged to abandon the Florida research. It seemed best to some university officials that we should continue directing our studies far away from Florida, beyond such disconcerting local and still unproven health concerns. Academic freedom! These were conflicting issues that resulted in very uncomfortable discussions.

Jenny Carolina and her team in Mexico, employing a habitat suitability risk model to assess the occurrence of white syndrome (WS) disease in *Pocillopora*, predicted WS could appear widely across the eastern tropical Pacific region (Rodríguez-Villalobos et al. 2022). Although not firmly established, a *Vibrio* bacterium might be the etiologic agent of this coral disease, which is manifest as a band of tissue loss above bleached tissue and bare skeleton at the colony base. Their predictions of the future prevalence and distribution of WS indicated occurrence primarily on coral reefs exposed to both high and low thermal anomalies (with high primary productivity and human impacts as additional variables of importance). Sea temperature extremes and variability influence disease pathogenesis directly by stressing and increasing the susceptibility of the coral host and indirectly by causing shifts in bacterial communities from mutualism to pathogenic associations. With increasing ocean temperatures globally, and associated ENSO extreme conditions, Jenny Carolina's team predicted that 70% of *Pocillopora* coral reefs will be susceptible to WS epizootics in the coming decades (Rodríguez-Villalobos et al. 2022).

The spread of biotic pathogens through the Panama Canal is of great concern. This issue was front and center in the 1970s during discussions of constructing a sea-level canal, which would have allowed the mixing of Pacific and Caribbean waters and their biotas (Jones 1972). The devastating *Diadema antillarum* epizootic that began on the Caribbean coast of Panama in 1983 was possibly introduced in ballast water by a ship transiting the canal (Lessios 1988). A biotic pathogen was

Oregon Department of Fish and Wildlife (ODFW) in Corvallis, Oregon. When Julie first joined the ODFW she was tasked with sampling salmon populations and spawning areas in streams in northeastern Oregon. This was challenging because it required snorkeling over extended periods to make accurate fish counts. In the winter months it was often necessary to break through ice to access the flowing stream habitat. Julie heroically accepted the challenge and developed new survey methods utilizing landscape models to identify spawning grounds in remote areas. More recently, Julie's work is focused on modeling the influence of climate change on aquatic habitats and salmon populations.

suspected to cause this regional mass mortality event.[17] Pathogenic microalgae, bacteria, and viruses have caused various coral diseases in the Caribbean but are not yet known to have invaded the eastern Pacific (Thurber et al. 2020). It is also necessary to be mindful of fouling organisms attached to ship's hulls. By towing Caribbean fouling organisms (and crustaceans, mollusks, and flatworms collected from the rocky intertidal zone), Bob Menzies reported that some species can survive the inter-ocean passage through the Panama Canal, across ~50 km of freshwater habitat (Menzies 1968).[18]

A minute gastropod ectoparasite, *Echineulima thanuumi* was sometimes found firmly attached to a sea urchin (*Diadema mexicanum*) and a sea cucumber (*Euapta godeffroyi*) on the Uva Island reef. This gastropod genus is a well-known parasite of echinoderms, infecting species in all major taxa. It inserts an extensible proboscis into the coelomic cavity where it ingests its host's tissues. In Mark Eakin's doctoral research at Uva Island, he reported a very low infestation rate of ~1% of the *Diadema* population. He also demonstrated parasitic castration, i.e., the consumption of gonadal tissues, from the dissection and examination of infected sea urchin organs (Eakin 1991). A Russian team, employing stable isotope analysis, reported that *Echineulima mittrei* in Vietnam also fed on the lipid-rich gonads of sea urchins (Dgebuadze et al. 2020). Is it possible that this ectoparasite could compromise sea urchin reproduction? Still unexplained are sharp declines of *Diadema* in Panama in the early 1990s (Eakin 2001). Could *Echineulima* have contributed to the decline of *Diadema* populations in the early 1990s?

Few multicellular organisms have been reported to parasitize alcyonarian and scleractinian cnidarians in the eastern Pacific. Only two cases of suspected (unconfirmed) fungi are listed as etiologic agents causing disease in alcyonarians (Rodríguez-Villalobos and Reyes-Bonilla 2019). Since focused disease studies were undertaken only recently in this region, the relative rarity of pathogens is likely attributed to a lack of research. Suspected coral "trematodiasis" is visible as distinct white or pink, 1–3 mm nodules on massive *Porites* and *Pavona* colonies. It has been reported in the Galápagos Islands and Clipperton Atoll (Vera and Banks 2009; Rodríguez-Villalobos and Reyes-Bonilla 2019; Pogoreutz et al. 2022). We have also

---

[17] A similar mass mortality of *D. antillarum*, beginning in the Caribbean region in 2022, was immediately investigated and resulted in definitive results. Combined molecular and histopathologic approaches, including fulfillment of Koch's postulates, revealed that a scuticociliate protozoan was the causative pathologic agent of the disease (Thurber and Hay 2023). It cannot be assumed that this parasite was responsible for the sea urchin mortality event in the early 1980s. Esther Peters performed histopathological observations on dying *D. antillarum* a couple of years after the 1983–84 mass mortality event but found no signs of the scuticociliate protozoans (Peters 2015). Establishing the identity of the etiologic agent(s) in suspected diseased species is an essential first step in the treatment and prevention of any epizootic.

[18] Bob duly noted that the towed specimens, wrapped in cheesecloth, planed and skipped over the water when the ship's speed approached 10 knots, thus interrupting complete immersion. This pioneering experiment would be worth repeating, sampling biota directly from the submerged hulls of ships, including diverse taxa (e.g., cryptic metazoans, and microorganisms) and allowing sufficient time to adequately assess recovery and survival.

observed these pink pustules on massive corals at our study sites in Costa Rica, Panama, and coastal Ecuador. Since these infections can depress coral calcification and reduce the competitive capacity of the coral host (Aeby 1992), it is imperative to confirm their genetic identity.[19] In the Hawaiian trematode life cycle (*Podocotyloides stenometra*), there are probably three hosts, a definitive fish corallivore, a suspected mollusk intermediate host (the intermediate hosts, harboring one or more sporocyst and redia stages of practically all trematodes), and *Porites*, the second intermediate host (harboring the cyst-like metacercaria of the trematode). The parasite's alteration of the coral host's polyps prevents their normal retraction rendering them more vulnerable to predation, thus enhancing transmission. The exposed polyps enhance the feeding of visual corallivores and the coral host by removing infected polyps. Based on evolutionary strategies (Dawkins 1982), Aeby suggested that the altered host's appearance (pinkish nodules) may represent a mutual symbiotic strategy by benefitting both the corallivore and the coral host. This is clearly a fascinating symbiosis in need of further study. What better place to conduct this investigation than in the eastern Pacific? Does a *Podocotyloides* trematode parasite cause the pink nodules? Does the life cycle include a definitive fish(es) host and two intermediate hosts, a mollusk and coral(s)? Does this parasitic symbiosis echo a mutualism? If a trematode parasite is involved, how widespread is this condition region wide? Which coral species are infected, what are the infection rates, and the ecological effects in coral communities?

Burrowing date mussels (bivalve mollusks, e.g., *Lithophaga* spp.) have long been regarded as parasites of reef-building corals, due chiefly to the damage they cause to coral skeletons and the nutritional benefit they receive by feeding on their host's mucus (Shafir and Loya 1983; Mokady et al. 1998). *Lithophaga* spp. are abundant internal bioeroders on massive *Porites* species in the eastern Pacific, with densities of 100 s to 1000 s of individuals m^{-2} (Scott et al. 1988). Evidence from continuing research suggests that this supposed case of parasitism is more akin to a true mutualism.

A possible benefit that *Lithophaga* is offering its coral host is an enhanced energy source. Mokady and team (1998) found that *Lithophaga* in the northern Red Sea release nitrogen (NH_3) enhancing zooxanthellae production, which in turn contributes to coral host tissue growth. The rates of ammonium production were found to satisfy a significant portion of the coral host's growth needs. An additional benefit was proposed by Paula Scott and co workers who hypothesized that date mussels can contribute to coral recovery via asexual propagation of coral fragments broken away through bioerosion (Scott et al. 1988). Absent sexual reproduction,[20] they pos-

[19] Since the encysted trematode larvae are ~0.3 mm (300 μm) in diameter, they would be visible to the naked eye and could be collected by careful dissection of the nodules in the field (Aeby 2015). Recall that spawned coral eggs are 0.1 mm (100 μm) in diameter and just visible to the naked eye under good lighting. Properly fixed (preserved), the larvae would then be suitable for histopathological and molecular analyses.

[20] At the time of Scott's study, it was generally held that asexual reproduction was the norm in eastern Pacific reef-building corals (Richmond 1985). Later research revealed that sexual reproduction occurred in all coral species, including *Porites* (see Chap. 6; Glynn et al. 1994).

ited that colony fragmentation could generate propagules that would help replenish coral populations following disturbance events such as El Niño. Thus, date mussels could promote coral host population recovery. This proposed example of mutualism involving bioerosion is logically weak because of the destruction of a large host coral colony that is structurally and reproductively important plus the increased vulnerability of *Lithophaga* to predators after coral fragmentation and exposure.

Concluding Remarks

Symbioses are ubiquitous in coral reef communities. The coral holobiont, comprising the coral host and associated microorganisms, manifests numerous commensal, mutualistic, and parasitic relationships, and is the essence, the *raison d'être*, for the very existence of coral reef ecosystems. Biodiversity and symbiosis are intimately related, the former the basis for the formation of a multitude of functional interdependent relationships. This vital connection is an important justification for protecting and sustaining biodiversity and the crucial symbiotic relationships that support the survival, growth, and persistence of coral reef ecosystems.

These are worrisome times concerning the numerous signs of coral stress, particularly regarding susceptible *Pocillopora* species, the predominant reef-building taxon in the eastern Pacific region. At present, we are largely ignorant of the causative agents of diseased corals.

While *Pocillopora* resilience and recovery from warming disturbances have been heartening (e.g., Glynn et al. 2009; Glynn et al. 2017a; Palacio-Castro et al. 2023; Turnham et al. 2023), projected thermal stress and El Niño heat waves will likely continue to alter conditions favorable to symbiotic coral pathogens that would become virulent with catastrophic mortality and reef loss.

References

Aeby GS (1992) The potential effect the ability of a coral intermediate host to regenerate has had on the evolution of its association with a marine parasite. In: Proceedings of the 7th international coral reef symposium, Guam, vol 2, pp 809–815

Aeby GS (2015) *Porites* trematodiasis. In: Woodley CM, Downs CA, Bruckner AW, Porter JW, Galloway SB (eds) Diseases of coral. Wiley, New York, pp 387–390. https://doi.org/10.1002/9781118828502.ch28

Alvarado JJ, Vargas-Castillo R (2012) Invertebrados asociados al coral constructor de arrecifes *Pocillopora damicornis* en Playa Blanca, Bahía Culebra. Costa Rica. Rev Biol Trop 60(Suppl 2):77–92

Auster PJ, Cortés J, Alvarado JJ, Beita-Jiménez Á (2019) Coordinated hunting behaviors of mixed-species groups of piscivores and associated species at Isla del Coco National Park (Eastern Tropical Pacific). Neotrop Ichthyol 17(1):2019 https://doi.org/10.1590/1982-0224-20180165

Auster P, Sánchez-Jiménez A, Rodríguez-Arrieta JA, Quesada AJ, Pérez C, Naranjo-Elizondo B, Blum S, Cortés J (2016) Facilitative behavioral interactions between deepwater piscivores at Isla del Coco National Park and Las Gemelas Seamount, Costa Rica. Revista de Biología Tropical. Int J Trop Biol 64(Suppl I):S187–S196

Baker AC (2003) Flexibility and specificity in coral-algal symbiosis: diversity, ecology, and biogeography of *Symbiodinium*. Annu Rev Ecol Syst 34:661–689

Bayer FM (1961) The shallow-water Octocorallia of the West Indian region. A manual for marine biologists. Studies on the fauna of Curaçao and other Caribbean islands. Martinus Nijhoff, The Hague, pp 1–373

Bavestrello G, Cerrano C, Cattaneo-Vietti R, Sarà M (1996) Relations between *Eudendrium glomeratum* (Cnidaria, Hydromedusae) and its associated vagile fauna. Sci Mar 60(1):137–143

Bermudes D, Margulis L (1987) Symbiont acquisition as neosome: origin of species and higher taxa. Symbiosis 4:185–198

Brattegard T (1968) On an association between *Acanthopleura granulata* (Polyplacophora) and *Dynamene* spp. (Isopoda). Sarsia 32(1):11–20

Brown BE (1987) Worldwide death of corals–natural cyclical events or man-made pollution? Mar Pollut Bull 18:9–13

Bruce AJ (1976) Shrimps and prawns of coral reefs, with special reference to commensalism. In: Jones OA, Endean R (eds) Biology and geology of coral reefs, vol 3. Academic Press, New York, pp 37–94

Castro P (1978) Movements between coral colonies in *Trapezia ferruginea* (Crustacea Brachyura), an obligate symbiont of scleractinian corals. Mar Biol 46:237–245. https://doi.org/10.1007/BF00390685

Castro P (1988) Animal symbioses in coral reef communities: a review. Symbiosis 5:161–184

Castro P (2015) Symbiotic Brachyura. In: Castro P, Davie PJF, Guinot D, Schram FR, von Vaupel Klein JC (eds) The Crustacea. Traité de Zoologie, vol. 9, Part C-1, Decapoda: Brachyura (Part 1). Brill, Leiden, Boston, pp 543–581

Castro P, Huber M (2024) Marine biology, 12th edn. McGraw Hill, New York, New York

Castro P, Ng PKL, Ahyong ST (2004) Phylogeny and systematics of the Trapeziidae Miers, 1886 (Crustacea: Brachyura), with the description of a new family. Zootaxa 643:1–70

Caves EM (2021) The behavioral ecology of marine cleaning mutualisms. Biol Rev 96(6):2584–2601

Chacón-Monge JL, Angulo A, Cortés J (2021) New hosts and morphological data for the star pearlfish *Carapus mourlani* (Ophidiiformes: Carapidae) from collections made in the North Pacific coast of Costa Rica. Rev Biol Trop 69:219–233

Cortés J (2003) Latin American coral reefs. Elsevier, Amsterdam, p 497

Côté IM (2000) Evolution and ecology of cleaning symbiosis in the sea. Oceanogr Mar Biol 38:311–355

Cushing PE (2012) Spider-ant associations: an updated review of myrmecomorphy, myrmecophily, and myrmecophagy in spiders. Psyche:151989. https://doi.org/10.1155/2012/151989

Dales RP (1957) Interrelations of organisms, A. Commensalism. In: Hedgpeth JW (ed) Treatise on marine ecology and paleoecology. Vol. 1 Ecology. The Geological Society of America. Waverly Press, Baltimore, pp 391–412

Dales RP (1966) Symbiosis in marine organisms. In: Henry SM (ed) Symbiosis. Volume 1, Associations of microorganisms, plants, and marine organisms. Academic Press, New York, pp 299–326

Dawkins R (1982) The extended phenotype. The gene as the unit of selection. W. H. Freeman, London, p 307

de Bary A (1879) Die Ersheinung der Symbiose. Verlag von Karl J. Trubner, Strassburg

Dgebuadze PY, Mekhova ES, Thanh NTN, Zalota AK (2020) Diet relationships between parasitic gastropods *Echineulima mittrei* (Gastropoda: Eulimidae) and sea urchin *Diadema setosum* (Echinoidea: Diadematidae) hosts. Mar Biol 167:150 (2020). https://doi.org/10.1007/s00227-020-03762-2

Eakin CM (1987) Damselfishes and their algal lawns: a case of plural mutualism. Symbiosis 4:275–288

Eakin CM (1988) Avoidance of damselfish lawns by the sea urchin *Diadema mexicanum* at Uva Island, Panama. In: Proceedings of the 6th international coral reef symposium, Townsville, vol 2, pp 21–26

Eakin CM (1991) The damselfish-algal lawn symbiosis and its influence on the bioerosion of an El Niño impacted coral reef, Uva Island, Pacific Panama. Ph.D. dissertation. University of Miami, p 158

Eakin CM (1992) Post-El Niño Panamanian reefs: less accretion, more erosion and damselfish protection. In: Proceedings of the 7th international coral reef symposium, Guam, vol 1, pp 387–396

Eakin CM (1996) Where have all the carbonates gone? A model comparison of calcium carbonate budgets before and after the 1982-1983 El Niño at Uva Island in the eastern Pacific. Coral Reefs 15:109–119

Eakin CM (2001) A tale of two ENSO events: carbonate budgets and the influence of two warming disturbances and intervening variability, Uva Island, Panama. Bull Mar Sci 69(1):171–186

Eibl-Eibesfeldt I (1955) Über Symbiosen, Parasitismus und andere besondere zwischenartliche Beziehungen tropischer Meeresfische. Z Tierpsychol 12:203–219

Feder HM (1966) Cleaning symbiosis in the marine environment. In: Henry SM (ed) Symbiosis. Vol 1, Associations of microorganisms, plants, and marine organisms. Academic Press, New York, pp 327–380

Firman JC (1995) Chronic toxicity of pesticides to reef-building corals: physiological, biochemical, cellular and developmental effects. Ph.D. dissertation. University of Miami, Coral Gables, Florida, p 135

Fong P, Glynn PW (1998) A dynamic size-structured population model: does disturbance control size structure of a population of the massive coral *Gardineroseris planulata* in the eastern Pacific? Mar Biol 130:663–674

Fong P, Glynn PW (2001) Population abundance and size-structure of an eastern tropical Pacific coral after the 1997-98 ENSO: a simulation model predicts field measures. Bull Mar Sci 69(1):187–202

Fong P, Smith TB, Muthukrishnan R (2017) Algal dynamics: alternate stable states of reefs in the eastern tropical Pacific. In: Glynn PW, Manzello DP, Enochs IC (eds) Coral reefs of the eastern tropical Pacific: persistence and loss in a dynamic environment. Coral reefs of the world 8. Springer, Dordrecht, pp 339–367. https://doi.org/10.1007/978-94-017-7499-4_11

Frankowiak K, Wang KT, Sigman DM, Gothmann AM, Kitahara MV, Mazur MJ, Meibom A, Stolarski J (2016) Photosymbiosis and the expansion of shallow-water corals. Sci Adv 2(11). https://doi.org/10.1126/sciadv.1601122

Glynn PJ, Glynn PW, Riegl B (2017a) El Niño, echinoid bioerosion and recovery potential of an isolated Galápagos coral reef: a modeling perspective. Mar Biol 164:146. https://doi.org/10.1007/s00227-017-3175-0

Glynn PJ, Glynn PW, Maté J, Riegl B (2020) Agent-based model of eastern Pacific damselfish and sea urchin interactions shows increased coral reef erosion under post-ENSO conditions. Ecol Model 423. https://doi.org/10.1016/j.ecolmodel.2020.108999

Glynn PW (1968) Ecological studies on the associations of chitons in Puerto Rico, with special reference to sphaeromid isopods. Bull Mar Sci 18(3):572–626

Glynn PW (1976) Some physical and biological determinants of coral community structure in the eastern Pacific. Ecol Monogr 46(4):431–456

Glynn PW (1983a) Increased survivorship in corals harboring crustacean symbionts. Mar Biol Lett 4:105–111

Glynn PW (1983b) Crustacean symbionts and the defense of corals: coevolution on the reef? In: Nitecki MH (ed) Coevolution. The University of Chicago Press, Chicago, pp 111–178

Glynn PW (1984) Widespread coral mortality and the 1982-83 El Niño warming event. Environ Conserv 11(2):133–146

Glynn PW (1987) Some ecological consequences of coral-crustacean guard mutualisms in the Indian and Pacific Oceans. Symbiosis 4:301–324

Glynn PW (1988) El Niño warming, coral mortality and reef framework destruction by echinoid bioerosion in the eastern Pacific. Galaxea 7(2):129–160

Glynn PW (1994) State of coral reefs in the Galápagos Islands: natural vs anthropogenic impacts. Mar Pollut Bull 29(1-3):131–140

Glynn PW, Colgan MW (1988) Defense of corals and enhancement of coral diversity by territorial damselfishes. In: Proceedings of the 6th international coral reef symposium, Australia, vol 2, pp 157–163

Glynn PW, D'Croz L (1990) Experimental evidence for high temperature stress as the cause of El Niño-coincident coral mortality. Coral Reefs 8:181–191

Glynn PW, Enochs IC (2011) Invertebrates and their roles in coral reef ecosystems. In: Dubinsky Z, Stambler N (eds) Coral reefs: an ecosystem in transition. Springer, Dordrecht, pp 273–325. https://doi.org/10.1007/978-94-007-0114-_18

Glynn PW, Wellington GM (1983) Corals and coral reefs of the Galápagos Islands. With an annotated list of the scleractinian corals of the Galápagos by John W. Wells. University of California Press, Berkeley, p 330

Glynn PW, Peters EC, Muscatine L (1985) Coral tissue microstructure and necrosis: relation to catastrophic coral mortality in Panama. Dis Aquatic Org 1:29–37

Glynn PW, Rumbold DG, Snedaker SC (1995) Organochlorine pesticide residues in marine sediment and biota from the Northern Florida reef tract. Mar Pollut Bull 30(6):397–402

Glynn PW, Wellington GM, Birkeland C (1979) Coral reef growth in the Galápagos: limitation by sea urchins. Science 203:47–49

Glynn PW, Bayer FM, Renegar DA (2014) *Coeloplana waltoni*, a new species of minute benthic ctenophore (Ctenophora: Platyctenida) from south Florida. Proc Biol Soc Wash 127(2):423–436

Glynn PW, Cortés J, Guzmán HM, Richmond RH (1988) El Niño (1982-83) associated coral mortality and relationship to sea surface temperature deviations in the tropical eastern Pacific. In: Proceedings of the sixth international coral reef symposium, Townsville, vol 2, pp 157–163

Glynn PW, Enochs IC, McCosker JE, Graefe AN (2008) First record of a pearlfish, *Carapus mourlani*, inhabiting the aplysiid opisthobranch mollusc *Dolabella auricularia*. Pac Sci 62:593–601

Glynn PW, Howard LS, Corcoran E, Freay AD (1984) The occurrence and toxicity of herbicides in reef building corals. Mar Pollut Bull 15:370–374

Glynn PW, Riegl B, Correa AMS, Baums IB (2009) Rapid recovery of a coral reef at Darwin Island, Galápagos Islands. Galapagos Res 66:6–13

Glynn PW, Coffman B, Primov KD, Moorhead SG, Vanderwoude J, Barrales RN, Williams MK, Roemer RP (2018) Benthic ctenophores (Platyctenida: Coeloplanidae) in south Florida: predator-prey interactions. Invert Biol 137(2):133–150

Glynn PW, Coffman B, Fuller MPC, Moorhead SG, Williams MK, Primov KD, Fortson TN, Barrales RN, Glynn PJ (2017b) Benthic ctenophores (Platyctenida: Coeloplanidae) in south Florida: environmental conditions, habitats, abundances, and behaviors. Invert Biol 136(4):379–393

Glynn PW, Colley SB, Eakin CM, Smith DB, Cortés J, Gassman NJ, Guzmán HM, Del Rosario JB, Feingold JB (1994) Reef coral reproduction in the eastern Pacific: Costa Rica, Panamá and Galápagos Islands (Ecuador). II. Poritidae. Mar Biol 118:191–208

Gochfeld DJ (2010) Territorial damselfishes facilitate survival of corals by providing an associational defense against predators. Mar Ecol Prog Ser 398:137–148

Goh NKC, Ng PKL, Chou LM (1999) Notes on the shallow water gorgonian-associated fauna on coral reefs in Singapore. Bull Mar Sci 65(1):259–282

Grutter AS (2005) Cleaning mutualism in the sea. In: Rohde K (ed) Marine parasitology. CSIRO Publishing, Collingwood, pp 264–278

Guzmán HM, Cortés J, Glynn PW, Richmond RH (1990) Coral mortality associated with dinoflagellate blooms in the eastern Pacific (Costa Rica and Panama). Mar Ecol Prog Ser 60:299–303

Hoeksema BW, García-Hernández JE (2020) Host-related morphological variation of dwellings inhabited by the crab *Domecia acanthophora* in the corals *Acropora palmata* and *Millepora complanata* (Southern Caribbean). Diversity 12(4):143

Honeycutt RN, Holbrook SJ, Brooks AJ, Schmitt RJ (2023) Farmerfish gardens help buffer stony corals against marine heat waves. PLoS One 18(3):e0282572. https://doi.org/10.1371/journal.pone.0282572

Janzen DH (1966) Coevolution of mutualism between ants and acacias in Central America. Evolution 20:249–275

Janzen DH (1974) Swollen-thorn acacias of Central America. In: Smithsonian contributions to botany, vol 13, pp 1–131

Jokiel PL, Coles SL (1974) Effects of heated effluent on hermatypic corals at Kahe Point, Oahu. Pac Sci 28:1–18

Jokiel PL, Coles SL (1977) Effects of temperature on the mortality and growth of Hawaiian reef corals. Mar Biol 43:201–208

Jones ML (1972) The Panamic biota: some observations prior to a sea-level canal. Bull Biol Soc Wash 2:270

Knudsen JW (1967) *Trapezia* and *Tetralia* (Decapoda, Brachyura, Xanthidae) as obligate ectoparasites of pocilloporid and acroporid corals. Pac Sci 21:51–57

Krediet CJ, Ritchie KB, Paul VJ, Teplitski M (2013) Coral associated microorganisms and their roles in promoting coral health and thwarting diseases. Proc R Soc B 280:20122328. https://doi.org/10.1098/rspb.2012.2328

Lessios H (1988) Mass mortality of *Diadema antillarum* in the Caribbean: what have we learned? Annu Rev Ecol Syst 19:371–393

Leung TLF, Poulin R (2008) Parasitism, commensalism, and mutualism: exploring the many shades of symbioses. Vie et Milieu · Life Environ 58(2):107–115

Limbaugh C (1961) Cleaning symbiosis. Sci Am 205:42–49

Limbaugh C, Pederson H, Chace FA Jr (1961) Shrimps that clean fishes. Bull Mar Sci Gulf Caribbean 11:237–257

Losey GS Jr (1987) Cleaning symbiosis. Symbiosis 4:229–258

Margulis L (1996) Archaeal-eubacterial mergers in the origin of Eukarya: phylogenetic classification of life. Proc Natl Acad Sci USA 93:1071–1076

Margulis L (1998) Symbiotic planet, a new view of evolution. Basic Books, New York, p 147

McKeon CS, Moore JM (2014) Species and size diversity in protective services offered by coral guard-crabs. PeerJ 2:e574. https://doi.org/10.7717/peerj.574

Menzies R (1968) Transport of marine life between oceans through the Panama Canal. Nature 220:802–803. https://doi.org/10.1038/220802a0

Mokady O, Loya Y, Lazar B (1998) Ammonium contribution from boring bivalves to their coral host–a mutualistic symbiosis? Mar Ecol Prog Ser 169:295–301

Mukherjee S (2022) The song of the cell, an exploration of medicine and the new human. Scribner, New York, p 473

Muscatine L (1961) Symbiosis in marine and fresh water coelenterates. In: Lenhoff HM, Loomis WF (eds) The biology of *Hydra* and some other coelenterates. University of Miami Press, Miami, pp 255–268

Norman MD, Finn J, Tregenza T (2001) Dynamic mimicry in an Indo-Malayan octopus. Proc R Soc London, Ser B: Biol Sci 268(1478):1755–1758

Ogden JC, Buckman NS (1973) Movements, foraging groups, and diurnal migratons of the striped parrotfish *Scarus croicensis* Bloch (Scaridae). Ecology 54. https://doi.org/10.2307/1935344

Oksanen A, Kärssin A, Berg RPKD, Koch A, Jokelainen P, Sharma R, Jenkins E, Loginova O (2022) Epidemiology of *Trichinella* in the Arctic and subarctic: a review. Food Waterborne Parasitol 28(2022):e00167

Palacio-Castro AM, Smith TB, Brandtneris V, Snyder GA, van Hooidonk R, Maté JL, Manzello D, Glynn PW, Fong P, Baker AC (2023) Increased dominance of heat-tolerant symbionts creates resilient coral reefs in near-term ocean warming. Proc Natl Acad Sci 120(8):e2202388120. https://doi.org/10.1073/pnas.2202388120

Paracer S, Ahmadjian V (2000) Heinrich Anton de Bary (1831–1888). In: Symbiosis, an introduction to biological associations, 2nd edn. Oxford University Press, New York, pp 231–233

Patton WK (1967) Studies on *Domecia acanthophora*, a commensal crab from Puerto Rico, with particular reference to modifications of the coral host and feeding habits. Biological Bull 132(1):56–67

Patton WK (1976) Animal associates of living reef corals. In: Jones OA, Endean R (eds) Biology and geology of coral reefs, vol 3, Biology 2. Academic Press, New York, pp 1–36

Pearse JS, Arch SW (1969) The aggregation behaivor of *Diadema* (Echinodermata, Echinoidea). Micronesica 5:165–171

Pearson RG, Endean R (1969) A preliminary study of the coral predator *Acanthaster planci* (L.) (Asteroidea) on the Great Barrier Reef, vol 3. Fishery Notes Department of Harbours and Marine, Queensland, pp 27–55

Peters EC (2015) Diseases of coral reef organisms. In: Birkeland C (ed) Coral reefs in the anthropocene. Springer, Dordrecht, pp 147–178

Peters EC, Gassman NJ, Firman JC, Richmond RH, Power EA (1997) Ecotoxicology of tropical marine ecosystems. Environ Toxicol Chem 16(1):12–40

Pogoreutz C, Clua EEG, Tortolero-Langarica JJA (2022) High live coral cover and incidence of a pink-spotted coral phenotype on remote reefs off Clipperton Island, tropical eastern Pacific. Mar Biol 169:115. https://doi.org/10.1007/s00227-022-04101-3

Pratchett MS (2001) Influence of coral symbionts on feeding preferences of crown-of-thorns starfish *Acanthaster planci* in the western Pacific. Mar Ecol Prog Ser 214:111–119

Pruitt JN, Keiser CN, Banka BT, Leidle JS, Brooks AJ, Schmitt RJ, Holbrook SJ (2018) Collective aggressiveness of an ecosystem engineer is associated with coral recovery. Behav Ecol 29:1216–1224. https://doi.org/10.1093/beheco/ary092

Quimbayo JP, Dias MS, Schlickmann ORC, Mendes TC (2016) Fish cleaning interactions on a remote island in the tropical eastern Pacific. Mar Biodiver. https://doi.org/10.1007/s12526-016-0493-2

Richmond RH (1985) Variations in the population biology of *Pocillopora damicornis* across the Pacific. In: Proceedings of the 5th international coral reef congress, Tahiti, vol 6, pp 101–106

Rinkevich B, Wolodarsky Z, Loya Y (1991) Coral-crab association: a compact domain of a multilevel trophic system. Hydrobiologia 216(217):279–284

Robertson R (1970) Review of the predators and parasites of stony corals, with special reference to symbiotic prosobranch gastropods. Pac Sci 24:43–54

Rodríguez-Villalobos JC, Reyes-Bonilla H (2019) History of perspectives on the study of coral disease in the eastern tropical Pacific. Dis Aquatic Org 136:243–253. https://doi.org/10.3354/dao03411

Rodríguez-Villalobos JC, Ayala-Bocos A, Calderon-Aguilera LE, Petatán-Ramírez D (2022) Potential distribution of White Syndrome in scleractinian corals in the eastern tropical Pacific. Mar Biol 169:46. https://doi.org/10.1007/s00227-022-04037-8

Rohwer F, Youle M (2010) Coral reefs in the microbial seas. Plaid, Granada Hills, CA, USA, p 201

Rouzé H, Leray M, Magalon H, Penin L, Gélin P, Knowlton N, Fauvelot C (2017) Molecular characterization reveals the complexity of previously overlooked coral-exosymbiont interactions and the implications for coral-guild ecology. Sci Rep 7:44923. https://doi.org/10.1038/srep44923

Sagan L (1967) On the origin of mitosing cells. J Theor Biol 14(3):225-IN6

Salas-Moya C, Vargas-Castillo R, Alvarado JJ, Azofeifa-Solano JC, Cortés J (2021) Decapod crustaceans associated with macroinvertebrates in Pacific Costa Rica. Mar Biodiver Rec 14:6. https://doi.org/10.1186/s41200-020-00199-w

Sánchez JA (2016) Diversity and evolution of octocoral animal forests at both sides of tropical America. In: Rossi S (ed) Marine animal forests. Springer, New York. https://doi.org/10.1007/978-3-319-17001-5_39-1

Scott PJB, Risk MJ, Carriquiry JD (1988) El Niño bioerosion and the survival of east Pacific reefs. In: Proceedings of the 6th international coral reef symposium, Townsville, vol 2, pp 517–520

Shafir A, Loya Y (1983) Consumption and assimilation of coral mucus by the burrowing mussel *Lithophaga lessepsiana*. Bull Ins Oceanograph Fish 9:135–140

Shmuel Y, Ziv Y, Rinkevich B (2022) *Trapezia* crabs that dwell in distinctive day/night canopy compartments of a marine animal forest forage on demersal plankton. J Mar Sci Eng 10(10):1522. https://doi.org/10.3390/jmse10101522

Snedaker SC, Glynn PW, Rumbold DG, Corcoran EF (1995) Distribution of n-alkanes in marine samples from southeast Florida. Mar Pollut Bull 30:83–89

Solano-Barquero A, Rojas A, Cortés J (2023) Metazoan marine parasites of Costa Rica: a review. Parasitologia 2023(3):116–141. https://doi.org/10.3390/parasitologia3020014

Sousa WP (2001) Natural disturbance and the dynamics of marine benthic communities. In: Bertness MD, Gaines SD, Hay ME (eds) Marine community ecology. Sinauer Associates, Sunderland, MA, pp 85–130

Stachowicz JJ (2001) Mutualism, facilitation, and the structure of ecological communities: positive interactions play a critical, but underappreciated, role in ecological communities by reducing physical or biotic stresses in existing habitats and by creating new habitats on which many species depend. Bioscience 51(3):235–246

Stambler N (2011) Zooxanthellae: the yellow symbionts inside animals. In: Dubinsky Z, Stambler N (eds) Coral reefs: an ecosystem in transition. Springer, Cham, pp 87–106

Stewart HL, Holbrook SJ, Schmitt RJ, Brooks AJ (2006) Symbiotic crabs maintain coral health by cleaning sediments. Coral Reefs 25:609–615. https://doi.org/10.1007/s00338-006-0132-7

Stimson J (1990) Stimulation of fat-body production in the polyps of the coral *Pocillopora damicornis* by the presence of mutualistic crabs of the genus *Trapezia*. Mar Biol 106:211–218

Styrsky JD (2014) An orb-weaver spider exploits an ant-acacia mutualism for enemy-free space. Ecol Evol 4(3):276–283. https://doi.org/10.1002/ece3.930

Thurber RV, Hay M (2023) Mystery solved? Disease detectives identify the cause of a mass die-off in the sea. Sci Adv 9(16):eadh5478

Thurber RV, Mydlarz LD, Brandt M, Harvell D, Weil E, Raymundo L, Willis BL, Langevin S, Tracy AM, Littman R, Kemp KM, Dawkins P, Prager KC, Garren M, Lamb J (2020) Deciphering coral disease dynamics: integrating host, microbiome, and the changing environment. Front Ecol Evol 8. https://doi.org/10.3389/fevo.2020.575927

Turnham KE, Aschaffenburg MD, Pettay DT, Paz-García DA, Reyes-Bonilla H, Pinzón J, Timmins E, Smith RT, McGinley MP, Warner ME, LaJeunesse TC (2023) High physiological function for corals with thermally tolerant, host-adapted symbionts. Proc Biol Sci 290(2003):20231021. https://doi.org/10.1098/rspb.2023.1021. Epub 2023 Jul 19

Van Oppen MJH, Mieog JC, Sanchez CA, Fabricius KE (2005) Diversity of algal endosymbionts (zooxanthellae) in octocorals: the roles of geography and host relationships. Mol Ecol 14(8):2403–2417

Vera M, Banks S (2009) Health status of the coral communities of the northern Galapagos Islands Darwin, Wolf and Marchena. Galapagos Res 66:65–74

Vernberg WB (1974) Symbiosis in the sea. University of South Carolina Press, Columbia, p 276

Weber JN, Woodhead PMJ (1970) Ecological studies of the coral predator *Acanthaster planci* in the South Pacific. Mar Biol (Berl.) 6:12–17

Webster NS, Reusch TBH (2017) Microbial contributions to the persistence of coral reefs. ISME J 11:2167–2174

Chapter 12
Retirement: A Subtropical Research Addendum

Abstract Teaching, student mentoring, and research were continued in southeastern Florida for several years after my formal retirement from the University of Miami in May 2016. Teaching ranged from academically advanced high school students in the Summer Scholars program (University of Miami) to a graduate level course in Coral Reef Ecology (Nova Southeastern University). Mentoring upper undergraduate students who desired an opportunity to engage in research was also enjoyable and a privilege. The occurrence of benthic ctenophore populations in Miami waters (two species), an under-studied taxon, provided an opportunity to add to our knowledge of these enigmatic invertebrates. The discovery of an exotic brittle star and its negative effects on ctenophores signal an invasive species status. A Florida Keys study demonstrated that forces from storm-generated breaking waves can generate giant, > 1 m diameter, mobile spherical coral colonies (coralliths). A study in Peru found that Late Pleistocene (~ 13,300 cal BP) *Pocillopora* fragments at the pre-historic Huaca Prieta settlement are likely a result of human transport and not indicative of an established local population. Recent studies have demonstrated significant long-term changes in coral communities in the Galápagos Islands (since 1975–76), and in a high intertidal community on the Central California coast (since 1959–61).

Introduction

In the early 2000s, my thoughts about retirement became more frequent and apprehensive. Teaching and research were the chief activities of my professional career with a focused interest on coral reefs. Coral reef ecosystems, their beauty, complexity, and ability to form, grow and often dominate tropical and subtropical seascapes, captivated my psyche. To investigate the environmental processes and species interactions of reef organisms and pass the findings on to others was an enjoyable and fulfilling endeavor. I was fearful of losing these pursuits upon retiring. There were still unpublished data, proposals, and projects—unfinished business—that needed attention. Associated with these undertakings were the camaraderie and collegiality

© The Author(s), under exclusive license to Springer Nature Switzerland AG 2025
P. W. Glynn, *Coral Reef Resilience in the Anthropocene*,
https://doi.org/10.1007/978-3-031-81421-1_12

involving scientific discussions and research with students and colleagues—e.g., questioning hypotheses, evidence, methodologies, and ideas for continuing investigations.

Academics

Happily, two opportunities allowed me to continue teaching into retirement, at least up to the COVID-19 pandemic. One of the courses was structured for the UM Summer Scholars Program (SSP), to serve academically advanced high school students. It was offered as a university, freshman-level marine science course with transferable credits. This course consisted of two parts, the first, *Life in the Sea*, was presented as class lectures, and the companion part, *Introduction to Marine Science Laboratory*, was designed to introduce students to field and laboratory activities in marine science with a special focus on unique tropical marine environments and biotas of South Florida. A survey of marine organisms, including examples from the six kingdoms of organisms, was presented as lectures in the morning and coordinated with laboratory exercises and field activities in the afternoon. Local species and communities were introduced, focusing on their biology, ecological relationships, and community dynamics. Coral reef studies involved overnight excursions in the Florida Keys. Professor Chris (Christopher) Langdon and I taught this course for two summers (2018, 2019), with enrollments of ~35 students each term. Teaching assistants from UM provided essential help with all assignments and needs. While challenging in terms of the rapid pace of instruction, Chris and I were greatly impressed with the student's enthusiasm, diligence, and discipline during this learning experience. In 2020, Chris alone taught the summer scholars course remotely via zoom. Our high regard for the performance of the SSP students was borne out by many who recently received undergraduate departmental science awards at UM.

The second teaching opportunity was kindly offered to me by Bernhard Riegl of the Nova Southeastern University (NSU) Oceanographic Center. This graduate-level course, *Coral Reef Ecology*, was presented by my stalwart teaching assistant, Nick (Nicholas) P. Jones,[1] and me at the Dania Beach, Florida campus. I structured the course around lectures on major coral reef topics, for example those presented in Sheppard et al. (2018). In addition, on-going research relevant to the Greater Caribbean region was presented as class lectures and discussions. Guest speakers were invited to discuss their research, in particular those topics related to coral species identifications, diseases, bleaching, ocean acidification, and evolutionary history. I am grateful to Andrew Baker, Ian Enochs, Nick Jones, Chris Langdon, Bill Precht, and Bernhard Riegl for their presentations. Further, each student was tasked

[1] Nick's broad interests and knowledge of coral reef ecology, population and community dynamics, and climate change biotic disturbances, added considerably to the quality and depth of discussions in our class (e.g., Jones et al. 2022).

with presenting a report, written and oral, on a coral reef research topic of their choosing. Due to the rapid spread of the COVID-19 virus, class instruction was converted to virtual mode via zoom presentations.

The cessation of class meetings, and laboratory and field exercises, was a necessary precaution limiting the rate of infection, yet, disappointing restrictions in many ways. In a remote lecture setting it is difficult to ascertain whether complicated topics are being fully understood. Facial expressions and body language of the students offer invaluable clues. In addition, organisms, their anatomy, identification and behavior, research equipment, and methods are most effectively introduced in person-to-person class demonstrations.

Research

In the early 2000s, grants to support coral research in the eastern tropical Pacific were not so readily available as in the 1980s and 1990s. Widespread coral reef bleaching and mortality, first reported in the eastern Pacific in 1982–83, was now occurring in all global reef regions and was in urgent need of attention and financial support for research. This caused a dilution of funding that was already in short supply. Therefore, it became necessary to restrict investigations to more accessible areas in southeastern Florida.

Ctenophores. The re-discovery of the benthic ctenophore, *Coeloplana waltoni*, near the RSMAES laboratory, and the interest of undergraduate students in this species, made it a perfect candidate for study (Fig. 11.3). Ted Bayer in 1960 was the first to collect and recognize this minute, flat, worm-like animal as a benthic ctenophore. It was commonly found clinging to the branches of various octocoral species, the chief subjects of Ted's research. Following Ted's death, Steve (Stephen) Cairns, Curator, Department of Invertebrate Zoology, Smithsonian National Museum of Natural History, came across his unfinished manuscript describing this ctenophore as a new species. Steve sent the manuscript to me and asked if I would search for the ctenophore to determine if it was still present in south Florida waters. I briefed the students in the invertebrate class I was teaching at the time on how we might search for the ctenophore on octocoral branches. It did not take long to find them. Our initial searches revealed numerous ctenophores on several species of octocorals. The ctenophores are small, 2–3 mm in length, and with translucent bodies that blend with their octocoral host's surface, making them generally difficult to observe in the field. We soon learned that the ctenophores and the branch tips of their octocoral hosts could be readily maintained in the laboratory. This was advantageous because the small size of the ctenophores required microscopical viewing. I had been searching for benthic ctenophores for decades, but without success. The discovery of this rare and enigmatic animal, likely a stem representative of the ancestral lineage of all metazoans, was a dream come true. Recent phylogenetic studies suggest that ctenophores rather than sponges are the sister lineage, i.e., the closest ancestral relations, to all other animals (e.g., Ryan et al. 2013; Whelan et al. 2017; Schultz et al. 2023).

The very first benthic ctenophore species discovered in American waters was also small, 4–6 mm in length; these were creeping individuals of *Vallicula multiformis* (Fig. 12.1). These animals, recognized as ctenophores by F. G. Walton Smith, were collected from aquaria at the UM Marine Laboratory in 1945.[2] Like *C. waltoni*, the occurrence of this ctenophore species in southeastern Florida was unstudied (and essentially forgotten?) for several decades. We began to wonder if *Vallicula* was still present in south Florida. One afternoon when I entered my laboratory, Joshua Levy, our laboratory manager, motioned for me to examine a large container chock-full of freshly collected algae. Much to my amazement, several individuals of *V. multiformis* were floating on the surface film of water. Joshua had spearheaded a search for *Vallicula*, with student volunteers, Samuel May, Ryan McMinds, and Joyce Yager, which was thankfully fruitful. When we began our quantitative surveys in 2012 and 2013, we were amazed by the abundances of the ctenophores. The mean densities of *Coeloplana* collected from octocoral stems ranged from 500 to 800 individuals per 100 ml DV (displacement volume)[3] of their microhabitat (Glynn et al. 2017a, b). Since we sampled only a few small branches from the octocoral colonies, it is likely that some individual colonies were host to thousands of benthic ctenophores. Like *Coeloplana*, the densities of *Vallicula* were also high. In the summer season it was common for collections to yield 100 to 200 individuals per 100 ml DV of a mix of algae and bryozoans. To our delight, we found that *Vallicula* could also be easily maintained in laboratory aquaria. This circumstance, along with their

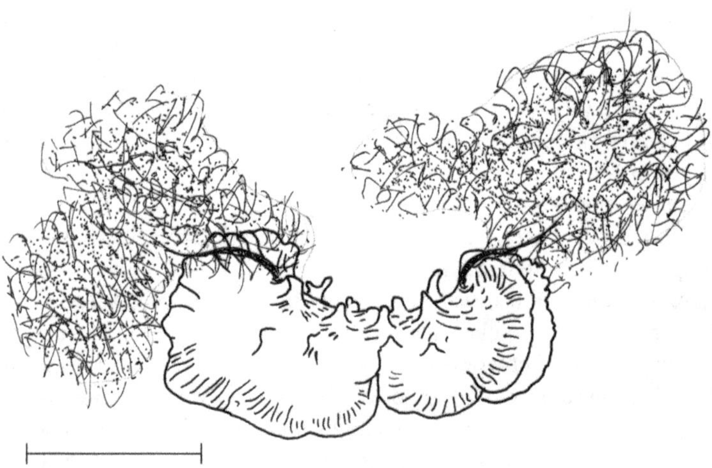

Fig. 12.1 *Vallicula multiformis*, defensive display, scale bar = 2 mm

[2] Prof. Smith founded the UM Marine Laboratory, which was the forerunner of RSMAES. This benthic ctenophore species was later described and studied in Jamaica by J. J. Rankin (1951, 1956).

[3] The displacement volume (DV) simply refers to the volume of sea water displaced by the ctenophore microhabitat, which was measured in graduated cylinders. The preferred habitat sampled and reported for *Coeloplana* was 10-cm long octocoral branches; for *Vallicula*, mixed fouling algal/bryozoan masses.

abundance and relative ease of collection, made the two species suitable subjects for comparative studies.

Both ctenophore species proved to be skillful predators and catholic in their selection of prey, feeding on a wide range of planktonic taxa (Glynn et al. 2018a). To prepare the ctenophores for feeding trials, each species with its octocoral host or algal mass habitat, was placed in a circular beaker with a magnetic stirrer to simulate natural current flow. The beaker was placed near a laboratory window to receive natural lighting, and the ctenophores were observed with a magnifying visor. This setup allowed a clear view of the ctenophores, including the movements of their tentacles. The tentacles, with numerous fine lateral branches (tentillae) bearing adhesive structures (colloblasts), were extended and retracted rapidly when the ctenophores were "fishing" (feeding). We were in constant awe observing the in-and-out movements of the tentacles and the effective prey capture behavior.

Both ctenophore species also quickly extended their tentacles when zooplankton was added to the beakers, resulting in the capture of adult copepods, and veliger and nauplii larvae (Fig. 12.1). Gut content analysis of field collected *Vallicula* revealed large crustacean limbs, indicating the ability to capture and ingest relatively large prey. Beyond the niche-diet interest of obscure benthic ctenophores, we found that both species readily consumed dolphinfish (mahi-mahi), *Coryphaena hippurus*, eggs and larvae in laboratory trials.[4] The captured and ingested egg sizes were 1 mm in diameter and the larvae 3.5–4.0 mm in length. The habitats of the two ctenophores, octocoral forests (*C. waltoni*) and biofouling algal communities, including floating *Sargassum* (*V. multiformis*),[5] are exposed to ocean currents in southeastern Florida carrying dolphinfish larvae. Dolphinfish occur in tropical and subtropical waters worldwide, are a sustainable food source, and important commercially and recreationally.

With an improved understanding of the ctenophore prey taxa and sizes, we next turned our attention to their predators. Considering the nature of the ctenophores, i.e., small, creeping, soft-bodied animals lacking an exoskeleton or other protective body covering, we expected the presence of numerous potential predators. We were not far off the mark. Two commonly observed predators of *C. waltoni* were the Flamingo Tongue (*Cyphoma gibbosum*), a gastropod mollusk, and the Caribbean Pygmy Angelfish (*Centropyge argi*).[6] The gastropod grazed on octocoral tissues and indirectly consumed ctenophores in its path; angelfish visually targeted ctenophores

[4] My RSMAES colleagues, Daniel Benetti and staff, Miami Experimental Hatchery, and Martin Grosell and team, Department of Marine Biology and Ecology, kindly supplied the young dolphinfish stages for the feeding trials. Martin and co-workers were investigating the effects of the Deepwater Horizon oil spill on the development of the fish larvae.

[5] *Vallicula multiformis* occurs in the Pacific, Atlantic, and Indian Oceans (Prasade et al. 2015). It is possible that it was introduced to the biofouling community in south Florida from the rafts of *Sargassun* that drifted into this region from the Sargasso Sea (Northern Atlantic Subtropical Gyre) during the summer season.

[6] Another predator of *C. waltoni*, observed less frequently, is a caprellid amphipod (*Caprella penantis*) that readily attacks and shreds ctenophores (Glynn et al. 2021).

and then plucked them from octocoral branches. Several other species, especially fishes, were observed feeding on octocoral colonies, but it was not possible to identify the prey due to their small size.

Numerous potential predators of *Vallicula multiformis* were observed feeding in the biofouling community, and several of these species readily consumed ctenophores in laboratory trials. The most frequently encountered predators, with widely different methods of prey capture, were an actinian cnidarian (*Exaiptasia pallida*), three species of brachyuran crabs [*Macrocoeloma trispinosum*, *Panopeus herbstii*, and *Portunus* (*P.*) *sayi*], and a juvenile snapper (*Lutjanus griseus*). Actinians ensnared ctenophores with their tentacles; crabs captured ctenophores with their pincers and shredded them with their mouthparts. Of the three fish species offered *Vallicula*, only the juvenile snapper consumed the ctenophore. In one trial the ctenophore was seized and quickly rejected, the snapper exhibiting apparent irritation around the mouth. This raised the possibility that the ctenophore may possess repellant properties. To evaluate this, *M. trispinosum* was fed a steady diet of *Vallicula* for over one month. The crab behaved normally and remained healthy during this period. Could it be that the ctenophore's tentacles, armed with sticky colloblasts, were acting as a physical deterrent to the snapper?

We began to suspect that ctenophore tentacles, besides their functional role in prey capture, could also serve a defensive function in protection against predators. During our routine handling of *Vallicula* in the laboratory we often observed agitated individuals rapidly extending their tentacles (Fig. 12.1). In simulating attacks by their natural predators, e.g., a sea anemone species and crabs, the ctenophores would often quickly extend both tentacles which presented dense tangles on both sides of their bodies (Glynn et al. 2018b). We hypothesized that this physical, sticky barrier would likely be sufficient to discourage attack by some small predators. Most of the larger predators in our feeding trials, however, were not deterred. Conversely, *Coeloplana* usually retracted its tentacles when agitated. It seemed to be well protected on its host, often observed seeking shelter alongside or within octocoral polyps. Its translucent body allowed the ctenophore to mimic its host's color and texture.

Initially puzzling were the large sizes of the tentacle tangles, which appeared considerably larger than the extended aggregate of fishing tentacles. The main tentacle filament and associated tentillae are armed with batteries of colloblast cells that, when stimulated, burst apart and discharge a strong adhesive substance (Brusca et al. 2016). Perhaps when the tentacles are stimulated simultaneously this effect is responsible for the enormous increase in tentacle volume?

The generally high abundances of both ctenophore species, juxtaposed with exposure to diverse predators, prompted our study of reproduction, i.e., the replenishment of population abundances following mortality. The principal topics we investigated were reproductive modes, seasonality, and recruitment (Glynn et al. 2019). Initially, we were apprehensive of this line of research, and fearful that little knowledge would be forthcoming since none of us had previous experience with ctenophore reproductive investigations. We were soon pleasantly surprised with the adaptations observed and the high level of sexual and asexual activities displayed by the ctenophores.

Both species are hermaphroditic brooders. During development, sexually generated cydippid larvae were often visible in the diaphanous bodies, with advanced stages actively rotating in brood chambers. Also, both species were reproductively active at an early age (paedogenesis), with gonadal development occurring at body sizes less than half the "adult" lengths. We were fortunate to witness the spawning of *Vallicula* larvae, as many as 30–40 individuals released during an event. Recently spawned larvae were without an apical sense organ and tentacles. These structures developed in two days, thus enabling zooplankton fishing and food procurement. Sexual reproduction in both ctenophore species was most active in the warm summer and fall months. Numerous pink adult individuals of *Coeloplana* appeared at this time. We suspected that the pink ctenophores were somehow involved in reproduction but were unable to perceive a function.

A life history trait shared by benthic ctenophores is their ability to reproduce asexually by fragmentation, known to occur in platyctenid species (e.g., Freeman 1967). Some ctenophores in the process of fragmentation reminded us of "molasses taffy pulling", with extreme stretching resulting in 2–3 times the original body length. The result: a tiny propagule, just a bit of tissue, and an adult ctenophore with no signs of injury (Fig. 12.2). Other ctenophores generated fragments by simply sloughing off peripheral bits of tissue. Fragmentation was highly variable timewise; some ctenophores showed no signs of fragmentation for weeks whereas others fragmented frequently, generating replicas or clones daily. Reflecting on the fate of clones, these recruited directly into the ctenophore's microhabitat, i.e. on to octocoral branches (*Coeloplana*) or in biofouling algal mats (*Vallicula*). They were not subject to the hazards of the plankton as were the released, sexually generated cydippid larvae. This style of recruitment is like asexual fragmentation in corals,

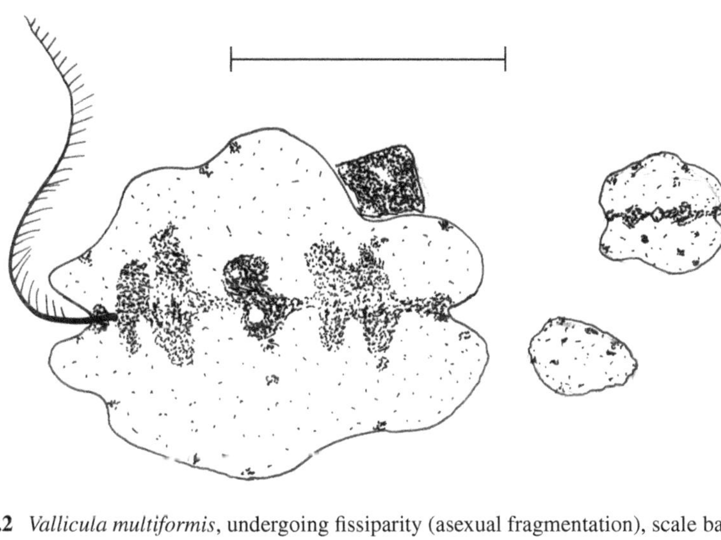

Fig. 12.2 *Vallicula multiformis*, undergoing fissiparity (asexual fragmentation), scale bar = 2 mm

which is a prevalent and successful mode of propagation in the eastern Pacific (Glynn et al. 2017a, b). In summary, the reproductive attributes observed—prolonged warm water spawning, larval brooding, rapid development, paedogenesis, and asexual fragmentation—would presumably allow these ctenophore species to recover quickly following predation or other mortality events.

A Non-native Brittle Star. In 2018 and 2019, we began encountering a non-native brittle star clinging to the octocoral stems inhabited by *Coeloplana waltoni* (Figs. 12.3 and 12.4). During this time, we also began noticing significant declines in the population density of *C. waltoni*. The brittle star, *Ophiothela mirabilis*, is an eastern Pacific endemic species described from the Pearl Islands, Panama (Verrill 1867). Its introduction to the western Atlantic region (Brazil, Caribbean Sea, south-eastern Florida), via the Panama Canal, has been attributed to shipping (e.g., Hendler and Brugneaux 2013; Araújo et al. 2018). Population densities of our beloved ctenophore, *Coeloplana waltoni*, revealed a precipitous decline in 2018 and 2019 at our Dania Beach study site during the sudden appearance of large numbers of brittle stars. This was deeply concerning. Is it possible that *Ophiothela* was in the process of displacing *Coeloplana*? This possibility changed the focus of our studies of the epizootic symbionts of octocorals. We then turned our attention to investigating the

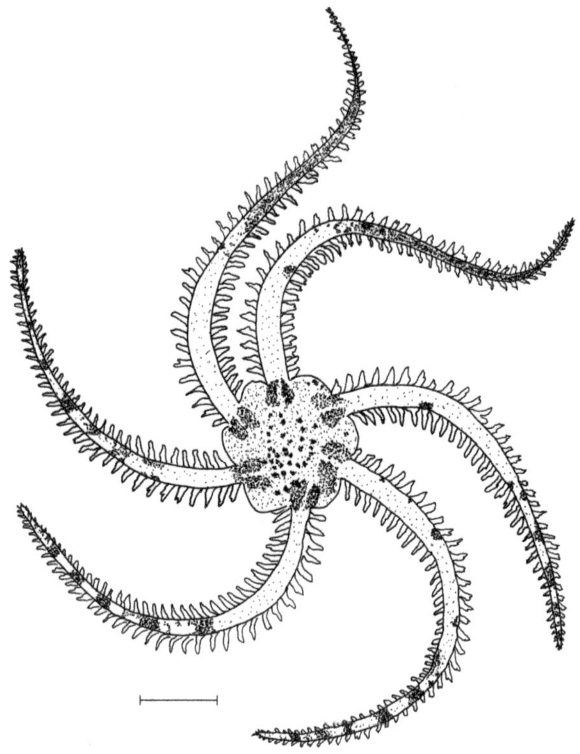

Fig. 12.3 *Ophiothela mirabilis*, scale bar = 1 mm

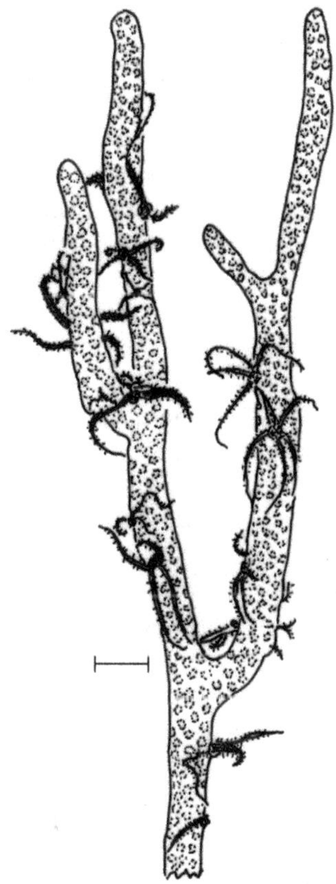

Fig. 12.4 *Ophiothela mirabilis*, several individuals clinging to an octocoral, scale bar = 1 cm

biology of *Ophiothela* and interactions with its octocoral hosts and co-occurring ctenophores[7].[8]

We hypothesized that the high densities of the brittle star, up to 25 individuals per 10 cm octocoral branch-length, in combination with their energetic movements and calcareous armature (spines, plates, and hooks), might be sufficient to physically irritate and encourage ctenophores to seek shelter elsewhere. Observations and experimental evidence supported this scenario. When *Ophiothela* meets *Coeloplana*, the ctenophore usually contracts and remains in place or moves away. The brittle stars made no assertive movements, i.e., predatory in nature, toward the ctenophores. In a 7-day experiment with both species cohabiting octocoral branches,

[7] Since two of our ctenophore studies were recognized as top cited papers from 2017–2018 and 2019–2020 in the journal *Invertebrate Biology*, we were gratified and encouraged to continue with the octocoral-symbiont investigations.

[8] We are grateful to Dean Roni Avissar and Associate Dean Prof. Michael C. Schmale for allowing us continued use of the Grosvenor Building research laboratory. Also, we thank Resource Manager Phillip Gillette for use of the laboratory in the "National Resource for *Aplysia*" facility.

ctenophore abundances declined by about 86% (Glynn et al. 2021). The few ctenophores remaining after the experiment were present at the branch tips of octocorals, relatively far away from the brittle stars.

Feeding observations showed that *Ophiothela* captured food with its arms extended in the water column. Gut content analysis revealed the presence of predominantly amorphous detritus and filamentous algae.[9] Few zooplankters were found. We therefore concluded that *Ophiothela* is negatively affecting ctenophore abundances through interference competition, greatly aided by its calcareous armature (Glynn et al. 2021). As noted above, a caprellid species readily attacked and consumed ctenophores. These predatory crustaceans are capable of directly impacting *Coeloplana* abundances. They are encountered sporadically, however, and presently cause only local declines in abundance.

During the initial phases of our studies, the basic color of *Ophiophila* was beige, however, orange-colored individuals began to appear in collections from mid-2019 and later. Molecular phylogenetic analyses indicated that the two color patterns represented two distinct lineages, possibly the result of at least two separate invasion events.[10] The orange morph from Florida is closely related to the Brazilian and Caribbean populations, suggesting a range expansion of the original invasion. The beige morph is genetically close to specimens from the Mexican Pacific and Peru, potentially representing a secondary and more recent introduction via shipping through the Panama Canal. It is perhaps not just coincidental that the first reports of the non-native brittle star were from localities in Brazil and Florida near seaports with heavy interoceanic ship traffic. It could be that *O. mirabilis* was transported on biofilms attached to the hull of a ship or in ballast water?

Since our investigations demonstrated negative effects of the brittle star on *Coeloplana* abundances, we could justifiably refer to *O. mirabilis* as an invasive species.[11] Is it possible that the brittle star could also adversely affect other native species, for example the octocoral hosts? Considering the widespread occurrence of *Ophiothela*, its high abundance, and active attachment behavior, several investigators expressed concern that the brittle star might have an adverse effect on octocoral hosts (e.g., Mantellatto et al. 2016; Rich et al. 2020). Some workers noted that octocoral polyps were not fully expanded and feeding normally in the presence of *Ophiothela* (Araújo et al. 2018; Ferry et al. 2020). Further, Ferry et al. (2020) observed dead octocoral patches where brittle stars were present. These reports need further study.

As the COVID-19 pandemic intensified, and student team members graduated, the octocoral-symbiont studies ended. Most of the undergraduate student volunteers

[9] Gut analyses were performed on preserved specimens at my home, in an improvised laboratory, at the height of the COVID-19 pandemic.

[10] We are indebted to two new team members with molecular genetic expertise for these findings: Renata Alitto, Instituto de Biologia, Universidade Estadual de Campinas, São Paulo, Brazil; Ana B. Christensen, Biology Department, Lamar University, Beaumont, Texas, USA.

[11] An introduced nonnative organism that can expand its range and has the potential to harm the environment, including any native organisms or the economy, is recognized as an invasive species.

were upper class third and fourth year marine science majors and remained on our team for 1–2 years. Some students entered my lab as first-year students or sophomores. It is fitting to recognize these pioneer and long-term team members: Brian Coffman, Joshua Dominguez, Michael P. C. Fuller, Nicolas Martinez, Shannon G. Moorhead, and Karim Primov. Some students became interested in particular research problems, enrolled in undergraduate research courses, and earned credit. These were exhilarating times with new scientific discoveries very much welcomed. I felt privileged mentoring these bright and motivated students, most of whom continued with biological sciences studies and earning advanced degrees.

Mobile coralliths. The discovery of giant, spherical corals (coralliths) captivated the interest of Bill (William F.) Precht[12] and me when we were hired as consultants, representing opposite sides of an environmental impact study, regarding construction of a boat dock in Key Largo, Florida. The shallow (1–3 m depth) construction site is an environmentally sensitive area with concern for dredging damage to sea grasses and corals. When Bill and I surveyed the area, we discovered numerous, large, spherical corals of several species. Coralliths are spherical, free-living (mobile), scleractinian colonies inhabiting present day and ancient coral reefs. They form by movement. i.e. rolling along the bottom through both biological (Glynn 1974) and physical (Hoeksema et al. 2018) action. The species and sizes of the largest coralliths we observed were *Solenastrea bournoni*, which ranged between 0.4 and 1.4 m in diameter (Fig. 12.5). One colony measuring 1.35 m in diameter revealed an estimated age of ~120 years.

Considering the shallow depth of the corallith habitat and the high frequency of tropical storms and hurricanes in south Florida, we hypothesized that the forces generated by shoaling, breaking storm waves would be sufficient to cause the formation of giant coralliths. We also noted the size advantage of large coralliths in resisting movement and surviving during turbulent storm events. Corallith species with small colonies must suffer more frequent movement and higher mortality than large colonies. Relatively frequent movement in different directions is key; all living colony surfaces must be exposed to sunlight and currents at some time. The large corals we observed revealed dead coral patches overgrown with algae, indicating prior contact with bottom sediments. One can surmise that the larger a corallith grows, the stronger a storm must be to initiate movement. Spoiler alert: Calculations based on hydrodynamic forces and field observations after a storm support the

[12] Bill stood out as an enthusiastic, bright student in the Coral Reef Ecology and Physiology course that Alina Szmant and I taught at RSMAES in 1987. Alina and I were impressed with his passion for coral reefs and impressive knowledge of early and current coral reef studies. While a graduate student at RSMAES, Bill worked in my laboratory as a research assistant for 1 year. He has exhibited a productive career in the areas of coral reef disturbances, including coral diseases, human impacts, and global climate change. Bill has been a visiting faculty member in Northeastern University's Three Seas Program for more than 30 years where he co teaches a class in coral reef ecology. He is also a visiting instructor at Columbia University where he co-teaches a course in coral reef geology. Bill is currently the Director of Marine and Coastal Programs for the environmental consulting firm Dial Cordy and Associates, Inc., based in Miami, Florida. Bill attained ICRS Fellow status in May 2024.

Fig. 12.5 *Solenastrea bournoni*, (**A**) upright living position; (**B**) rotated to show embedded, dead colony bottom. Colony diameter 45 cm, water depth ~ 2 m. Islamorada, Florida Keys (2006). Photographs courtesy of William Precht

efficacy of storm-generated waves and surge to move and thus assist in the formation of giant coralliths. This study, "Giant mobile coralliths from the Florida Keys", is now finalized and will be submitted to a scientific journal for review and hopefully acceptance for publication.[13]

***Pocillopora* occurrences in Peru.** Another subject requiring attention was the occurrence of zooxanthellate corals in northern Peru, referred to in Chap. 4. This matter began in 2014 when I received a letter from Prof. Patricia Netherly of Vanderbilt University, requesting help in the identification of two species of coral. I was anxious to help since so little is known of the corals in the Panamic-Peruvian transition zone. A small coral fragment recovered from the Huaca Prieta cultural mound (Fig. 12.6), one of three specimens, was *Pocillopora damicornis*. This specimen was heavily encrusted as originally excavated but revealed a well preserved skeleton when cleaned (Fig. 12.7). A second larger specimen, collected offshore by a local fisher at Máncora, Peru, I identified as *Pocillopora capitata*. These identifications were based on morphological characteristics. Considering recently attained coral genetic evidence and species-specific endosymbiont identities of these morphospecies, it is likely the two colonies are synonymous and belong to the taxon *Pocillopora grandis*.

The small fragment, radiocarbon dated with an age of ~13,300 BP, is from the late Pleistocene epoch and in remarkably good condition. Since the marine thermal climate during the late Pleistocene was cooler and likely limited the occurrence of zooxanthellate scleractinian corals at Huaca Prieta's latitudinal position, the question of its origin was key. If not present at Huaca Prieta as an established population, could it have dispersed southwards from Ecuador via planktonic larvae during an El

[13] Latest development: Our study was submitted to the Bulletin of Marine Science, passed peer review, and is now published (Precht et al. 2024). Mark R. Patterson, Northeastern University, joined our study to add insight of the hydrodynamic forces affecting corallith movements.

Fig. 12.6 General view of the Huaca Prieta prehistoric settlement mound near the Pacific Ocean shore. The arrows point to the deep excavation trenches on the east side of the mound. Modified after Glynn et al. (2022)

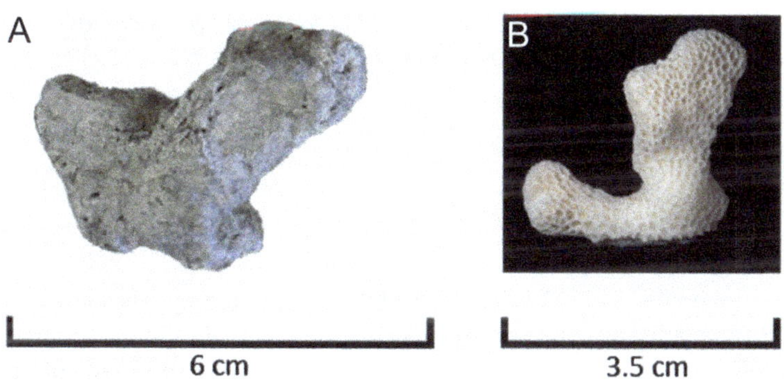

Fig. 12.7 *Pocillopora damicornis* terminal branch fragment excavated from the Huaca Prieta settlement mound. (**A**) uncleaned fragment as originally found; (**B**) cleaned skeleton. Adapted after Glynn et al. (2022)

Niño warming event? An alternative explanation would involve the transport of the coral down the coast from tropical Ecuadorian waters by terminal Pleistocene people. This explanation fits a later pattern between 8000 and 4000 years ago whereby warm water marine species such as *Spondylus* shells (bivalve mollusks) appear at the Huaca Prieta site in middle Holocene time (Dillehay 2017). These occurrences correspond to the mid Holocene thermal optimum period (Wanner et al. 2008). We tentatively accepted the north-south overland transport hypothesis (Glynn et al. 2022).

We are uncertain of the significance of the coral fragments recovered from the late Pleistocene cultural deposits at Huaca Prieta. Malcolm Shick, in an engaging narrative, has reviewed the historical cultural uses of corals, ranging from aesthetic curiosities and votive offerings to aides in the afterlife (Shick 2018). The Peruvian fragments could have had a symbolic meaning or were offered as an exotic curiosity as were fossils and minerals at several later sites in north Peru (Dillehay 2011). Elsewhere in the Americas the burial of coral offerings by the Mexica people in the 15th and 16th centuries is believed to have had a major religious importance (Medina-Rosas et al. 2021). Recent excavations in the sacred precinct of Tenochtitlan in Mexico City have revealed that corals were the most frequently buried gifts. The corals were among a diverse assemblage of cnidarians, including 15 species of soft, black, and fire corals from the Atlantic and Pacific coasts of Mexico. Surprisingly, the two coral morphospecies from Peru, *Pocillopora damicornis* and *P. capitata*, were among the species excavated in Mexico.

In addition to the question regarding the origin of the Huaca Prieta coral fragments, there is the matter of the *Pocillopora* occurrence reported by a fisher at Máncora. Máncora lies ~440 km north of Huaca Prieta and has a marine climate conducive to coral growth (monthly mean sea temperature = 23.17 °C; Glynn et al. 2022). No fossil or contemporary reef-building corals are known from the marine environment at Huaca Prieta where the monthly mean sea temperature today is about 2 °C cooler than at Máncora. The nearshore hard-bottom banks off Máncora should be surveyed for corals. Is it possible that *Pocillopora* corals exist in this area, and if so, are they present as ephemeral, El Niño-related introductions or established populations?

Galápagos Islands

A second invitation arrived from Pat Hutchings (Australian Museum Sydney) in 2018, 25 years after my first review reporting on the condition of Galápagos coral communities (Glynn 1994; Glynn et al. 2018c). This invitation could not be refused because it offered an opportunity for a manuscript submission to the widely read journal *Marine Pollution Bulletin*, a special issue in honor of Charles Sheppard.[14] Additionally, it allowed our eastern Pacific coral team to dedicate the review to Jerry Wellington, who pioneered coral research and conservation in the Galápagos Islands in the early 1970s. Much has changed since the devastating effects of the 1982–83 El Nino, with new insights on coral condition emerging from the molecular

[14] This special issue was dedicated to Prof. Charles Sheppard for his stalwart commitment to coral reef research, management, and conservation, and service as Chief Editor of the Marine Pollution Bulletin for 25 years. Sadly, Charles passed suddenly in early 2024. His kindly manner and breadth of knowledge will be truly missed.

identities of coral endosymbionts, ocean acidification, and extended ecological research. More recently, Ana Palacio-Castro et al. (2023) and colleagues, focusing on the genetic identity of heat tolerant coral symbionts surviving three El Niño heatwaves, offer a ray of hope, at least provisionally. This information is presented in Chap. 13, in the context of the broader eastern Pacific region and the condition of global coral reefs.

Pacific Grove Revisited

In early 2023, I was surprised to receive a draft manuscript of a long-term ecological study of a high intertidal community on the Central California coast. This study, 'Six Decades of Biodiversity Change in an Intertidal Community', is the same community and location (intertidal shoreline at the Hopkins Marine Station, Pacific Grove, California) where I conducted my doctoral research from 1959–1961, under the guidance of Donald P. Abbott. The lead author of the current study, Jamie (Samuel J.) Leonard, kindly sent the manuscript to me for review and comments (Leonard et al. ms.). The primary goal of my 3-year study was to quantitatively characterize the community structure and trophic interactions of the algal-barnacle zone (Glynn 1965). This intertidal assemblage has exhibited many fundamental changes over a six-decade period, particularly in biodiversity and intertidal range shifts. Its presence in the high intertidal zone at the Hopkins Marine Station campus has attracted students and researchers who have documented notable ecological changes over the years (e.g., Barry et al. 1995; Sagarin et al. 1999).

Significant declines in biodiversity and changes in community structure have occurred. Overall, species abundance declined by 59.5% across all invertebrate taxa over the study period. The decline or disappearance of foundation species, which provide the physical shelter within which associated species live, was largely responsible for the degradation and change in community structure. *Musculus pygmaeus*, a minute bivalve mollusk, was one such species that sheltered in algal holdfasts and vacant barnacle tests, habitat forming species now uncommon or absent from this community (Fig. 12.8).[15] Such relationships, involving negative effects to foundation species, have been amply demonstrated in diverse marine communities, e.g. sea grasses, salt marshes, mangrove forests, and coral reefs (Bruno and Bertness 2001). This is a sad reminder of the pervasive effects of climate change with the capacity to negatively affect biotic communities worldwide.

[15] After extensive searching in the molluscan literature, I failed to find a named species that fit the description of this minute bivalve. With specimens in hand, I visited Prof. Angeline Myra Keen at Stanford University with the hopes that she would agree with my assessment and describe the species. A world renown malacologist and invertebrate paleontologist, I was confident that Prof. Keen would be willing to help with this task. She was blunt, paraphrasing, "Yes, it does appear to be an unnamed species, but I will not describe it; you will describe it, and I will oversee your study." Hence, my first venture with describing a new species (Glynn 1964).

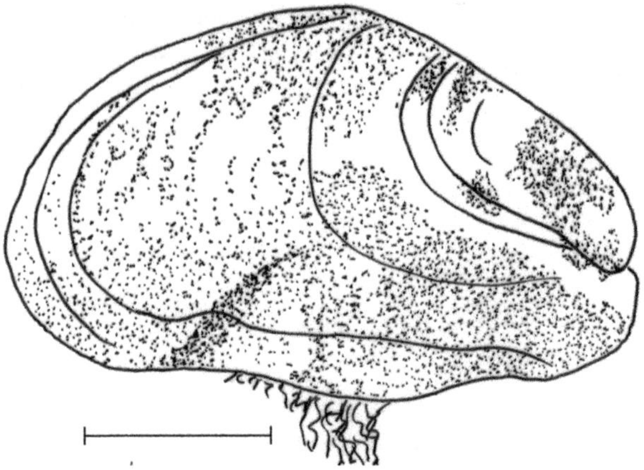

Fig. 12.8 *Musculus pygmaeus*, sale bar = 1 mm

References

Araújo JT, de Oliveira Soares M, Matthews-Cascon H, Monteiro FAC (2018) The invasive brittle star *Ophiothela mirabilis* Verrill, 1867 (Echinodermata, Ophiuroida) in the southwestern Atlantic: filling gaps of distribution, with comments on an octocoral host. Latin Am J Aquat Res 46:1123–1127. https://doi.org/10.3856/vol46-issue5fulltext-25

Barry JP, Baxter CH, Sagarin RD, Gilman SE (1995) Climate-related, long-term faunal changes in a California rocky intertidal community. Science 267(5198):672–675

Bruno JF, Bertness MD (2001) Habitat modification and facilitation in benthic marine communities. In: Bertness MD, Gaines SD, Hay ME (eds) Marine community ecology. Sinauer Associates, Sunderland, Massachusetts, pp 201–218

Brusca RC, Moore W, Shuster SM (2016) Invertebrates, 3rd edn. Sinauer Associates, Sunderland, Massachusetts, p 1104

Dillehay TD (ed) (2011) From foraging to farming in the Andes: new perspectives on food production and social organization. Cambridge University Press, London

Dillehay TD (2017) Where the land meets the sea: fourteen millenia of human history at Huaca Prieta, Peru, 1st edn. University of Texas Press, Austin

Ferry R, Hubert L, Philippot V, Priam F, Smith J (2020) First record of the non-indigenous brittle star species *Ophiothela mirabilis* Verrill, 1867 (Echinodermata: Ophiuroidea), off Martinique Island, French Lesser Antilles. Bioinvas Rec 9:228–238

Freeman G (1967) Studies on regeneration in the creeping ctenophore, *Vallicula multiformis*. J Morphol 123:71–84. https://doi.org/10.1002/mor.1051230107

Glynn PW (1964) *Musculus pygmaeus*. Nov., a minute mytilid from the high intertidal zone at Monterey Bay, California (Mollusca: Pelecypoda). Veliger 7(2):121–128

Glynn PW (1965) Community composition, structure, and interrelationships in the marine intertidal *Endocladia muricata—Balanus glandula* association in Monterey Bay, California. Beaufortia 12(148):1–198

Glynn PW (1974) Rolling stones among the Scleractinia: mobile coralliths in the Gulf of Panamá. In: Proceedings of the 2nd international coral reef symposium, Brisbane, vol 2, pp 183–198

Glynn PW (1994) State of coral reefs in the Galápagos Islands: natural vs anthropogenic impacts. Mar Pollut Bull 29(1–3):131–140

Glynn PW, Colley SB, Carpizo-Ituarte E, Richmond RH (2017b) Coral reproduction in the eastern Pacific. In: Glynn PW, Manzello DP, Enochs IC (eds) Coral reefs of the eastern tropical Pacific: persistence and loss in a dynamic environment. Coral reefs of the world 8. Springer, Dordrecht, pp 435–476

Glynn PW, Dillehay TD, Netherly PJ, Riegl BM (2022) Prehistoric reef-building coral occurrence in North Peru. J Island Coastal Archaeol 19:420–432. https://doi.org/10.1080/15564894.2022.2072026

Glynn PW, Coffman B, Primov KD, Moorhead SG, Vanderwoude J, Barrales RN, Williams MK, Roemer RP (2018a) Benthic ctenophores (Platyctenida: Coeloplanidae) in South Florida: predator-prey interactions. Invertebr Biol 137(2):133–150

Glynn PW, Coffman B, Vanderwoude J, Martínez N, Domínguez JH, Gross JM, Renegar DA (2018b) Antipredatory escape behaviors of two benthic ctenophores in South Florida. Ecology 100(1):1–3. https://doi.org/10.1002/ecy.2497

Glynn PW, Coffman B, Dettloff K, Dominguez J, Gillette PR, Martinez N, Jones NP, Riegl BM (2021) Non-native brittle star interactions with native octocoral epizoites: an endemic benthic ctenophore in peril? Mar Biol 168:142. https://doi.org/10.1007/s00227-021-03927-7

Glynn PW, Coffman B, Fuller MPC, Moorhead SG, Williams MK, Primov KD, Fortson TN, Barrales RN, Glynn PJ (2017a) Benthic ctenophores (Platyctenida: Coeloplanidae) in South Florida: environmental conditions, habitats, abundances, and behaviors. Invertebr Biol 136(4):379–393

Glynn PW, Coffman B, Primov KD, Renegar DA, Gross J, Blackwelder P, Martinez N, Dominguez J, Vanderwoude J, Riegl BM (2019) Benthic ctenophore reproduction (order Platyctenida) reproduction, recruitment, and seasonality in South Florida. Invertebr Biol 138. https://doi.org/10.1111/ivb.12256

Glynn PW, Feingold JS, Baker A, Banks S, Baums IB, Cole J, Colgan MW, Fong P, Glynn PJ, Keith I, Manzello D, Riegl B, Ruttenberg BI, Smith TB, Vera-Zambrano M (2018c) State of corals and coral reefs of the Galápagos Islands (Ecuador): past, present and future. Mar Pollut Bull 133:717–733. https://doi.org/10.1016/j.marpolbul.2018.06.002

Hendler G, Brugneaux SJ (2013) New records of brittle stars from French Guiana: *Ophiactis savignyi* and the alien species *Ophiothela mirabilis* (Echinodermata: Ophiuroidea). Mar Biodivers Rec 6:e113

Hoeksema BW, Hassell D, Meesters EHWG, van Duyl FC (2018) Wave-swept coralliths of Saba Bank, Dutch Caribbean. Mar Biodivers 48(4):2003–2016. https://doi.org/10.1007/s12526-017-0712-5

Jones NP, Ruzicka RR, Colella MA, Pratchett MS, Gilliam DS (2022) Frequent disturbances and chronic pressures constrain stony coral recovery on Florida's coral reef. Coral Reefs 2022(41):1665–1679. https://doi.org/10.1007/s00338-022-02313-z

Mantellatto MC, Vidon LF, Silveira RB, Menegola C, Moreira da Rocha R, Creed JC (2016) Host species of the non-indigenous brittle star *Ophiothela mirabilis* (Echinodermata: Ophiuroidea): an invasive generalist in Brazil? Mar Biodivers Rec 9:8. https://doi.org/10.1186/s41200-016-0013-x

Medina-Rosas P, López-Luján L, Zúñiga-Arellano B (2021) Corales para los dioses: ofrendas marinas en Tenochtitlan. Arqueología Mexicana 169:20–29

Palacio-Castro AM, Smith TB, Brandtneris V, Snyder GA, van Hooidonk R, Maté JL, Manzello D, Glynn PW, Fong P, Baker AC (2023) Increased dominance of heat-tolerant symbionts creates resilient coral reefs in near-term ocean warming. Proc Natl Acad Sci 120(8):e2202388120. https://doi.org/10.1073/pnas.2202388120

Prasade A, Apte D, Kale P, Oliveira OMP (2015) *Vallicula multiformis* Rankin, 1956 (Ctenophora, Platyctenida): first record from the Indian Ocean. Check List J Biodivers Data 11(1):1544. https://doi.org/10.15560/11.1.1544

Precht WF, Patterson MR, Glynn PW (2024) Giant mobile coralliths from the Florida keys. Bull Mar Sci 100(1):95–104. https://doi.org/10.5343/bms.2023.0051

Rankin JJ (1951) A new platyctenid ctenophore from Jamaica. Nature 4285:1947

Rankin JJ (1956) The structure and biology of *Vallicula multiformis* gen. et sp. nov., a platyctenid ctenophore. Journal Linnean Society London. Zoology 43(289):55–71

Rich LP, Dennis MM, Freeman FA (2020) New record of the non-native *Ophiothela mirabilis* (Verrill 1867) in St. Kitts, West Indies. Adv Oceanogr Mar Biol 2(1):1–4. https://doi.org/10.33552/AOMB.2020.02.000526

Ryan JF, Pang K, Schnitzler CE, Nguyen A-D, Moreland RT, Simmons DK, Koch BJ, Francis WR, Havlak P, Smith SA, Putnam NH, Haddock SHD, Dunn CW, Wolfsberg TG, Mullikin JC, Martindale MQ, Baxevanis AD (2013) The genome of the ctenophore *Mnemiopsis leidyi* and its implications for cell type evolution. Science 342(6164). https://doi.org/10.1126/science.124259

Sagarin RD, Barry JP, Gilman SE, Baxter CH (1999) Climate-related change in an intertidal community over short and long time scales. Ecol Monogr 69(4):465–490

Schultz DT, Haddock SHD, Bredeson JV, Green RE, Simakov O, Rokhsar DS (2023) Ancient gene linkages support ctenophores as sister to other animals. Nature 2023:110–117. https://doi.org/10.1038/s41586-023-05936-6

Sheppard CRC, Davy SK, Pilling GM, Graham NAJ (2018) The biology of coral reefs, 2nd edn. Oxford University Press, Oxford, p 370

Shick JM (2018) Where corals lie, a natural and cultural history. Reaktion Books, London, p 304

Verrill AE (1867) Notes on Radiata in the museum of Yale College with descriptions of new genera and species. No. 2. Notes on the echinoderms of Panama and the west coast of America, with descriptions of a new genus. Trans Connect Acad Arts Sci 1(2):251–322

Wanner H, Beer J, Bütikofer J, Crowley TJ, Cubasch U, Flückiger J, Goosse H, Grosjean M, Joos F, Kaplan JO et al (2008) Mid- to late Holocene climate change: an overview. Quat Sci Rev 27:1791–1828. https://doi.org/10.1016/j.quascirev.2008.06.013

Whelan NV, Kocot KM, Moroz TP, Mukherjee K, Williams P, Paulay G, Moroz LL, Halanych KM (2017) Ctenophore relationships and their placement as the sister group to all other animals. Nat Ecol Evol 1(11):1737–1746. https://doi.org/10.1038/s41559-017-0331-3

Chapter 13
Uncharted Waters, the Fate of Eastern Pacific Coral Reef Ecosystems

Abstract As climate change and global ocean warming proceed and become increasingly stressful, the Earth's coral reef ecosystems would become maladjusted with the eventual extinction of innumerable species. This projected decline is now occurring worldwide, more so in some areas (western Atlantic, Arabian Gulf) than others (Indian Ocean, eastern Pacific). Eastern Pacific coral reefs have experienced severe bleaching and mortality in recent decades, however, the predominant reef-building coral taxon, *Pocillopora grandis*, consists of bleaching resistant populations hosting thermally tolerant, coevolved dinoflagellate symbionts (*Durusdinium*, *Cladocopium*) that greatly alleviate impacts and potentially allow for rapid recovery and persistence. Increasing local scale extirpations of corals, coincident with recent record-breaking high ocean temperatures, are expected to result in mass extinctions by the end of the present century. The 2023–24 El Niño event, superimposed on an already stressful warming ocean, resulted in widespread coral bleaching and mortality, the fourth global coral bleaching event since 2000. Although the survival of the world's coral reef ecosystems as we know them today is in doubt, the majority of reef investigators still maintain hope that greenhouse gas emissions can be reduced. Various methods of coral reef protection and restoration are currently underway to preserve reef building coral stocks to sustain some vestige of this ecosystem in the future.

Introduction

Over the past 600 million years, organic reefs have experienced several mass extinctions, exhibiting a "collapse-rebuild model" (Stanley Jr. 2001). During the past 200 million years, modern scleractinian coral reefs existed for tens of millions of years during Triassic-Cretaceous time and in the Cenozoic Era. Over geologic time, all reefs have experienced global collapse, however, today's coral reefs are in the throes of rapid degradation and likely on the cusp of world-wide loss (e.g., Hughes et al. 2017, 2018; Hoegh-Guldberg et al. 2023). While a plethora of hypotheses has been advanced to explain coral ecosystem demise (e.g., volcanic outgassing, sea

© The Author(s), under exclusive license to Springer Nature Switzerland AG 2025

P. W. Glynn, *Coral Reef Resilience in the Anthropocene*,
https://doi.org/10.1007/978-3-031-81421-1_13

level change, climate deterioration, over nutrification), evidence for the causes of modern reef demise is credible. Triassic reef building ceased during a brief icehouse period; Jurassic reefs were left high and dry during sea-level regression; and Cretaceous reefs disappeared suddenly due to the Chicxulub bolide impact (Riegl et al. 2009). Coral reef recovery and expansion occurred rapidly in Tertiary time and during the Pleistocene experienced species extinctions and diversification to the present (Budd 2000).

Paradoxically, two periods of prolific coral reef expansion occurred in Cretaceous and PETM (Paleocene-Eocene Thermal Maximum) hothouse environments when global mean surface temperatures (GMSTs) were 28–38 °C compared with Earth'coldhouse (11–18 °C) conditions today (Judd et al. 2024). Perhaps the key to this dilemma is the rate of temperature change.[1] If biosphere temperatures increase slowly this would allow time for evolutionary adaptive changes and time for organisms to migrate to habitats (refugia) supportive of life. Aware of robust coral reef expansion during past hothouse conditions, a couple of my reef scientist colleagues were not overly concerned with the present state of coral mortality and reef loss. I do not share this indifferent attitude.

Ocean warming and acidification are wide-reaching conditions impacting coral reefs today. Several other disturbances, such as coral diseases, human disruptions, and storm damage, usually more spatially restricted, also have direct negative effects that can exacerbate thermal and acidification impacts. Here, I summarize the condition of present day eastern Pacific coral reefs and note environmental hazards in the context of global coral reef decline. These effects are considered in previous chapters and are now discussed in the context of accelerating climate change, global warming, and coral mortality. As a matter of urgency, efforts to protect eastern Pacific coral reefs and methods of restoration are also examined. Finally, a perspective on the way forward, with respect to the projected demise of coral reefs, concludes my 60 year journey of study and dissemination of knowledge of one of Planet Earth's most diverse, dynamic, and vibrant ecosystems. These are truly unprecedented times in human history, with global environmental conditions rapidly degrading and impacting all ecosystems, both aquatic and terrestrial, as well as human civilization.

Status of Eastern Pacific and Global Coral Reefs

It is paradoxical that coral reefs in the eastern tropical Pacific, long recognized as impoverished and feeble compared with reef development in other regions, have demonstrated resilience and persistence concurrent with ocean warming and coral bleaching during the past several decades. Despite widespread coral bleaching and

[1]The rate of heating during Cretaceous time was 0.000025 °C per 100 years, and in the PETM period 0.025 °C per 100 years. Modern global warming is 1–4 °C per 100 years (Kump 2011).

mortality associated with elevated sea temperatures accompanying El Niño events (1982–83, 1997–98, and 2015–16), impacted coral reefs have demonstrated notable recovery across large swaths of the eastern Pacific region. Focusing on the status of eastern Pacific coral reefs, 'the good, the bad, and the ugly',[2] Reyes-Bonilla et al. (2021) presented at the International Coral Reef Symposium 2021 Bremen an optimistic summary of the overall good (healthy) condition of reefs in this region. This assessment was based on long term monitoring of reefs on the mainland and offshore islands by research teams.[3] For 20 years, live coral cover showed a 'good' positive trend at 62% of monitored sites. The 'bad and ugly' (a negative trend), was reported for 35% of sites; 4% of sites demonstrated stability. A 44-year data set (1970–2014) exhibited temporary declines in coral cover following major ENSO events but no overall regional decline (Romero-Torres et al. 2020). Unlike the significant coral reef declines and losses in the Caribbean and Indo-Pacific regions, eastern Pacific coral assemblages often exhibit 30–50% live coral cover and recover from bleaching events in 10–15 years.

The positive trends in live coral cover were reported for tropical coastal Mexico, Central America, and the Galápagos, Cocos, and Malpelo islands. The Gulf of California, Revillagigedo Islands, and Clipperton Atoll reported stable trends in coral cover. Eastern Panama, Colombia, and coastal Ecuador were treated as contiguous areas showing a questionable negative trend in live coral cover. The status of this southern coastal area should be re-evaluated based on new information from eastern Panama and Colombia, as noted below.

Research by Victoria M. Glynn in the Pearl Islands (Gulf of Panama) during 2022 revealed high coral cover and robust coral reef development, similar to pre-ENSO disturbance communities 20 years ago (Glynn VM 2023; Glynn VM et al. 2023).[4] Coral reefs at Gorgona Island, three insular reef sites representing Colombia, also demonstrated a positive trend in live coral cover over 10 years (2011–2020). Mean live coral cover at the Gorgona Island reef sites ranged between 56% and 70% (Zapata and Rodríguez-Moreno 2021). This positive trend is worth pointing out due to the location of Gorgona Island within a marine protected area (MPA), the Parque Nacional Natural (PNN) Gorgona (Gorgona Natural National Park). Gorgona Island receives constant protection from human impacts, such as fishing, extraction of reef organisms, and pollution, *inter alia*. Recurrent low tidal exposures and ENSO thermal extremes are natural disturbance events impacting Colombia's coral reefs.

---

[2] "The Good, the Bad, and the Ugly" is an Italian Western (Spaghetti Western) epic 1966 film directed by Sergio Leone and starring Clint Eastwood. The film's haunting and unique soundtrack by Ennio Morricone has entered popular and classical music venues.

[3] Souter et al. (2021) of the Global Coral Reef Monitoring Network, compiled survey results. These were available at https://gcrmn.net/2020-report/

[4] Victoria Glynn conducted her doctoral field studies in Panama as a Fulbright U.S. Student Grant awardee at STRI. She is a Ph.D. candidate at McGill University (Montréal, Quebec) and a pre-doctoral Fellow at STRI. See Glynn VM 2023 for the original publication, and the article featuring this study via the *Journal of Biogeography*'s Early Career Researcher blog (Glynn VM et al. 2023). Victoria is often asked, "Are you related to PW Glynn?", and her response is always "Perhaps".

Protecting herbivorous sea urchins and fishes and fish corallivores, keystone species that facilitate coral recovery, is key to sustain the observed health of Gorgona's PNN coral reefs.

Considering the southern-most eastern Pacific coral outpost, Rapa Nui (Easter Island), the predominant coral cover, to depths of ~40 meters, can reasonably be assigned a positive trending state. Historically, coral communities recovered rapidly after a record rainfall coral bleaching event in 1980 (Cea Egaña and DiSalvo 1982) and prolonged elevated sea temperature stress in 2000 (Wellington et al. 2001). Scant information is available on the condition of Rapa Nui corals during the past 5–6 years due to COVID-19 and a mandatory quarantine imposed on the island by the government of Chile. Evie Wieters, with whom we collaborated earlier on coral research at Rapa Nui, has offered some positive information.[5] Her summary assessment of recent coral condition, to about 40 m depth, follows. "The predominant corals (*Pocillopora verrucosa* and *Porites lobata*) have continued to remain overall healthy, and local coral cover and structure within most sites continues to be high. For our monitoring sites dominated by corals, mean total coral cover = 76% ± 1.84. Coral cover has generally proven capable of recovering from spatially heterogeneous and acute, low/mid-intensity disturbances (e.g., storms/wave assault, low-level 'background' bleaching, and overgrowth in some years of seasonally ephemeral outbreaks of the green alga *Codium*)." A report of active sexual reproduction of the dominant Rapa Nui corals has offered hope regarding recovery processes (Buck-Wiese et al. 2018).

Mesophotic coral buildups (*Leptoseris* spp.) at depths of 40–350 m, recently discovered by investigators at Universidad Católica del Norte, Coquimba, Chile, revealed overgrowth of dead coral colonies by filamentous algal mats (Sellanes et al. 2021). These communities are located near the settlement of Hanga Roa and were possibly affected by human eutrophication via seeping sewage (Fig. 2.2). Deep coral communities are relatively rare in the eastern Pacific. Monitoring these communities is challenging considering the technical problems of deep diving and the cost.

Summarizing, the vistas of beautiful,[6] golden brown/green *Pocillopora* reefs with their teeming life were very much in evidence and admired in the eastern

---

[5] Evie is a Pontificia Universidad Católica de Chile, Santiago professor, in the Departamento de Ecología. She spends most of her time researching and teaching at the marine laboratory in Las Cruces (Estación Costera de Investigaciones Marinas) with her husband-marine ecologist and colleague, Sergio Navarrete. She joined our research activities in 1999 and 2000, and quickly formed a Chilean team of investigators to pursue long-term ecological studies of Rapa Nui coral communities. An overview of some of their research is presented in section, 5.3.14, Rapa Nui (Easter and Salas y Gómez Islands, Chile), Chap. 5, in Glynn et al. 2017.

[6] Is there beauty in a coral reef? When viewing a healthy coral reef, I would occasionally take pause and try to comprehend the panoply of diverse on-going biological and ecological functions—microbial coral host interactions, photosynthesis, calcification, reproduction, trophic interrelationships, symbioses, competition, among other things—of the living reef community, admittedly an impossible task. How are these processes organized and successfully interact to result in a vibrant, living and growing coral reef? Lamentably, global coral reef decline today could be likened to a dysfunctional orchestra, like Haydn's farewell symphony No. 45, where musicians (analogous to ecological functions) retreat consecutively until the music ends in silence.

Pacific as early as the 1940s and 1950s (Steinbeck and Ricketts 1941; Steinbeck 1951; Squires 1959), during the 1960s–1970s (Glynn 1972; Brusca and Thompson 1975; von Prahl et al. 1979), and have continued to offer enjoyment to humankind to the present day. Disturbances causing coral mortality occurred in the eastern tropical Pacific for hundreds and even thousands of years, and when strong and long-lasting have interfered with or halted coral and reef growth over varying time intervals (Colgan 1990; Dunbar et al. 1994; Cole and Tudhope 2017; Toth et al. 2012, 2017). These disturbances can be classified as natural or anthropogenic, a result primarily of natural causes or human disruptions. Common-occurring natural disturbances in the eastern Pacific include extreme upwelling events, tidal exposures, cyclonic storms, and ENSO activity (see Chap. 5). Anthropogenic disturbances, such as climate change, ocean acidification, and nutrient pollution, also can result from natural causes but with heightened and accelerated effects during the Anthropocene, in close step with the growing human population (e.g., Dodson et al. 2020; Riegl and Glynn 2020). The further back in time one attempts to identify the causes of disturbances, the more complex the task. The study of past disturbances requires the application of special methods, e.g., geochemical analyses, description of sedimentary rock structures, detection of variations in paleoclimates, and examination of fossils and their assemblages.

Coral diseases are also of great concern due to their lethality and potentially rapid regional propagation. A coral disease, black band disease (BBD), was first recognized by Arnfried Antonius in Florida in the early 1970s (Antonius 1973, 1977). The Caribbean-W Atlantic and Indo-Pacific regions have suffered enormous coral losses from different diseases since the 1980s, often associated with heat-induced coral bleaching events (Harvell et al. 1999; Porter 2001; Bruno et al. 2007). In discussions with my colleagues, Rich (Richard) Aronson and Bill Precht, who documented a 3000-year record of continuous *Acropora* reef growth in Belize (Aronson et al. 2002), we lamented the current inability to detect the presence of coral diseases in the fossil record. Could today's emergent and diverse coral diseases possibly be causally related to Anthropocene environmental deterioration?[7] The stony coral tissue loss disease (SCTLD), first recognized by Bill Precht and colleagues, has precipitated mass coral mortality throughout the Greater Caribbean region (Precht et al. 2016; Alvarez-Filip et al. 2022). Between the devastating effects of SCTLD and the current, record-breaking high sea temperatures (see below), the fate of coral populations in the western Atlantic region (and globally) is in extreme peril.

From regional assessments of coral recovery (percent changes in live coral cover), we began to adopt an optimistic outlook on the improving health condition of eastern Pacific coral reefs. A regional, 44-year (1970–2014) assessment of live

[7] Ancient DNA sequencing and protein signatures in the teeth of plague victims have successfully identified the Black Death bacillus (*Yersenia pestis*) (Haensch et al. 2010), offering hope for application to coral disease research. If this line of research is pursued it is imperative that the fossil corals selected for analysis have not been subject to trace-effacing bioerosion following their demise.

coral cover exhibited temporary reductions following major ENSO events but no overall long-term decline (Romero-Torres et al. 2020). While highly variable in rate and spatial extent, by the early 2000s, more than one-third of 21 sites in the equatorial eastern Pacific demonstrated positive increases in live coral cover (Wellington and Glynn 2007; Baker et al. 2008). As noted above, positive trends in eastern Pacific coral cover have doubled (from ~33% to 62%) in recent surveys (Reyes-Bonilla et al. 2021). Nearly all *Pocillopora* patch reefs in the southern Galápagos Islands have disappeared, chiefly from coral mortality followed by sea urchin (*Eucidaris*) bioerosion, with little to no coral recruitment evident. A few sites are showing surprising recovery, however, for example at Concha y Perla, near Villamil, Isabela Island (Baums et al. 2014), Devil's Crown, Floreana Island (Feingold and Glynn 2014; also Feingold, Fig. 10 in Glynn et al. 2019), and the northeast anchorage at Santa Fe Island (Ellen Prager, personal communication). Asexual fragmentation is responsible for the rapid recovery at Concha y Perla, whereas sexually produced colonies are prominent at Devil's Crown and Santa Fe Islands. Perhaps circulation differences influence this contrast—reducing colony fragmentation and dispersion or increasing sexual recruitment. Concha y Perla is a partially enclosed pond whereas the two island sites experience frequent open ocean currents likely transporting coral larvae.

As discussed earlier regarding coral reproduction (Chap. 6), our studies have demonstrated that sexual and asexual reproduction contribute significantly to coral recovery under different circumstances (Glynn et al. 2017a, b). Coral recovery can depend critically on the sizes of corals that survive severe disturbance events. The survival of large colonies of a given species, which are highly fecund, can contribute substantially more to sexual reproduction and recruitment than small colonies. Thus, the sizes and age structure, including the reproductive capacity of surviving post-disturbance corals, are critical in influencing recovery processes. The demographic structure of coral populations, and how this relates to population growth, was recognized early by Joe Connell (1973). More recently, workers investigating the dynamics of coral recovery are urging the inclusion of demographic methods in recovery studies (Edmunds and Riegl 2019; Dietzel et al. 2020). This sort of information on the size structure of populations is far superior to coral cover metrics alone as a predictor of recovery; however, it requires a much greater sampling effort to acquire. Also, in some areas coral cover is so low that in field studies the variance is greater than the observable change without greatly increasing sample sizes—this has forced ecologists to consider coral ind/m^2 instead of % cover.

The impact of coral bleaching and mortality, and the extent of recovery in the world's major coral reef regions has been of pressing interest following the El Niño events of 1982–83 and 1997–98. Do certain regions demonstrate more resistance and resilience to thermal disturbances than others? These are important considerations because they may influence the capacity for coral reef survival under rapidly deteriorating environmental conditions. In light of the considerable physical and biological differences between inter-ocean coral reef ecosystems, a comparison of the rate and extent of recovery is fraught with biases. Nonetheless, regional recovery differences might reveal insights into critical drivers involved in coral community resilience. By 2007, observations were sufficient in number to allow for global inter-regional comparisons of coral recovery (Baker et al. 2008).

Numerous Indian Ocean coral reefs revealed robust recovery in as little as 2 years. The dominant coral cover at several surveyed sites was of *Acropora* and *Pocillopora* species, suggesting a post-disturbance pulse of sexual recruits. All but one site in the Western Atlantic (Belize, Reef 2) demonstrated coral cover decline. This decline resulted from high temperature stress followed by coral disease-induced mortality. Recent information available for the Arabian Gulf reveals a significant decline in coral cover. Coral bleaching from high thermal stress is followed by diseases and mortality resulting in <10% live coral cover across the southeastern Arabian Gulf (Burt, J. E., and Riegl, B. Jr., pers. comm.; Riegl et al. 2018; Aeby et al. 2020). Although highly variable spatially, coral recovery in the Central/Southern/Western Pacific (CSWP) and eastern Pacific regions amounted to ~60% at all surveyed sites. Heat-induced bleaching initiated most coral mortality events across the Pacific Ocean with subsequently varied disturbances, such as corallivory, algal competition, diseases, and cyclones, causing additional mortality in the CSWP region. In the Eastern Pacific, bioerosion was often dominant and interfered with recovery. Recurrent ENSO activity and heat waves have interfered with recovery in all major coral reef regions worldwide (Hoegh-Guldberg et al. 2023).

A meta-analysis of 48 global coral reefs, investigating the rates of live coral recovery, examined ecological factors such as coral recruitment rates, structural complexity of reefs, functional diversity of reef-associated fishes, and degree of isolation; in this analysis eastern Pacific reefs were reported to be the slowest to recover (Graham et al. 2011).[8] This result is not in line with the 'good' condition of eastern Pacific reefs reported by Reyes-Bonilla et al. (2021) and other recent studies supporting robust recovery (Romero-Torres et al. 2020; Palacio-Castro et al. 2023). Also, Turnham et al. (2024) have offered a rare, optimistic outlook of the persistence and potential spread of thermally tolerant host-symbiont combinations (*Pocillopora grandis* hosting species of *Durusdinium* and *Cladocopium* zooxanthellae) during ocean warming and irradiance extreme conditions. The survival of these colonies following stress events would be expected to accelerate reef recovery. In an analysis of 14 survey sites, Baker et al. (2008) found that one-half demonstrated robust recovery and one-half did not, perhaps reflecting the mix of thermally tolerant and non-tolerant coral holobiont combinations. Baird et al. (2009) indicated that differences in the susceptibility of reef-building coral species to stressful conditions can also be influenced by coral host physiological mechanisms, e.g., fluorescent pigments, antioxidant systems, and heat shock proteins. Both members of the coral holobiont need to be considered.

Sea urchin bioerosion is a critical driver of reef framework destruction in the Galápagos Islands and is largely responsible for the slow recovery of reefs in the

[8] In the meta-analysis of Graham et al. (2011), the coral recovery data from the eastern Pacific was very limited. Only coral cover in Costa Rica was considered, including just three survey sites on two islands. Their study did not consider the changes in coral cover of numerous *Pocillopora* reefs in Mexico and the equatorial eastern Pacific.

central and southern archipelago (Reaka-Kudla et al. 1996; Glynn et al. 2019). Sea urchin abundance and bioerosion are significantly reduced on the Wellington Reef at Darwin Island, thus allowing time for coral recovery and continued reef accretion after ENSO disturbance events (Glynn et al. 2017a). Fish invertivores are abundant at Wellington Reef, the only continuously surviving coral reef since 1982–83; these predators are assumed responsible for limiting the population growth of sea urchins and their bioerosion.

Coral Species Extinctions

With repeated bouts of coral bleaching, mortality, reduced recruitment and population recovery, vulnerable coral species will begin to experience local, regional, and ultimately global extinctions. In the aftermath of the 1982–83 coral bleaching event in the equatorial eastern Pacific, about one-third of zooxanthellate coral species experienced severe declines in abundance and local to regional-scale extinctions (Glynn 1997). The disappearance of two fire coral species (*Millepora* spp.) turned our attention to the possibility of extant global coral extinctions. Our documentation of the fire coral disappearances was the first reported incident of zooxanthellate coral loss in current times (Glynn and de Weerdt 1991).

Fearful of not documenting the extinction of an undescribed zooxanthellate species, I contacted Wallie H. de Weerdt, a fire coral systematist, and invited her to join us in a search effort to find fire corals that had disappeared in Panama. She enthusiastically accepted my invitation, and we were soon underway for the Gulf of Chiriquí.[9] We found numerous dead fire corals, but not a single living colony of the suspected undescribed species (later named *Millepora boschmai*). After 8 years of searching (including 204 diver hours) in the known former range of *Millepora* in the Gulf of Chiriquí, we tentatively concluded that the two fire coral species were no longer present in Panama (Glynn and de Weerdt 1991).

Carpenter et al.'s 2008 report jolted the coral reef community of researchers with their assessment that 231 of 704 (32.8%) of reef-building coral species faced an elevated risk of extinction associated with rising sea temperatures, disease incidence, and local-scale anthropogenic disturbances. Further, 407 corals in the threatened and near-threatened categories amounted to 57.8% of the world's coral fauna. A recent worldwide assessment indicates that extinction threats to major marine taxa continue unabated, driven by climate change/variability and human-related

[9] At that time, Wallie investigated the taxonomy of *Millepora* spp. in the western Atlantic region. She followed in the footsteps of Hilbrandt Boschma, a distinguished hydrocoral systematist of the Leiden Museum, Netherlands, who first recognized one of the Panamanian *Millepora* colonies as an undescribed species. Continuing surveys in the search for surviving fire corals, greatly aided by Joshua Feingold, Mark Eakin, and Juan Maté, revealed some live colonies of *M. boschmai*, which died soon after discovery. In the absence of ENSO activity and *Acanthaster* predation, how these colonies died was mystifying. Further, nearby scleractinian corals were alive and healthy. A complete account of the eastern Pacific *Millepora* storyline is published in Chap. 7 (Glynn 2012) in Lee Hannah's book, *Saving a Million Species*.

pollution (Nikolaou and Katsanevakis 2023). Most reports of coral species losses are examples of local extinctions, which is not surprising considering the difficulty of searching all suitable habitats across large areas. There is always the possibility that a remnant cluster of corals exist in some unsearched, remote corner of a species' range. In our experience, multiple surveys of identical areas have revealed the presence of specific colonies or their absence in subsequent surveys. Variable sea conditions and visibility can interfere with thorough searches. Still, some disappearances have remained a mystery. Surprisingly, estuarine habitats of limited underwater visibility (~ 1 m) in the Gulf of Chiriquí have revealed unbleached and surviving coral species after El Niño events. Such communities, present in seemingly hostile environments, could serve as 'refuge' habitats in easily overlooked areas. Port Miami, with numerous coral species surviving the 2023 bleaching event, might also qualify as a refuge habitat (Rubin et al. 2021; Enochs et al. 2023).

If sea temperatures continue to increase worldwide, as projected (and realized to date, 2023), coral mortality will occur over large areas, including the entire ranges of numerous species, thus resulting in mass extinction events (Hoegh-Guldberg 1999; b, 2018; Eakin et al. 2022). Along with corals, which provide habitat and sustenance for numerous microbial, invertebrate, and fish taxa (Glynn 2011; Glynn and Enochs 2011; Brandl et al. 2018), cascading mortality and extinctions will severely impact all coral reef ecosystems. Many cryptic taxa depend on dead coral rubble as a habitat, and without living, calcifying corals, this critical resource would quickly vanish (Enochs 2012). Absent migration and adaptation, widespread global extinctions would inevitably occur. The likelihood of such a scenario is very real in the present century, with global warming exceeding the upper temperature tolerance limits of coral holobionts. I proposed this outcome as a hypothesis of high temperature as the cause of mass coral extinctions at the Lyell Meeting on 'Organism-Environment Feedbacks in Carbonate Platforms and Reefs' held in the year 2000 at the Geological Society of London, United Kingdom (Glynn 2000a).

If increasing greenhouse conditions today lead to mass coral extinctions, this would not be unprecedented in Earth's history. Today's rate of change is much faster, decades instead of 100,000 to millions of years in the past. For example, warming in Upper Paleozoic time occurred contemporaneously with tabulate and rugose coral extinctions (Dickins 1984); and in Early Cretaceous time, coral-algal reefs were replaced by rudist bivalve mollusks in the thermally elevated equatorial zone of the Tethys Sea due to sustained super-greenhouse conditions (Kauffman and Johnson 1997). The early notion that rudists displaced corals by competitive replacement has shifted to the view that corals were less resistant to global warming than rudists (Pandolfi and Kiessling 2014). Littler et al. (2011) have found that sea surface temperatures in earliest Cretaceous time, > 32 °C at 15–20°N, were significantly higher than modern day values at equivalent latitudes (see also Wilson and Norris 2001). Cenozoic coral extinctions in the Caribbean during late Pleistocene time are extensively documented by Ann F. Budd and co-workers (Budd et al. 1993).[10]

[10] A two-day Colloquium of case histories of global coral reefs' health, hazards, and history was conceived and organized by R. N. Ginsburg (Chairperson, Organizing Committee), and convened at RSMAES in 1993. The motivation for this meeting stemmed from the ever growing concern of the coral reef scientific community for the accelerating degradation of global reefs (and in com-

Climate Change and Global Warming

In the 1980s, data demonstrating global ocean warming were of concern regarding the devastating effects a slight increase in temperature could cause coral reefs. Increasing sea surface temperature trends were revealed by satellite data (Strong 1989) and instrumental records beginning as early as 1880 (Shein 2006). High and prolonged temperature anomalies, mainly in the eastern Pacific region, offered convincing evidence of severe threats to coral reefs (Glynn 1984; Glynn and D'Croz 1990). Ove Hoegh-Goldburg (1999) convincingly put this potential global catastrophe affecting coral reef ecosystems in perspective in the Anthropocene Epoch with several more following warnings (e. g., Veron et al. 2009; b). Most of Earth's warming over the past 60+ years is attributed to anthropogenic forcings, namely increasing greenhouse gases and deforestation (IPCC 1992, 2001).

Sea surface and deeper water temperatures of the world's oceans have been increasing at an alarming rate based on observations and forecast modeling studies by the IPCC. In early April 2023, the average global ocean temperature was 21.1 °C, a new record and 0.1 °C higher than the previous record in March 2016 (Jones 2023). Record high SSTs occurred in the ETP and wider Caribbean region in the summer of 2023, resulting in severe coral heat stress (Hoegh-Guldberg et al. 2023). Projected sea temperature increases to 2100 are expected to diverge around 2040 into moderate and high trajectories of continued warming (Hoegh-Guldberg et al. 2014). Simulated 'moderate' annual average sea surface temperatures to 2100 are projected to increase over a range of 0.25–1.0 °C (5th–95th percentile range) from the present; simulated 'high' annual average temperature increases are projected to reach 2.5–4.0 °C by 2100. These climate model scenarios are based on Representative Concentration Pathways,[11] which consider varying rates of greenhouse gas emissions. Moderate rates of temperature increase reflect a degree of pollution abatement whereas high rates denote routine and continuing excessive high-level anthropogenic emissions. These scenarios can be considered as 'best-case' and 'worst-case' impacts on coral reefs. Both outcomes result in coral reef losses, the latter more rapidly than the former.

Acute and deadly heating, rarely if ever experienced worldwide by humankind, has catastrophic effects on living organisms and contributing to extreme weather events (wildfires, violent storms, floods, and droughts). Heat dome, a seldom-noted meteorological phenomenon, began to appear commonly in 2023 summer weather reports. Rising hot ocean air becomes trapped in a high-pressure system forming a heat dome, which when prolonged, leads to heat waves. Four heat domes formed in 2023, over Northern Mexico-Southwestern USA (including California), the North

memoration of F. G. Walton Smith on the 50th anniversary of the founding of the Rosenstiel marine school).

[11] Representative Concentration Pathways (RCPs) are greenhouse gas concentration trajectories adopted by the IPCC for use in climate model runs. RCPs are four graded sets of scenarios that attempt to capture fossil fuel use versus adoption of renewable energy sources (van Vuuren et al. 2011).

Atlantic, North Africa, and South Asia. Coral reefs and other coastal marine ecosystems were (and are still) stressed by the extreme, prolonged, elevated temperatures generated in these domes. Brian McNoldy, RSMAES steadfast custodian of NOAA weather and climatology data, has made available extreme high sea water temperature records for Virginia Key, Miami. Fifteen daily maximum water temperature records, ranged from 32.8 °C to 33.6 °C from 5 July to 12 August, 2023.[12] This warm water period greatly exceeded the optimal temperatures for corals (26°-28 °C) in Southeastern Florida.

Cascading stress effects and mortality can result from the heated and oxygen-depleted waters. In S Florida, bleaching and mortality of symbiotic corals and other cnidarians occurred during July and August 2023. Fishes and diverse invertebrates also succumbed in Biscayne Bay and other shallow water coastal areas. From observations of past summer-season disturbance events, sea grasses (submerged aquatic vegetation [SAVs] including algae) could soon experience mortality; and turtles, manatees, and other organisms that depend on sea grasses for shelter and nourishment. Elevated concentrations of nutrients, resulting from death and decay, may unleash red tides or water blooms that could further exacerbate the degradation of coastal ecosystems. *Karenia brevis*, a common red tide dinoflagellate in American waters, produces brevetoxin, a neurotoxin that can harm birds, dolphins, turtles, manatees, and humans (particularly those persons with preconditions such as emphysema or asthma).[13] A comprehensive review of the effects of the extreme 2023 warming event on marine life in Florida would be of high value in assessing the impact of this disturbance.

Heat stress alone can have devastating health impacts on humans. In addition to adverse outcomes in diverse natural ecosystems, there are steadily increasing reports of heat mortality associated with rising global temperatures. The Center for Disease Control and Prevention reports 600–700 heat deaths annually in the USA. These numbers are indeed underestimates as reporting is inconsistent and heat-induced causation of death is difficult to determine. It is heart wrenching to witness essential outdoor laborers (e.g., farmworkers, roofers, construction workers, *inter alia*) struggling in their tasks during the unrelenting heat of the day.

Increasing acidification of the world's oceans is also harming coral reefs and associated species, however, at present at a somewhat less impactful rate than global warming. The world's oceans have absorbed about 30% of Anthropogenic CO_2, decreasing pH by approximately 0.1 unit over 100 years. This rate of change is unprecedented within the last 65 Ma and is now impacting coral calcification, skeletogenesis of diverse marine organisms, and has an accelerating effect on

[12] The temperature recording site is on the RSMAES dock at Bear Cut, south end of Virginia Key (NOAA Co-ops site 8,723,214). The temperature sensor is at a depth of 1.5 meters and exposed to strong tidal currents flushing Biscayne Bay.

[13] Several invertebrate suspension feeding species, such as mussels, oysters, and clams, accumulate neurotoxins when feeding. The consumption of such species can cause neurotoxic shellfish poisoning, which can be fatal. In my hometown (Coronado, California), several members of an Asian family died by consuming contaminated mussels during a shellfish poisoning event.

bioerosion (see Chaps. 4 and 8). A doubling of pre-industrial atmospheric CO_2 concentration, from 280 to 560 ppm, results in about a 10–50% decrease in the calcification rate of reef-building corals and coralline macroalgae (Kleypas and Langdon 2006). Nearly all tropical seas will shift to net dissolution values of predicted CO_2 concentration (450 ppm) and Ω_{arag} by 2040 (Hoegh-Guldberg et al. 2007). By the mid-twenty-first Century, the low pH oceans will have dissolution effects mainly on the dead skeletons of corals, coralline macroalgae, and other calcifying organisms that have succumbed earlier from heat stress. In addition, experiments simulating high CO_2 conditions have shown that sea grasses and euendolithic boring microorganisms[14] can benefit in an acidic milieu (Kleypas and Yates 2009). Acidification could greatly exacerbate the demise of coral reefs through competitive displacement by sea grasses and likely increasing bioerosion by microorganisms as ocean acidification progresses. Recent studies have shown, however, that seagrasses are also at risk from increasing SSTs (Marbà et al. 2022).

In early 2023 two juxtaposed news items seemed at odds in our local Miami Herald newspaper relative to the dilemma of climate change. One piece noted how Australian children were psychologically affected by the reality of climate change, namely, with symptoms of anxiety, depression, and disempowerment. The second news item was an upbeat piece about the success of the local Grand Prix 'Formula One' motor racing event in Miami. This race involved 20+, 1000 horsepower cars that powered around 57 laps (308 km) travelling up to 322 km/hr. I wondered if any of the spectators, with a record attendance of 270,000, considered the enormous amount of fossil fuel consumed and entering the atmosphere as greenhouse gasses? In addition to the pollution created from the race itself, to fairly measure the total impact on our atmosphere, it would also be necessary to add the carbon emitted during the qualifying trials and spectator's travels (domestic and international) to and from the event. Such energy-consuming sporting events are not conducive to slowing the accelerating pace of climate change and global warming. One segment of our society is seemingly unconcerned with the harm that is presently befalling humankind, all of Earth's biosphere and the peril facing future generations. And the Grand Prix F1 races continue. An estimated 275,000 spectators were in attendance in Miami during the 3–5 May 2024 event. According to the World Meteorological Organization, an El Niño warming event has a 60% chance of developing between May and July, and up to an 80% chance of emerging by October 2023. El Niño has arrived (July), triggering havoc in the eastern tropical Pacific (see **2023/24 El Niño Projection** below). El Niño activity contributes to the development of abnormally high sea temperatures.

[14] Euendolithic cyanobacteria are abundant and wide-ranging organisms that live within and actively erode carbonate skeletons (Tribollet et al. 2009).

2023/24 El Niño Projection

On May 16, 2023, the NOAA Coral Reef Watch (coralreefwatch.noaa.gov) outlook predicted a 90% probability of heat stress associated with the development of an El Niño event. From an unusually large amount of warm water that accumulated in the western Pacific basin during a 3-year La Niña event, Tao Lian and co-workers predicted the development of an intense El Niño event at the end of 2023 to early 2024 (Lian et al. 2023). This was unwanted news but not entirely unexpected.[15] Derek Manzello, coordinator of the Coral Reef Watch program, urgently contacted coral reef researchers in the equatorial eastern Pacific region to be alert for coral heat stress and to report incidences of bleaching and mortality. The Degree Heating Weeks heat stress indices (DHW ≥4 = beginning of bleaching; DHW ≥8 = mortality expected) have occurred at several coral sites for 2 months (mid-May to mid-July) from southern Mexico to coastal Ecuador and the Galápagos Islands. The extreme and record-breaking temperatures of the northern hemispheric summer noted above result from El Niño warming, superimposed on warm oceans, and overlapping in occurrence with the warm summer season (Fig. 13.1).

In one sense, it is uplifting to witness the flood of responses regarding Derek's request for field verification of coral bleaching. Researchers up and down the eastern Pacific region, from Mexico to Ecuador, including the Galápagos and other offshore islands, began messaging colleagues about the condition of corals in their study areas and beyond. This response was also greatly facilitated by a recently created network, the Eastern Tropical Pacific (ETP) Node.[16] This network created a time series database of eastern Pacific coral reefs for regional status reports to be submitted to the Global Coral Reef Monitoring Network (GCRMN). During the first 2 months (mid-May to mid-July), preliminary coral bleaching observations were reported at Huatulco (severe) in southern Mexico; Los Cóbanos (severe), in El Salvador; Golfo Dulce (severe) and Cocos Island (light, 1% coral cover), in Costa Rica; Malpelo Island (widespread bleaching), in Colombia. No bleaching was reported at Bahía Banderas in Mexico, in the Gulf of Papagayo, Costa Rica, in Panama[17] or the Galápagos Islands. Detailed information will become available at

[15] An increase in ENSO SST variability is projected under greenhouse warming (Cai et al. 2018), and stronger warm anomalies begin earlier in the eastern Pacific, by around 2030, than previously observed (Geng et al. 2022).

[16] The second meeting of the ETP Node was organized by Fernando Zapata of the Universidad del Valle, Cali, Colombia, and Sonia Bejarano from the Leibniz Center for Tropical Marine Research in Bremen, Germany. The first meeting, organized by Franz Smith, was held at the Charles Darwin Research Station (Charles Darwin Foundation) in the Galápagos Islands in February 2019, with David Obura serving as the facilitator. The second meeting convened in June 2023 in Cali, during the initial development of the predicted El Niño 2023/24 event.

[17] A message received from Juan Maté (22 July 2023) contained the unwelcome news of extensive reef-coral bleaching at Iguana Island in the western sector of the Gulf of Panama. This bleaching incident was documented by Marco Díaz of Pedasí, Panama, on 21 July 2023. The prominent species on the Iguana reef exhibited bleaching, *Pocillopora damicornis* (cf. *P. grandis*) and *Porites lobata*. In the YouTube video of coral condition, https://youtu.be/gYahPStggM0., the upper branches of *Pocillopora* colonies were bleached and the basal branches still revealed normal pig-

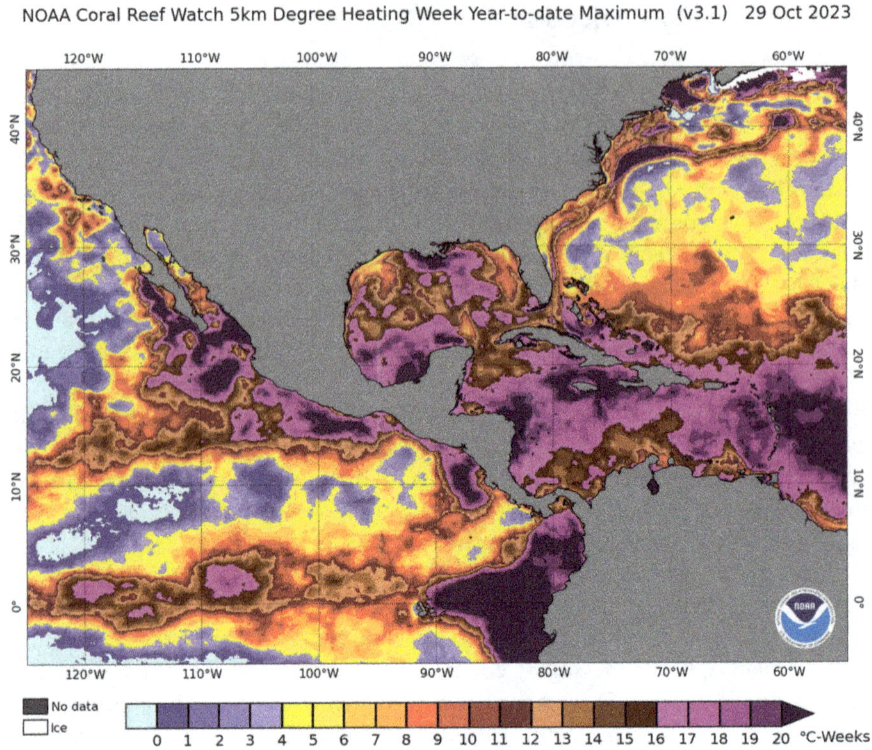

Fig. 13.1 Year-to-date maximum degree heating weeks (DHW) in tropical eastern Pacific and western Atlantic waters as of 29 October 2023. Image from Coral Reef Watch website, https://coralreefwatch.noaa.gov/. Significant coral reef bleaching is expected at DHW >4 (shades of yellow), and coral mortality is expected to begin at DHW >8 (shades of orange), especially for the most heat sensitive species. During past bleaching events, DHW values in excess of 20 have been associated with catastrophic coral mortality (89–99%). Black areas denote DHW > 20, white-light and dark purple areas indicate no data

the NOAA Coral Reef Watch site and the Global Coral Reef Monitoring Network in due course.

Nearly the exact opposite occurred in Florida, however—Derek Manzello and his group released Bleaching Alert warnings in May and June—by early July it was evident that Florida was heading into a disastrous situation—Derek and Bill Precht were interviewed on local and national media including the televised weather channel. By the end of July, the earliest and worst bleaching episode in Florida was

---

mentation. Some colonies exhibited normal pigmentation, showing no signs of bleaching. Marco reported reef sea water temperatures ranging between 31.9 and 32.4 °C. To follow future coral responses to this developing El Niño event, the reader should keep abreast of the U.S. government (NOAA, USA) and NGO postings

Fig. 13.2 Uniform bleaching of *Pocillopora* spp. at Isla del Canal de Afuera (7.6910°N, 81.6241°W), Gulf of Chiriquí, Panamá, 10 October 2023. No *Pocillopora* tissue mortality observed at this time. Most massive corals (*Porites*, *Pavona*) with bleached and dead tissues on upper colony surfaces (courtesy of Juan L. Maté)

already killing corals—the response from the agencies, and most of the coral nurseries involved in restoration, were caught off guard.

If development of this El Niño results in a heat wave affecting coral reefs in the Gulf of Chiriquí, it could be an excellent natural test of the heat tolerance of *Pocillopora* populations that survived the 2015–16 El Niño stress event. Earlier observations have shown that certain colonies of *Pocillopora* that survived major bleaching events demonstrated higher heat resistance in later events (Baker et al. 2004; Romero-Torres et al. 2020; see Fig. 13.2).[18] Genetic analyses, spearheaded by Ana Palacio-Castro, revealed one *Pocillopora* lineage that increased association with a thermotolerant algal symbiont (*Durusdinium glynnii*) during the 2015–16 stress event. This *Pocillopora-Durusdinium* symbiotic relationship experienced reduced bleaching and mortality compared with another *Pocillopora* lineage that did not possess *Durusdinium* (Palacio-Castro et al. 2023). Another factor relating to the heat resistance and abundance of *Pocillopora* species is their thermal

[18] In April 2024, 6 months after widespread *Pocillopora* bleaching in the Gulf of Chiriquí, the usual golden-brown pigmentation of colonies had returned at all of several reef sites surveyed (J. L. Maté, B. M. Riegl, and P. J. Glynn, personal communication.

history—populations that have experienced heat stress are probably better prepared for future heat waves. This is because they survived heat episodes and thus would contribute to increased frequencies of thermotolerant associations. A significant question: Do present-day reefs consist of sufficient abundances of heat resistant *Pocillopora* such that major structural portions of reefs will remain intact? This hopeful prospect must be tempered by the reality of global warming, which by mid-century (2040–2050) will present a stark challenge to the existence of any symbiotic corals.

Coral bleaching and mortality have also been reported at several western Atlantic sites contemporaneously with eastern Pacific coral disturbances. This appears to result from abnormally high atmospheric and ocean temperature records experienced in the northern hemisphere in the early summer of 2023 (Hausfather 2024). Light to moderate bleaching and low-level mortality of zooxanthellate cnidarians also occurred across the Isthmus of Panama during the 1982–83 El Niño event and at other Caribbean localities (Glynn 1984; Lasker et al. 1984; Coffroth et al. 1990). This bleaching was a harbinger of the shocking worldwide coral reef devastation. Coral bleaching in 2023, including colonies of branching and massive species, was observed on reefs at Bocas del Toro (Panama), on the Belizean barrier reef, on the Gulf of Mexico and Caribbean sides of Yucatán, and in the Florida Keys. Sea water temperatures in the Florida Keys have been at record levels this summer (Hoegh-Guldberg et al. 2023). For example, a reef bottom temperature reported at Sombrero Key was 34 °C. Also, seawater temperatures along the Keys, from Miami to Key West, have consistently ranged between 32–34 °C during July. In addition to scleractinian corals, other zooxanthellate cnidarians, such as gorgonians (sea fans and sea whips) and zoanthids, also experienced bleaching. The Florida Keys National Marine Sanctuary reported bleaching and rapid tissue loss in *Acropora* corals at nursery sites. The documentation of coral bleaching and mortality will follow in other regions as the heat wave continues. We must remember that coral bleaching and mortality are indicators of and responses to global warming, a human-caused disturbance that impacts all of the world's oceans.

Coral City, an Urban Coral Community

It has been possible for Miamians (and distant others with internet access) to follow remotely the fate of "urban corals" exposed to the extreme summer 2023 warming in July and August. Courtesy of Coral Morphologic (CM),[19] an underwater camera

---

[19] This internet site, "Miami's underwater livestream" can be accessed at <coralcitycamera.com>. The Coral Morphologic (CM) enterprise was founded in 2007 by marine biologist Colin Foord and musician J. D. McKay to bring awareness to the public of the beauty and plight of reef-building corals. Their mission statement elegantly captures the goals of the CM venture: "a multi-faceted platform for the development of symbiosis between humans and coral", its "unique methodology

is installed over a coral community ('Coral City') at the harbor entrance of the Port of Miami. The camera is live-streaming, launched online in February 2020, allowing continuous (24/7) viewing of corals, associated fishes, motile invertebrates, and occasional tidal-borne trash. This real-time viewing of a natural coral community is available and enjoyed by the public and also serves as an informative resource for managers and scientists. Urban corals are reef-building, zooxanthellate corals that live in and tolerate harsh, urbanized environments. At 'Coral City', 13 coral species and 139 fish species have been identified as members of this urban coral community (Enochs et al. 2023).[20] Unfortunately, when the DHW coral bleaching threshold was exceeded in 2023, this caused alarm and sorrow (Fig. 13.3).

The greater Miami area is fortunate to have environmentalists who advocate for protecting our water ways and marine ecosystems from pollution and climate change impacts. Louis Aguirre, a multiple award-winning environmental journalist and WPLG Local 10 TV anchor, is actively involved in coastal cleanup efforts via Channel 10's station-wide outreach program titled "Don't Trash Our Treasure". The ecowarriors involved in this program are young and adult community volunteers. Miami Waterkeeper (MWK), a non-governmental organization, also advocates for the protection of south Florida's waterways, from the Everglades to Biscayne Bay, including wetlands, mangroves, sea grasses, and coral communities. Rachel Silverstein, Executive Director of MWK, was one of Andrew Baker's earlier doctoral students. Rachel has received numerous conservation awards; her doctoral research centered on coral holobiont responses to environmental stressors (e.g., Silverstein et al. 2017; Cunning et al. 2018). MWK's mission involves advocating for clean water, ecosystem protection, and sea-level rise resiliency, grounded in science and driven by community. This is structured according to six community outreach programs involving environmental education and practical conservation activities. The Rescue-a-Reef coral conservation program was developed at the University of Miami's Rosenstiel School in 2015 to raise awareness and promote action on coral conservation via educational and experimental activities. This initiative arose from Diego Lirman who has been engaged in south Florida coastal

blends science and art in a way that enamors popular culture with the beauty of coral while inspiring the next generation to restore the reefs and protect the planet." A non-profit "Coral City Foundation" (CCF), via the Key Biscayne Community Foundation, was recently launched by CM. The goal of the CCF is to introduce the Coral City live feed into as many public schools in Miami-Dade County as possible. This can be a powerful tool to stimulate children's interest in science and the health of the ocean's ecosystems. The Coral City live feed is currently being played at the Biscayne Nature Center with the hope that the teachers there will be encouraged to develop a curriculum around it. Colin Foord and Sam (Samuel A.) May, two UM undergraduate students in 2002–2004, kindly assisted in setting up the first coral aquaculture system at RSMAES. This involved colonies of *Pocillopora* transported to Miami from Panama in 2002. Panamanian *Pocillopora* transported to RSMAES in 2005 are still viable, thanks to Tom (Thomas) Capo and Phil (Phillip R.) Gillette, both Resource Managers of the "National Resource for *Aplysia*" facility. These colonies readily produce clones, through asexual fragmentation, and have provided valuable research material for RSMAES' students and faculty over the years (e.g., Connelly et al. 2022).

[20] More recently, as of September 2024, Colin Foord has identified 207 fish species at the "Coral City" site.

Fig. 13.3 Still frame images of the condition of branching *Acropora cervicornis* colonies at the Port of Miami, Florida, urban Coral City Camera (CCC) monitoring site during the extreme ocean warming event of 2023. (**A**)—Colonies normally pigmented, 30.60 °C, 30.26–30.86 °C, 0.34 DHW (mean, minimum-maximum daily temperatures, Degree Heating Weeks. (**B**)—Bleached colonies, 31.28 °C. 30.86–31.78 °C, 9.67 DHW. (**C**)—Colonies dead, 31.06 °C, 30.56–31.68 °C, 17.89 DHW. (**D**)—Colonies algal-covered and bioeroded, 26.01 °C, 25.51–26.39 °C, 2.99 DHW. *Pseudodiploria strigosa* (front center) bleached (C) then recovered (D). Credits: CCC images, Colin Foord, Coral Morphologic; temperature records (HOBO Pendant, 64 K, Onset), Richard Karp, Coral Program, NOAA AOML; Rescue a Reef UM; Coral Reef Futures Laboratory UM

ecology and coral reef ecosystem rehabilitation since the early 1990s. Andrew Baker and Diego Lirman are now collaborating on coral restoration research at the University of Miami with the goal of scaling up the production of genetically diverse corals whose climate resilience is enhanced through a variety of new interventions, like selective breeding, sourcing parents from outside Florida, and rearing coral recruits with heat-tolerant symbionts prior to out-planting.

These conservation initiatives are to be praised and supported to protect and sustain Miami's urban coral communities. Pollution is a constant threat, requiring high vigilance. For example, a recent discovery of toxic sediments in Biscayne National Park indicates the importance of continuing clean-up and restoration action (May et al. 2022).

Two iconic corals that adorned the Coral City community were *Acropora cervicornis* and *Acropora palmata*, reef-building species that have suffered extreme mortality (95–98%) during the past few decades and are now classified as critically endangered on the IUCN Red List of Threatened Species. Colonies of these species were transplanted to Coral City by the "Rescue-a-Reef" conservation team to help promote community diversity and avoid local extinctions. It was encouraging to observe the growth of these threatened species when viewing Coral City, at least up to the early summer of 2023. Despite record high water temperatures, the *Acropora* colonies still retained their healthy, golden-brown coloration until the end of July. Then suddenly, and not unexpectedly, nearly all colonies bleached and appeared bone white. By the end of August, the bleached colonies had turned to dull shades of brown, an indication of tissue death and the overgrowth of benthic algae. The dead coral skeletons had virtually disappeared by the end of 2023 due to intense fish bioerosion. The recent significant decline of *Acropora* species and other structurally important species is concerning and likely spells the end of reef-building in Florida and the greater Caribbean region (Toth et al. 2019).

Coral Reef Protection and Restoration

Due to the near certainty of continuing and intensifying climate change stressors in the present century, particularly extreme high temperatures and increasing ocean acidification, consideration of coral rescue and protective measures is advised to encourage action to aid in the survival and persistence of threatened reef-building corals. Effective, science-based measures must be enacted now. Failure is temporary, regrets over inaction can linger forever. Here I note proposed and actively practiced local-scale methods of coral reef protection and restoration emphasizing efforts in the eastern Pacific. These direct, small-scale endeavors should proceed in step with efforts to reduce and control greenhouse gas emissions, the root cause of climate change (global warming). Global-scale methods, such as space-based geo-engineering and carbon sequestration, are under increasing consideration and briefly reviewed in Kulkarni (2022). The engineering expertise and cost of these developments are an enormous challenge and can be addressed only at national and

Fig. 13.4 Nohora Galvis at the United Nations Ocean Conference in New York where she defended the protection of a coral reef with ~500 year old colonies from channel dredging in Varadero, Colombia (2017). Her proposal was approved by the President of Colombia who granted the requested protection. Nohora is the recipient of several conservation prizes, including the 2023 ICRS World Reef Award

international levels. Local intervention strategies, however, can be effective in limiting coral mortality and are necessary in attempts to restore the structural integrity of degraded structural reefs. While all these methods may prove to be effective in relatively small areas, affecting 100 s to 1000 s of coral colonies, the challenge will be to up-scale restoration to reef locations and regions. In recent years, a notable increase in coral reef protection and restoration activities has paralleled the escalation of eastern Pacific coral reef investigations. In-country researchers, conservationists, environmentalists, and local fishers have performed yeoman's service in supporting the protection of coral reefs. A pioneer in coral reef conservation and management in Colombia, Nohora Galvis, was awarded the ICRS (International Coral Reef Society) 2023 Coral Conservation Award (Fig. 13.4).[21] These efforts must follow scientific protocol with documentation of procedures and results, and dissemination to interested practitioners for adoption where feasible.

[21] The challenges of environmental protection in some Latin American countries came into focus through Nohora when I served as a co-advisor during her Colombian case studies at RSMAES in the early 1990s. Nohora has completed over 35 years of implementing coral reef protection and conservation measures in the Pacific and Caribbean coasts and the Colombian oceanic islands. Nohora has devoted considerable effort in protecting Pacific coral reefs and mangroves in the Gulf of Tribugá and the Gulf of Panama, and Caribbean reefs at Capurgana—Cabo Tiburón and the Varadero Reef in Cartagena Bay (Varadero is the Spanish word for shipyard or drydock, not to be confused with the resort city of Varadero, Cuba). Some of these expansion projects required great care and skill in marshaling scientific support opposing politically powerful developer consortia. Lamentably, she experienced nefarious personal threats in supporting conservation measures to protect mangrove and coral habitats. In consideration of Nohora's unrelenting defense of sensitive coastal habitats and her environmental leadership she received a medal from Colombia's Ministry of Environment in 2019.

With accelerating world-wide coral reef degradation and mounting public concern, the US National Academies NGO initiated a comprehensive study of interventions to limit coral mortality under climate change.[22] A committee of 12 coral reef scientists, chaired by Stephen R. Palumbi of Stanford University, was convened to synthesize current knowledge and provide a roadmap for informed decisions regarding coral reef conservation (National Academies of Sciences, Engineering, and Medicine. 2019a, b). For this study, 23 areas of intervention were examined under four general categories: (a) genetics and reproduction, (b) physiology, (c) coral populations and communities, and (d) environment. The detailed information associated with each tested method of intervention is critical and useful for those methods under present and future consideration for specific habitats and conditions. Each intervention stratagem is accompanied with a discussion of current feasibility, potential scale of focus, limitations, and risks. These are all critical factors that must be carefully considered in coral reef management. A structured, adaptive management framework for decision making was proposed as a rational approach for introducing intervention measures.

When we first began investigations on the Uva Island reef in the early 1970s, trawler boats from the shrimp fishing fleet commonly anchored near the reef, which would occasionally result in anchor dragging and significant damage to the reef framework. We brought this matter before Dr. Carlos Arellano, director of the Panamanian fisheries bureau, and requested help to prevent further damage to the Uva reef and other reefs in the Gulf of Chiriquí. Carlos, an environmentalist, was sympathetic to our concerns and ordered the Uva Island area off limits to all trawler fishing vessels. This was our first attempt to help create a marine protected area (MPA), with beneficial results. To safeguard this 11 meter, vertically thick 6000 year old coral reef required little effort. With growing recognition of the pristine marine and terrestrial communities of high biodiversity in this area, the Coiba National Park (including Uva Island) was established in 1991. At >2000 km^2, this is Panama's largest, no-take reserve and is now recognized as a UNESCO World Heritage site.

Héctor M. Guzmán is credited with promoting the establishment of several MPAs with reef-building corals in Panama.[23] He also initiated reef restoration

[22] The National Oceanic and Atmospheric Administration, and the Paul G. Allen Family Foundation provided funds for the consensus study report.

[23] Héctor was the first Master's (M.Sc.) student of Jorge Cortés at the University of Costa Rica, San Pedro, San José. He completed his Ph.D. degree at Newcastle University, United Kingdom and joined the scientific staff of STRI in 1994. He continues with coral reef studies with an emphasis on conservation biology. For example, he and colleagues provided detailed information on the occurrence of coral communities and coral reefs across all sectors of the Coiba National Park (Guzmán et al. 2004), and the Pearl Islands Archipelago (Guzmán et al. 2008), the largest and second largest MPAs of Pacific Panama, respectively. Beginning in 2000, he began gorgonian systematic studies in the eastern Pacific, the latter in collaboration with Odalisca Breedy of the University of Costa Rica, San Pedro. Odalisca and Héctor have revealed a notably diverse gorgonian fauna in the eastern tropical Pacific with 111 species reported from Mexico to Peru (Cortés et al. 2017). Unlike the usual occurrence of gorgonians on coral reefs in the western Atlantic, in the eastern Pacific gorgonians occur more commonly in reef-coral communities on rocky substrates. Héctor, an early participant in the MigraMar conservation program, has also advanced research

practices by transplanting *Pocillopora* fragments onto degraded reef frameworks in Costa Rica in the 1980s (Guzmán 1991). This pioneering intervention method greatly assisted reef recovery and resulted in Héctor's receiving a Rolex Conservation Award.

Other approaches to protect corals entail selecting areas that minimize bleaching risk, such as habitats that are naturally protected from bleaching, e.g. upwelling centers, islands with turbulent circulation, and substrates partially shielded from continuous sunlight (Salm et al. 2001). Also, marine protected areas (MPAs) that prevent or reduce pollution, overfishing, and structural damage by humans would minimize such secondary stressors that occur in unprotected areas. The effectiveness of MPAs in limiting coral reef stress and decline is not absolute, however, due to the intense negative effects of climate change and associated elevated sea temperature and other pivotal conditions (Bruno et al. 2019). A notable recent increase of MPAs in the eastern Pacific, with coral habitats or coral reefs, is reassuring. Existing and proposed MPAs have increased from 28 (Glynn 2000b) to 46 (Alvarado et al. 2017) in less than 20 years.

Intervention strategies noted in Baker et al. (2008) propose reef shading, polyp feeding, electrochemical stimulation, wave-powered artificial upwelling, and symbiont inoculation. Some of these methods have been evaluated in the Indo-Pacific region with promising results. The motivation for creating upwelling conditions to limit coral bleaching and mortality originates from observations of natural cooling that greatly tamped down El Niño warming in the Gulf of Panama in 1997–98 (Glynn et al. 2001) and the Galápagos Islands in 2015–16 (Riegl et al. 2019). Genetic analyses of surviving *Pocillopora* colonies during the 2015–16 El Niño heat stress event revealed an increased association of *Durusdinium glynnii* in coral hosts, a thermotolerant algal symbiont (Palacio-Castro et al. 2023; see also Ch.10, Biodiversity, ["Symbiodiniaceae Zooxanthellae"]). A more recent study also supports the positive role of thermally resilient symbionts in protecting corals during ocean warming events (Turnham et al. 2024). Coral probiotics, or the introduction of resilient endosymbionts into corals, is a rapidly developing procedure under intense study in the broader Pacific region (Peixoto et al. 2021; Peixoto and Voolstra 2023; Voolstra et al. 2023; Guibert 2024)

For an upwelling center—with surfacing, cool, nutrient-rich waters that would lessen or prevent bleaching—the timing of upwelling must coincide with the occurrence of the stressful warm water event, which is not always the case (Chollett et al. 2010; Smith et al. 2017). Upwelling did not occur in the Gulf of Panamá during the 1982–83 El Niño event; coral bleaching and mortality were severe (Glynn 1990). Further, as of this writing (July 2023), the San Agustín *Pocillopora* reef in the upwelling center of Huatulco (Gulf of Tehuantepec, southern Mexico), exhibited severe El Niño-related bleaching with all corals stark white (López-Pérez et al.

---

and protection to highly migratory eastern Pacific megafauna such as whale sharks, whales, sharks, and turtles (Guzmán et al. 2018, 2022).

2023, 2024). Upwelling normally begins in this region in October or November, possibly too late to prevent mass coral mortality?

Current coral restoration projects have met with notable success in the eastern Pacific in advancing effective intervention methods and enhancing coral community recovery. Transplantation methods employed in Mexico have demonstrated the effectiveness of this approach (Martínez-Castillo et al. 2022; Gómez-Petersen et al. 2023; Nava et al. 2023; Tortolero-Langarica et al. 2023). The Nava et al. (2023) study is noteworthy because molecular analysis was employed to identify colonies hosting bleaching-resistant *Durusdinium* symbionts. In Costa Rica, transplantation and out planting (from nurseries) of branching and massive coral species in non-upwelling and upwelling habitats have proved successful in the rehabilitation of degraded coral reefs (Chomitz et al. 2023; Fabregat-Malé et al. 2023). Finally, out planting of *Pocillopora* in Colombia demonstrated a clear advantage in that acquired arborescent colony morphology in nursey-grown fragments revealed higher survival and growth rates than reef-derived fragments (Ishida-Castañeda et al. 2020).

It is instructive to consider the "Reefs of Hope" coral adaptation and restoration paradigm in Fiji and other Pacific Island sites. This initiative, under the direction of Austin Bowden-Kirby of the NGO "Corals for Conservation," aspires to rescue coral diversity as global warming continues to decimate coral reefs. In a nut shell, the procedure, modelled on tropical forest restoration, incorporates the following steps: (a) bleaching-resistant corals[24] from hot, nearshore reefs are translocated[25] to cooler, protected reef areas (nurseries) where they serve as bleaching resistant brood stock; (b) multiple species are grown, maintained and harvested for coral reef restoration and climate change adaptation; (c) genetically diverse patches of bleaching resistant corals are established on cooler outer reefs to restore sexual reproduction by facilitating crosses between bleaching-resistant and cool-adapted strains; (d) patches of bleaching resistant corals are established among bleaching sensitive corals to promote potential sharing of resistant algal symbionts, thus spreading resilience; (e) patches of bleaching resistant corals are introduced to dead reefs that could then release a strong settlement signal to attract dispersing symbiont-free coral larvae and bleaching resistant symbiotic algae, thus accelerating natural adaptation processes. It is hypothesized that the restoration of resistant coral community patches could serve as a habitat to attract coral-associated fishes, invertebrates, and other reef organisms. Importantly, these patches would expand and enhance natural

[24] The resistant holobionts include both coral host and symbiont selected/adaptive genotypes (Morikawa and Palumbi 2019).

[25] Removing and transporting corals from their aquatic habitat requires care to avoid desiccation, and excessive exposure to sunlight and rain. We have transported live corals from various sites in the eastern Pacific to our laboratory and aquaculture facility in Miami (never for transplantation to open Atlantic waters) with virtually zero bleaching or mortality. Small colonies (2–8 cm maximum length) of *Pocillopora* spp., *Millepora intricata*, and *Diaseris distorta* have been transported by carefully wrapping them in seawater-saturated nylon stockings or cotton cloth. Care is also taken to avoid abrasion or hard knocks in transport. The corals survive airline flights of 8–10 h and are quickly revived, with polyp expansion, when immersed in seawater.

recruitment, recovery, and functional and adaptation processes. Research supporting this approach, and the pros and cons of other adaptation coral rescue initiatives (e.g., the "50 Reef Initiative", Beyer et al. 2018) are discussed in Bowden-Kerby 2023). The common goal is to **buy time** by saving threatened species, hoping that climate change will eventually come under control. The "Reefs of Hope" protocol would seem well suited for the frequently disturbed eastern Pacific region by accelerating natural adaptation and recovery processes.

There is an opportunity for amateur and professional coral aquarists to assist in safeguarding the genetic diversity of corals and their symbiotic algae. The expertise of Colin Foord and Julian Sprung,[26] coral aquarium authorities in Miami, were immensely helpful in offering advice and lending a hand in establishing and maintaining corals in my laboratory. The Panamanian *Pocillopora* aquaculture system, designed by Colin, was set up in 2002 in Tom Capo's *Aplysia* wet lab at RSMAES. It consisted of plastic egg trays set in 2–3 m long by 0.5 m wide fiberglass tanks illuminated with 'power compact' fluorescent light bulbs (before the arrival of LED lighting). Colin's current coral nursery and studio/laboratory, in a refurbished warehouse in the Allapattah neighborhood, is like discovering a buried gemstone, unexpected and inspiring (Figs. 13.5 and 13.6). During one of Charlie Veron's visits to Miami, he and I were invited to Julian's home on Miami Beach to inspect (and admire) his reef aquaria. One of Julian's aquariums was a living replica of a natural coral reef ecosystem, with a miniature living mangrove shoreline and simulated shallow and deep-water coral assemblages. In mimicking common patterns of coral zonation, Julian arranged branching coral colonies near the shallow shoreline and dome-shaped and encrusting colonies in deeper parts of the aquarium. We mentioned to Julian that branches from some of the shallow colonies were growing at and slightly above the aquarium's waterline. He remarked that frequent branch-trimming was necessary due to the rapid growth rates of these corals. The success of maintaining and propagating corals in aquaria, long-term (indefinitely?), has been abundantly demonstrated. In terms of preserving coral diversity for restoration purposes, careful record-keeping (e.g., recording in detail the coral hosts and *Symbiodinium* species, genomes, and provenances) and coordination with donor aquarists are essential (Carlson 1999). Besides serving a critical role in safeguarding threatened coral species, public aquaria offer enjoyment, awareness, and knowledge of ecosystem function. Many coral reef scientists have vivid childhood memories of the joy and fascination they experienced in visiting aquariums and natural history museums.

The addition of recent interventions to the coral health toolbox and an improved understanding of immune responses can potentially benefit coral conservation efforts, for example: (a) cryopreservation (Lin et al. 2022; Powell-Palm et al. 2023); (b) genetic editing, CRISPR-Cas 9 (Cleves et al. 2020); (c) inoculation of

[26] Julian, known locally as the "Reef King," is the proprietor of Two Little Fishies, founded in 1991 to promote the reef aquarium hobby. An informative source on the necessary physical-chemical conditions, available and suitable species, and feeding of coral reef organisms in aquaria is accessible in Delbeek and Sprung (2005).

Fig. 13.5 Colin Foord, cofounder (with J. D. McKay) of Coral Morphologic, beside cultured corals at the Allapattah, Miami aquarium facility. Stepping into the dim, blue LED lighting (420–470 nm wavelength) of the aquarium room creates an other-worldly sensation. The coral's fluorescent proteins are activated by blue light wavelengths thereby creating the vibrant greens/oranges/reds display of the coral's soft tissues. This indoor lighting is close to the natural color spectrum below about 20 m depth. Colin's shirt features a print of a fossil brain coral, probably a species in the genus *Pseudodiploria*. Photograph courtesy of Alfonso Duran, October 2022

Fig. 13.6 Bouquet of fluorescing corals under LED lighting, an assortment of mostly hard and soft IndoPacific taxa (*Lobactis scutaria*, *Zoanthus* spp., *Acropora* spp., *Montipora spongodes*, *Pachyseris* sp., *Goniopora* sp., *Favites pentagona*, *Goniastrea* sp., *Caulastrea furcata*, *Cespitularia* sp., *Acanthastrea undata*, *Pectinia* sp., *Oxypora* sp., and *Duncanopsammia axifuga*. Coral Morphologic, courtesy of Colin Foord

thermotolerant algal symbionts (Buerger et al. 2020; Scharfenstein et al. 2022); (d) antibiotics (Forrester et al. 2022; Studivan et al. 2023); (e) innate immunity (Snyder et al. 2021; Connelly et al. 2022). Concerning the ability to perform coral immunity, according to Nikki Traylor Knowles (pers. comm.), "we are not there yet". The interaction of the coral's immune system with probiotics, however, may be the best present avenue of research into this subject.[27] Some of these methods demonstrate promise in limited areas (e.g., aquaria, natural coral community patches, and particular reef sites), but their effectiveness at larger reef scales may be insurmountable.

Afterword

The "window of opportunity to secure a livable and sustainable future" is "rapidly closing." "Wake up! This is your last chance, humanity." (IPCC 2023).

> Earth is now well outside of the safe operating space for humanity. (Richardson et al. 2023).

It would be an injustice to deny the advances that fossil fuel use has allowed in the genesis and development of our world's civilizations. This energy source was essential for advancing science, technology, human health, and associated comforts. The chief driver of climate change, however, is the increasing concentration of CO_2 from fossil fuels, which is causing intolerable high temperature stress to all life. Fossil fuels are "incompatible with human survival." (Antonio Guterres, Secretary General of the United Nations, June 15, 2023). While increasing atmospheric CO_2 concentration and ocean temperature have received primary attention throughout this narrative, due to the sensitivity of coral reefs to thermal stress, all marine and terrestrial ecosystems are subject to additional multifarious disturbances. Many of these impactful factors are a result of unprecedented human population growth and the resulting disruptions this brings, e.g. exploitation of natural resources, pollution, extinction of species, declines of forested land, release of synthetic chemicals, *inter alia* (Dodson et al. 2020; Riegl and Glynn 2020; Richardson et al. 2023).[28]

---

[27] Nikki Traylor Knowles, who joined the RSMAES faculty (and coral reef research team) when I retired, has brought genomics expertise and an appreciation of the value of biological collections. Nikki has introduced a new academic track, "Marine Invertebrate Biology and Evolution", which is designed to give students experience in molecular techniques, phylogenetics, taxonomy, and museum/collection science. Upon arrival at RSMAES, Nikki quickly collaborated with Nancy Voss, then Director of the Voss Marine Invertebrate Collection (VMIC). Nikki is promoting renewed interest and support for VMIC, internationally recognized for comprehensive collections of tropical marine invertebrate species from the western Atlantic, west central Africa, and eastern Pacific regions. The VMIC collections are CITES-certified, with coral and mollusk taxa particularly well represented. CITES, Convention on International Trade in Endangered Species (of Wild Fauna and Flora), is the legal and correct way to exchange endangered species for scientific study.

[28] An emerging school of scholarship shows that population number and growth are not the driving factors of over-exploitation but rather the over consumption of resources by a small fraction of top income earners. Jason Hickel and colleagues argue that Earth's current climate disruption and

Few of my colleagues have lost hope in stemming the decline of coral reefs but are working diligently across a broad range of disciplines for their safe keeping (e.g., Kleypas et al. 2021; Knowlton et al. 2021; Arias-González et al. 2022; Van Oppen and Aranda Lastra 2022; Bowden-Kerby 2023; Kaufman et al. 2023).[29] Something loved and marveled at is not easily abandoned. Even though frightening predictions of increasing environmental extreme conditions, anathema to supporting earth's biota, are regularly recognized by dedicated researchers, the future is unknowable. Is it possible that the forces driving present day climate change and global warming could stabilize at a tolerable level by humankind's ingenuity (Schubert 2021) or change naturally by some climate stabilizing event (Wunderling et al. 2023). I urge the scientific community, climate engineers, environmentalists, and all concerned citizens to continue working toward altering the course of climate change to sustainability to protect our planet's biosphere, including a safe and just future for humanity.

Today's generation is aware of our planet's deteriorating environmental conditions and sterile future due to climate change. There is a movement among our youthful heirs to step forward to take action to slow or hopefully reverse the ongoing devastation. Greta T. E. E. Thunberg, a 15-year old (in 2018) Swedish environmental activist, has forcefully pressed world leaders to act on climate change mitigation. In the USA, young activists in Montana recently won a land-mark lawsuit (Held v. State of Montana, June 2023): that the state's failure to consider climate change when approving fossil fuel projects was unconstitutional. The Montana constitution guarantees a clean and healthy environment for its citizens. Similar lawsuits have been filed against Hawaii, Utah, Virginia, and Oregon governments. In Miami in November 2023, on the heels of record breaking extreme ocean warming and coral mortality, the Frost Science Museum hosted the inaugural ReeFlorida Symposium dedicated to conserving and restoring coral reefs (Kaufman et al. 2023). Of the 159 meeting registrants, 43 were students, many of whom spoke at the meeting. A common mantra prevailed: Follow science, advocate for a healthy, life-supporting environment, and keep the pressure on the policymakers. Planet Earth's biosphere is fragile and perhaps unique in our Milky Way galaxy and beyond. To this end, all nations should quickly commit to the best of their abilities and resources, to support and accelerate action for a healthy environment, greater equity among people, and the protection of natural ecosystems and biodiversity.

---

ecological crisis result from economies that depend upon ever increasing growth and exploitation of resources. These workers propose an economy that tends toward an equilibrium state and is globally just and in balance with the natural world (e.g., Hickel 2020; Rammelt et al. 2022; Gupta et al. 2023; Hickel and Sullivan 2023).

[29] Some of my colleagues whose hope has significantly waned feel that coral reef researchers are too narrowly focused on conservation and restoration actions instead of on over-consumption and drivers, and human equality measures, i.e., striving toward 'a safe and just' future for humanity (Obura et al. 2023).

References

Aeby GS, Howells E, Work T, Abrego D, Williams GJ, Wedding LM, Caldwell JM, Moritsch M, Burt JA (2020) Localized outbreaks of coral disease on Arabian reefs are linked to extreme temperatures and environmental stressors. Coral Reefs 39:829–846

Alvarado JJ, Aburto-Oropeza O, Abad R, Barraza E, Brandt M, Cantera J, Estrada P, Gaymer CF, Guzmán-Mora AG, Herlan JJ, Maté JL (2017) Coral reef conservation in the eastern tropical Pacific. In: Glynn PW, Manzello DP, Enochs IC (eds) Coral reefs of the eastern tropical Pacific: persistence and loss in a dynamic environment. Coral reefs of the world 8. Springer, Dordrecht, pp 565–591

Alvarez-Filip L, González-Barrios FJ, Pérez-Cervantes E, Molina-Hernández A, Estrada-Saldívar N (2022) Stony coral tissue loss disease decimated Caribbean coral populations and reshaped reef functionality. Commun Biol 5(1):440

Antonius A (1973) New observations on coral destruction in reefs. Association of Island Marine Laboratories of the Caribbean. In: 10th Meeting, Mayagüez, Puerto Rico, Abstract, p 3

Antonius A (1977) Coral mortality in reefs: a problem for science and management. In: Proceedings of the 3rd international coral reef symposium, Miami, vol 2, pp 617–623

Arias-González JE, Baums IB, Banazak AT, Prada C, Rossi S, Hernández-Delgado EA, Rinkevich B (2022) Coral reef restoration in a changing world: science-based solutions. Front Mar Sci 9-2022. https://doi.org/10.3389/fmars.2022.919603

Aronson RB, Macintyre IG, Precht WF, Thaddeus JT, Murdoch TJT, Wapnicki CM (2002) The expanding scale of species turnover events on coral reefs in Belize. Ecol Monogr 72(2):233–249

Baird AH, Bhagooli R, Ralph PJ, Takahashi S (2009) Coral bleaching: the role of the host. Trend Ecol Evol 24(1):16–20

Baker AC, Glynn PW, Riegl B (2008) Climate change and coral reef bleaching: an ecological assessment of long-term impacts, recovery trends and future outlook. Estuar Coast Shelf Sci 80:435–471. https://doi.org/10.1016/j.ecss2008.09.003

Baker AC, Starger CJ, McClanahan TR, Glynn PW (2004) Corals' adaptive response to climate change. Nature 430:741

Baums IB, Devlin-Durante M, Laing BAA, Feingold J, Smith T, Bruckner A, Monteiro J (2014) Marginal coral populations: the densest known aggregation of *Pocillopora* in the Galápagos Archipelago is of asexual origin. Front Mar Sci 1:59. https://doi.org/10.3389/fmars.2014.00059

Beyer HL, Kennedy EV, Beger M, Chen CA, Cinner JE, Darling ES, Eakin CM, Gates RD, Heron SF, Knowlton N et al (2018) Risk-sensitive planning for conserving coral reefs under rapid climate change. Conserv Lett 11. https://doi.org/10.1111/conl.12587

Bowden-Kerby A (2023) Coral-focused climate change adaptation and restoration based on accelerating natural processes: launching the "Reefs of Hope" paradigm. Oceans 4:13–26. https://doi.org/10.3390/oceans4010002

Brandl SJ, Goatley CH, Bellwood DR, Tornabene L (2018) The hidden half: ecology and evolution of cryptobenthic fishes on coral reefs. Biol Rev 93(4):1846–1873

Bruno JF, Côté IM, Toth LT (2019) Climate change, coral loss, and the curious case of the parrotfish daradigm: why don't marine protected areas improve reef resilience? Annu Rev Mar Sci 11:307–334

Bruno JF, Selig ER, Casey KS, Page CA, Willis BL, Harvell CD, Sweatman H, Melendy AM (2007) Thermal stress and coral cover as drivers of coral disease outbreaks. PLoS Biol 5(6):e124. https://doi.org/10.1371/journal.pbio.0050124

Brusca RC, Thompson DA (1975) Pulmo reef: the only "coral reef" in the Gulf of California. Cienc Mar 2:37–53

Buck-Wiese H, Burgués I, Medrano A, Navarrete-Fernandez T, García M, Wieters EA (2018) Patterns in sexual reproduction of the dominant scleractinian corals at Rapa Nui (Easter Island): *Pocillopora verrucosa* and *Porites lobata*. Aquat Biol 27:1–11. https://doi.org/10.3354/ab00691

Budd AF (2000) Diversity and extinction in the Cenozoic history of Caribbean reefs. Coral Reefs 19:25–35

Budd AF, Johnson KG, Stemann TA (1993) Plio-Pleistocene extinctions and the origin of the modern Caribbean reef-coral fauna. In: Ginsburg RN (ed) Proceedings of the colloquium on global aspects of coral reefs: health, hazards and history. Rosenstiel School of Marine and Atmospheric Science, University of Miami, pp 7–13

Buerger P, Alvarez-Roa C, Coppin CW, Pearce SL, Chakravarti LJ, Oakeshott JG, Edwards OR, van Oppen MJH (2020) Heat-evolved microalgal symbionts increase coral bleaching tolerance. Sci Adv 6:eaba2398

Cai W, Wang G, Dewitte B, Wu L, Santoso A, Takahashi K, Yang Y, Carréric A, McPhaden MJ (2018) Increased variability of eastern Pacific El Niño under greenhouse warming. Nature 564(7735):201–206. https://doi.org/10.1038/s41586-018-0776-9

Carlson BA (1999) Organism responses to rapid change: what aquariums tell us about nature. Am Zool 39(1):44–55

Carpenter KE, Abrar M, Aeby G, Aronson RB, Banks S, Bruckner A, Chiriboga A, Cortés J, Delbeek JC, DeVantier L, Edgar GJ (2008) One-third of reef-building corals face elevated extinction risk from climate change and local impacts. Science 321(5888):560–563

Cea Egaña A, DiSalvo LH (1982) Mass expulsion of zooxanthellae by Easter Island corals. Pac Sci 36:61–63

Chollett L, Mumby P, Cortés J (2010) Upwelling areas do not guarantee refuge for coral reefs in a warming ocean. Mar Ecol Prog Ser 416:47–56

Chomitz BR, Kleypas JA, Cortés J, Alvarado JJ (2023) Succession of the sessile benthic community at a coral reef restoration site. Rev Biol Trop 71:e54881. https://doi.org/10.15517/rev.biol.trop..v71iS1.54881

Cleves PA, Tinoco AI, Bradford J, Perrin D, Bay LK, Pringle JR (2020) Reduced thermal tolerance in a coral carrying CRISPR-induced mutations in the gene for a heat-shock transcription factor. Proc Natl Acad Sci 117(46):28899–28905. https://doi.org/10.1073/pnas.1920779117

Coffroth MA, Lasker HR, Oliver JK (1990) Coral mortality outside of the eastern Pacific during 1982-1983: relationship to El Niño. In: Glynn PW (ed) Global ecological consequences of the 1982–83 El Niño Southern Oscillation. Elsevier oceanography series 52. Elsevier, Amsterdam, pp 141–182

Cole J, Tudhope AW (2017) Reef-based reconstructions of eastern Pacific climate variability. In: Glynn PW, Manzello DP, Enochs IC (eds) Coral reefs of the eastern tropical Pacific: persistence and loss in a dynamic environment. Coral reefs of the world 8. Springer, Dordrecht, pp 535–548

Colgan MW (1990) El Niño and the history of eastern Pacific reef building. In: Glynn PW (ed) Global ecological consequences of the 1982–83 El Niño-Southern Oscillation. Elsevier oceanography series 52. Elsevier, Amsterdam, pp 183–232

Connell JH (1973) Population ecology of reef-building corals. In: Jones OA, Enderan R (eds) Biology and geology of coral reefs, vol 2, Biology 1. Academic, New York, pp 205–247

Connelly MT, McRae CJ, Liu P-J, Martin CE, Traylor-Knowles N (2022) Antibiotics alter *Pocillopora* Symbiodiniaceae-bacteria interactions and cause microbial dysbiosis during heat stress. Front Mar Sci 8:814124. https://doi.org/10.3389/fmars.2021.814124

Cortés J, Enochs IC, Sibaja-Cordero J, Hernández L, Alvarado JJ, Breedy O, Cruz-Barraza JA, Esquivel-Garrote O, Fernández-García C, Hermosillo A, Kaiser KL, Medina-Rosas P, Morales-Ramírez A, Pacheco C, Pérez-Matus A, Reyes-Bonilla H, Riosmena Rodríguez R, Sánchez-Noguera C, Wieters EA, Zapata FA (2017) Marine biodiversity of eastern tropical Pacific coral reefs. In: Glynn PW, Manzello DP, Enochs IC (eds) Coral reefs of the eastern tropical Pacific: persistence and loss in a dynamic environment. Coral reefs of the world 8. Springer, Dordrecht, pp 203–250

Cunning R, Silverstein RN, Baker AC (2018) Symbiont shuffling linked to differential photochemical dynamics of *Symbiodinium* in three Caribbean reef corals. Coral Reefs 37:145–152

Delbeek JC, Sprung J (2005) The reef aquarium: science, art, and technology, vol 3. Two Little Fishies, Ricordia, Florida, p 680

Dickins JM (1984) Evolution and climate in the upper Paleozoic. In: Brenchly PJ (ed) Fossils and climate. Wiley, Chichester, pp 317–327

Dietzel A, Bode M, Connolly SR, Hughes TP (2020) Long-term shifts in the colony size structure of coral populations along the Great Barrier Reef. Proc R Soc B 287:20201432. https://doi.org/10.1098/rspb.2020.1432

Dodson JC, Dérer P, Cafaro P, Götmark F (2020) Population growth and climate change: addressing the overlooked threat multiplier. Sci Total Environ 748(2020):141346. https://doi.org/10.1016/jscitotenv.2020.141346

Dunbar RB, Wellington GM, Colgan MW, Glynn PW (1994) Eastern Pacific sea surface temperature since 1600 A.D.: the $\delta^{18}O$ record of climate variability in Galápagos corals. Paleoceanography 9(2):291–315

Eakin CM, Devotta D, Heron S, Connolly S, Liu G, Geiger E, Cour JDL, Gomez A, Skirving W, Baird A, Cantin N, Couch C, Donner S, Gilmour J, Gonzalez-Rivero M, Gudka M, Harrison H, Hodgson G, Hoegh-Guldberg O, Manzello D (2022) The 2014-17 global coral bleaching event: the most severe and widespread coral reef destruction. https://doi.org/10.21203/rs.3.rs-1555992/v1

Edmunds P, Riegl B (2019) Urgent need for coral demography in a world where corals are disappearing. Mar Ecol Prog Ser 635:233–242

Enochs IC (2012) Motile cryptofauna associated with live and dead coral substrates: implications for coral mortality and framework erosion. Mar Biol 159(4):709–722

Enochs IC, Studivan MS, Kolodziej G, Foord C, Basden I, Boyd A, Formel N, Kirkland A, Rubin E, Jankulak M, Smith I, Kelble CR, Manzello DP (2023) Coral persistence despite marginal conditions in the port of Miami. Sci Rep 13:6759. https://doi.org/10.1038/s41598-023-33467-7

Fabregat-Malé S, Mena S, Alvarado JJ (2023) Nursery-reared coral outplanting success in an upwelling-influenced area in Costa Rica. Rev Biol Trop 71:e54879. https://doi.org/10.15517/rev.biol.trop..v71iS1.54879

Feingold JS, Glynn PW (2014) Coral research in the Galápagos Islands, Ecuador. In: Denkinger J, Vinueza L (eds) The Galápagos marine reserve, social and ecological interactions in the Galápagos Islands, pp 3–22. https://doi.org/10.1007/978-3-319-02769-2_1

Forrester GE, Arton L, Horton A, Nickles K, Forrester LM (2022) Antibiotic treatment ameliorates the impact of stony coral tissue loss disease (SCTLD) on coral communities. Front Mar Sci 9:859740. https://doi.org/10.3389/fmars.2022.859740

Geng T, Cai W, Wu L, Santoso A, Wang G, Jing Z, Gan B, Yang Y, Li S, Wang S, Chen Z, McPhaden MJ (2022) Emergence of changing Central-Pacific and Eastern-Pacific El Niño-Southern Oscillation in a warming climate. Nat Commun 13:6616. https://doi.org/10.1038/s41467-022-33930-5

Ginsburg RN (1993) Global aspects of coral reefs: health, hazards, and history. Rosenstiel School of Marine and Atmospheric Science, University of Miami

Glynn PJ, Glynn PW, Riegl B (2017a) El Niño, echinoid bioerosion and recovery potential of an isolated Galápagos coral reef: a modeling perspective. Mar Biol 164:146. https://doi.org/10.1007/s00227-017-3175-0

Glynn PW (1972) Observations on the ecology of the Caribbean and Pacific coasts of Panamá. In: Jones ML (ed) The Panamic biota: some observations prior to a sea-level canal. Bull Biol Soc Wash 2:13–30

Glynn PW (1984) Widespread coral mortality and the 1982–83 El Niño warming event. Environ Conserv 11(2):133–146

Glynn PW (1990) Coral mortality and disturbances to coral reefs in the tropical eastern Pacific. In: Glynn PW (ed) Global ecological consequences of the 1982–83 El Niño Southern Oscillation. Elsevier oceanography series 52. Elsevier, Amsterdam, pp 55–126

Glynn PW (1997) Eastern Pacific reef coral biogeography and faunal flux: Durham's dilemma revisited. In: Proceedings of the 8th international coral reef symposium, Panama, vol 1, pp 371–378

Glynn PW (2000a) El Niño-Southern Oscillation mass mortalities of reef corals: a model of high temperature marine extinctions? In: Insalaco E, Skelton PW, Palmer TJ (eds) Carbonate platform systems: components and interactions. Geological Society, London, pp 117–133

Glynn PW (2000b) Eastern Pacific coral reef ecosystems. In: Seeliger U, Kjerfve B (eds) Coastal marine ecosystems of Latin America. Ecological studies 144. Springer, Berlin, pp 281–305

Glynn PW (2011) In tandem reef coral and cryptic metazoan declines and extinctions. Bull Mar Sci 87(4):767–794. https://doi.org/10.5343/bms.2010.1025

Glynn PW (2012) Global warming and widespread coral mortality: evidence of first coral extinctions. In: Hannah L (ed) Saving a million species: extinction risk from climate change. Island Press, Washington, DC, pp 103–119

Glynn PW, D'Croz L (1990) Experimental evidence for high temperature stress as the cause of El Niño-coincident coral mortality. Coral Reefs 8:181–191

Glynn PW, de Weerdt WH (1991) Elimination of two reef-building hydrocorals following the 1982–1983 El Niño warming event. Science 253:69–71. https://doi.org/10.1126/science.253.5015.69

Glynn PW, Enochs IC (2011) Invertebrates and their roles in coral reef ecosystems. In: Dubinsky Z, Stambler N (eds) Coral reefs: an ecosystem in transition. Springer, Dordrecht, pp 273–325

Glynn PW, Manzello DP, Enochs IC (2017) Coral reefs of the eastern tropical Pacific: persistence and loss in a dynamic environment. Coral reefs of the world 8, Springer, Dordrecht, p 657

Glynn PW, Colley SB, Carpizo-Ituarte E, Richmond RH (2017b) Coral reproduction in the eastern Pacific. In: Glynn PW, Manzello DP, Enochs IC (eds) Coral reefs of the eastern tropical Pacific: persistence and loss in a dynamic environment. Coral reefs of the world 8. Springer, Dordrecht, pp 435–476

Glynn PW, Maté JL, Baker AC, Calderón MO (2001) Coral bleaching and mortality in Panama and Ecuador during the 1997-1998 El Niño-Southern Oscillation event: spatial/temporal patterns and comparisons with the 1982-1983 event. Bull Mar Sci 69(1):79–109

Glynn PW, Feingold JS, Baker A, Banks S, Baums IB, Cole J, Colgan MW, Fong P, Glynn PJ, Keith I, Manzello D, Riegl B, Ruttenberg BI, Smith TB, Vera Zambrano M (2019) State of corals and coral reefs of the Galápagos Islands (Ecuador): past, present and future. Mar Pollut Bull 133:717–733. https://doi.org/10.1016/j.marpolbul.2018.06.002

Glynn VM (2023) Bouncing back in a warming world. Smithsonian Voices. https://www.smithsonianmag.com/blogs/smithsonian-institution-office-fellowships-and-int/2023/01/30/bouncing-back-in-a-warming-world/

Glynn VM, Vollmer SV, Kline DI, Barrett RDH (2023) Environmental and geographical factors structure cauliflower coral's algal symbioses across the Indo-Pacific. J Biogeogr 50(4):669–684. https://doi.org/10.1111/jbi.14560

Gómez-Petersen P, de Tortolero-Langarica JA, Rodríguez-Troncoso AP, Cupul-Magaña AL, Ortiz M, Ríos-Jara E, Rodríguez-Zaragoza FA (2023) Testing the effectiveness of natural and artificial substrates for coral reef restoration at Isla Isabel National Park, vol 71. Revista Biología Tropical, Mexico, p e54738

Graham NAJ, Nash KL, Kool JT (2011) Coral reef recovery dynamics in a changing world. Coral Reefs 30:283–294. https://doi.org/10.1007/s00338-010-0717-z

Guibert I (2024) Assisted evolution of algal symbionts to enhance coral reef bleaching tolerance: a success story. Glob Chang Biol 30:e17150. https://doi.org/10.1111/gcb.17150

Gupta J, Liverman D, Prodani K, Aldunce P, Bai X, Broadgate W, Ciobanu D, Gifford L, Gordon C, Hurlbert M (2023) Earth system justice needed to identify and live within earth system boundaries. Nat Sustain 6:630–638. https://doi.org/10.1038/s41893-023-01064-1

Guzmán HM (1991) Restoration of coral reefs in Pacific Costa Rica. Conserv Biol 5:189–195

Guzmán HM, Collatos CM, Gómez CG (2022) Movement, behavior, and habitat use of whale sharks (*Rhincodon typus*) in the tropical eastern Pacific Ocean. Front Mar Sci 9:793248. https://doi.org/10.3389/fmars.2022.79324

Guzmán HM, Guevara CA, Breedy O (2004) Distribution, diversity and conservation of coral reefs and coral communities in the largest marine protected area of Pacific Panama (Coiba Island). Environ Conserv 31:111–121

Guzmán HM, Benfield S, Breedy O, Mair JM (2008) Broadening reef protection across the marine conservation corridor of the eastern tropical Pacific: distribution and diversity of reefs in Las Perlas Archipelago, Panama. Environ Conserv 35:46–54

Guzmán HM, Gomez CG, Hearn A, Eckert SA (2018) Longest recorded trans-Pacific migration of a whale shark (*Rhincodon typus*). Mar Biodiv Rec 11:8. https://doi.org/10.1186/s41200-018-0143-4

Haensch S, Bianucci R, Signoli M, Rajerison M, Schultz M, Kacki S, Vermunt M, Weston DA, Hurst D, Achtman M, Carniel E, Bramanti B (2010) Distinct clones of *Yersinia pestis* caused the black death. PLoS Pathog 6(10):e1001134

Harvell CD, Kim K, Burkholder JM, Colwell RR, Epstein PR, Grimes DJ, Hofmann EE, Lipp EK, Osterhaus ADME, Overstreet R, Porter JW, Smith G, Vasta GR (1999) Emerging marine diseases—climate links and anthropocene factors. Science 285:1505–1510

Hausfather Z (2024) State of the climate: 2023 smashes records for surface temperature and ocean heat, published in the Carbon Brief on January 12, 2024. https://www.carbonbrief.org/state-of-the-climate-2023-smashes-records-for-surface-temperature-and-ocean-heat/. Accessed Jan 13, 2024

Hickel J (2020) Less is more: how degrowth will save the world. Penguin Random House, London

Hickel J, Sullivan D (2023) Capitalism, global poverty, and the case for democratic socialism. Mon Rev 75(3):99–113

Hoegh-Guldberg O (1999) Climate change, coral bleaching and the future of the world's coral reefs. Mar Freshw Res 50(8):839–866. https://doi.org/10.1071/MF99078

Hoegh-Guldberg O, Skirving W, Dove SG, Spady BL, Norrie A, Geiger EF, Liu G, De La Cour JL, Manzello DP (2023) Coral reefs in peril in a record-breaking year. Science 382(6676):1238–1240

Hoegh-Guldberg O, Cai R, Poloczanska ES, Brewer PG, Sundby S, Hilmi K, Fabry VJ, Jung S (2014) The Ocean. In: Barros VR, Field CB, Dokken DJ, Mastrandrea MD, Mach KJ, Bilir TE, Chatterjee M, Ebi KL, Estrada YO, Genova RC, Girma B, Kissel ES, Levy AN, MacCracken S, Mastrandrea PR, White LL (eds) Climate change 2014: impacts, adaptation, and vulnerability. Part B: regional aspects. Contribution of working group II to the fifth assessment report of the intergovernmental panel on climate change. Cambridge University Press, Cambridge, UK, NY, USA, pp 1655–1731

Hoegh-Guldberg O, Mumby PJ, Hooten AJ, Steneck RS, Greenfield P, Gomez E, Harvell CD, Sale PF, Edwards AJ, Caldeira K, Knowlton N, Eakin CM, Iglesias-Prieto R, Muthiga N, Bradbury RH, Dubi A, Hatziolos ME (2007) Coral reefs under rapid climate change and ocean acidification. Science 318:1737–1742. https://doi.org/10.1126/science.1152509

Hughes TP, Barnes ML, Bellwood DR, Cinner JE, Cumming GS, Jackson JB, Kleypas J, Van De Leemput IA, Lough JM, Morrison TH, Palumbi SR (2017) Coral reefs in the anthropocene. Nature 546(7656):82–90

Hughes TP, Anderson KD, Connolly SR, Heron SF, Kerry JT, Lough JM, Baird AH, Baum JK, Berumen ML, Bridge TC et al (2018) Spatial and temporal patterns of mass bleaching of corals in the anthropocene. Science 359:80–83

IPCC (1992) Climate change 1992: the supplementary report to the IPCC scientific assessment. In: Houghton JT, Callander BA, Varney SK (eds) Intergovernmental panel on climate change. Cambridge University Press, Cambridge

IPCC (2001) In: Houghton JT, Ding Y, Driggs DJ, Noguer M, van der Linden PJ, Dai X, Maskell K, Johnson CA (eds) Climate change 2001: the scientific basis. Contribution of working group I to the third assessment report of the Intergovernmental Panel on Climate Change. Cambridge University Press, Cambridge

IPCC (2023) Climate change 2023: synthesis report. A report of the Intergovernmental Panel on Climate Change

Ishida-Castañeda J, Pizzaro V, López-Victoria M, Zapata FA (2020) Coral reef restoration in the eastern tropical Pacific: feasibility of the coral nursery approach. Restor Ecol 28(1):22–28

Jones N (2023) The ocean is hotter than ever: what happens next? Nature, News Explorer. https://doi.org/10.1038/d41586-023-01573-1

Judd EJ, Tierney JE, Lunt DJ, Montañez IP, Huber BT, Wing SL, Valdez PJ (2024) A 485-million-year history of earth's surface temperature. Science 385:1316

Kauffman EG, Johnson CC (1997) Ecological evolution of Jurassic-Cretaceous Caribbean reefs. In: Proceedings of the 8th international coral reef symposium, Panamá, vol 2, pp 1669–1676

Kaufman L, Precht WF, Auster P, Glynn PW (2023) Breathing new life into Florida's coral reef. Reef Encounter 38(2):10–11. https://doi.org/10.53642/OGHN7156

Kleypas JA, Langdon C (2006) Coral reefs and changing seawater chemistry. In: Phinney JT, Skirving W, Kleypas J, Hoegh-Guldberg O (eds) Coral reefs and climate change: science and management. American Geophysical Union, Washington, DC, pp 73–110

Kleypas JA, Yates KK (2009) Coral reefs and ocean acidification. Oceanography 22(4):108–117

Kleypas J, Allemand D, Anthony K, Baker AC, Beck MW, Hale LZ, Hilmi N, Hoegh-Guldberg O, Hughes T, Kaufman L (2021) Designing a blueprint for coral reef survival. Biol Conserv 257(2021):109107

Knowlton N, Grottoli AG, Kleypas J, Obura D, Corcoran E, de Goeij JM et al (2021) Rebuilding coral reefs: a decadal grand challenge in International Coral Reef Society and Future Earth Coasts, p 56. https://doi.org/10.53642/NRKY9386

Kump LR (2011) The last great global warming. Sci Am 305(1):56–61

Kulkarni S (2022) Reversing climate change with geoengineering. Harvard University, SITN. https://sitn.hms.harvard.edu/flash/2022/reversing-climate-change-with-geoengineering/

Lasker HR, Peters EC, Coffroth MA (1984) Bleaching of reef coelenterates in the San Blas Islands, Panama. Coral Reefs 3:183–190

Lian T, Wang J, Chen D, Liu T, Wang D (2023) A strong 2023/24 El Niño is staged by tropical Pacific Ocean heat content buildup. Ocean-Land-Atmo Res 2(0011). https://doi.org/10.34133/olar.0011

Lin C-C, Li H-H, Tsai S, Lin C (2022) Tissue cryopreservation and cryobanking: establishment of a cryogenic resource for coral reefs. Biopreserv Biobank 20(4):409–411. https://doi.org/10.1089/bio.2021.0089

Littler K, Robinson SA, Bown PR, Nederbragt AJ, Pancost RD (2011) High sea-surface temperatures during the early Cretaceous epoch. Nat Geosci 4(3):169–172. https://doi.org/10.1038/ngeo1081

López-Pérez A, Ramirez E, Granja Fernández R, Rodriguez Zaragoza FA, Valéncia-Méndez O, Antonio Perez V (2023) El Niño 2023 event in the southern Mexican Pacific: a preliminary report of an imminent disaster. Reef Encounter 38(2):64–66

López-Pérez A, Granja Fernández R, Ramírez-Chávez E, Valencia-Méndez O, Rodríguez-Zaragosa FA, González-Mendoza T, Martínez-Castro A (2024) Widespread coral bleaching and mass mortality of reef-building corals in southern Mexican Pacific reefs due to 2023 El Niño warming. Oceans 5:196–209. https://doi.org/10.3390/oceans5020012

Marbà N, Jordà G, Bennett S, Duarte CM (2022) Seagrass thermal limits and vulnerability to future warming. Front Mar Sci 9:860826

Martínez-Castillo V, Rodríguez-Troncoso AP, Mayfield AB, Rodríguez-Zaragoza FA, Cupul-Magaña AL (2022) Coral recovery in the central Mexican Pacific 20 years after the 1997-1998 El Niño event. Oceans 3:48–59. https://doi.org/10.3390/oceans3010005

May LA, McDonald EM, Kothera RT, Toline CA, McDonough V, Moffitt ZJ, Miller CV, Woodley CM (2022) Assessment of sediment porewater toxicity in Biscayne National Park with sea urchin (*Lytechinus variegatus*) embryos. PLoS One 17(12):e0728695. https://doi.org/10.1371/journal.pone.0278695

Morikawa MK, Palumbi S (2019) Using naturally occurring climate resilient corals to construct bleaching-resistant nurseries. Proc Nat Acad Sci 116:10586–10591

National Academies of Sciences, Engineering, and Medicine (2019a) A research review of interventions to increase the persistence and resilience of coral reefs. The National Academies Press, Washington, DC. https://doi.org/10.17226/25279

National Academies of Sciences, Engineering, and Medicine (2019b) A decision framework for interventions to increase the persistence and resilience of coral reefs. The National Academies Press, Washington, DC. https://doi.org/10.17226/25424

Nava H, González-Rodríguez A, Narchi NE, Méndez-Medina AC, Maldonado-López Y, Cárdenas-Alvarado MA, Figuero-Camacho AG, Drouet-Cruz HT, Corona-Morales N (2023) Towards reef restoration in Zihuatanejo, Guerrero, México: lessons learned. Rev Biol Trop 71:e54792

Nikolaou A, Katsanevakis S (2023) Marine extinctions and their drivers. Reg Environ Chang 23:88. https://doi.org/10.1007/s10113-023-02081-8

Obura DO, DeClerck F, Verburg PH, Gupta J, Abrams JF, Bai X, Bunn S, Ebi KL, Gifford L, Gordon C (2023) Achieving a nature- and people-positive future. One Earth 6:105–117. https://doi.org/10.1016/j.oneear.2022.11.013

Palacio-Castro AM, Smith TB, Brandtneris V, Snyder GA, van Hooidonk R, Maté JL, Manzello D, Glynn PW, Fong P, Baker AC (2023) Increased dominance of heat-tolerant symbionts creates resilient coral reefs in near-term ocean warming. Proc Natl Acad Sci 120(8):e2202388120. https://doi.org/10.1073/pnas.2202388120

Pandolfi JM, Kiessling W (2014) Gaining insights from past reefs to inform understanding of coral reef response to global climate change. Curr Opin Environ Sustain 7:52–58. https://doi.org/10.1016/j.cosust.2013.11.020

Peixoto RS, Voolstra CR (2023) The baseline is already shifted: marine microbiome restoration and rehabilitation as essential tools to mitigate ecosystem decline. Front Mar Sci 10:1218531. https://doi.org/10.3389/fmars.2023.1218531

Peixoto RS, Sweet M, Villela HDM, Cardoso P, Thomas T, Voolstra CR, Høj L, Bourne DC (2021) Coral probiotics: premise, promise, prospect. Annu Rev Anim Biosci 9:265–288. https://doi.org/10.1146/annurev-animal-090120-115444

Porter JW (ed) (2001) The ecology and etiology of newly emerging marine diseases. Kluwer Academic, Dordrecht, p 228

Powell-Palm MJ, Henley EM, Consiglio AN et al (2023) Cryopreservation and revival of Hawaiian stony corals using isochoric vitrification. Nat Commun 14:4859. https://doi.org/10.1038/s41467-023-40500-w

Precht WF, Gintert BE, Robbart ML, Fura R, Van Woesik R (2016) Unprecedented disease-related coral mortality in southeastern Florida. Sci Rep 6(1):31374

Rammelt CF, Gupta J, Liverman D, Scholtens J, Ciobanu D, Abrams JF, Bai X, Gifford L, Gordon C, Hurlbert M (2022) Impacts of meeting minimum access on critical earth systems amidst the great inequality. Nat Sustain 6:212–221. https://doi.org/10.1038/s41893-022-00995-5

Reaka-Kudla ML, Feingold JS, Glynn PW (1996) Experimental studies of rapid bioerosion of coral reefs in the Galápagos Islands. Coral Reefs 15:101–107

Reyes-Bonilla H, Smith F, Arriaga A, Alvarado JJ, Barraza JE, Cortés J, Olán M, Olivier D, Quimbayo JP, Rivera F, Zapata F, Glynn PW, López A, Cupul A (2021) Status of eastern Pacific coral reefs: the good, the bad, and the ugly. Theme 7: scalable observations and technologies. In: Proceedings of the 14th international coral reef symposium, Bremen, 19-23 July 2021

Richardson K, Steffen W, Lucht W, Bendtsen J, Cornell SE, Donges JF, Drüke M, Fetzer I, Bala G, von Bloh W et al (2023) Earth beyond six of nine planetary boundaries. Sci Adv 9:eadh2458

Riegl BM, Glynn PW (2020) Population dynamics of the reef crisis: consequences of the growing human population. Adv Mar Biol 87:1–30. https://doi.org/10.1016/bs.amb.2020.08.010

Riegl B, Bruckner A, Coles SL, Renaud P, Dodge RE (2009) Threats and conservationin an era of global change. Ann N Y Acad Sci 1162:136–186. https://doi.org/10.1111/j.1749-6632.2009.04493.x

Riegl B, Johnston M, Purkis S, Howells E, Burt J, Steiner SC, Sheppard CR, Bauman A (2018) Population collapse dynamics in *Acropora downingi*, an Arabian/Persian gulf ecosystem-engineering coral, linked to rising temperature. Glob Chang Biol 24(6):2447–2462

Riegl B, Glynn PW, Banks S, Keith I, Rivera F, Vera-Zambrano M, D'Angelo C, Wiedenmann J (2019) Heat attenuation and nutrient delivery by localized upwelling avoided coral bleaching mortality in northern Galapagos during 2015/2016 ENSO. Coral Reefs 38:773–785. https://doi.org/10.1007/s00338-019-01787-8

Romero Torres M, Acosta A, Palacio-Castro AM, Treml EA, Zapata FA, Paz-García DA, Porter JW (2020) Coral reef resilience to thermal stress in the eastern tropical Pacific. Glob Chang Biol 26:3880–3890. https://doi.org/10.1111/gcb.15126

Rubin ET, Enochs IC, Foord C, Mayfield AB, Kolodziej G, Basden I, Manzello DP (2021) Molecular mechanisms of coral persistence within highly urbanized locations in the port of Miami, Florida. Front Mar Sci 8:695236

Salm RV, Coles SL, West JM (2001) Coral bleaching and marine protected areas. In: Proceedings of the workshop on mitigating coral bleaching impact through MPA design, Asia Pacific coastal marine program report (vol 102)

Scharfenstein HJ, Chan WY, Buerger P, Humphrey C, van Oppen MJH (2022) Evidence for de novo acquisition of microalgal symbionts by bleached adult corals. Multidiscip J Micro Ecol 16:1676–1679. https://doi.org/10.1038/s41396-022-01203-0

Schubert J (2021) Engineering the climate: science, politics, and visions of control. Mattering Press, Manchester, p 280

Sellanes J, Gorny M, Zapata-Hernández G, Alvarez G, Muñoz P, Tala F (2021) A new threat to local marine biodiversity: filamentous mats proliferating at mesophotic depths off Rapa Nui. PeerJ 9:e12052. https://doi.org/10.7717/peerj.12052

Shein KA (ed) (2006) State of the climate in 2005. Bull Am Meteorol Soc 87:S1–S102

Silverstein RN, Cunning R, Baker AC (2017) Tenacious D: *Symbiodinium* in clade D remain in reef corals at both high and low temperature extremes despite impairment. J Exp Biol 220 (Pt 7):1192–1196. https://doi.org/10.1242/jeb.148239. Epub 2017 Jan 20

Smith TB, Maté JL, Gyory J (2017) Thermal refuges and refugia for stony corals in the eastern tropical Pacific. In: Glynn PW, Manzello DP, Enochs IC (eds) Coral reefs of the eastern tropical Pacific: persistence and loss in a dynamic environment. Coral reefs of the world 8. Springer, Dordrecht, pp 501–515

Snyder GA, Eliachar S, Connelly MT, Talice S, Hadad U, Gershoni-Yahalom O, Browne WE, Palmer CV, Rosental B, Traylor-Knowles N (2021) Functional characterization of Hexacorallia phagocytic cells. Front Immunol 12:662803. https://doi.org/10.3389/fimmu.2021.662803

Souter D, Planes S, Wicquart J, Logan M, Obura D, Staub F (2021) Status of coral reefs of the world: 2020. Global Coral Reef Monitoring Network (GCRMN) and International Coral Reef Initiative (ICRI). https://gcrmn.net/2020-report/

Squires DF (1959) Reports of the Puritan-American Museum of Natural History Expedition to Western Mexico, 7. Corals and coral reefs in the Gulf of California. Bull Am Mus Nat Hist 118(art. 7):367–432

Stanley GD Jr (2001) Introduction to reef ecosystems and their evolution. In: Stanley GD Jr (ed) The history and sedimentology of ancient reef systems. Kluwer Academic/Plenum, New York, pp 1–39

Steinbeck J (1951) The log from the Sea of Cortez. Penguin Books, New York, p 288

Steinbeck J, Ricketts EF (1941) Sea of Cortez, a leisurely journal of travel and research. The Viking Press, New York

Strong AE (1989) Greater global warming revealed by satellite-derived sea-surface temperature trends. Nature 338:642–645

Studivan MS, Eckert RJ, Shilling E, Soderberg N, Enochs IC, Voss JD (2023) Stony coral tissue loss disease intervention with amoxicillin leads to a reversal of disease-modulated gene expression pathways. Mol Ecol 00:1–20. https://doi.org/10.1111/mec.17110

Tortolero-Langarica JJA, Rodríguez-Troncoso AP, Alvarez-Filip L, Cupul-Magaña AL, Carricart-Ganivet JP (2023) The contribution of assisted coral restoration in calcium carbonate production in eastern Pacific reefs. Rev Biol Trop 71:e54849

Toth LT, Macintyre IG, Aronson RB (2017) Holocene reef development in the eastern tropical Pacific. In: Glynn PW, Manzello DP, Enochs IC (eds) Coral reefs of the eastern tropical Pacific: persistence and loss in a dynamic environment. Coral reefs of the world 8. Springer, Dordrecht, pp 177–201. https://doi.org/10.1007/978-94-017-7499-4_6

Toth LT, Stathakopoulos A, Kuffner IB, Ruzicka RR, Colella MA, Shinn EA (2019) The unprecedented loss of Florida's reef-building corals and the emergence of a novel coral-reef assemblage. Ecology 100(9):e02781

Toth LT, Aronson RB, Vollmer SV, Hobbs JW, Urrago DH, Cheng H, Enochs IC, Combosch DJ, van Woesik R, Macintyre IG (2012) ENSO drove 2500-year collapse of eastern Pacific coral reefs. Science 337:81–84

Tribollet A, Godinot C, Atkinson MJ, Langdon C (2009) Effects of elevated pCO_2 on dissolution of coral carbonates by microbial euendoliths. Glob Biogeochem Cycles 23:GB3008. https://doi.org/10.1029/2008GB003286

Turnham KE, Aschaffenburg MD, Pettay DT, Paz-García DA, Reyes-Bonilla H, Pinzón J, Timmins E, Smith RT, McGinley MP, Warner ME, LaJeunesse TC (2024) High physiological function for corals with thermally tolerant, host-adapted symbionts. Proc R Soc B 290:20231021. https://doi.org/10.1098/rspb.2023.1021

Van Oppen MJH, Aranda Lastra M (eds) (2022) Coral reef conservation and restoration in the omics age, vol 15. Springer, Cham

van Vuuren DP, Edmonds J, Kainuma M, Riahi K, Thomson A, Hibbard K, Hurtt GC, Kram T, Krey V, Jean-Francois L, Masui T, Meinshausen M, Nakicenovic N, Smith JS, Rose SK (2011) The representative concentration pathways: an overview. Clim Change 109(5). https://doi.org/10.1007/s10584-011-0148-z

Veron JE, Hoegh-Guldberg O, Lenton TM, Lough JM, Obura DO, Pearce-Kelly PAUL, Sheppard CR, Spalding M, Stafford-Smith MG, Rogers AD (2009) The coral reef crisis: the critical importance of < 350 ppm CO_2. Mar Pollut Bull 58(10):1428–1436

von Prahl H, Guhl F, Grogl M (1979) Gorgona. Futura Grupo Editorial Ltd., Bogotá, Colombia, p 279

Voolstra CR, Peixoto RS, Ferrier-Pagès C (2023) Mitigating the ecological collapse of coral reef ecosystems: effective strategies to preserve coral reef ecosystems. EMBO Rep 24:e56826. https://doi.org/10.15252/embr.202356826

Wellington GM, Glynn PW (2007) Responses of coral reefs to El Niño-Southern Oscillation sea-warming events. In: Aronson RB (ed) Geological approaches to coral reefs. Ecological studies 192. Springer, pp 342–385

Wellington G, Glynn PW, Strong AE, Navarrete SA, Wieters E, Hubbard D (2001) Crisis on coral reefs linked to climate change. Eos Trans AGU 82(1):5–5

Wilson PA, Norris RD (2001) Warm tropical ocean surface and global anoxia during the mid-Cretaceous period. Nature 412(6845):425–429

Wunderling N, von der Heydt A, Aksenov Y, Barker S, Bastiaansen R, Brovkin V, Brunetti M, Couplet V, Kleinen T, Lear CH, Lohmann J, Roman-Cuesta RM, Sinet S, Swingedouw D, Winkelmann R, Anand P, Barichivich J, Bathiany S, Baudena M, Bruun JT, Chiessi CM, Coxall HK, Docquier D, Donges JF, Falkena SKJ, Klose AK, Obura D, Rocha J, Rynders S, Steinert NJ, Willeit M (2023) Climate tipping point interactions and cascades: a review. EGUsphere [preprint] 2023. https://doi.org/10.5194/egusphere-2023-1576

Zapata FA, Rodríguez-Moreno M (2021) Dinámica temporal de la cobertura bentónica y abundancia de herbívoros durante una década (2011-2020) y el estado de conservación de los arrecifes coralinos del Parque Nacional Natural Gorgona. In: Technical Report, Monitorea Arrecifes Coralinos PNN Gorgona 2011–2020, Cali, Colombia

Index

© The Editor(s) (if applicable) and The Author(s), under exclusive license to Springer Nature Switzerland AG 2025

P. W. Glynn, *Coral Reef Resilience in the Anthropocene*,
https://doi.org/10.1007/978-3-031-81421-1

O

P

T

GPSR Compliance
The European Union's (EU) General Product Safety Regulation (GPSR) is a set of rules that requires consumer products to be safe and our obligations to ensure this.

If you have any concerns about our products, you can contact us on

ProductSafety@springernature.com

In case Publisher is established outside the EU, the EU authorized representative is:

Springer Nature Customer Service Center GmbH
Europaplatz 3
69115 Heidelberg, Germany

www.ingramcontent.com/pod-product-compliance
Lightning Source LLC
Chambersburg PA
CBHW070400090825
30826CB00059B/73
* 9 7 8 3 0 3 1 8 1 4 2 0 4 *